军事计量科技译丛

痕量分析手册
Handbook of Trace Analysis
Fundamentals and Applications

［波兰］伊雷娜·巴拉诺夫斯卡（Irena Baranowska） 主编

赵 华 李 颖 巩 琛 黄 辉 李本涛 译
赵 辉 万晓东

国防工业出版社

·北京·

著作权合同登记　图字:01-2023-0336 号

图书在版编目(CIP)数据

痕量分析手册/(波)伊雷娜·巴拉诺夫斯卡主编；赵华等译. —北京:国防工业出版社,2024.4
（军事计量科技译丛）
书名原文：Handbook of Trace Analysis：Fundamentals and Applications
ISBN 978-7-118-12901-4

Ⅰ.①痕… Ⅱ.①伊…②赵… Ⅲ.①痕量分析—手册 Ⅳ.①O656.21-62

中国国家版本馆 CIP 数据核字(2023)第 137224 号

First published in English under the title
Handbook of Trace Analysis：Fundamentals and Applications
edited by Irena Baranowska
Copyright © Springer International Publishing Switzerland，2016
This edition has been translated and published under licence from Springer Nature Switzeriand AG.
本书简体中文版由 Springer 授权国防工业出版社独家出版。
版权所有,侵权必究。

※

国防工业出版社出版发行
(北京市海淀区紫竹院南路 23 号　邮政编码 100048)
雅迪云印（天津）科技有限公司印刷
新华书店经售

*

开本 710×1000　1/16　插页 1　印张 26½　字数 475 千字
2024 年 4 月第 1 版第 1 次印刷　印数 1—1500 册　定价 198.00 元

(本书如有印装错误,我社负责调换)

国防书店：(010)88540777	书店传真：(010)88540776
发行业务：(010)88540717	发行传真：(010)88540762

译者序

痕量分析技术在生态环境、化学工程、食品化工、生物医学和药物化学等多个领域发挥着重要作用。本书是痕量分析专业著作,不仅系统讲述了样品处理、基体分离、仪器分析、干扰识别、数据分析与处理等痕量分析全过程的知识,而且详细介绍了痕量分析技术在环境、药物、食品、法医学、毒理学等方面的应用。本书是近年来化学分析学科领域不可多得的兼顾理论与实践知识的专著。

初次拜读英文原著是因工作需要,书中的痕量分析前处理方法和数据统计知识给予译者很大帮助与启发,此后,译者经常翻阅此书,从中学到了很多实用的痕量分析知识与技巧,收益良多。

译者认为这是一本教科书式的工具书,截至目前国内未见相关内容的图书,因此,从个人及专业角度,都很有必要翻译成中文并出版这部图书,从而为更多科研工作者提供帮助。

本书对痕量分析技术特点、特性及应用进行了全面细致的论述,从基本概念入手,即便初涉此领域的研究人员也能很快了解痕量分析的概念、目的和方法。书中还介绍了痕量分析技术的发展历程,以及其随时代变革而发生的技术进步。同时详述了测量误差分类、不确定度评估、标准物质选择等常见问题的解决办法。因此,对于化学分析和计量测试的工作人员而言,本书是一本非常实用的工具书,具有很好的指导意义。

本书在痕量分析的应用方面,介绍了有机化合物分析、药物杂质分析、食品研究中的痕量元素分析、医学诊断和选择药物监控以及法医学分析,作者借鉴引用了大量文献,希望通过不同角度的知识应用给予大家更多启迪。另外,本书还结合丰富的实例对痕量分析的特殊应用进行了解析,对从事相关工作的科研人员有很强的指导性,可以大大提高工作效率。

本书对痕量分析技术,既有现状分析,也有前景展望,系统全面地向人们展示了这种分析方法的广泛用途,相信本书会为更多的人指点迷津。

本书翻译工作由赵华负责。第1~3章由赵华翻译;第4~6章由黄辉翻译;第7~9章由巩琛翻译;第10~11章由万晓东翻译;第12~14章由李颖翻译;第15章由赵辉翻译。

本书审校工作由赵华负责。第1~6章、第11~12章由赵华校对;第7、8、14章由李艳玲校对,第9、10、13章由吕辉校对;第15章由林敏校对。

由于译者水平有限,书中难免存在不妥和疏漏之处,恳请读者批评指正。

译者

2023年10月26日

目录

第一部分 概 论

第1章 痕量分析的特点和特性 003
1.1 概述 003
1.2 痕量分析中使用的浓度单位 006
1.3 痕量分析在化学分析方法发展中的应用 008
1.4 检测限的概念及其测量方法 011
参考文献 013

第2章 分析结果的质量：误差分类与测量不确定度的评定 015
2.1 概述 015
2.2 测量误差 016
2.3 不确定度 019
2.4 小结 020
参考文献 020

第3章 痕量分析中的校准问题 022
3.1 分析校准 022
3.2 校准方法 023
  3.2.1 标准曲线法 023
  3.2.2 标准加入法 024
  3.2.3 内标法 026
  3.2.4 间接法 027
  3.2.5 稀释法 028
3.3 流动分析中的校准 029
  3.3.1 流动分析的特点 029
  3.3.2 流动分析的校准方法 031

V

3.4 痕量分析领域的校准实例 ·· 033
3.5 小结 ·· 037
参考文献 ·· 038

第4章 无机痕量分析中的标准物质 ·· 039

4.1 概述和基本概念 ··· 039
4.2 痕量分析的特征 ··· 040
4.3 测量准确度的检验方法 ·· 043
4.4 标准物质和有证标准物质 ··· 043
4.5 标准物质的种类 ··· 045
4.6 无机痕量分析用化学成分标准物质和有证标准物质 ············ 047
4.7 无机痕量化学成分标准物质的制备和认定 ······················· 048
4.8 有证标准物质的选择 ··· 051
4.9 有证标准物质的应用 ··· 052
4.10 有证标准物质的有效性 ··· 054
4.11 小结 ··· 056
参考文献 ·· 056

第5章 无机痕量元素分析中的样品分解技术 ······························ 061

5.1 概述 ·· 061
5.2 参考书目 ·· 062
5.3 样品分解技术 ·· 062
  5.3.1 湿法化学分解 ··· 063
  5.3.2 燃烧分解 ··· 083
  5.3.3 熔融分解 ··· 087
5.4 结论与发展趋势 ··· 087
参考文献 ·· 090

第6章 痕量分析中的萃取方法 ·· 102

6.1 概述 ·· 102
6.2 萃取方法的分类及分析性能 ·· 103
6.3 提取方法的选择 ··· 104
  6.3.1 固相萃取 ··· 104
  6.3.2 膜萃取 ·· 107
  6.3.3 微波萃取 ··· 109

  6.3.4 超声萃取 ·· 112
  6.3.5 顺序分离 ·· 113
  6.3.6 生物酶反应 ··· 115
  6.3.7 浊点提取 ·· 117
  6.3.8 超临界流体萃取 ·· 119
  6.3.9 QuEChERS 技术 ··· 121
 6.4 小结 ·· 122
 参考文献 ··· 122

第二部分　痕量分析的应用

第 7 章　选择性有机化合物的痕量分析 ·· 131
 7.1 概述 ·· 131
 7.2 致癌物 ··· 132
 7.3 类雌激素活性化合物 ·· 132
  7.3.1 雌激素 ·· 132
  7.3.2 双酚 A ·· 134
  7.3.3 双酚 A 衍生物 ··· 134
  7.3.4 烷基酚 ·· 136
  7.3.5 金属雌激素 ··· 137
  7.3.6 苯甲酮 ·· 138
 7.4 亚硝胺 ··· 142
 7.5 阻燃剂 ··· 145
 7.6 小结 ·· 147
 参考文献 ··· 148

第 8 章　药物杂质分析 ·· 155
 8.1 概述 ·· 155
 8.2 药品杂质控制条例 ··· 156
 8.3 药物杂质特征 ··· 158
 8.4 药物杂质的仪器分析方法 ·· 160
  8.4.1 分离方法 ·· 161
  8.4.2 光谱方法 ·· 163
  8.4.3 串联方法 ·· 164
  8.4.4 无机杂质分析方法 ··· 165

Ⅶ

8.4.5　挥发性溶剂残留物的分析方法 ……………………………… 167
8.5　小结 ……………………………………………………………………… 168
　参考文献 …………………………………………………………………… 168

第9章　食品研究中的痕量元素分析 …………………………………… 175

9.1　概述 ……………………………………………………………………… 175
9.2　分析样品的前处理 ……………………………………………………… 176
9.3　食品中元素含量的测定 ………………………………………………… 178
9.4　元素的形态分析 ………………………………………………………… 184
9.5　评估痕量分析结果的化学计量技术 …………………………………… 189
　参考文献 …………………………………………………………………… 190

第10章　痕量分析在医学诊断和选择药物监控中的应用 …………… 216

10.1　代谢物组学 …………………………………………………………… 216
10.2　系统生物学中的代谢组学 …………………………………………… 217
10.3　代谢组学工具 ………………………………………………………… 218
　　10.3.1　样品预处理 …………………………………………………… 218
　　10.3.2　分析测定 ……………………………………………………… 219
　　10.3.3　获得数据集的生物信息学分析 ……………………………… 220
10.4　代谢组学的研究思路 ………………………………………………… 221
10.5　代谢组学在低剂量代谢物分析中的应用实例 ……………………… 222
　　10.5.1　人体红细胞糖酵解循环代谢物代谢概况的电泳分析 ……… 222
　　10.5.2　肝癌患者尿液中新陈代谢概况的测定 ……………………… 223
　　10.5.3　利用核磁共振对肌萎缩性脊髓侧索硬化症患者的血浆样
　　　　　　本进行代谢分析 ……………………………………………… 224
　　10.5.4　利用高效液相色谱与串联质谱法测定潜在癌症标记物尿核苷
　　………………………………………………………………………… 225
10.6　小结 …………………………………………………………………… 226
10.7　在生物体液中所选药物及其代谢物的测定 ………………………… 227
　　10.7.1　引言 …………………………………………………………… 227
　　10.7.2　多变量诊断学 ………………………………………………… 227
　　10.7.3　生物分析在个性化治疗中的作用 …………………………… 228
　　10.7.4　药物代谢动力学和生物药物可利用率 ……………………… 228
　　10.7.5　药物与其他化合物的相互作用 ……………………………… 229
　　10.7.6　分析方法的选择 ……………………………………………… 231

  10.7.7 操作规范和分析技巧 ·············· 233

 参考文献 ·············· 245

第11章 法医学分析 ·············· 253

 11.1 刑事学分析 ·············· 253

  11.1.1 引言 ·············· 253

  11.1.2 微观痕迹的特点 ·············· 254

  11.1.3 犯罪痕迹检测的难点 ·············· 255

  11.1.4 痕迹分析方法 ·············· 256

  11.1.5 测量结果的利用 ·············· 260

  11.1.6 微观痕迹的种类 ·············· 260

  11.1.7 小结 ·············· 272

 11.2 毒理学痕量分析 ·············· 272

  11.2.1 引言 ·············· 272

  11.2.2 中毒的类型 ·············· 273

  11.2.3 原始毒药 ·············· 273

  11.2.4 现代毒药 ·············· 273

  11.2.5 中毒的途径 ·············· 277

  11.2.6 中毒症状 ·············· 277

  11.2.7 毒理学分析结果的解释 ·············· 278

 11.3 现代毒理学分析 ·············· 280

  11.3.1 筛选方法 ·············· 281

  11.3.2 确认方法 ·············· 283

  11.3.3 识别系统 ·············· 284

  11.3.4 分析方案 ·············· 285

  11.3.5 小结 ·············· 286

 参考文献 ·············· 287

第三部分 痕量分析的特殊应用

第12章 无机形态和生物无机形态分析的问题和前景 ·············· 295

 12.1 形态和形态分析 ·············· 295

 12.2 取样和储存 ·············· 297

 12.3 元素总量的测定 ·············· 300

 12.4 固体样品的萃取和浓缩 ·············· 302

12.5 应用联用技术测定与鉴别元素形态 ………………………………… 305
    12.5.1 气相色谱联用技术 ………………………………………… 305
    12.5.2 液相色谱联用技术 ………………………………………… 308
    12.5.3 超临界流体色谱技术 ……………………………………… 313
    12.5.4 毛细管电泳技术 …………………………………………… 313
12.6 小结 …………………………………………………………………… 314
参考文献 ……………………………………………………………………… 315

第13章 生物和环境样品中贵金属的定量分析 ……………………………… 329

13.1 概述 …………………………………………………………………… 329
13.2 样品制备 ……………………………………………………………… 330
    13.2.1 临床样品 …………………………………………………… 330
    13.2.2 环境样品 …………………………………………………… 331
13.3 铂族金属元素仪器检测技术 ………………………………………… 334
13.4 临床样品中的铂和钌 ………………………………………………… 336
    13.4.1 临床样品中铂的测定 ……………………………………… 338
    13.4.2 临床样品中钌的测定 ……………………………………… 339
13.5 环境样品中的铂、钯和铑 …………………………………………… 340
    13.5.1 植物 ………………………………………………………… 340
    13.5.2 大气颗粒物和粉尘 ………………………………………… 341
    13.5.3 土壤和沉积物 ……………………………………………… 342
13.6 质量控制 ……………………………………………………………… 343
参考文献 ……………………………………………………………………… 343

第14章 挥发性有机化合物的浓缩和分析 …………………………………… 362

14.1 概述 …………………………………………………………………… 362
14.2 样品制备 ……………………………………………………………… 364
14.3 挥发性有机化合物测定方法 ………………………………………… 373
    14.3.1 气相色谱质谱法 …………………………………………… 373
    14.3.2 质子-转移-反应质谱法 …………………………………… 374
    14.3.3 选择离子流动管质谱法 …………………………………… 375
    14.3.4 离子迁移谱法 ……………………………………………… 377
14.4 小结 …………………………………………………………………… 378
参考文献 ……………………………………………………………………… 378

第15章 放射性核素分析 ································ 389

15.1 概述 ·· 389
- 15.1.1 放射性及其伴随现象 ····························· 389
- 15.1.2 天然和人工放射性 ································ 389
- 15.1.3 辐射测量方法 ······································· 390
- 15.1.4 γ能谱法 ·· 391
- 15.1.5 β能谱法 ·· 391
- 15.1.6 α能谱法 ·· 393
- 15.1.7 中子活化分析法 ···································· 393

15.2 放射分析方法在环境中的应用:分析方面 ·········· 394
- 15.2.1 ^{40}K 活度测定 ······································· 394
- 15.2.2 活化产物 ^{55}Fe、^{60}Co、^{63}Ni 的活度测定 ···· 394
- 15.2.3 ^{137}Cs 活度测定 ······································ 395
- 15.2.4 ^{90}Sr 活度测定 ······································· 396
- 15.2.5 天然和人工放射性核素活度的测定 ·········· 398
- 15.2.6 ^{241}Pu 活度测定 ····································· 402

参考文献 ·· 403

著者列表 ·· 410

第一部分
概　　论

第1章
痕量分析的特点和特性

1.1 概述

化学中"痕量"的概念大概源于19世纪对化学试剂纯度的描述。当不需要或不可能更准确地描述化学试剂纯度时,将会使用"痕量"来描述试剂的次要成分。在毒理学领域有一个非常重要的"马什试验"用于测定微量元素砷,便是一个早期的痕量测量的例子,该方法起源于1836年。

"痕量分析"(德语"Spurenalyse",法语"analyse de traces")一词是在20世纪开始在分析化学领域开始使用的。当时的研究表明,材料中非常少的物质可以严重影响材料(如金属和半导体)的某些重要的性能。20世纪四五十年代,随着核技术的出现,该领域的研究人员需要对含量低于($10^{-2}\% \sim 10^{-5}\%$)的微量成分以及杂质进行定量分析,这使得痕量分析技术得到了巨大的发展。同时,人类活动的很多方面也对极低浓度的元素和化合物的定量分析提出要求,特别是医学、生物学、生物化学和环境科学等领域中涉及的生物活性成分分析,很多试验研究的最终结论都是依据研究对象中低含量,有时甚至低于$10^{-8}\%$含量(低于$10^{-8} g/g$)的元素和/或分子的定性和定量分析数据[1-6]来确定的。

显而易见,一个可靠的痕量分析过程需要使用与之品质相当的试剂。在大多数情况下痕量分析不能使用主成分含量90%~95%内的"技术"级试剂,即使是99.0%~99.9%主成分含量的"纯试剂"也无法满足痕量分析要求。由此出现了试剂等级的描述,比如,"分析纯"是指主成分含量为99.9%~99.99%,"化学纯"具有99.99%~99.999%纯度,"光谱纯"具有99.999%~99.9999%纯度,"原子纯"主成分含量高于99.9999%。这些术语标志着在极低浓度定量分析可行性方面技术的进步。

在生物学研究领域,对存在于活体生物中的各种元素在生物反应中发挥作用的研究与认知促进了动植物生物体中极低浓度元素分析方法的发展。"微量元素"一词被普遍使用,微量元素不仅浓度低,而且在许多自然系统中有特殊作用。

在20世纪末21世纪初,人们对于痕量元素测量的需求,出现了越来越多的痕量分析方法研究的出版物,促进了痕量分析的发展。

极低浓度物质在分析化学方面的进展,催生了许多具有通用性质的科学组织,以及一些专门讨论痕量分析特殊方面的会议召开,并公开出版了部分专著[1-6]。但需要注意的是"痕量分析"这一术语的意义已经发生了变化,在19世纪30年代,它描述的是10^{-4} g/g数量级浓度。20年后,这一限度变成了10^{-7} g/g。不久以后,分析家引入了微量$10^{-8} \sim 10^{-6}$ g/g、极微量$10^{-11} \sim 10^{-9}$ g/g以及亚微量$10^{-14} \sim 10^{-12}$ g/g的定义。然而,上述对于痕量范围的划分是相当随意的,因为所有低于10^{-4} g/g的浓度均属于"痕量"范畴。

在现代复杂基体的痕量分析中,第一步通常是对被分析物进行分离和预富集等预处理,这可以实现被分析物成分的准确测量。然而,对被分析物的处理过程中,每个操作过程都有可能引入误差,误差可能是正的(如偶然引入的杂质成分),也可能是负的[4-5](如被分析物成分的损失)。在无机痕量分析中,由于被分析物中化学成分(元素)通常具有较稳定的化学性质,在预处理过程中不会发生改变,因此可以采用多种比较剧烈的技术手段对其进行预处理。以前通常依据经验来判断被分析物中化学成分种类的数量,但目前随着分析方法和手段的不断进步,被分析物的定性、定量分析方法都逐步被开发,应用范围也越来越广泛。有机痕量分析与无机痕量分析有很大不同,理论上有机分析物的化学成分种类是非常多的,而且不断地增加。充分了解有机样品来源显然对成分分析非常重要,同时,在对有机分析物分析的过程中,分析人员需要密切注意一些难以预料的成分对分析过程的干扰。

有机分析与无机分析的一个重要不同是,许多有机组分在条件发生变化时是不稳定的,这就限制了分析物前处理过程中所使用的方法。有机组分的转化通常在化学反应发生条件下,产生诸如水解、氧化还原和酶催化或微生物学的过程,转化条件也可以受温度和辐射等物理因素的影响。因此,有机分析需要对有机物的性质有比较全面的了解,当然无机痕量分析也同样需要。全面了解有机物的性质对推论得出分析结果至关重要,决定了如何采取分析方案、确定分析程序包括分析过程中的每个分析操作步骤,甚至能够帮助分析判断或评估样品在存储期间可能发生的变化。

当有机分析中被分析物中包含化学性质相似的组分时,问题就更加复杂。除非有特定的操作规程,分析步骤通常首先从分离开始,其中首选的是各种有效的色谱分离方法,它构成了整个分析过程的重要组成部分。因此,20世纪后半叶有机痕量分析的迅速进步是以色谱技术的发展以及其他物理方法识别物质种类等技术的发展为基础的。

测量给定产品中的已知化合物是相对简单的一种情况,例如,食品中抗氧化剂、维生素或着色剂的测定,或特定农作物中庄稼防护剂等的测定。当分析物在样品中有可能进行化学转化时情况就变得复杂,如测定尿液、血液或其他生物样品中的药物、毒素或麻醉剂。在这种情况下必须考虑在特定介质中可能发生的生化过程,也需要考虑分析物与事先未知的成分形成的不明代谢物。

由于很多有机物被人为无节制地引入环境生态体系中,因此痕量有机化合物的测量变得特别重要。由于这些物质具有很高的活性而被使用,有害作用可能在后期才会凸显。例如,杀虫剂二氯二苯基三氯乙烷(DDT)在自然界中不能迅速分解,因而在许多生物体中积累产生危害;再如,在许多工业生产过程中形成的其他活性物质,包含多环芳烃和二噁英(二苯并二噁英的氯化衍生物),现在这两种高毒性物质几乎无处不在。从上述例子中可知,有些待测物有许多同源物、异构体、同系物等相似结构的化合物,化学性质也相似,这些物质的共存导致对特定单一待测物进行测量变得很困难。

这些都表明有机痕量分析是有难度的,同时,相对于无机痕量分析通常可使用有证标准物质,有机痕量分析物因其多样性以及化学不稳定性,导致定量分析更加困难。

无论是有机痕量分析还是无机痕量分析,都会面临需要进行物质形态分析的问题,几十年来人们逐渐意识到形态分析的重要性。经研究证实,同一元素的各种衍生物能发挥完全不同的化学和生物活性,因此形态研究就成了热点。形态分析需要测定各种环境样本中处于痕量水平的金属和非金属有机化合物,这些化合物(如有机锡衍生物)可以是人工合成的,也可以是元素及其无机衍生物与自然存在的有机物相互作用形成的。这些测量需求促进了联用技术(或耦合技术)的发展。这些联用技术将分离技术(如气相色谱)与非常灵敏的仪器检测技术(如质谱)相结合,具有优异的灵敏度,使研究痕量新物质的演变成可能[6]。

痕量分析结果的准确性不仅与量值水平有关,而且与最终使用检测仪器的灵敏度有关。分析过程中的每个操作程序都会在最终结果中引入一个不确定度分量,虽然有些影响因素不是最主要的,但必须考虑。因此,相对于常规简单的测量结果,痕量分析结果具有更大的不确定度,分析物浓度越小,最终结果的不确定度就越大。这通常是以下因素导致的。

(1) 样品中有含量水平较高的其他成分;
(2) 分析程序中包含多个操作步骤;
(3) 样品污染而产生的绝对误差;
(4) 分析物有损失的可能,如吸附或降解;
(5) 其他共存组分引起的干扰;
(6) 测定高含量分析物时,仪器分析比化学分析,精度更低。

综合许多实验测量结果研究表明,当分析物浓度降低时,测量精密度随之降低,这种相关性用霍维茨图表示[7],如图 1-1 所示。图 1-1 表明,当浓度降低两个数量级时,最终结果的相对标准偏差是原来的 2 倍。相关报告仅提到结果的精密度,并没有说明结果的准确度。应该强调的是,即使当分析物浓度从 10^{-4}% 降至 10^{-6}% 时,标准偏差增大了 20%,也不能认为测量结果不准确度。相反,主成分测量结果的标准偏差虽然仅有百分之几,但如果在计算中有错误或结果质量控制出现异常趋势,也不意味着测量结果准确。

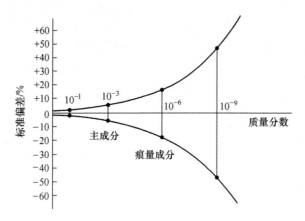

图 1-1　霍维茨图表明标准偏差与分析物含量的关系[7]

1.2　痕量分析中使用的浓度单位

在痕量分析中,用于组分含量的单位应该与浓度数量级相对应。最合适的表示方法是在微量元素(B)的质量分数(w_B)SI 单位前加前缀头,如 m(10^{-3})、μ(10^{-6})、n(10^{-9})、p(10^{-12})、f(10^{-15})或 a(10^{-18})(表 1-1)。成分含量与整个样品的质量以 g 或 kg 表示,通常使用缩写给出样品中 10^{-6}(ppm)或 10^{-9}(ppb)部分痕量组分的数目。这个方法是普遍使用的,需要注意的是,在美国术语中"billion"一词表示 10 亿,而在欧洲术语中,它意味着万亿,是 10 亿的 1000 倍,即"European billion"相当于"American trillion"。因此,ppb(parts per billion)可能会有不同理解,同样的误解也可能出现在 ppq(parts per quadrillion)中。国际纯粹与应用化学联合会(IUPAC)最终接受了美国的体系,即 1ppb = $1/10^9$,但如此简短的单位描述也会带来风险,即"ppt"可能被误为千分之几[8]。

如果人们不理解这些表达式必须严格用于质量单位而不是体积单位时,就会产生混淆。只有水溶液时这种差别可以忽略不计,因为 1L 实际上约等于 1kg;但是,其他液体体系时这种差别不应忽略,应该使用质量浓度单位(c_B或γ_B),即分析物质量与液体样品体积的比值。当固体样品的密度与整个体系明显不同,或没有说明分析物质量是指整个样品的质量还是样品的体积时,这种方法的错误性尤其严重。

表1-1 痕量分析用含量单位

质量分数ω	百分含量/%	质量比		质量浓度	不建议使用单位
10^{-3}	10^{-1}	1mg/g	1g/kg	1g/L	
10^{-6}	10^{-4}	1μg/g	1mg/kg	1mg/L	1ppm
10^{-9}	10^{-7}	1ng/g	1μg/kg	1μg/L	1ppb
10^{-12}	10^{-10}	1pg/g	1ng/kg	1ng/L	1ppt
10^{-15}	10^{-13}	1fm/g	1pg/kg	1pg/L	1ppq
10^{-18}	10^{-16}	1ag/g	1fg/kg	1fg/L	

特别是比较各种方法的检测能力时,质量分数的单位以对数标度来表示。例如,用负对数分别为6.00和5.30来代替质量分数0.000001和0.000005。就像电位测量一样,在分析信号是浓度的对数函数时尤其方便。

当评估气体,特别是空气中气体杂质的检测限时,所测物质的含量通常以体积单位给出,例如,毫升每立方米(mL/m^3),相当于ppm(体积单位)。以质量单位(mg/m^3)表示气体组分的含量不太方便,因为它受温度和压力影响,并且应该在标准条件下给出,如20℃,1013mbar,$1bar=10^5Pa$。

由于痕量分析的主要目的是测定待测样品中给定组分的质量,虽然可以用摩尔数表示,但很少使用摩尔浓度,但电化学分析是例外,其分析信号(如电流)是摩尔浓度的直接函数[9]。因此,在伏安法中,检测限通常以摩尔浓度单位给出(表1-2)。

表1-2 电化学法的检测限[10]

方 法	检测限	
	mol/L	g/mL
恒流极谱法(DME)	10^{-5}	$5×10^{-7}$
脉冲极谱法(DME)	$5×10^{-7}$	$2.5×10^{-8}$
微分脉冲极谱法(DME)	10^{-8}	$5×10^{-10}$
方波极谱法(DME)	10^{-8}	$5×10^{-10}$
反伏安法(HMDE)	10^{-10}	$5×10^{-12}$

注:摩尔质量人为设定为50 g/mol;DME—滴汞电极;HMDE—悬汞电极。

由此,在反向伏安法测镍(摩尔质量 $M=58.7\text{g/mol}$)时,检测限表示为 6×10^{-9} mol/L,如用质量分数则为 $3.5\times10^{-7}\text{g/dm}^3$ 或用百分比表示为 $0.35\times10^{-7}\%$。在分光光度法中,当以摩尔单位给出浓度时,也使用摩尔吸光度。例如,摩尔吸光度 $\varepsilon=5\times10^{-4}\text{L}/(\text{mol}\cdot\text{cm})$ 对应于(摩尔质量 $M=58.7\text{g/mol}$)摩尔吸光度 $a=5\times10^4/(58.7\times10^3)\text{mL}/(\text{g}\cdot\text{cm})$,即 $0.85\text{mL}/(\text{g}\cdot\text{cm})$。

1.3 痕量分析在化学分析方法发展中的应用

分析化学技术和方法的发展为痕量分析提供了更多的可能性。在20世纪上半叶,由于检测能力有限,微量/痕量的测定需要增加样品量。基于质量法和体积法的传统化学分析都不能有效地进行痕量分析,直到光谱光度法发展起来以后才真正地推动了痕量分析的发展,痕量分析是建立在发射光谱以及满足其灵敏度的光谱光度试剂基础上的。

20世纪50年代,直流电弧作为激发源的发射光谱可以测量固体样品中 10^{-5}g/g 含量级别的元素,在特定条件下,甚至可以低至 10^{-8}g/g。这种测量方法一般被认为是半定量方法,但在金属和矿物样本的研究中发挥着重要的作用。用火焰光度法测定液体中的碱金属和碱土金属元素也可以得到类似的检出限。

使用显色剂的分光光度法(光度法或比色法)可以测定许多 $10^{-7}\sim10^{-6}\text{g/g}$ 含量水平的金属元素。由于测量设备易于操作,因此在样品相对容易溶解的情况下,该方法被广泛使用[11-12]。随着专用有机试剂的物理化学合成知识逐渐被人们掌握,分光光度法在20世纪60年代得到普及。但受到摩尔吸光度大小的限制,该方法检测能力无法进一步显著提高。

在经典的电化学方法中,极谱法能够测定溶液中 10^{-5}mol/L 含量水平的离子,该离子浓度相当于(当 $M=60\text{g/mol}$)$6\times10^{-4}\text{g/L}$(或 $6\times10^{-7}\text{g/g}$)。根据20世纪中叶的测量标准,溶解固体样品至少需要10倍的稀释液,经典极谱法检测限不具备测量痕量离子能力。相比之下,基于原位富集的溶出伏安法可以测定比经典极谱法低3个甚至4个数量级的浓度[9-10]。脉冲或方波极谱法使测定浓度显著低于经典极谱法,而且不局限于无机物分析。因此,即使不做预富集处理,电化学方法在没有干扰的情况下也已成功地应用于痕量分析。

20世纪前几十年,原子光谱法是一种非常重要的痕量分析方法(表1-3)。由于新技术的发明,在20世纪60年代原子光谱法再次体现了重要的地位。石墨炉原子化技术的发展使得原子吸收光谱法的测量下限达到 10^{-9}g/g 的水平。通过生成挥发性的化合物可以获得更低的测量下限,由于一般挥发性化合物为氢化物,因此这种方法仅适用于少数元素。电感耦合等离子体发展了原子化和原子激发技术,虽然检测限不能达到 $10^{-10} \sim 10^{-9}$ g/g,但能显著拓宽金属和非金属元素的分析范围。只有以电感耦合等离子体为离子源,采用质谱检测器,才能达到 10^{-11} g/g 的检测限[2, 13],仪器结构的改进和电离新技术的采用使某些无机物和有机物的检测限达到了 10^{-12} g/g。

表 1-3 用光谱方法测定元素大约的检出限

检出限/(ng/g)	FAAS	ICP-OES	GF-AAS	ICP-MS
100	Ag, Se, Al, Sb, V, Pb	K, Pt, Ce, Pt, Pd, Se, Na, Pb, Sb, Sn, Al, As, Ni, Au, W	Hg	
10	Ba, Co, Au, Ni, Fe, Cr, Mn, Ca, Cu	Co, Ni, Cr, Ag, V, B, Cu, Fe, Cd, Zn, Li, Ti	Se, Li, As, Pt, Sn, V	K, Ca
1	Ag, Zn, Cd, Li	Be, Bi, Sr, Ca, Mn, Ba	Sb, Au, Bi, Ni, Ca, Pb, Ba	Fe, Mg, P, Br, Se, Zn, Cr
0.1	Mg	Mg	Co, Fe, Cu, Al, Mn, Cr, Ag, Mg, Cd	B, Na, Hg, Co, Mn, Li, Be, Au, Pb
0.01			Zn	Cd, Mg, Al, Ni, As, Hg, Ag, Cu, Pd, Ba, Mn, Pt
0.001				Bi, La, Ce, In

注:对于 GFAAS,假设样品体积为 20 μL;FAAS—火焰原子化器原子吸收光谱;ICP-OES—电感耦合等离子体原子发射光谱;GFAAS—石墨炉原子化器原子吸收光谱;ICP-MS—电感耦合等离子体质谱。

在分子光谱和原子光谱学中,利用激光激发荧光可以有效提高检测能力。在原子光谱中,利用激光激发荧光进行电热原子化可将检测限提高 2 个数量级;在分子荧光光谱法中,被激发的主要是具有有机配体的金属配合物,检测能力提升主要依靠增加激光强度,同时降低背景强度,然而激光激发荧光适用于有限数量的分析物[14]。化学荧光试剂和色谱检测器选择性和灵敏度的改善,使检测的最低限度达到 10^{-14} g/g[15]。中子活化分析能够在非常宽的范围内测定固体中 $10^{-11} \sim 10^{-10}$ g/g 浓度水平的某些元素,而无须样品溶解[16]。

只有少数的分析技术可以在没有化学预处理的情况下对样品进行分析。大多数情况下,光谱和电化学测试过程需要先对被测物进行溶解,这种处理方式可以通过分析物预浓缩提高检测能力。然而预处理过程常常会导致待测元素的损失以及污染物的引入,从而产生分析误差。

有机物中痕量有机成分的测定存在更复杂的情况。在分析物测定之前,首先要进行分析样品的纯化、分离和预浓缩等多个程序的预处理。色谱法和其他用于有机物分离技术的发展使有机分析有了显著进步,有机化学和生物化学的先进研究成果促进了许多先前未知物理化学性质的化合物定量检测技术的进步。气相色谱法能够检测用经典化学方法无法检测的微量成分。20世纪60年代,杀虫剂DDT用经典化学方法只能测定μg/g量级,采用气相色谱法使检测限降低5个数量级浓度,有力地提高了人们对杀虫剂危害环境和生物体的认知。

如果最终检测选择性不够,痕量有机物的测定就需要选择更合适的方法。例如,在某些情况下选择性免疫化学方法不需要对被测物进行复杂的分离处理。如果检测器是非选择性的,则可以选择现代色谱方法进行分离。色谱法的优势是可以同时测定几种具有相似分子结构的分析物,但如果要测定数十种不同形式基体物质共存的环境样品中的ng/g量级的多核芳烃、多氯联苯和其他化合物,通常需要先去除样品中的基体物质。

测量痕量成分的化学物质种类和物理性质主要依赖气相色谱仪器和液相色谱仪器。许多实验室工作人员根据分离分子的性质、结构和大小,对固定相和流动相的结构和组成进行开发和优化,研究它们的理化性质与色谱参数的相关性。许多商品化仪器和新开发的仪器主要针对某类典型的应用。

为了测量微量分析物,不同的色谱检测器用来选择性检测,它们或者是在一些化合物的官能团上具有选择性,或者是能同时测定不同种类的物质。例如,火焰离子化检测器对于有机化合物是通用的,但其检测能力只有ng/g量级。电子捕获检测器只对卤素有机物有选择性,而且灵敏度高出2个数量级。一般来说,荧光检测器具有较好的灵敏度。以质谱仪作为检测器的气相色谱和液相色谱,将有机物检测限显著降低至pg/mL。

在大多数情况下,采用色谱法进行痕量分析需要对样品前处理。为此,除了传统的液-液萃取之外,还开发了前置的特殊萃取系统,这些系统使色谱对微量成分的定量测量具有很低的检测限和很好的选择性。

在各种分析技术中同位素稀释质谱法发挥着独特的作用。这种测量方法最初用于无机分析物的测定,在20世纪的后几十年中它们也开始用于有机物分析。作为权威方法[14],同位素稀释质谱法需要使用同位素标记的化合物,并具有非常良好的精密度和准确度。

现代分析化学的进步常常与工作流程优化、简化相关,并不总是以检测限度的提高为目的。复杂的自动识别电子系统可以提供关于ng/g量级分析物的定性信息。

人们对生物系统中化学反应过程的研究与日俱增,痕量分析在生物化学技术领域的应用越来越多。经分析检测研究显示,pg/mL量级的酶和免疫反应仍具有较高的特异性和适应性。各种样品中有机物的测定需求越来越多,如种类繁多的环境样本(水、土壤和空气)、典型农用化学品和食品,这些领域中有毒有害成分的最高限量都有法定要求,这些都促进了检测技术的进步。另一个具有较高关注度的重要领域是监测生物和环境样品中的药物残留,与之类似,犯罪学和天然物种演变研究领域的痕量分析也具有重要意义。这些新兴、活跃的领域将持续催生出各种新的痕量分析方法。

1.4 检测限的概念及其测量方法

检测能力是选择痕量分析方法的一项基本参数,检测限(DL、LOD、L_D)是检测能力的定量特征,它定义为可检测到的特定分析物的最小量或浓度,该最小量或浓度通常以给定的概率在实验中给出。只有给出其评价过程的详细实验条件时检测限的量值才有意义。有时使用术语"仪器检测限",其通常是指针对特定的仪器,在被测物的"纯"标准溶液且不存在其他可能会直接干扰分析信号强度的物质的条件下进行测量。"仪器检测限"通常用于普通教科书中的多种技术比较或用于仪器的宣传。因为即便采用同一分析仪器进行不同分析物(元素、化合物)检测,检测限也会有很大差别,"仪器检测限"不会描述测量检出限时的详细实验条件,通常是指仪器在最理想状态下获得的最佳测量能力。当讨论"方法检测限"时,通常指的是在最优或最差条件下得到的数据。在一些技术中(如电化学),检测限与被测物的性质几乎不相关(表1-2)。

相对来说"方法检测限"更有参考和应用价值,它是指在分析方法实际操作条件下进行测量,要考虑分析者在进行测量时遇到的实际状况,包括测量干扰、预处理(如萃取、沉淀和蒸馏等)不完全、被测物的损失,这些因素均能直接影响测量结果的准确度和精密度。

若对不同痕量分析方法的检测限进行比较,必须说明"方法检测限"的测量条件和计算方法。采用同一分析方法,"纯"溶液分析的检测限和生物样本或食品分析时的检测限会有1个数量级的差别,与无机材料分析时的检测限会有2个数量级的差别,与矿物或地质样本分析时的检测限甚至会达到3个数量级的差别[13]。

单次测量结果都有一个随机误差,当测量次数足够多时,测量结果呈正态分布(高斯分布),检测限的信号响应非常小,此时不能满足正态分布,这时测量结果用背景水平的标准偏差 s_B 来描述。因为准确地测定检测限通常会比较困难,一般假定 s_B 与检测限相近,不会有很大的差异,其计算公式为

$$s_B = \sqrt{\frac{(\gamma_B - \overline{\gamma})^2}{n-1}} \tag{1-1}$$

式中:γ_B 为检测结果;$\overline{\gamma}$ 为 n 次测量的平均值。

检测限计算：

$$L_D = \bar{\gamma} + ks_B \qquad (1-2)$$

式中：k 为置信水平的扩展系数，通常 $k=3$。

在这里测量结果如果以响应信号的单位给出（如计数、吸收值等），那么获得的结果需要除以方法的灵敏度。方法的灵敏度定义为响应信号与分析物浓度的比值，可以用分析曲线的斜率来度量。

检测限的另一种计算方法是在空白溶液（原则上不含被测物）中，加入已知量值的被测物标准溶液，用校准曲线 $y = bc + a$ 的外推部分与分析物浓度（含量）交叉点的标准偏差来计算检测限。因为空白溶液与真实样品存在很大差别，所以这种检测限计算方法会有一定的偏差。

由于很低的浓度测量结果不一定符合正态分布，因此给出的检测限应不超过两位有效数字。即使是测量次数很多，检测限结果也可能存在严重的误差。

如果分析方法中包含预富集（如固体样品的分离或溶剂蒸发）过程，那么检测限会出现明显的降低。例如，原子吸收光谱法可以直接测定溶液中 0.1ng/mL 含量的金属，经分离或预浓缩后浓度增加了 20 倍，整个方法的检出限会下降到 0.005ng/mL。因此只有对两个检测限的测量方法进行准确描述，这两个数值才具有实际意义。

同一种被测物的不同分析方法的检测限有时是不能比较的。例如，火焰原子吸收光谱法或极谱分析法这类分析方法，其信号强度不依赖引入仪器中的溶液数量；相反，如石墨炉原子化器原子光谱这类分析方法中，检测能力与加入仪器中的样品量相关。在这些情况下必须准确描述所有的测定条件，尤其是当样品量与分析信号强度非线性相关时。那么当被测物产生的信号以一个给定的概率能和背景值区开时，可以用产生这个信号的样品量来描述检测能力。例如，可以将一个检出限为 1μg/L 的 20 μL 样品写成 20pg。

在色谱这类分析技术中，分析物响应信号会被记录为保留时间的函数，检测能力依赖响应信号、保留时间和峰面积，检测器的灵敏度用质量和时间表示。火焰光度检测器色谱的检测限为 10^{-11} g/s，火焰离子化检测器色谱的检测限为 10^{-12} g/s，氮磷检测器色谱的检测限为 10^{-13} g/s[13,18]。在对测定的全部条件了解后，对某一特定样品这些值可以被重新计算为质量单位或被测物含量。在实际应用时，检测限定义为产生 2 倍检测器噪声信号的物质的最小量。信号可以用电压、电流或者是吸光度单位表示，也可以是所有大于 1s 的脉冲信号。

分析信号的特性及其与分析物含量的函数关系也影响了检测能力的描述。以离子选择电极检测为例，在很宽的范围内，指示电极的电位 E（实际为电解池的电动势）是电极反应物浓度 c 的对数函数，这种定量关系可以绘制成 E 与 $\log c$ 的坐标图（图 1-2）。从分析图上看，信号响应低于平行于浓度轴的直线（恒定离子强度)的分析物浓度均低于方法的检测限。该方法检测限通常是校准曲线的两个线性部分交点的横坐标,这一检测限的可靠性取决于绘制分析图所用点的标准偏差。

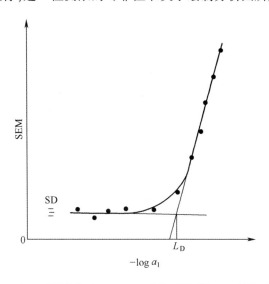

图 1-2　离子选择电位法(SEM-电动势)检测限(L_D)计算示意图

还有一种比较特殊的情况,如表面分析方法,检测限以单原子层的被测分析物的百分含量表示。

检测限是一个估算值,其测量及计算过程都会对检测限有显著影响,如果涉及特殊元素和化合物就更要关注检测限的计算过程。同时,需要强调的是不同的分析人员给出的检测限会有差别,采用不同的仪器得到的检测限也会存在很大差别。与检测能力紧密相关的参数还有定量限(LOQ、LQ),其是指在"可接受的精密度"下能被测定的最小浓度(量值、含量等),其中"可接受的精密度"与检测目的相关联。定量限与检测限在概念上有微小差异,经常认为检测限应该是接近于空白水平的标准偏差的 3 倍,而定量限是标准偏差的 10 倍,定量限可接受的相对精密度假定为 10%水平[18]。

在一些分析方法中,会存在低浓度范围内的响应分析图不是线性的,随浓度增大响应曲线才变成直线。在这种情况下通常认定直线部分的起点是定量限,因为从该点开始的精密度足以用于定量测定。

参考文献

[1] Koch, O.G., Koch-Dedic, G.A.: Handbuch der Spurenanalyse. Springer, Berlin (1964)

[2] Vandecasteele, C., Block, C.B.: Modern Methods for Trace Element Determination. Wiley, Chichester (1993)

[3] Minczewski, J., Chwastowska, J., Dybczyński, R.: Separation and Preconcentration Methods in Inorganic Trace Analysis. E. Horwood, Chichester (1982)

[4] Dulski, T.R.: Trace Elemental Analysis of Metals. Methods and Techniques. Marcel Dekker, New York (1999)

[5] Beyermann, K.: Organic Trace Analysis. E. Horwood, Chichester (1984)

[6] Szpunar, J., Łobiński, R.: Hyphenated Techniques in Speciation Analysis. Royal Society of Chemistry, Cambridge (2003)

[7] Horwitz, W.: Evaluation of analytical methods used forregulation of food and drugs. Anal. Chem. 54, 67A-76A (1982)

[8] Inczedy, J., Lengyel, T., Ure, A.M.: Compendium of Analytical Nomenclature. Definitive Rules 1997. Blackwell Science, Oxford (1998)

[9] Kemula, W., Kublik, Z.: Application de la goutte pendant de mercure à la determination de minimes quantités de différents ions. Anal. Chim. Acta 18, 104–111 (1958)

[10] Wang, J.: Analytical Electrochemistry, 3rd edn. Wiley, Chichester (2006)

[11] Sandell, E.B.: Colorimetric Determination of Traces of Metals. Interscience, New York (1959)

[12] Marczenko, Z., Balcerzak, M.: Separation, Preconcentration and Spectrophotometry in Inorganic Analysis. Elsevier, Amsterdam (2000)

[13] Dybczyński, R.: Neutron activation analysis and its contribution to inorganic trace analysis. Chem. Anal. 46, 133–160 (2001)

[14] Dybczyński, R., Wasek, M., Maleszewska, H.: A definitive method for the determination of small amounts of copper in biological materials by neutron activation analysis. J. Radioanal. Nucl. Chem. 130, 365–388 (1989)

[15] Prichard, E., Mackay, G.M., Points, J.: Trace Analysis: A Structural Approach to Obtaining Reliable Results. Royal Society of Chemistry, Cambridge (1996)

[16] Westmoreland, D.G., Rhodes, G.R.: Analytical techniques for trace organic compounds -II. Detectors for gas chromatography. Pure Appl. Chem. 61, 1147–1151 (1989)

[17] Thompson, M., Ellison, S.L.R., Wood, R.: Harmonized guidelines for single-laboratory validation of methods of analysis. Pure Appl. Chem. 74, 835–855 (2002)

[18] Pritchard, E.: Analytical Measurement Technology. Royal Society of Chemistry, Cambridge (2001)

第2章
分析结果的质量：误差分类与测量不确定度的评定

2.1 概述

分析结果最重要的是其数据的可靠性，其数值不是一个不变的常数，每个分析结果必须评估测量数据的误差和不确定度。

获得可靠的分析结果必须选择适当的测量方法，同时从取样到最终测定的全部测量过程都尽可能准确可靠。

确保分析结果的准确性，并能够反映分析物在样品中的真实含量或"真值"，必须满足两个基本条件：

(1) 用来分析的样品组成应代表分析对象的组成；
(2) 测量结果应反映被分析样品中分析物的真实含量，测量数据准确可靠。

分析化学的主要发展趋势是被分析对象的基体越来越复杂，同时需要测量的分析物浓度越来越低，处于痕量的水平。痕量分析面临的挑战主要体现在以下五个方面[1]：

(1) 分析物浓度越来越低；
(2) 样本基体的组成越来越复杂；
(3) 计量学原理的一些新概念对分析化学提出了新要求；
(4) 需要研究分析结果的溯源性，并对分析结果的不确定度进行评估；
(5) 随着全球化，不同实验室分析结果的可比性要求。

分析结果通常不是对整个分析对象的测量而得到，而是通过检测适当的样品来获得，因此只有当测量样品能够真实地、精确地代表整个分析对象时，分析结果才是可靠的。因此，抽样、样品预处理是非常重要的，也是非常复杂的。这给分析人员带来了巨大的挑战，因为分析人员既要关注所得结果的质量保证(QA)，又要关注结果的质量控制(QC)[2]。

"可靠的数据"与测量的质量密切相关,质量控制和质量保证测量结果更加可靠。

"质量"在不同领域和不同用途的分析中具有特定的意义。国际计量学词汇(VIM)[3]将"质量"定义为对特定要求的实现程度,其中"要求"包括质量控制体系的标准和物质可接受的要求。

化学分析结果的测量质量是指所获得结果与公认的假定结果的一致性[3],可以分为以下四个方面:

(1) 数据的质量;
(2) 方法的质量;
(3) 仪器的质量;
(4) 工作程序的质量。

测量结果的质量取决于分析人员所使用的分析程序以及实验过程中所涉及的各个方面。测量结果的质量的表征参数是测量误差和不确定度评估值。

2.2 测量误差

误差的定义是测量值和期望(真)值之间的差值。因此,误差的计算方法是用测量值减去期望(真)值[3]。测量误差是用来评价测量结果及分析程序的准确性的。

准确度是指单个测量结果与参考量值之间的一致性[3]。正确度是指无限次的重复测量值的平均值与参考量值之间的一致性[3]。这两个参数都与误差的估计密切相关。

由于测量误差的存在,单个测量结果的值通常不等于期望(真)值,根据误差对测量结果的影响类型,将测量误差分为三种基本类型:

(1) 粗大误差;
(2) 系统误差也称偏差;
(3) 随机误差。

根据测量误差的表示方式不同,测量误差可以分为绝对误差和相对误差。

绝对误差以相关性表示:

$$d_{x_i} = x_i - \mu_x \tag{2-1}$$

式中:x_i 为测量结果的值;μ_x 为期望(真)值。

相对误差可表示如下:

$$e_{x_i} = \frac{d_{x_i}}{\mu_x} \tag{2-2}$$

根据误差的来源不同,测量误差可以分为方法误差、仪器误差、人为误差。

单个测量结果的误差可分为三个分量,即[2]

$$d_{x_i} = x_i - \mu_x = \Delta x_{sys} + \Delta x_i + \delta x_i \qquad (2-3)$$

式中:d_{x_i} 为测量结果的总误差;Δx_{sys} 为系统误差;Δx_i 为随机误差;δx_i 为粗大误差。

粗大误差是一个偶然影响因素对测量结果的影响,使测量结果偏离平均值。粗大误差通常以随机变量的形式出现在一些测量中。这种误差最容易检测到,也最容易消除。如果对同一样品进行平行测量,则测量次数越多,越利于消除粗大误差,从而减小检测结果带有粗大误差的概率。

计算粗大误差有许多方法,只是不同方法的适用条件不同[2]。通常以多次平行测量结果的平均值作为最终测量结果,如果没有粗大误差,最终测量结果就只受系统误差和/或随机误差的影响。总误差为:

$$d_{x_m} = x_m - \mu_x = \Delta x_{sys} + \Delta x_m \qquad (2-4)$$

式中:d_{x_m} 为最终测量结果的总误差;x_m 为平行测量结果的平均值;Δx_m 为随机误差。

随机误差是指实验中的典型波动引起的误差。对同一材料中同一分析物的多次测量,随机误差将随着测量次数增大而减小。对于单个结果,无法评估计算随机误差。虽然随机误差的数值较小,但它是计算精密度的基础,是分析结果测量不确定度的一个组成部分[2]。

系统误差是指在相同条件下进行多次测量时保持不变的误差。在真值未知的情况下,无法计算系统误差。系统误差的数值应该很小,这决定了测量的准确性。系统误差可以是单个测量或一个分析方法的参数,在这种情况下也称"偏倚"[2]。

测定系统误差是评判分析方法正确性的一种方法。当测量次数足够多时,随机误差相对于系统误差小到可以忽略不计,即当 $n \to \infty$,则 $s \to 0$(s 为标准偏差),则下面的等式关系成立[2]:

$$d_{x_{met}} = E(x_{met}) - \mu_x - \Delta x_{sys} \qquad (2-5)$$

式中:$d_{x_{met}}$ 为应用分析方法测定结果的总误差;$E(x_{met})$ 为给定分析方法的期望值。

系统误差的出现使得相同条件下的一系列重复测量结果与期望值的差有一个固定值。通常有以下两种类型:

(1) 恒定偏倚 a_{sys}:不依赖分析物浓度水平的值。

(2) 可变偏倚 $b_{sys}\mu_x$:取决于分析物浓度水平,通常呈线性变化。

总的系统误差可以通过下式来描述:

$$\Delta x_{sys} = a_{sys} + b_{sys}\mu_x \qquad (2-6)$$

假设随机误差与系统误差相比可以忽略不计,则可以呈现以下相关性:

$$x_m = \mu_x + \Delta x_{sys} = \mu_x + a_{sys} + b_{sys}\mu_x = a_{sys} + (1 + b_{sys})\mu_x \qquad (2-7)$$

不同类型的误差对最终测量结果的影响如图 2-1 所示。

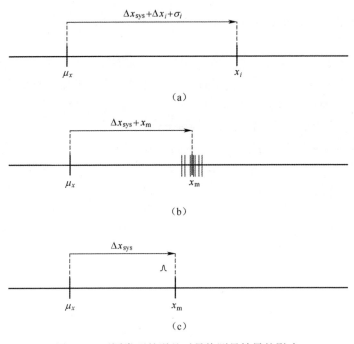

图 2-1　不同类型的误差对最终测量结果的影响

Δx_{sys}—系统误差；Δx_i—单次测量的随机误差；σ_i—粗大误差；μ_x—期望(真)值；x_i—单次测量结果；
x_m—系列测量结果的平均值；Δx_m—多次平行测量的随机误差。

在剔除粗大误差和恒定偏差(通常为校准结果的修正值)之后，测量结果仍然包含随机误差。随机误差的值与测量结果的精密度相关。

精密度定义为在规定条件下，对同一或类似的被测对象重复测量所得的示值或测量值之间的一致程度[3]。它与随机误差相关，表示了围绕平均值的离散程度，通常用标准偏差表示。根据得到一系列测量结果的测量条件，测量精密度可用于定义以下三种。

(1) 重复性：在一组可重复测量条件下的测量精密度[3]；在包括相同的实验室、分析人员、测量仪器、试剂等重复测量条件下获得结果的精密度。

(2) 期间精密度：在同一个实验室内，长期测量过程中测量结果的精密度。期间精密度是比重复性通用的概念，对于重复性而言具有更多可能变化的参数[3]。

(3) 复现性：不同实验室间不同分析人员使用同一测量方法获得测量结果的精密度[3]。

2.3 不确定度

不确定度是每个测量结果的基本属性,存在于分析过程的每个阶段,必须区分以下两类不确定度:

(1) 测量不确定度:基于所使用的信息表征被测量量值分散性的非负参数。

(2) 定义不确定度:对被测量定义描述有限而引起的测量不确定度分量。

常规分析测试结果的不确定度来源见表2-1[4]。

表2-1 常规分析测试结果的不确定度来源[4]

个人原因	仪器原因
测量不准确或不精确的定义	应用的测量仪器分辨率不够
抽样过程可能缺乏代表性	所用标准和/或参考材料相关的不确定性
不适当的测定方法	在单次测量中需要引用并用于最终结果计算的常数不确定度,如物理化学常数
模拟信号读取中的个人偏差	
未知外界因素对测量结果的影响	所用测量方法以及所用仪器相关的近似和假设
与所用测量仪器校准有关的不确定度	在近似相同测量条件下,测量仪器仪表在重复测量过程中的波动

评估测量不确定度可以增加测量结果的可靠性,有利于实验室间测量结果的比对,同时评价测量质量也非常重要。分析程序中的所有步骤都会对最终测量不确定度有贡献[5-9]。因此,应详细说明每个分析程序各个阶段的数值来源和测量不确定度分量[10-12]。

不确定度评估有以下程序[10-13]。

(1) 分析来源:基于测量不确定度的各个来源的识别、量化和组合。

(2) 确定数学模型:基于与参数的定义匹配的函数关系的确定,它采用代数表达式的形式,并描述了不确定度与分析物含量之间的关系。

(3) B类评定:基于实验室间研究获得的精密度数据。

(4) A类评定:基于实验室或实验室内部确认过程,明确测量结果所涉及的精密度、正确度、校准结果、检测限、稳健性等。

(5) 确定包含概率:基于实验室间测试的稳健性检验。

因此,分析的最终结果包括[13]以下三方面。

(1) 测量值及其单位;

(2) 测定结果及扩展不确定值的结果及其单位,用 $y \pm U$ 表示,其中 y 是测量结果，U 是扩展不确定度；

（3）包含因子 k 的值，由此计算扩展不确定度。

不确定度是测量结果的必要组成部分。测量过程的每个阶段都会产生不确定度，这是测量的基本属性，因此，评估不确定度是测量程序中不可或缺的部分。

测量误差和不确定度之间存在差异，测量误差是测定值和期望值之间的差值，不确定度是测量值以某一个概率出现的量值区间范围，因此不确定度不能用于修正测量结果。

2.4 小结

在分析结果的质量保证和质量控制过程中，遇到的主要难题是如何选择以及如何获得足够的信息选用统计工具。最重要的应该是使用统计工具的依据和方法，这是计量的核心问题。

分析测量结果是化学分析工作人员的工作成果。如同生产的产品一样，对测量结果进行质量评价是非常必要的，同时测量结果质量也是分析测量工作本身的质量保证要求。作为产品特性的测量质量是将测量所获得的结果量值与参考值进行比较，其中参考值包括量值期望值、标准值、规范规定值和质量要求值等。为了使测量结果与参考值具有可比性，同时也具有权威性和可靠性，必须对分析测量过程进行高质量的记录，并对这些记录进行保存和维护。首先必须保证测量结果的质量，然后才能类似检验产品质量那样来评价测量结果的质量。

应当注意的是，表征测量结果质量的基本参数和必要参数是溯源性和测量不确定度。不能提供溯源性声明记录和测量不确定度，测量结果通常很难谈得上质量，这两个参数是确保分析结果可靠的基本要求。

误差和不确定度的量值大小很大程度上取决于分析物含量或浓度的水平。在分析物含量处于百分数水平时，测量结果误差和不确定度量值较高是不可接受的，在痕量量值水平下是可接受的。

参考文献

[1] Mermet, J.M., Otto, M., Valcárcel, M. (eds.): Analytical Chemistry: A Modern Approach to Analytical Science. Wiley, Weinheim (2004)

[2] Konieczka, P., Namieśnik, J.: Quality Assurance and Quality Control in the Chemical Analytical Laboratory: A Practical Approach. CRC Press/Taylor and Francis, Boca Raton, FL (2009)

[3] International Vocabulary of Metrology: Basic and General Concepts and Associated Terms

(VIM), Joint Committee for Guides in Metrology, JCGM 200 (2012)

[4] Konieczka, P.: The role of and place of method validation in the quality assurance and quality control (QA/QC) system. Crit. Rev. Anal. Chem. 37, 173–190 (2007)

[5] Paneva, V.I., Ponomareva, O.B.: Quality assurance in analytical measurements. Accred. Qual. Assur. 4, 177–184 (1999)

[6] Populaire, A., Campos, G.E.: A simplified approach to the estimation of analytical measurement uncertainty. Accred. Qual. Assur. 10, 485–493 (2005)

[7] Roy, S., Fouillac, A.-M.: Uncertainties related to sampling and their impact on the chemical analysis of groundwater. Trends Anal. Chem. 23, 185–193 (2004)

[8] Meyer, V.R.: Measurement uncertainty. J. Chromatogr. A. 1158, 15–24 (2007)

[9] Kadis, R.: Evaluating uncertainty in analytical measurements: the pursuit correctness. Accred. Qual. Assur. 3, 237–241 (1998)

[10] Conti, M.E., Muse, O.J., Mecozzi, M.: Uncertainty in environmental analysis: theory and laboratory studies. Int. J. Risk. Assess. Manag. 5, 311–335 (2005)

[11] Love, J.L.: Chemical metrology, chemistry and the uncertainty of chemical measurements. Accred. Qual. Assur. 7, 95–100 (2002)

[12] Armishaw, P.: Estimating measurement uncertainty in an afternoon. A case study in the practical application of measurement uncertainty. Accred. Qual. Assur. 8, 218–224 (2003)

[13] ISO Guide to the Expression of Uncertainty in Measurement (GUM), Geneva (1993)

第3章
痕量分析中的校准问题

3.1 分析校准

不论是定性分析还是定量分析,一个完整的分析程序都是由几个相对独立的工作阶段组成的。如果是定量分析过程,那么必然包括分析校准这个基本的操作程序。

对某一组分进行定量分析通常需要借助专用测量仪器,根据分析物在仪器中产生信号响应来计算其浓度,因此需要首先建立信号响应与分析物浓度之间的定量相关性。校准的关键就是定义这种相关性,从而利用它来定量测定样品中的分析物。由此可见,校准是定量分析的前提条件和重要程序。

建立相关性的校准很难,甚至不可能依据某种理论或采用半经验方法获得,而是完全通过实验的方法来进行,通常以校准曲线的形式建立实验相关性。通过测定已知分析物浓度的标准溶液,将响应信号对应已知分析物浓度来绘制校准曲线,从而计算出分析物浓度。

校准曲线是否能反映分析物在测量仪器或测量系统中的响应特性,取决于被分析样品中分析物的化学环境。一些样品中的基体成分可能会影响分析信号,造成"干扰效应"。但基体成分能够干扰分析物的响应信号就会改变校准曲线的斜率和曲线形式,如图3-1所示。有时干扰效应会显现出浓度乘积或加法形式的正干扰特征影响,这通常在分光光度测量中遇到。在实际应用中,由于样品可能包含以不同方式和不同强度影响分析信号的各种成分,因此校准曲线可能会出现更加复杂的非线性变化。

如果所使用的分析方法存在干扰效应,则需要设计实验方案以便对原有经验的校准曲线进行重建,这可能是一项艰巨的任务。如果校准曲线不能正确重建,仪器检测到样品中分析物响应信号会因为"错误"的校准曲线而偏离样品中的真实分析物浓度,导致不准确的测量结果。

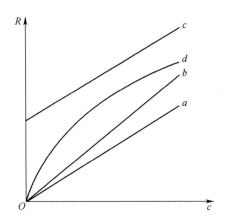

图 3-1 干扰效应对校准曲线的影响
a—理想校准曲线;b—浓度相关的正干扰;c—加法平移正干扰;d—复杂的非线性。

样品基体中成分种类越多,干扰效应越大;干扰物相对于分析物浓度比例越大,干扰效应也越大。在痕量分析中,干扰物通常显著地高于分析物浓度,干扰效应可能会导致较大的分析误差。因此,干扰效应是分析过程中必须考虑的因素。如有必要,则应该采用不同方法来帮助消除或降低干扰效应。常通过沉淀和提取等方法或利用色谱、电泳等分离技术将分析体系中的干扰物和分析物分离。有时,可以给样品加入与干扰物反应的特殊物质,从而对干扰物进行化学消除。如果在预处理之后仍然存在不可忽视的干扰效应,则可以尝试采用具有选择性的校准方法来精确地重建校准曲线。

3.2 校准方法

在分析化学实验室中有几种校准方法获得了广泛使用[1],它们中大多数可用于痕量分析,这些方法的不同:①校准溶液的制备;②测量结果的解释和校准图的构建;③最终分析结果的计算。本节介绍化学分析中最常用的校准方法,并指出了它们的优缺点,这对于痕量分析尤其重要。

3.2.1 标准曲线法

标准曲线法是最普通、最常规的校准方法,也是任何其他校准方法的基础。该方法是制备分析物浓度逐级增加的序列标准溶液,并将样品中分析物预期浓度包含在这个范围内。如图 3-2 所示,基于对这些溶液测量的响应信号来建立校准图,然后在相同的实验条件下测量样品的响应信号并将其与校准图相关联,以插值法计算分析结果。

在这种校准曲线绘制中,标准溶液通常仅包含分析物,如果不存在干扰效应,则分析测量的结果是准确的。然而,如果样品含有一些干扰物,分析物测定结果会引入一个系统误差,如图3-3所示。

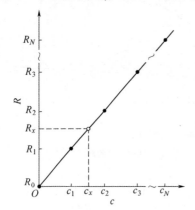

图3-2 标准曲线法校准图
$R_0 \sim R_N$—序列标准溶液的响应信号;
$c_1 \sim c_N$—标准溶液中的分析物浓度;
R_x—样品的响应信号;c_x—样品中分析物浓度。

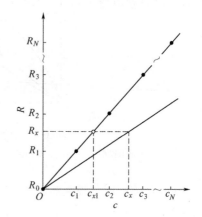

图3-3 标准曲线法中的系统误差校准图
c_x—分析物实际浓度;c_{x1}—由于干扰效应导致校准曲线偏离而测得的分析物浓度。

当干扰效应存在时,通常采用两种方法来确保应用标准曲线法时获得准确的分析结果:一是将存在于样品中所有干扰成分以相同的量引入标准中,显然这是很难做到的,只有在样品中包含少量可识别干扰成分时才可能实现;二是将一些特殊试剂添加到样品中,这种方法相对更实用和更可靠。这些试剂的作用是用化学法消除干扰物对分析物的影响,或者将这些试剂同步添加到标准和样品中,使样品和标准中的干扰效应保持在同一水平上。例如,在原子吸收光谱分析中,通过加入过量的碱金属降低电离干扰。然而,这种方法需要识别各种分析物-干扰系统中具有干扰效应的试剂数量和种类。

标准曲线法显著的优点是简单和高效,一旦绘制校准曲线图,就可以用于许多样品中分析物的测定,这也是受到分析人员普遍欢迎的主要原因。然而,当干扰效应不可忽视时,特别是在痕量分析中该方法必须慎重使用。因此,在经验证明干扰效应可忽略的情况下可使用标准曲线法;否则,首先要证实样品中没有干扰物或明确其消除方法后,然后才能使用。

3.2.2 标准加入法

与标准曲线法不同在标准加入方法中,先将分析物浓度递增的序列标准溶液添加到等体积的样品溶液中,留一个样品不添加标准;再将所有溶液稀释到相同体积后测量响应信号,响应信号值是样品和添加标准的总分析物浓度的响应强度。由于在每个溶液中的总分析物浓度仍然是未知的,测量数据呈现为随标准溶液中分析物浓度递增的函数(图3-4)。因此,必须在一定的已知标准浓度范围内绘制校准图,并且通过将该图外推到零信号值来计算分析结果。

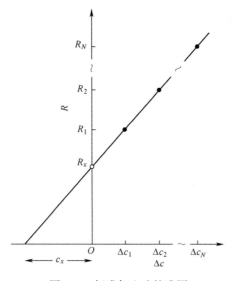

图 3-4 标准加入法校准图

$R_1 \sim R_N$—用标准加入后的样品响应信号;$\Delta c_1 \sim \Delta c_N$—标准加入后分析物浓度增加值;
R_x—样品测量的响应信号;c_x—计算样品中分析物的浓度。

这种校准方法显著特点是所有校准溶液中的分析物都处于相同样品组分环境,包括潜在的干扰物。因此,即使存在干扰效应,也可以通过校准曲线获得准确的测量结果。从这个角度来看,标准加入法是消除或补偿干扰效应的一种方法,而且不需要考虑干扰物种类、数量和在样品中的浓度。这些优势使标准加入法在痕量分析中得到应用。

该方法不足之处[2]:一是在相同的实验条件下,标准加入法会比标准曲线法引入更大的随机误差;二是由于标准加入法是基于外推得到的测量结果,因此在某些情况下会导致较严重的系统误差。

标准加入法在外推校准曲线时,外推部分与实验部分的测量曲线共用相同的函数,这个函数关系通常是线性的。事实上,由于样品中有更多的基体干扰物,所以外推部分与实验部分的函数关系不完全相同,外推部分函数关系通常是非线性的。当样品中分析物的化学形式不同于加入样品中标准溶液分析物的化学形式时,也会出现类似的问题。如果样品和标准溶液中的分析物都受到干扰物的影响,但干扰效应的程度可能是完全不同的,那么两部分的校准曲线事实上具有不同的函数关系。如图 3-5 所示,在这两种情况下通过外推法得到的分析物浓度与预期结果相去甚远。

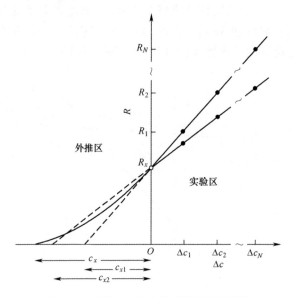

图 3-5 标准加入法中的系统误差校准图

c_x—当外推区的校准相关性为非线性时分析物浓度;c_{x1}—当外推部分和实验部分的函数关系为同一线性关系时测得的浓度;c_{x2}—样品中分析物与添加到样品中的分析物的化学形式不同时测得的浓度。

由于存在上述问题,因此标准加入法在痕量分析中应该谨慎使用。即便存在基体干扰的差异化,但是相对于目前的其他方法而言,标准加入法在补偿干扰效应方面仍然有明显优势并具有非常广泛的应用。

3.2.3 内标法

内标法需要测量两个组分的分析响应信号。校准溶液制备时,首先将已知恒定浓度的内部标准加入分析物浓度递增的标准溶液和被测定的样品中,然后在分析物和内部标准的测量条件下测量所有校准溶液。内标法校准曲线图(以插值的方式计算测量结果)如图 3-6 所示。

由于响应信号比单个信号有更好的测量精度,因此内标法的基本优点是增加了分析物测量结果精度。内标物应尽可能选择与分析物的信号响应特性相同的物质,因此内标物的选择是非常关键的,最好预先开展必要的筛选实验或根据文献数据进行选择。此外,内标物不能是样品中存在的或能够与样品成分发生反应的物质。

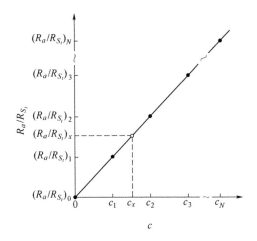

图 3-6 内标法校准曲线图

$(R_a/R_{S_t})_0 \sim (R_a/R_{S_t})_N$—在一组标准溶液中测量的分析物和内标物的响应信号比值;

$c_1 \sim c_N$—标准溶液中的分析物浓度;$(R_a/R_{S_t})_x$—样品中的分析物和内标物

的响应信号比值;c_x—计算的样品中的分析物浓度。

内标法的缺点是其对抗干扰能力小于标准曲线法。样品基体成分不仅会影响分析物的响应信号,也会影响内标物的响应信号。特别是在痕量分析中,这两个方面的影响因素完全相同或相似的概率非常小。因此,只有用标准曲线法消除或减小干扰效应后才能利用内标法。

3.2.4 间接法

当所使用的分析仪器不能直接产生分析物的信号时,可用能够与分析物发生反应的物质对分析物进行间接测定。间接试剂应该相对于样品分析物最高浓度过量加入样品和标准溶液中,以确保样品中和所有标准溶液中的分析物能够反应完全。间接法校准曲线图如图 3-7 所示。

由于间接法实际上测定的是与分析物的反应的间接物质,因此间接法是与标准曲线法互补的校准方法。例如,间接法经常用于原子吸收光谱法测定许多痕量阴离子[3]。有趣的是间接法提供了"积极地"利用干扰效应的机会。将分析物像标准曲线法中的干扰物那样通过化学反应进行了去除后测量干扰物的响应信号,此时干扰物和分析物在这两种方法中似乎交换了角色。

在间接法中干扰效应仍然是非常严重的问题,因为样品基体成分不仅直接影响间接试剂的响应信号,而且可能间接地通过与分析物的反应来改变响应信号。

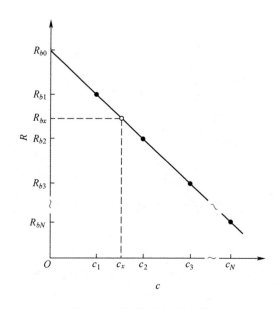

图 3-7　间接法校准曲线图

$R_{b0} \sim R_{bN}$—在序列浓度的标准溶液中响应信号；$c_1 \sim c_N$—标准溶液中分析物浓度；

R_{bx}—样品的响应信号，c_x—样品中分析物浓度。

由于标准溶液和样品是独立制备和测量的，因此除非采取适当的措施来消除存在的基体干扰，否则都会产生测量结果的系统误差。

3.2.5　稀释法

稀释法是使用最少的一种校准测量方法，这种方法是对样品和标准溶液的进行逐级稀释并对每一步稀释后的分析物表观浓度进行测量和内插计算（图 3-8(a)）。因为表观浓度的集合在统计学原则上是彼此相等的，所以分析结果以它们的平均值计算。

样品中的干扰物对分析物的影响随逐级稀释而发生变化，同时表观浓度也发生变化，并在一定浓度值下接近非线性，这被假定为最终分析结果（图 3-8(b)）。该值通过对与实验点相符的非线性函数的外推来计算。这种方法的最大的缺点是分析精度和准确度较差；另外，样品稀释过程中干扰效应通常会减弱，分析物有可能不受干扰影响，因此不需要做干扰消除也可以获得准确的测量结果。

在痕量分析中，稀释法的问题是随着样品的逐级稀释原本含量很低的初始分析物浓度变得更低。因此，只有当分析物具有足够高的灵敏度能允许样品至少稀释几次也能被测量时才能应用该方法。

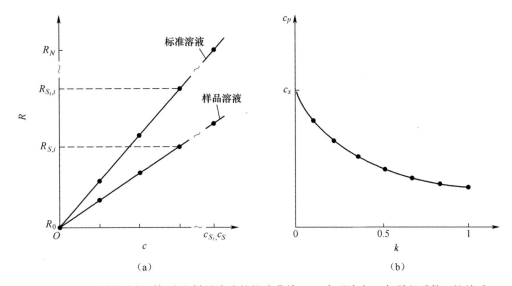

图 3-8 （a）稀释法得到标准和样品溶液的校准曲线；（b）表观浓度 c_P 与稀释系数 k 的关系。c_{S_t} 和 c_S—标准溶液和样品溶液中的初始分析物浓度；$R_{S_t,i}$ 和 $R_{S,i}$—稀释到相同程度的标准溶液和样品溶液的信号；c_x—计算的分析物浓度在样品中的浓度。

3.3 流动分析中的校准

校准的正确实施在很大程度上依赖制备标准溶液和测量样品的正确性和细致程度，这在痕量分析中尤其重要。因为校准曲线绘制时，微小的随机误差或系统误差也能显著地影响测量结果的精密度和准确性。

从实际分析过程来看，制备一系列标准溶液的操作是非常耗时、耗力的，如通常要考虑添加用于消除干扰效应等相关物质，向测量样品中添加外标或内标，确保所用试剂和溶液纯度满足要求等。当需要分析的样品数量较多，或者样品基体较复杂时，操作的复杂性就成为非常突出的问题。

使用流动分析技术有助于克服上述困难。流动分析技术可以正确和有效地制备标准溶液，从而在痕量分析中获得可靠的分析结果。这些技术已被成功地应用于一系列分析领域，同时也代替传统操作方法广泛用于各种校准程序中。

3.3.1 流动分析的特点

在流动分析中，溶液通过内径 1mm 的聚四氟乙烯管并经流动泵流入检测仪器中(图3-9)。如果需要将样品溶液、试剂等多种溶液同时引入,则首先是分别通过单独的进样管,然后在检测模块之前混合在一起。溶液的流动动力通常依靠蠕动泵或注射器提供,并能够实现溶液流速的分别调节和控制。在流动过程中,溶液可以充分混合或进行消解、提取等更加复杂的处理。在流动系统中还可以通过安装机械或电磁阀门来实现溶液从一个管到另一个管的流动。这些阀门也可用于引入体积几十微升的溶液,也可以将一种溶液注入另一种溶液,这就是流动注射分析技术。根据分析操作的实际需求,流动系统可以增加附加的管路和扩展其他功能模块。

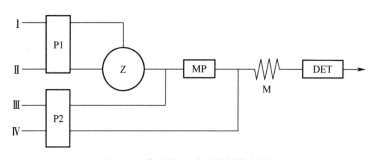

图3-9 典型的流动系统结构框图
Ⅰ、Ⅱ、Ⅲ、Ⅳ—为溶液,如样品溶液、标准溶液及试剂等;P1、P2—蠕动泵;
Z—阀门;MP—样品制备模块;M—混合线圈;DET—检测系统。

就痕量分析而言,用封闭的管道将溶液引入测量仪器,能够有效消除分析过程中溶液被污染的风险。而且使用窄口径管和加样阀,所用溶液的体积可比传统方法所用溶液的体积小得多。流动进样对流入检测系统的被测样品进行各种制备处理的可能性大大提高,并且加快了分析进程。由于流动系统中的所有过程都具有程序性和同步性,因此使预处理及进样过程具有非常好的重复性和再现性,从而提高了分析结果的精密度。

如图3-9所示的流动系统适用于建立校准曲线的工作。例如,将样品引入一个管中,与另一个管中水等溶剂混合,经历消解或紫外线辐射等样品处理后,再与消除干扰效应的试剂混合,最后进入检测系统。这个过程也可以用传统的方法在预先准备好所有的校准溶液后进行仪器测量,虽然流动系统的样品处理比较程序化,有些步骤有时会显得多余,但是所有溶液在同等条件下进行相同处理有助于获得精确的分析结果。

采用流动系统可以收集用于绘制标准曲线校准图的测量数据。在流动系统的阀门后面安装标准溶液引入管和前面的样品溶液混合可实现标准加入校准曲线的绘制。同样地,样品溶液、含内标物的标准溶液以及间接法需要的反应试剂都可以在流动系统中实现有效混合。与传统模式相比,流动注射技术有助于更高效和更经济地制备复杂的校准溶液。

3.3.2 流动分析的校准方法

几十年来,流动分析在各个领域都有了很大的发展,并衍生出一些适用于流动分析模式,不同于分析人员日常使用的各种校准方法[4-5]。相对于传统分析,它们不仅可以更有效、更自动化地实现校准和全部分析程序,而且可以更有效地利用分析信号获得更丰富的测量信息。

在流动分析中使用一个标准溶液就可以获得序列浓度的测量结果,使校准更容易完成。网络校准方法就是一个典型的例子,将不同长度的进样管彼此连接成网络,并连通检测器构成流动系统(图3-10(a))[6]。在该系统中,标准溶液的注入段被分成几个部分(本图中为3个部分),因为每个进样路径长度不同,所以各路溶液到达检测器的时间不同。通过调整管路中的流速和长度,可以使原始段的特定部分彼此交叉重叠,可获得由3个信号的重叠峰,测量图像可以区分5个特征点(图3-10(b))。将样品加入相同参数的系统后,可以获得对应的5个信号,该信号与标准信号相比成比例降低。

上述校准方法是标准曲线的一个实现方案,其中分析信号不是通过使用含有不同浓度的分析物标准溶液来获得的,而是通过单个标准溶液流动条件变化来获得的,因此,其对测量结果的处理也不同,如图3-11所示。对于标准溶液的不同时间段获得的信号值对应分析物浓度,可以建立"两点"直线校准图。然后,将同

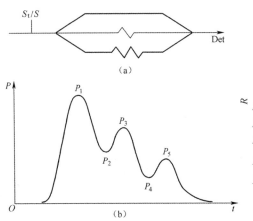

图3-10 网络校准方法

S_t—标准溶液,S—样品溶液,Det—检测系统;
$P_1 \sim P_5$—用于建立校准图的特征点,t—注射后时间。

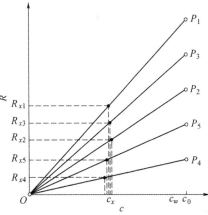

图3-11 网络校准方法获得的样品测量数据

$P_1 \sim P_5$—标准溶液获得的信号;$R_{x1} \sim R_{x5}$—样品获得的信号;c_w—标准溶液中分析物浓度;c_x—样品中分析物浓度。

样条件下样品信号与校准图相比,就可以得到样品中分析物浓度的 5 个估计值,其平均值作为最终分析结果的测量值。

近年来,研究人员开发了由多个泵或多个注射器组成的多泵多通道流动系统,如标准加入法可以使用带有注射阀和三向电磁阀的多通系统[7],该系统通过连续地将样品与 3 种不同浓度分析物标准混合而自动制备一系列校准溶液。笔者认为,该方法特别适用于样品中的分析物浓度低于检测器的线性响应下限时的痕量分析,如电位滴定法测定氟离子选择性电极的线性响应下限浓度 1/10 的氟离子时,该方法非常合适。

在流动注射分析中,有些校准溶液信号以不对称峰的形式出现,图 3-12 是截取的注入载体溶液中部分待测物的浓度测量图像。假定两个不同浓度的待测物溶液,按先后顺序注入同一个系统中,在仪器条件相同时,它们的底部峰宽相同,即注射后出峰时间相同。

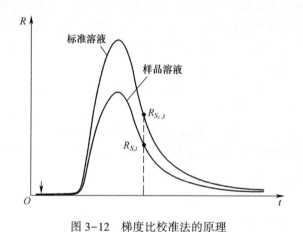

图 3-12 梯度比校准法的原理

$R_{S_t,i}$ 和 $R_{S,i}$——标准溶液和样品溶液在相同条件下注入载流后同一时间点的信号值。

用上述程序进行校准的方法称为梯度比校准法(gradient ratio calibration method, GRCM)[8],这是一种在流动分析中实现稀释法的方法。

在这种方法中,样品溶液和标准溶液按先后顺序注入流动系统中,获得相互重叠的峰(图 3-12),从两个峰较缓一侧的最大点开始逐点向后分析。根据在同一时间点上的标准和样品对应的测量信号值,可以在样品图上各个时间点测定待测物的浓度。对于样品中待测物最终的浓度,当它们的测定值彼此独立、不相关时,可以取这些浓度的平均值。

在 GRCM 中,由于存在干扰效应,因此需要假定样品图的各个时间点的待测物浓度彼此不同,并且与真实浓度也不相同。然而,在被考察的一侧(最高点之后)记录的样品峰实际上是一个逐渐稀释的样品曲线图,随着样品稀释过程中干扰的减小,稀释浓度的计算值逐渐接近样品中真实待测物的浓度。实际上,通过将稀释浓度的值表示为样品稀释的函数,并将该函数外推至无限大稀释样品的值,就可以确定该样品的浓度(图3-8(b))。

3.4 痕量分析领域的校准实例

综上所述,曲线校准决定着痕量分析结果的可靠性,因此,对于一个分析程序,记录关键校准程序及测量的步骤是非常重要的。校准过程最重要的是确保所有校准溶液在相同的实验条件下测量,包括相同的仪器测量参数和化学处理过程。因此,这些所有溶液的制备应该在尽可能短的时间内完成。此外,尽可能让校准溶液和样品具有相同的制备程序,使分析物的任何不可预测的损失或增加在所有溶液中等同地发生。如果实现上述制备通常是复杂的或耗时的,那么建议通过在流动模式中改变所选择的分析程序来实现。

原则上不能使用与样品分析不同日期绘制的校准曲线,即便在连续分析的情况下,一天内的周期性地重复校准也是必要的,因为在仪器参数的微小变化也会使分析信号发生很大的变化,先进高质量的分析仪器时也是如此。所以不应该认为仪器的先进性、价格或制造商声誉等就能够带来可靠的测量结果,这是常见的误解。

为了获得良好精密度和最优定量限,应选择实验条件以获得样品中分析物的最大信号。同时,所建立的分析方案和操作过程应尽可能简单,因为样品的每一步物理或化学处理步骤都必然会带来随机误差。在痕量分析中尤其要避免样品中分析物的大量损失,如不合适的萃取条件等;同时尽可能避免对样品的大比例稀释,如大体积加入处理试剂等。

测量结果的准确度首先取决于分析物测量过程是否受到干扰效应的影响。因此,判断样品是否含有干扰物是至关重要的。如果存在干扰效应并且无法消除,则可以考虑用标准加入法进行分析。当然,也可以通过一些实验来评估干扰效应。

1. 干扰效应评估

评估干扰效应最常用的方法非常简单,就是分别单独测量已知浓度的分析物和在干扰物存在下的分析物。这种方法可以逐一检测组成样品的所有物质对分析物的影响。当然,这是繁琐的且耗时的,并且不能确保考虑到所有样品中共存成分。应该注意的是,校准所用标准溶液中分析物的化学形式可能不同于其在样品中的化学形式,导致干扰成分与分析物的相互作用与在实际样品中的作用不相同。

上述方法还有一个问题,即多种干扰物共存在样品中时不能简单地就某一种成分与分析物的相互作用得出结论。如果潜在干扰组分数目不是很多,那么最好在包含不同浓度的所有组分的合成溶液中评估分析物引发的信号变化。在设计实验方案的基础上选择溶液的组成,以使用相对较少的溶液得到关于干扰物与分析物相互作用的丰富信息[9]。

还有一种常见的识别干扰效应的方法,它基于对单独的样品溶液、一定浓度的分析物标准溶液以及添加同浓度标准溶液的样品进行测量。在这3个结果的基础上,可以制备两个校准曲线,如图3-13所示。如果样本包含产生干扰效应的成分,则这些图具有不同的斜率;否则,这两个曲线彼此平行。用这种方式可以评估样品中存在影响因子的所有干扰物的干扰效应,而不用考虑它们的数量、种类和浓度。同时,用所得到的结论可以判断样品测量适合采用标准曲线法还是标准加入法。

测量包含已知分析物浓度 c_{St} 的标准溶液信号响应 R_{St}、样品响应 R_x 和添加标准的样品信号响应 R_{x+St},通过获得的两个校准线的斜率的比较来考察干扰效应。

互补稀释法(CDM)[10]是上述干扰效应检验方法的衍生。它需要根据图3-14给出的方案初步准备6种溶液,所有溶液中分析物的测量结果可以绘制4个校准曲线(图3-15)。由于分析物存在于不同浓度水平的相应溶液中,因此图中每个曲线的相互斜率都是不同的。图形诠释的不是它们的相互交叉,而是通过比较从图中获得的4个分析结果。

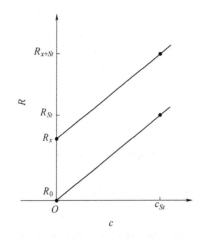

图3-13 干扰效应评估曲线

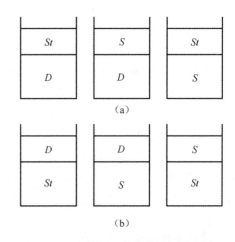

图3-14 根据互补稀释法制备的溶液组
D—含有稀释剂;St—标准溶液;S—样品。

图3-15中的4个结果中有两个是通过标准曲线法的插值获得的,另外两个是通过标准加入法外推获得的,都是样品分析物浓度的估计值。因此,这些结果不同,干扰效应以很高的概率存在。与内插计算的结果相比,通过外推得到的结果更接近样品中的实际分析物含量。

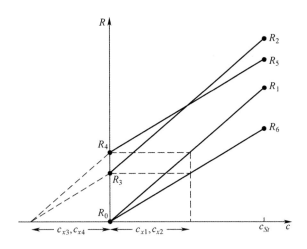

图 3-15 基于图 3-14 所示的解获得的信号 $R_0 \sim R_6$ 的校准图

注:通过内插(c_{x1},c_{x2})和外推(c_{x3},c_{x4})计算的样品中分析物浓度。

可以使用原始流动注射歧管[11]的流动模式改进上述实验过程,样品和单一标准溶液被引入系统中并可以逐级稀释,可以针对不同稀释度对样品中的分析物浓度进行内插和外推估计。这个过程也可以评估可能存在的干扰,而且是一种综合标准曲线法、标准加入法和稀释法的校准方法,实验证明,上述任何校准方法都可以与歧管集成使用。

2. 确定校准条件

在实际分析中只要建立有效的校准方法,特别是正确制备校准溶液和应用恰当的校准方案,即使是在干扰成分存在的情况下,也能获得可靠的分析结果。

在绝大多数情况下,校准过程就是先制备标准溶液,它独立于样品溶液。制备原则是在标准溶液中尽可能地复制样品的基体组成,但是向标准溶液中引入附加组分有带入一定量分析物的风险。因此,当样品中的一种成分相对于分析物极大过量时,没有必要进行标准和样品的化学成分精确匹配,在标准中只包括主要成分足够。

标准物质,即已知成分组成(通常通过实验室间测试定值)的物质,对正确制备标准溶液有很大帮助。它们通常用于确认给定分析方法的正确度和所获得分析结果的准确度。如果参考物质的组成与所分析的样品相匹配,并且待分析的物质是其组分之一,那么该材料可通过添加已知量的分析物来制备一系列标准溶液的方式来校准。

当待分析的一系列样品具有相似的化学组成,至少在干扰物的组成方面相似,并已知的某些组分以相似的量存在于所有样品中,难以预估其干扰效应且对分析结果的可靠性有影响时,可以使用另一种非常重要的且相对简单的方法,即可以选定一个样品用标准加入法进行建立校准曲线,剩余样品中的样品用内插法进行测定(图3-16)。这样得到的结果通常比对所有样品都使用标准曲线法更准确,而且分析速度比用标准加入法分析所有样品时的速度要快。

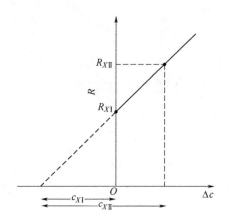

图 3-16 组合校准方法

R_{XI} 和 R_{XII}——具有相似基体成分的两个样品(Ⅰ和Ⅱ)获得的信号;

c_{XI} 和 c_{XII}——用外推和内插法测得的样品Ⅰ和Ⅱ中的分析物浓度。

当干扰效应恒定时,干扰效应和样品中分析物浓度与干扰物浓度之比无关。例如,当存在于样品中的干扰物浓度很高,即使它们相对于分析物大量减少时也不会改变干扰物的影响。在这种情况下,标准加入法可以与校准溶液的逐级稀释结合使用。最简单的方法是按照"顺序标准加入校准"(S-SAC)进行,就是将固定体积的标准溶液连续添加到样品中的方法[12]。另一种为"标准加入和指示性稀释法"(SAIDM),样品用单一标准加标,用中性稀释剂连续稀释溶液,直到测量的信号等于未稀释样品产生的信号[13]。

在相同的情况下可以采用"内插标准加入法"(ISAM)[14],使用流动注射技术进行校准。一系列标准溶液被相继注入连续流向检测器的样品中。由样品产生稳定基线信号,而标准溶液使该信号以峰的形式发生变化(图3-17(a))。分析物浓度高于样品的标准溶液产生"正"峰,低于样品的标准溶液产生"负"峰。计算这些峰的最大点或最小点与稳态信号之间的差值并绘制校准图,其表示信号差异对标准溶液中分析物浓度的相关性(图3-17(b))。在图3-17(b)中,只有当标准和样品中的分析物浓度相等即信号差值为零时,标准中的分析物浓度等于样品中的分析物浓度,标准测量的信号不会引起稳态信号的变化,然后从校准曲线和浓度轴的交叉点确定分析结果。

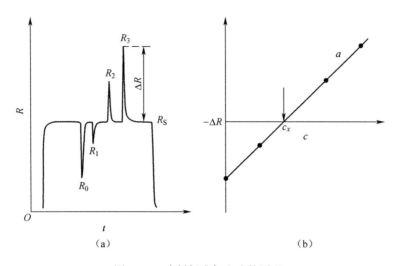

图 3-17 内插标准加入法的原理

R_S—样品的测量信号;$R_0 \sim R_3$—标准溶液的测量信号;c_x—样品中计算的分析物浓度。

3.5 小结

综上所述,有多种校准方法可应用于化学分析中,方法的选择取决于分析问题的类型和分析过程中预期的随机误差来源。很难说哪种方法特别适用于痕量分析,由于痕量分析的特殊性,需要特别关注校准方法的选择及在校准过程每一步的可靠性。

本书介绍的是最常用的校准方法,有些方法也可以用于痕量分析,但很少使用或只在某阶段实施,因此,不能与已建立的校准方法相提并论,也很难确认是否满足痕量分析的要求。由于分析校准在化学分析中具有非常重要的作用,因此校准方法和程序的开发、实施和普及是分析人员最重要和最紧迫的任务之一。

校准方法的发展和普及在很大程度上取决于流动分析领域中分析实验室仪器和方法的先进程度。如前面所指出的,流动技术有助于高效和经济地制备校准溶液,也为获得非典型但有用的测量结果创造了条件。流动分析催生了许多新的校准方法,将分析人员从常规的思维方式和操作中解放出来,而且会继续发挥积极的作用。希望随着时间的推移,流动技术将在实验室中得到越来越多的认可和应用。

总之,从校准的角度来看,流动技术在校准溶液制备阶段提供更大的自由度和操作的可能性,分离技术为消除干扰效应提供了帮助。而质谱仪、ICP 光谱仪等仪器实现了不同分析物同步高灵敏度检测的能力。这二者的结合形成了完美的分析系统。

参考文献

[1] Kościelniak, P.: Calibration methods-nomenclature and classification. In: Namieśnik, J. (ed.) New Horizons and Challenges in Environmental Analysis and Monitoring. CEEAM, Gdańsk (2003)

[2] Welz, B.: Abuse of the analyte addition technique in atomic absorption spectrometry. Fresen. Z. Anal. Chem. 325, 95-101 (1986)

[3] Pinta, A. (ed.): Atomic Absorption Spectrometry. Halstead, New York (1975)

[4] Kościelniak, P., Kozak, J.: Review and classification of the univariate interpolative calibration procedures in flow analysis. Crit. Rev. Anal. Chem. 34, 25-37 (2004)

[5] Kościelniak, P., Kozak, J.: Review of univariate standard addition calibration procedures in flow analysis. Crit. Rev.Anal. Chem. 36, 27-40 (2006)

[6] Tyson, J.F., Bysouth, S.R.: Network flow injection manifolds for sample dilution and calibration in flame atomic absorption spectrometry. J. Anal. At. Spectrom. 3, 211-215 (1988)

[7] Galvis-Sánchez, A.C., Santos, J.R., Rangel, A.O.S.S.: Standard addition flow method for potentiometric measurements at low concentration levels: application to the determination of fluoride in food samples. Talanta 133, 1-6 (2015)

[8] Fan, S., Fang, Z.: Compensation of calibration graph curvature and interference in flow injection spectrophotometry using gradient ratio calibration. Anal. Chim. Acta 241, 15-22 (1990)

[9] Kościelniak, P., Parczewski, A.: Empirical modelling of the matrix effect in atomic absorption spectrometry. Anal. Chim. Acta 153, 111-119 (1983)

[10] Kościelniak, P., Kozak, J., Herman, M., Wieczorek, M., Fudalik, A.: Complementary dilution method-a new version of calibration by the integrated strategy. Anal. Lett. 37, 1233-1253 (2004)

[11] Kościelniak, P., Wieczorek, M., Kozak, J., Herman, M.: Versatile flow injection manifold for analytical calibration. Anal. Chim. Acta 600, 6-13 (2007)

[12] Brown, R.J.C., Mustoe, C.L.: Demonstration of a standard dilution technique for standard addition calibration. Talanta 122, 97-100 (2014)

[13] Kościelniak, P., Kozak, J., Wieczorek, M.: Calibration by the standard addition and indicative dilution method in flame atomic absorption spectrometry. J. Anal. At. Spectrom. 26, 1387-1392 (2011)

[14] Tyson, F.J.: Low cost continuous flow analysis. Anal. Proc. 18, 542-545 (1981)

第4章
无机痕量分析中的标准物质

4.1 概述和基本概念

20世纪后半叶和21世纪初,化学科学中的一个特征现象是痕量分析化学的快速发展。原子能时代的到来催生了新材料的纯度标准,核反应堆燃料的铀和核技术中使用的材料要求必须尽可能完全地去除镝(Dy)、钆(Gd)等某些杂质元素,这些元素对于热中子的吸收具有较高的横截面,这些元素在材料中的浓度不应超过几毫克每千克。这些控制要求促进了发射光谱法、分光光度法和中子活化分析(NAA)等技术,以及元素分离和预富集技术的快速进展。电子技术和半导体技术的进步也提出对材料科学中高纯度的要求。同时,生物医学领域的进展也引起了人们对生命必需的微量元素(如 Co、Cu、Fe、I、Mn、Mo、Ni、Se、Sn 等)或有毒元素(如 As、Cd、Hg、Pb 等)的关注,测定组织、体液中的这些元素,通常是医学诊断和治疗的基础。生态保护运动也开始关注全球环境污染,使得监测水、空气、土壤和食品中的多种微量化学成分和元素已成为实验室的日常工作。目前,测量某一元素总浓度已经远远不能满足要求,还需要具有对元素的各种物理和化学形式的形态分析能力。在全球化时代,国际贸易快速增长,国际标准被认可使用,许多行政、医疗、商业、科技和环境保护相关的领域的重要决策都依赖化学分析的结果。

只有依据准确、可靠的分析结果,才可以采取正确的决策。下面分析化学领域的基本术语和定义[1]:

测量准确度:被测量的测得值与真值之间的一致程度。

测量精密度:在规定条件下,对同一或类似被测对象重复测量所得示值或测量值之间的一致程度。

图4-1给出了准确度和精密度的概念说明[2]。

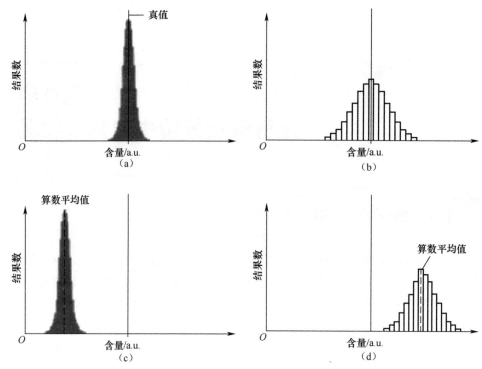

图 4-1 准确度和精密度的概念说明
注:(a)方法准确且精密度高;(b)方法准确,但精密度不高;
(c)方法不准确但精密度高;(d)方法不准确且精密度不高。

4.2 痕量分析的特征

痕量分析产生粗大误差的可能性比微量成分分析要大得多。参加国际实验室间比对(ILC)的实验室报送的分析结果的分散程度令人吃惊,如表 4-1 所列。应该注意的是,数量级差异的结果不仅发生在不限实验室范围的开放性比对中,而且同时出现在具有良好的声誉和经验的优选实验室的比对数据中。

表 4-1 过去 40 年国际实验室比对中分析结果的量值范围

材料	元素	实验室数量/个	结果范围	参考文献
燃油	Ag	4[①]	0.006~0.1mg/kg	von Lehmden et. al. 1974[3]
	Hg	4[①]	0.005~0.4mg/kg	
煤	K	8[①]	20~22000mg/kg	von Lehmden et. al. 1974[3]

续表

材料	元素	实验室数量/个	结果范围	参考文献
奶粉 IAEA A-11	Cd	8	1.10~1660ng/g	Dybczyński et al. 1980[4]
	Co	18	0.004~51.5mg/kg	
	Cr	16	0.016~1160mg/kg	
奶粉 IAEA A-11	Co	7①	3.7~40ng/g	Byrne et al. 1987[5]
	Ni	4①	22.1~500ng/g	
	Pb	4①	43~300ng/g	
乳清粉 IAEA-155	As	7	1.33~3893ng/g	Zeiller et al. 1990[6]
	Cd	25	0.73~38000ng/g	
	Co	22	3.41~4980ng/g	
	Cs	11	0.051~20.1mg/kg	
	Hg	12	0.75~54333ng/g	
	Pb	21	0.025~6.53mg/kg	
东方烟叶 CTA-OTL-1	Cr	43	0.038~11.6mg/kg	Dybczyński et al. 1993, 1996[7-8]
	Cs	17	0.117~315mg/kg	
	Na	43	48.2~8083mg/kg	
	Ni	36	0.005~22.7mg/kg	
	Pb	40	0.051~19.5mg/kg	
波兰混合草药 INCT-MPH-2	Al	33	4.25~1095mg/kg	Dybczyński et al. 2002,2004[9-10]
	As	37	0.027~1.74mg/kg	
	Ca	75	465~15243mg/kg	
	Fe	93	78.7~825mg/kg	
	Na	53	170~20700mg/kg	
	Pb	69	0.077~14.6mg/kg	
	Se	14	9.6~720ng/g	
玉米粉 INCT-CF-3	As	19	0.367~220ng/g	Polkowska-Motrenko et al. 2006, 2007[11, 12]
	Cr	27	10~5710ng/g	
	Fe	72	9.8~1034mg/kg	
	Pb	31	11.6~7455ng/g	
	Se	7	1.0~160ng/g	
东方芳香型烟叶	Cr	46	0.071~19.52mg/kg	Samczyński et al. 2011[13]
	Fe	63	410.15~2006.7mg/kg	
	Pb	53	0.127~9.53mg/kg	
	Sc	19	1~753ng/g	
	Se	9	0.045~1.44mg/kg	

续表

材料	元素	实验室数量/个	结果范围	参考文献
波兰弗吉尼亚烟叶 INCT-PVTL-6	Ni	36	0.294~68.2mg/kg	Samczyński et al. 2011[14]
	Pb	20	0.066~79.9mg/kg	
	Th	11	64.14~12485ng/g	
	U	8	9.37~132.5ng/g	

注：①基于良好声誉和经验选择的实验室。

当测量典型样品的微量成分时，分析人员通常需在一定程度上了解某些成分的近似浓度。当出现与预期显著不同的结果时，会警觉并对分析过程进行评判，进行重复测定。一般来说，在痕量分析中，很难预见某一样品中痕量成分的浓度应该是多少，因此出现跨数量级的结果是可能的。现代仪器分析技术的发展以及先进仪器的商品化推广应用，为痕量分析能力提供了巨大的潜能，可以弥补痕量分析的经验不足等负面影响。有些人认为花费了数十万美元购买的高度自动化和计算机化的分析仪器就应该给出理想的测量结果，但是由于存在基体效应和各种类型的干扰，昂贵的设备也不能自动消除产生粗大误差的可能性。分析的过程开始于取样阶段，随后是包含分离或预浓缩步骤的样品制备，最后才是仪器测量，粗大误差可能在这些分析过程的第一阶段就已经发生。所测定的分析物或元素也可能在样品制备或分离/浓缩过程中发生挥发、吸附等而损失，也可能环境空气、所用实验室容器和试剂，以及溶解样品的酸中等都会对样品造成污染。尽管同一实验室报出的结果也会存在较大差异，不仅包括"难测"的痕量元素（As、Cd、Co、Pb、Se），而且包括高浓度的普通元素（Ca、Fe、K、Na）（表4-1）。痕量分析中粗大误差的可能来源如表4-2所列[15]。误差的大小在很大程度上取决于分析物的浓度。对于某种类型的样品，当成分含量在mg/kg量级时通常能获得令人满意的准确度和精密度，但对于ng/g量级来说测量结果就会相当差[15-16]。

表4-2 无机痕量分析误差来源[15]

取样不足	
痕量分析试验样品的不均匀性	
样品中分析物可能损失的情况：	样品中分析物可能污染的情况：
挥发，吸附，不完全溶解，未能将分析物转化为所需的化学形式，预浓缩和/或分离步骤中的非定量回收	试剂、酸、溶剂等，取样装置、反应容器、玻璃器皿、油管等，离子交换树脂类，萃取剂，吸附剂，水，实验室空气

续表

仪器误差
基体干扰
空白问题
计算误差
人为的小错误(如在报告结果中写入错误的质量或浓度单位)

4.3 测量准确度的检验方法

如前所述,分析人员必须关注分析结果的可靠性,即准确度和精密度。使用某一方法获得良好的测量精密度,并不能充分说明这是一个满意的结果。有时实验室尽管提供了精密度非常高的结果,但与正确值相去甚远[8, 11, 15]。虽然精密度可以很容易评估,例如通过多次重复测定,并计算标准偏差、置信水平等。但检验准确性就相对困难,以下几种方法是可行的:

(1) 采用两种或多种不同的原理的分析方法对样品进行测量;
(2) 参与实验室间比对;
(3) 与权威的或基准的公认准确的方法的测量结果进行比对;
(4) 使用与分析样品相似基体和分析物浓度相匹配的有证标准物质(CRM)。

不是所有实验室都能采用具有相应的良好的测定限和不确定度的办法来检验内部结果的准确性(方法(1)),参与实验室间比对是非常有用和有建设性的途径(方法(2)),但比对结果的阐述和发布可能持续数月或更长时间,因此发现并验证实验室的结果是否偏离及为什么偏离等问题不是很及时。使用国际标准化组织(ISO)[1]的基准参考测量程序(方法(3))是验证结果准确度的最可靠手段,遗憾的是只有高度专业化的实验室才具有基准测量方法的技术能力。

使用有证标准物质(方法(4))对于大多数实验室是可行的。如果分析人员测定分析物(如 CRM 中的元素)获得正确的结果(与认定值一致),则可以预期其他样品中该元素的测定结果也是可靠的。然而,需要分析的样品的基本成分和所确定的元素浓度与 CRM 中的相似。

4.4 标准物质和有证标准物质

在第三版《国际计量学词汇—通用、基本概念和相关术语》(VIM3)[1]中给出标准物质(RM)和有证标准物质(CRM)的最新定义。

标准物质是具有足够均匀和稳定的确定特性的物质,其特性被证实适用于测量中或标称特性检查的预期用途。

注:(1) 标称特性检查提供一个标称特性值及其不确定度,该不确定度不是测量不确定度。

(2) 赋值或未赋值的标准物质都可用于测量精密度控制,只有赋值的标准物质才可用于校准或测量准确度控制。

(3) 标准物质"既包括具有量的物质,也包括具有标称特性的物质。

具有量的标准物质举例:

(1) 规定了纯度的水,其动力学黏度用于校准黏度计;

(2) 没有对胆固醇浓度的量赋值的人类血清,仅用作测量精密度控制;

(3) 阐明了所含二噁英的质量分数的鱼尾性纸巾,可用作校准物。

具有标称特性的标准物质举例:

(1) 一种或多种指定颜色的色图;

(2) 含有特定的核酸序列的 DNA 化合物;

(3) 含 19-雄烯二酮的尿。

(4) 标准物质与特定装置是一体化的。

例如:三相点瓶中已知三相点的物质;置于投射滤光器支架上已知光密度的玻璃;安装在显微镜载玻片上尺寸一致的小球。

(5) 有些标准物质的量值溯源到 SI 制外的某个测量单位,这类物质包括由世界卫生组织指定的国际单位(IU)的疫苗。

(6) 在某个特定测量中,所给定的标准物质只能用于校准或质量保证两者中的一种用途。

(7) 对标准物质的说明应包括该物质的溯源性,指明其来源和加工过程[17]。

(8) 国际标准化组织/标准物质委员会有类似定义[17],但采用术语"测量过程"意指"检查"(ISO 15189:2007,3.4),它既包含了量的测量,又包含了对标称特性的检查。

有证标准物质是指附有由权威机构发布的文件,提供使用有效方法获得的具有不确定度和溯源性的一个或多个特性量值的标准物质。

示例:在所附证书中,给出胆固醇浓度值及其测量不确定度的人体血清,用作校准器具或测量正确度控制的物质。

注:(1) "文件"以"证书"的形式给出(见 ISO Guide 31:2000[28])。

(2) 有证标准物质制备和颁发证书的程序是有规定的(如 ISO Guide 34[31]和 ISO Guide 35[32])。

(3) 在定义中"不确定度"包含了"测量不确定度"和"标称特性值的不确定

度",这样做是为了一致和连贯。"溯源性"既包含"量值的计量溯源性",也包含"标称特性值的溯源性"。

(4) 有证标准物质的特性量值要求附有测量不确定度的计量溯源性[17]。

(5) 国际标准化组织/标准物质委员会有类似定义[17],但使用修饰词"metrological"和"metrologically"来表示量和标称特性。

这里应该提到的是,自从第一份 ISO/REMCO 出版物[18]出版以来,标准物质和有证标准物质的定义经历了明显的演变,相信这种演变应该不会结束。VIM2[19]给出的 CRM 定义中,溯源性要求被描述为"通过规定的不间断的校准链,将测量结果与参照对象联系起来的测量结果的特性,校准链中的每项校准均会引入测量不确定度"。这表明,如果严格按照这一要求,目前实际上不存在用于无机痕量分析的固体有证标准物质。这些标准物质通常需要使用样品溶解的分析技术进行量值认证,这可以理解为破坏比较链条。因为这些定义是不了解化学测量独有特性的传统计量学家给出的,显然在化学分析领域必须加以修改和调整。在VIM3 中,"计量溯源链"定义为"用于将测量结果与参考标准相关联的测量标准和校准序列"[1],通过使用有证标准物质来实现溯源性的方法仍在讨论中[20]。在现实中,尽管人们普遍承认溯源性对于认定值的重要性,但是在以往的实践及其文献中,怎样才算确保了溯源性并没有普遍的共识[20]。标准物质与测量研究所(IRMM)、德国国家材料研究与测试研究所(BAM)和英国政府化学家实验室(LGC)是欧洲三大有证标准物质生产商,他们最近针对有证标准物质生产建立了合作,并就标准物质的制备、认证和如何在证书中记录溯源性等方面达成了共识[21]。

随着时间的推移,有证标准物质的定义变得越来越长,这是因为添加了越来越多的详细注释,充分考虑了许多不同属性的存在。例如,"名义属性"包括在VIM3 中,用于定义有证标准物质,其他例子还包括"人的性别"或"ISO 中两个字母的国家代码"。但当定义变得过于宽泛时,广大实际用户对定义的理解就会不一致。

4.5 标准物质的种类

标准物质可以分为以下四类[2, 22-23]。

(1) 化学成分标准物质,如钢铁、合金、金属中的气体、植物和动物来源的生物材料、地质材料、矿石、临床分析材料、玻璃、陶瓷等;

(2) 同位素含量标准物质;

(3) 物理特性标准物质,如离子活度、光学特性、放射性、计量学、导电性、磁性等；

（4）工程和技术特性标准物质，如标准橡胶、标准筛等。

另外，还有特殊的标准物质是标准气体混合物和化学表面改性硅胶或玻璃，这种标准物质在受控热解过程中会释放出一定的挥发性化合物[23]。也有基于液体（如天然水）制备的标准物质[24]。

按照标准物质的用途，其可以分为以下六种：

（1）科技和工业材料领域的有证标准物质，如纯金属、合金、汽车用催化剂、半导体、矿物燃料、玻璃、塑料等；

（2）用于健康保护的医学和生物学研究、法医毒理学、生物技术的有证标准物质，如体液、代表个体的基因等；

（3）用于评价物质的纯度和组成的有证标准物质，如固体、液体有机或无机化合物，化学元素等；

（4）具有认定的物理特性，如热性能（如热导系数）、光学热性（如吸光度、波长）和力学性能（如硬度）等的有证标准物质；

（5）用于研究食物和饲料的基体型有证标准物质，如饮用水和饮料、基因修饰生物体、蛋白质、脂肪、硝酸盐、氨基酸、纤维等；

（6）环境研究用有证标准物质，如水、土壤、植物、作为生物指示剂的动物组织等。

本章的主题是化学成分标准物质，主要用于无机痕量分析领域。

早期的文献记录，甚至在一些最近的出版物以及通俗读物中，标准物质通常被称为"标准"，或者这两个术语等同使用[2]。以 ILC 为基础的生物或地质类型的标准物质，有时被称为天然基体标准，这个词显然是较早具有某个象征意义的标准物质生产者所用的传统术语，并被保留至今，比如美国国家标准与技术研究院（NIST），或是以前的国家标准局（NBS）。NIST 发布的有证标准物质仍然称为"标准参考物质"（SRM）。SRM 的最初定义是"批量生产的具有良好表征的材料，用于校准测量系统以保证国家测量的一致性"[25]。

除了历史因素，"标准"一词目前仍用以表示最高计量质量。例如，精确化学计量的高纯金属和高纯化合物可以被认为是基准测量标准，可用于标准溶液的制备。

正如在 4.3 节中提到的，有证标准物质的基本功能是在全球范围内传递准确量值和确保分析测量的一致性和溯源性。当有证标准物质被实验室用于质量保证时，这些实验室的结果可追溯到 SI 或其他国际单位。因此，有证标准物质本身也必须满足严格的质量标准。ISO 指南（30~35 系列）[26-32]描述了有证标准物质的制备、认证使用以及生产的要求。有证标准物质具有多种类型和用途，因此很难或不可能制定出对每一种有证标准物质都适用的国际规则。

ISO 指南 30 及附件[26-27]给出了与有证标准物质相关的大多数定义和术语。ISO 指南 31[28]描述了应该在证书和标签上给出的数据。在 ISO 指南 32 和 ISO 指南 33[29-30]中给出了有证标准物质的使用原理和建议。ISO 指南 34 和 ISO 指南 35 规定了有证标准物质生产者的要求,特别是需要建立质量体系的要求[31]。在 ISO 指南 35[32]中讨论了赋值的方法,包括统计方法、不确定度的评估和溯源性保证等。在 ISO 指南 34[31]的最新版本中,还涉及在研究有证标准物质的均匀性、稳定性和定值的过程中实现溯源性的要求。ISO 指南 34[31]与 ISO 17025[33]是有证标准物质生产者认可过程中依据的基本文件,符合国际实验室认可合作组织(ILAC)2004 的决议。欧洲主要有证标准物质生产商,如 IRMM(2004/2005)和 LGC(2006)均依据上述文件获得了认可。

4.6 无机痕量分析用化学成分标准物质和有证标准物质

只要对痕量组分和最小取样量进行了认定且保证了材料的均匀性,适合于在实验室分析方法的有证标准物质都适用于痕量化学成分分析。按照标准物质的基体类型,将无机痕量分析的化学成分标准物质分为以下五种:
(1) 地质标准物质,如矿物、岩石、矿石、土壤等;
(2) 生物标准物质,如动植物组织、体液、食物和饲料等;
(3) 环境标准物质,如水、沉积物、生物指示剂、土壤、灰烬和灰尘等;
(4) 工业/科技标准物质,如金属、合金、结构材料、半导体、玻璃、陶瓷等;
(5) 标准溶液和气体混合物。

表 4-3 列出了用于痕量分析的有证标准物质示例。

认定证书中给出了认定值和信息值。认定值总是与它们的不确定度一起引用。信息值被引用为没有不确定度的数值,它只用于表明分析物浓度的近似水平,不能用于评价测量方法的准确性。

表 4-3 痕量分析用有证标准物质示例

有证标准物质	分类	生产商	具有认定值的分析物	具有信息值的分析物
SRM 688 巴氏岩	地质	NIST	Al_2O_3、FeO、Fe_2O_3、K_2O、MnO、Na_2O、P_2O_5、SiO_2、TiO_2、Cr、Rb、Sr、Th、Pb	CaO、CO_2、F、MgO、Ce、Co、Eu、Hf、Co、Lu、Sc、Ba、Tb、U、Yb、Cu、Ni、V、Sm、Zn
BCR 464 金枪鱼中总汞和甲基汞	生物	IRMM	总 Hg;CH_3Hg^+	

续表

有证标准物质	分类	生产商	具有认定值的分析物	具有信息值的分析物
ERM BD150 脱脂奶粉	生物	IRMM	Ca, Cd, Cl, Cu, Fe, Hg, I, K, Mg, Mn, Na, P, Pb, Se, Zn	
BCR 482 地衣	生物/环境	IRMM	Al, As, Cd, Cr, Cu, Hg, Ni, Pb, Zn	
IAEA-375 土壤	环境	IAEA	^{106}Ru, ^{125}Sb, ^{129}I, ^{134}Cs, ^{137}Cs, ^{232}Th, ^{40}K, ^{90}Sr, Th	^{228}Th, ^{234}U, ^{238}U, ^{238}Pu, $^{239+240}$Pu, ^{241}Am, As, Ba, La, Ni, Rb
LGC6189 河水沉积物-可提取金属①	环境	LGC	As, Cd, Cr, Cu, Mn, Mo, Ni, Pb, Zn	Ba, Se
D 271-1 不锈钢	工业	BAM	O, N	
ERM-EB504 汽车催化剂	工业	BAM	Pd, Pt, Rh	
SRM 2614a 空气中一氧化碳	环境、气体混合物	NIST	CO	

注：①可通过规定参考程序提取的金属含量。

认定证书应该包含有证标准物质用户需要的所有信息，认定证书包含的细节数量取决于材料的性质和认定报告的有效性。当没有认定报告时，证书应该有足够详细的内容使用户以此正确判断有证标准物质预期应用。例如，波兰华沙核化学与技术研究所(INCT)提供的化学成分有证标准物质证书中，包含对所采用的制备方法和认定方法的详尽描述[34]。

4.7 无机痕量化学成分标准物质的制备和认定

标准物质的制备和认定是一项复杂、多阶段且耗时的工作，因为任何阶段的工作失败都会破坏先前的成果，使有证标准物质无效，因此必须关注所有阶段的细节。这项工作需要许多经验丰富的分析人员的协作，同时需要工作协调者对整个过程自始至终的持续监督，这些都是成功的必要条件。

表4-4[35-36]列出了INCT研制和采用的无机痕量分析基体有证标准物质制备的通用方法。关于标准物质制备的大量细节的讨论超出了本书的范围，有关有证标准物质候选物的处理、均匀性检验、建立测定方法、稳定性考核、辐射灭菌和ILC组织的各个阶段的综合信息可以参见文献[9-14, 35-44]。

表 4-4 INCT 研制和采用的无机痕量分析用基体有证标准物质的制备和认定程序[35-36]

有证标准物质制备与认定方案
材料类型的选择
收集适量的材料
制备,包括采用粉碎、研磨、筛分、适当粒径部分的分离、均质化等工作
稳定性初检
选择和购买合适的容器;标签的设计和印刷
均匀性初检
材料主要成分含量的表征(可选)
粒度分布的测定
材料在容器中的分布
最终均匀性检验
辐射灭菌(对生物材料是强制性项目)
测定方法的建立
短期和长期稳定性试验的规划和实施
实验室间比对的组织
结果的统计评价,包括异常值的检测和剔除、计算均值、计算标准偏差、确定置信区间、合成并扩展不确定度等
基于先前制备和完整测定过程给出认定证书和信息值
打印证书
发放和销售的有证标准物质
全程储存期的长期稳定性监测

实践和经验证明,标准物质的认定量值方式大致可分为以下两类。

(1) 单家权威实验室(如 NIST)使用基准方法或至少两个独立的参考方法。

(2) 基于实验室间比对(ILC)结果的认证:

① 指定实验室的比对——仅通过选择的专业实验室网络提供结果(如 BCR);

② 开放实验室的比对——邀请多家实验室自愿参与的比对,如 ICT、IAEA、美国地质调查局(USGS)、加拿大有证标准物质项目(CCRMP)和国际公约组织——国际工作组(GIT-IWG)。

关于如何阐述和解释 ILC 的结果,也有大量的争论。

INCT 一直采用基于全球 ILC 结果的原始认定方法。参与实验室提供的比对数据,通过基于异常值剔除的方法进行统计评估,该方法同时在 0.05 的显著水平,进行狄克逊、格拉布斯、偏度和峰度的统计检验,在离群值剔除后,进行总体算术平均值、标准偏差、标准误差和置信区间的计算。判断离群值剔除过程的正确性可以使用已知元素成分或放射性核素含量真实值的测试材料与所得的 ILC 数据进行验证[45],得到的平均值与真实值一致。在实际验证案例中,利用先前制定的专业公认准则,将总体均值确定为认定值或信息值,或判定为不能进行任何认定。这些准则综合考虑单侧置信区间与总体平均值的比值、参加定值实验室所用方法、总体分析技术的数量以及不同分析技术获得的结果之间的一致性[37-39]。只有当所有情况都满足准则要求时,才可以给出认定值;当所有情况不完全满足准则要求,但结果满足要求略低的标准时,给出信息值。方法评价的可靠性可通过测定有证标准物质(CTA-OTL-1)的认定值和置信区间与ISO[31]推荐的胡贝尔稳健方法和置信区间进行比较验证[37]。

在量值认定过程中,量值认定的主导实验室通常会向参加定值实验室提供有证标准物质,同时也会要求定值实验室使用权威方法或指定的分析程序。这些辅助措施有利于保证量值认定过程的质量和认定值的溯源性。

在最近的一些ILC中,参加实验室分析了一个有证标准物质候选物,以及一个由主导实验室提供的未识别的有证标准物质。这样就能够创建两个数据组:第一个数据组包含由参加实验室提供的有证标准物质候选物中的所有元素的全部测量结果;第二个数据组是在第一个数据组基础上创建的,是参加实验室的元素测量值,相对有证标准物质其是有效的测量结果。第二组数据仅包含参加实验室的元素测量值的平均置信极限与有证标准物质的置信极限重叠部分的结果。两个数据组的数据都按照前述统计方法进行计算,尽管两个数据组的总体平均值非常接近,但大部分的元素都通过对第二个数据组的统计评估结果来认定。

2000—2005年,通过实施ILC并向参加比对的实验室发放有证标准物质作为被测物,所获得的平均值与标准值(ILC 1991)的良好一致性证明了所采用的认定程序的正确性和可靠性,也证明了根据INCT程序制备的有证标准物质的稳定性[37,40]。

文献[46-48]对几种元素基于放射化学如中子活化法(NAA)等的权威基准方法(RPRMP)进行了详细阐述。作为具有最高计量品质的方法,它们是用于独立确定认定值的可靠方法。若采用前文所述方法获得的认定值与RPRMP所获得的结果有很好的一致性,则认定值以下形式呈现:

$$X \pm U \tag{4-1}$$

式中:X 为认定值;U 为扩展不确定度($U = 2u_c$)。

合成标准不确定度由下式给出:

$$u_c = \sqrt{u_{interlab}^2 + u_{lstab}^2 + u_{inhom}^2 + u_m^2} \tag{4-2}$$

式中:$u_{interlaber}$ 为总体平均值的估计标准偏差;u_{lstab} 为从长期稳定性研究中估计的标准不确定度;u_{inhom} 为同均匀性研究估计的标准不确定度;u_m 为其他影响测定结果的标准不确定度。

有时,短期稳定性 u_{shstab} 也被考虑在式(4-2)的右边部分,但其值通常非常小,可以忽略不计。

4.8 有证标准物质的选择

市场上销售的有证标准物质数量可能不能满足大量科学团体对标准物质的实际需要,然而这并不意味着追求质量保证的分析人员可以随意使用有证标准物质。针对特定的分析任务在选择合适优质的有证标准物质时有以下五个通用标准[15, 35]。

(1) 可以提供大量痕量元素的认定值。目前的痕量分析方法如 NAA、ICP-OES 和 ICP-MS 可以同时测定多种元素。测定一个或两个元素是相对较少的要求,因此,选用具有多种微量元素认定值的有证标准物质有助于节省时间和资金。

(2) 具有良好的均匀性,即便痕量元素在非常低浓度水平。均匀性是指在给定材料中某一性质或物质随机分布的程度[49]。在分析化学的术语中,这意味着从该材料获得的某一质量的样品在适当置信极限内具有相同的平均成分[50]。类似的定义[32]指出,如果材料的不同部分中表征性质的值没有差异,则对该性质来说材料是完全均匀的。在实际中,当材料不同部分的特性值的差异与特性值不确定度相比可以忽略不计时,该材料被认为是均匀的。

值得注意的是,地质、生物、环境等天然基体的固体有证标准物质在微观尺度上总是不均匀的,因为它们来源于某一地质材料或某一生物组织。表观均匀性是通过对单个微粒或颗粒的粉碎和精确混合来实现的。有时,使用英格米尔采样常数 K_s [51-52]表征材料的均匀性(更确切地说是不均匀性):

$$K_s = R_s^2 \times m_s \tag{4-3}$$

式中:R_s^2 为采样方差(百分比);m_s 为用于被分析的样品的质量。

一些工作人员还使用 $K_s^{1/2}$ 来表征材料的均匀性[51, 53]。对于一个双组分混合物,其中被测痕量元素 A,100% 仅存在于颗粒中,取样常数为:

$$K_s \approx 10^4 \frac{m_{particle}}{C_A} \tag{4-4}$$

式中:$m_{particle}$ 为材料的单个微粒或颗粒的质量;C_A 为痕量元素的浓度(g/g)。

由此可见,当痕量组分的浓度较低且材料的粒径较大时,材料中该组分出现不均匀的可能性变大[35, 50]。因此,有证标准物质生产者应该表明材料用于分析的最小质量。如今现代仪器方法允许对非常少量的样品进行分析,然而不均匀性的误差可以随着分析的有效质量的减少而急剧增加,并且用于微量分析的有证标准物质是稀缺的,甚至是不存在的[42-43]。

(3) 短期稳定性和长期稳定性。有证标准物质必须在整个特定时间内保持稳定,因为只有这样认定值与其不确定度才是有效的,稳定性的前提是材料存储条件是合适的。因此,分析人员应该注意材料的保质期,这应该在证书中注明。在 ISO 的文献[26-32]中描述了在实验室中进行有证标准物质的稳定性试验和储存期限的要求。稳定性试验是通过类似于制药中使用的方法进行的,它们包括在特定条件下存储有证标准物质并测量此期间的特定量值变化[54-58]。这些测试通常持续 2 年,并将其结果外推至将来。一般来说,稳定性试验要研究在运输条件下的短期稳定性和在储存条件下的长期稳定性。无机痕量分析有证标准物质稳定性试验基于单个元素浓度的变化是时间的函数,测量结果使用趋势分析进行统计计算,类似于均匀性研究。重要的是在稳定性试验中使用的分析方法应具有较小的不确定度。如果不同时间的测量结果重复性与测量方法重复性在统计上无显著性差异,则表明材料是足够稳定的,可以用作有证标准物质。浓度变化作为时间的函数可以直线方程的形式表示,稳定性的标准不确定度等于直线斜率的标准偏差[54, 58-59]。如果有证标准物质用于实验室在很长一段时间的工作质量监控,则其稳定期是特别重要的。

（4）与被分析样品有相似的基体和痕量元素浓度水平。这一要求基于假设,即如果存在基体效应、光谱干扰等引起的系统误差,则它们对痕量元素测量结果的影响在有证标准物质和分析样品中是相同的或至少是相似的。

（5）能够简便地脱水干燥。是否需要进行水分的精确测定需要参考元素浓度认定值与"干质量"基体材料的关系,在大多数情况下认定值以有证标准物质的每"干质量"单位表示。在实际中,重要的往往不是材料中水分的精确测定,而是设计一种干燥方法,从而重复获得具有认定值的"干质量"有证标识物质。同时,所设计的干燥方法应简单易行且适用于大多数实验室。

有证标准物质所有重要的数据和说明都应该在证书中予以说明[28]。

4.9 有证标准物质的应用

有证标准物质在分析实验室的质量保证体系中发挥着重要的作用,特别是在痕量分析中显得尤为重要。通常,有证标准物质主要用于以下几方面:

（1）分析结果的准确度、精密度和可靠性的鉴定;
（2）开发新的分析程序;
（3）分析方法的验证;
（4）实验室或分析人员能力考核;
（5）不同分析方法的结果比对;
（6）内部质量控制(如使用休哈特控制图);
（7）测量溯源性的建立;

（8）测量设备的校准；

（9）实现和维持认证认可。

一个普遍推荐且被公认的检查准确度的方法是将有证标准物质与常规分析样本一起分析。显然用于此目的的标准物质 S 在基体类型和测定元素的含量水平应尽可能与分析样本相似,这样就可以假设由基体效应、光谱干扰等可能引起的系统误差在有证标准物质和分析样品中是相同的或至少是相似的。然后将 $X_m \pm U_m$ 的测量结果与认定值 $X_{ref} \pm U_{ref}$ 进行比较,其中 U_m 和 U_{ref} 分别为分析样品测量结果和认定值的扩展不确定度。测量结果与认定值吻合较好,表明分析正确。定量评估为：

$$|X_m - X_{ref}| \leq 2\sqrt{u_m^2 + u_{ref}^2} \tag{4-5}$$

式中：u_m 和 u_{ref} 分别为测量结果和认定值的标准不确定度($k=1$)。

$u_\Delta = \sqrt{u_m^2 + u_{ref}^2}$ 的数值用于估计与测量偏差相关的测量不确定度的组成[61],即当测量偏差很小时,所引入的不确定度与合成不确定度相比可以忽略不计。

示例：NBS1547 桃树叶中总 Se 的测定结果 $X \pm U(k=2)$ 为(127 ± 5) ng/g[62],认定值为(120 ± 10) ng/g。因此,标准偏差 $u_m = 5/2 = 2.5$(ng/g),$u_{ref} = 10/2 = 5$(ng/g),$u_\Delta = \sqrt{2.5^2 + 5^2} = 5.59$,$U_\Delta = 2u_\Delta = 11.2$。$|X_m - X_{ref}| = |127 - 120| = 7$,数值小于 U_Δ。因为两个值相差不大,所以该方法被认定为没有系统误差。

通常情况下,有证标准物质用于检验一个实验室得出的分析结果的正确度、精密度和可靠性,即用于检查实验室日常工作的质量。有证标准物质在特定的时间间隔内进行分析,所得到的结果用于绘制控制图(如休哈特图表)[63]。这实现了对出现的测量系统和系统误差等进行直观评估。因为有证标准物质具有均匀性和稳定性,以及通过与认定值进行比对来评估实验室获得结果的准确度的能力,所以其在控制图的构建中的应用是有利的。

如果证明所用的分析方法是准确的,相当于表明样品在包括溶解、萃取等制备阶段不会引起系统误差,否则这个阶段可能会破坏溯源链。这样测量结果就可以溯源到认定值表示的单位。

有证标准物质也可用于测量仪器的校准。例如,纯金属或合金有证标准物质用于工业实验室中的 X 射线光谱仪(XRF)和其他光谱仪器的校准。由于与认定值相关的不确定度 u_{ref} 对测量的总不确定度有贡献,因此 u_{ref} 应尽可能小。由此,在可能的情况下用来校准仪器的有证标准物质应该是纯物质或溶液。因为一般基体有证标准物质认定值的不确定度通常高于纯物质或溶液的不确定度,所以不推荐使用基体有证标准物质来进行测量仪器的校准。此外,使用这种方法分析人员不需要对所得结果正确度进行单独验证,是有证标准物质的主要应用优势。

使用有证标准物质是依据 ISO/IEC 17025:2005(条款 5.6.3)[33]获得并维持实验室认可的要求之一。有证标准物质在 ILC 中也可用作试验材料,然而这不是常见的做法,会使比对成本大幅增加。

4.10 有证标准物质的有效性

有关有证标准物质的信息可从多个来源获得,许多供应商会在他们的网站上提供数据库。在欧洲公认的有证标准物质的生产者是 IRMM、LGC 和 BAM,他们所生产的有证标准物质(统称为欧洲参考物质(ERM)的信息)可以在欧洲参考物质网站上找到。欧洲参考物质分为以下六个不同的类别。

(1) 工业和工程材料的成分认定;
(2) 健康相关基体材料的成分认定;
(3) 物理性能材料的认定;
(4) 食品/农业及相关基体材料的成分认定;
(5) 非基体材料纯度、浓度或活性的认定;
(6) 环境和相关的基体材料的成分认定。

图 4-2 展示了不同类别的有证标准物质的不同分类及占比。

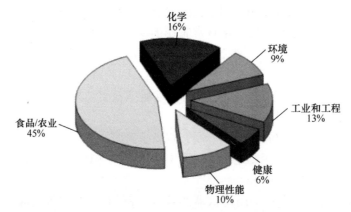

图 4-2 有证标准物质的不同分类及占比

波兰的无机痕量分析有证标准物质生产者 INCT 已发布了 10 种有证标准物质(表 4-5)。这些有证标准物质的详细内容可以在其网站上找到。

世界领先的有证标准物质生产者是 NIST、加拿大国家研究委员会(NRCC)和国际原子能机构(IAEA)。关于 NIST 生产的有证标准物质的信息可以在 NIST 网站找到。

表 4-5 INCT 生产的有证标准物质

CRM	具有认定值的元素	具有信息值的元素
CTA-AC-1 磷灰石精矿	Ba, Ca, Ce, Co, Cu, Eu, Gd, Hf, La, Lu, Mn, Na, Nd, Sc, Si, Sm, Ta, Tb, Th, Ti, U, V, Y, Yb, Zn	Al, Cr, Dy, Er, Fe, Ho, K, Mg, Ni, Pr, Sr, Zr
CTA-FFA-1 精制粉煤灰	Al, As, Ba, Ce, Co, Cr, Cs, Cu, Dy, Er, Eu, F, Fe, Gd, Hf, La, Li, Lu, Mn, Na, Nd, Ni, P, Pb, Rb, Sb, Sc, Si, Sm, Sr, Ta, Tb, Th, Tm, U, V, W, Y, Yb, Zn	Be, Ca, Cd, Ga, In, K, Mg, Mo, Se, Ti
CTA-OTL-1 东方烟叶	Al, As, Ba, Br, Ca, Cd, Ce, Co, Cr, Cs, Cu, Eu, K, La, Li, Mg, Mn, Na, Ni, P, Pb, Rb, S, Se, Sm, Sr, Tb, Th, V, Zn	Au, Cl, Fe, Hf, Hg, Mo, Na, Sb, Sc, U, Yb
CTA-VTL-2 弗吉尼亚烟草	As, Ba, Br, Ca, Cd, Ce, Cl, Co, Cr, Cs, Cu, Fe, Hf, Hg, K, La, Li, Mg, Mn, Mo, Ni, P, Pb, Rb, Sb, Sm, Sr, Tb, Th, U, V, W, Zn	Al, Eu, Na, S, Sc, Se, Si, Ta, Ti, Yb
INCT-TL-1 茶叶	Al, As, Ba, Br, Ca, Cd, Ce, Cl, Co, Cr, Cs, Cu, Eu, Hg, K, La, Lu, Mg, Mn, Na, Ni, Pb, Rb, S, Sc, Sm, Sr, Tb, Th, Tl, V, Yb, Zn	B, Fe, Hf, Nd, P, Sb, Se, Ta, Ti, Tm
INCT-MPH-2 波兰混合草药	Al, As, Ba, Br, Ca, Cd, Ce, Cl, Co, Cr, Cs, Cu, Eu, Hf, K, La, Lu, Mg, Mn, Nd, Ni, Pb, Rb, S, Sb, Sc, Sm, Sr, Ta, Tb, Th, V, Yb, Zn	Fe, Mo, Na, P, Ti, Tl, U, W
INCT-CF-3 玉米粉	B, Br, Cl, Cu, Fe, K, La, Mg, Mn, Mo, Ni, P, Rb, S, Sc, Zn	Al, As, Ba, Ca, Cd, Co, Cr, Cs, Hg, Na, Pb, Sb, Sr, Ti
INCT-SBF-4 大豆粉	Al, B, Ba, Br, Ca, Cl, Co, Cs, Cu, Fe, K, La, Mg, Mn, Mo, Ni, P, Rb, S, Sr, Th, Zn	Cd, Cr, Hg, Na, Pb, Sc, Sm, Ti, V
INCT-OBTL-5 东方芳香型烟叶	Ag, Al, As, B, Ba, Br, Ca, Cd, Ce, Co, Cs, Cu, Er, Eu, Hf, Hg, K, La, Mg, Mn, Mo, Nd, Ni, P, Pb, Rb, S, Sb, Sc, Sm, Sr, Ta, Tb, Th, V, Yb, Zn	Au, Be, Cl, Cr, Dy, Fe, Gd, Ho, Li, Lu, Na, Pr, Ti, Tl, Tm, U, Y
INCT-PVTL-6 波兰弗吉尼亚烟草	Ag, Al, As, B, Ba, Br, Ca, Cd, Ce, Co, Cu, Er, Eu, Hg, Hf, K, La, Li, Mg, Mn, Mo, Nd, Ni, P, Pb, Rb, S, Sb, Sc, Sm, Sr, Ta, Tb, Th, V, Zn	Bi, Cl, Cs, Fe, Na, Pr, Sn, Ti, Tl, U, Y, Yb

目前,国际原子能机构(IAEA)专门研制放射性同位素含量的有证标准物质,但也为无机痕量分析、有机分析和稳定同位素的测定提供有证标准物质。信息可以在国际原子能机构主页上找到。

ISO-ReMCO 在 1984 年建立了 COMAR 数据库[64]。目前,COMAR 数据库包括来自 20 多个国家的超过 200 个生产者的 1 万多个 RMS/CRM 的信息[65]。

由 IRMM 研制的 CRM 的信息可以在其网站上找到,BAM 的产品相关网站找到。

参考材料虚拟研究所(VIRM)通过其主页提供了广泛的质量保证和质量控制

工具,参考物质生产者列表、有证标准物质的大型数据库、通讯录等。登录 CRM 数据库是免费的,VIRM 提供的服务需要订阅费。

关于可用 CRM 的信息和他们选择的建议也可以从 LGC 标准的主页获得。

4.11 小结

CRM 是分析实验室质量保证和质量控制的重要工具,可用于正确度和精密度的验证,即验证测量结果和分析程序的可靠性、建立测量溯源性和测量设备的校准。CRM 的使用是 ISO/IEC 17025 标准推荐的,因此,对于希望获得和保持实验室认可的人来说是强制性的。

由于研制过程相当费力和耗时,因此 CRM 通常价格昂贵,个人选择使用时应该慎重考虑。值得注意的是,只有当 CRM 和测试样本的基体类型和测定元素含量水平尽可能相似时,基于 CRM 分析得出的结果准确性的结论才是完全可靠的。

还应特别注意标准物质证书中对有关制备过程、采用的认定方法、认定值及其相关的不确定度、材料的均匀性和稳定性、水分测定方法、溯源性说明、失效日期,以及如何使用标准物质的说明信息。

参考文献

[1] ISO/IEC Guide 99:2007: International vocabulary of metrology-Basic and general concepts and associated terms (VIM3). ISO, Geneva (2007)

[2] Dybczyński, R.: Reference materials and their role in quality assurance in inorganic trace analysis: problems of quality of trace analysis in the investigations of natural environment. In: Kabata-Pendias, A., Szteke, B. (eds.) (in Polish) Wyd. Edukacyjne Zofii Dobkowskiej, Warsaw (1998)

[3] von Lehmden, D.J., Jungers, R.H., Lee Jr., R.E.: Determination of trace elements in coal, fly ash, fuel oil, and gasoline. Preliminary comparison of selected analytical techniques. Anal. Chem. 46, 239–245 (1974)

[4] Dybczyński, R., Veglia, A., Suschny, O.: Milk powder (A-11): A new IAEA reference material for trace and other element analysis. In: Bratter, P., Schramel, P. (eds.) Trace Element Analytical Chemistry in Medicine and Biology, pp. 657–674. de Gruyter, New York (1980)

[5] Byrne, A.R., Camara-Rica, C., Cornelis, R., De Goeij, J.J.M., Iyengar, G.V., Kirkbright, G., Knapp, G., Parr, R.M., Stoeppler, M.: Results of a co-ordinated programme to improve the certification of IAEA milk powder A-11 and animal muscle H-4 for eleven "difficult trace elements". Fresen. Z. Anal. Chem. 326, 723–729 (1987)

[6] Zeiller, E., Strachnov, V., Dekner, R.: Intercomparison study IAEA-155 on the determination of inorganic constituents in whey powder IAEA/AL/034. IAEA, Vienna (1990)

[7] Dybczyński, R., Polkowska-Motrenko, H., Szopa, Z., Samczyński, Z.: New Polish certified reference materials for multielement inorganic trace analysis. Fresen. J. Anal. Chem. 345, 99–104 (1993)

[8] Dybczyński, R., Polkowska-Motrenko, H., Samczyński, Z., Szopa, Z.: Preparation and certification of the Polish reference material oriental tobacco leaves (CTA-OTL-1) for inorganic trace analysis. Raporty IChTJ Seria A nr. 1/96 (1996)

[9] Dybczyński, R., Danko, B., Kulisa, K., Maleszewska, E., Polkowska-Motrenko, H., Samczyński, Z., Szopa, Z.: Preparation and certification of the Polish reference material: mixed Polish herbs (INCT-MPH-2) for inorganic trace analysis. Raporty IChTJ Seria A, nr 4/2002 (2002)

[10] Dybczyński, R., Danko, B., Kulisa, K., Chajduk-Maleszewska, E., Polkowska-Motrenko, H., Samczyński, Z., Szopa, Z.: Final certification of two new reference materials for inorganic trace analysis. Chem. Anal. (Warsaw) 49, 143–158 (2004)

[11] Polkowska-Motrenko, H., Dybczyński, R., Chajduk, E., Danko, B., Kulisa, K., Samczyński, Z., Sypuła, M., Szopa, Z.: Polish reference material: corn flour (INCT-CF-3) for inorganic trace analysis - preparation and certification. Raporty IChTJ, Seria A, nr, 3/2006 (2006)

[12] Polkowska-Motrenko, H., Dybczyński, R. S., Chajduk, E., Danko, B., Kulisa, K., Samczyński, Z., Sypuła, M., Szopa, Z.: New Polish certified reference materials for inorganic trace analysis: corn flour (INCT-CF-3) and soya bean flour (INCT-SBF-4). Chem. Anal. (Warsaw) 52, 361–376 (2007)

[13] Samczyński, Z., Dybczyński, R.S., Polkowska-Motrenko, H., Chajduk, E., Pyszynska, M., Danko, B., Czerska, E., Kulisa, K., Doner, K., Kalbarczyk, K.: Preparation and certification of the new Polish reference material oriental basma tobacco leaves (INCT-OBTL-5) for inorganic trace analysis. Institute of Nuclear Chemistry and Technology, Warsaw (2011)

[14] Samczyński, Z., Dybczyński, R.S., Polkowska-Motrenko, H., Chajduk, E., Pyszynska, M., Danko, B., Czerska, E., Kulisa, K., Doner, K., Kalbarczyk, K.: Preparation and certification of the new Polish reference material Polish Virginia tobacco leaves (INCT-PVTL-6) for inorganic trace analysis. Institute of Nuclear Chemistry and Technology, Warsaw (2011)

[15] Dybczyński, R.: Considerations on the accuracy of determination of some essential and/or toxic elements in biological materials. Chem. Anal. (Warsaw) 47, 325–334 (2002)

[16] Versieck, J.: Collection and manipulation of samples for trace element analysis: Quality assurance considerations. In: Quality assurance in biomedical neutron activation analysis. IAEA-TECDOC-323. IAEA, Vienna, pp. 71–82 (1984)

[17] Emons, H., Falgelj, A., van der Veen, A.M.H., Watters, R.: New definitions on reference materials. Accred. Qual. Assur. 10, 576–578 (2006)

[18] ISO Guide 30:1981: Termsand definitions used in connection with reference materials. ISO, Geneva, pp. 1-5 (1981)

[19] ISO Guide 99:1993: International vocabulary of basic and general terms in metrology (VIM2). ISO, Geneva (1993)

[20] Koeber, R., Linsinger, T.P.J., Emons, H.: An approach for more precise statements of metrological traceability on reference material certificates. Accred. Qual. Assur. 15, 255-262 (2010)

[21] Emons, H.: Policy for the statement of metrological traceability on certificates of ERM certified reference materials. ERMhttp://www.erm-crm.org/ERM_products/policy_on_traceability/Documents/erm_traceability_statement_policy_final_5_may_2008.pdf. Accessed 15 May 20015

[22] Taylor, J.K.: Standard reference materials. In: Handbook for SRM users, NBS Special Publication. U.S. Government Printing Office, Washington, DC (1993)

[23] Konieczka, P., Namieśnik, J.: Quality assurance and quality control in the analytical chemical laboratory: a practical approach. CRC, Boca Raton (2009)

[24] Gawlik, B.M., Linsinger, T., Kramer, G.N., Lamberty, A., Schimmel, H.: Organic contaminants in water - conceptual considerations for the production of liquid reference materials in support of the new water framework directive. Fresen. J. Anal. Chem. 371, 565-569 (2001)

[25] Cali, J.P., Plebanski, T.: Standard reference materials: guide to United States reference materials. National Bureau of Standards Special Publication 260 - 57. U.S. Government Printing Office, Washington, DC (1978)

[26] ISO Guide 30:1992: Terms and definitions used in connection with reference materials. ISO, Geneva (1992)

[27] ISO Guide 30:1992: Amendments 1:2008, Revision of definitions for reference materials. ISO, Geneva (2008)

[28] ISO Guide 31:2000: Contents of certificates and labels. ISO, Geneva (2000)

[29] ISO Guide 32:1997: Calibration in analytical chemistry and use of CRMs. ISO, Geneva (1997)

[30] ISO Guide 33:2000: Uses of CRMs. ISO, Geneva (2000)

[31] ISO Guide 34:2009: General requirements for the competence of reference material producers. ISO, Geneva (2009)

[32] ISO Guide 35:2006: Reference materials - general and statistical principles for certification. ISO, Geneva (2006)

[33] ISO/IEC 17025:2005: General requirements for the competence of testing and calibration laboratories. ISO, Geneva (2005)

[34] www.ichtj.waw.pl

[35] Dybczyński, R.: Preparation and use of reference materials for quality assurance in inorganic trace analysis. Food Addit. Contam. 19, 928-938 (2002)

[36] Dybczyński, R., Polkowska-Motrenko, H., Samczyński, Z.: History, achievements and present time of production of CRMs for inorganic trace analysis in Poland. In: Proceedings of the 1st Scientific Conference, Reference materials in measurement and technology, part 1. (Federal A-

gency on Technical Regulating and Metrology) 10-14 Sept 2013, Ekaterinburg, Russia (2013)

[37] Polkowska-Motrenko, H., Dybczyński, R., Chajduk, E.: Certification of reference materials for inorganic trace analysis, the INCT approach. Accred. Qual. Assur. 15, 245-250 (2010)

[38] Dybczyński, R., Polkowska-Motrenko, H., Samczyński, Z., Szopa, Z.: Two new Polish geological-environmental reference materials: apatite concentrate (CTA-AC-1) and fine fly ash (CTA-FFA-1). Geostand Newslett 15, 163-185 (1991)

[39] Dybczyński, R., Danko, B., Kulisa, K., Maleszewska, E., Polkowska-Motrenko, H., Samczyński, Z., Szopa, Z.: Preparation and certification of the Polish reference material: tea leaves (INCT-TL-1) for inorganic trace analysis. Raporty IChTJ Seria A, nr 3 (2002)

[40] Polkowska-Motrenko, H., Dybczyński, R.: Activities of the INCT, Warsaw, in the domain of quality assurance for inorganic trace analysis. J. Radioanal. Nucl. Chem. 269, 339-345 (2006)

[41] Dybczyński, R., Danko, B., Polkowska-Motrenko, H.: NAA study on homogeneity of reference materials and their suitability for microanalytical techniques. J. Radioanal. Nucl. Chem. 245, 97-104 (2000)

[42] Dybczyński, R., Kulisa, K., Polkowska-Motrenko, H., Samczyński, Z., Szopa, Z., Wasek, M.: Neutron activation analysis as a tool for checking homogeneity of certified reference materials. Chem. Anal. (Warsaw) 42, 815-825 (1997)

[43] Dybczyński, R., Polkowska-Motrenko, H., Samczyński, Z., Szopa, Z.: Virginia tobacco leaves (CTA-VTL-2) - new Polish CRM for inorganic trace analysis including microanalysis. Fresen. J. Anal. Chem. 360, 384-387 (1998)

[44] Samczyński, Z., Dybczyński, R.S., Polkowska-Motrenko, H., Chajduk, E., Pyszynska, M., Danko, B., Czerska, E., Kulisa, K., Doner, K., Kalbarczyk, K.: Two new reference materials based on tobacco leaves: certification for over a dozen of toxic and essential elements. Scientific World Journal 2012, Article ID 216380 (2012). doi:10.1100/2012/216380

[45] Dybczyński, R.: Comparison of the effectiveness of various procedures for the rejection of outlying results and assigning consensus values in interlaboratory programs involving determination of trace elements or radionuclides. Anal. Chim. Acta 117, 53-70 (1980)

[46] Dybczyński, R. S., Danko, B., Polkowska-Motrenko, H., Samczyński, Z.: RNAA in metrology: a highly accurate (definitive) method. Talanta 71, 529-536 (2007)

[47] Dybczyński, R. S., Polkowska-Motrenko, H., Chajduk, E., Danko, B., Pyszynska, M.: Recent advances in ratio primary reference measurement procedures (definitive methods) and their use in certification of reference materials and controlling assigned values in proficiency testing. J. Radioanal. Nucl. Chem. 302, 1295-1302 (2014). doi:10.1007/s10967-014-3607-y

[48] Dybczyński, R.S.: 50 years of adventures with neutron activation analysis with the special emphasis on radiochemical separations. J. Radioanal. Nucl. Chem. 303, 1067-1090 (2015).doi: 10.1007/s10967-014-3822-6

[49] Kratochvil, B., Wallace, D., Taylor, J.K.: Sampling for chemical analysis. Anal. Chem. 56, 113R-129R (1984)

[50] Dybczyński, R., Danko, B., Polkowska-Motrenko, H.: Some difficult problems still existing in the preparation and certification of CRMs. Fresen. J. Anal. Chem. 370, 126–130 (2001)

[51] Ingamells, C.O., Switzer, P.: A proposed sampling constant for use in geochemical analysis. Talanta 20, 547–568 (1973)

[52] Ingamells, C.O.: A further note on the sampling constant equation. Talanta 25, 731–732 (1978)

[53] Stoeppler, M., Kurfürst, U., Grobecker, K.H.: Der Homogenitätsfaktor als Kenngrösse für pulverisirte Festproben. Fresen. Z. Anal. Chem. 322, 687–691 (1985)

[54] Linsinger, T.P.J., Pauwels, J., Van der Veen, A., Schimmel, H., Lamberty, A.: Homogeneity and stability of reference materials. Accred. Qual. Assur. 6, 20–25 (2001)

[55] Ellison, S.L.R., Burke, S., Walker, R.F., Heydorn, K., Mansson, M., Pauwels, J., Wegscheider, W., te Nijenhuis, B.: Uncertainty for reference materials certified by interlaboratory study: recommendations of an international study group. Accred. Qual. Assur. 6, 274–277 (2001)

[56] van der Veen, A.M.H., Linsinger, T., Schimmel, H., Lamberty, A., Pauwels, J.: Uncertainty calculations in the certification of reference materials. 4. Characterisation and certification. Accred. Qual. Assur. 6, 290–294 (2001)

[57] Pauwels, J., Lamberty, A., Schimmel, H.: Quantification of the expected shelf-life of certified reference materials. Fresen. J. Anal. Chem. 361, 395–399 (1998)

[58] Linsinger, T., Pauwels, J., Lamberty, A., Schimmel, H., Van der Veen, A., Siekmann, L.: Estimating the uncertainty of stability for matrix CRMs. Fresen. J. Anal. Chem 370, 183–188 (2001)

[59] Gellert, W., Kustner, H., Hellwich, M., Kasner, H. (eds.): Mathematics at a glance. Compendium. VEB Bibliographisches Institute, Leipzig (1975)

[60] Linsinger, T.: Comparison of a measurement result with the certified value. ERM application note 1 (2005). ERM, Geel. https://ec.europa.eu/jrc/sites/default/files/erm_application_note_1_en.pdf. Accessed 15 May 2015

[61] Ellison, S.L.R., Williams, A. (eds.): Eurachem/CITAC guide: quantifying uncertainty in analytical measurement, 3rd edn. http://www.eurachem.org (2002)

[62] Chajduk, E., Polkowska-Motrenko, H., Dybczyński, R.S.: A definitive RNAA method for selenium determination in biological samples. Uncertainty evaluation and assessment of degree of accuracy. Accred. Qual. Assur. 13, 443–451 (2008)

[63] Bulska, E.: Chemical metrology. the art of carrying out measurements (in Polish), p. 191. Malamut, Warsaw (2008)

[64] COMAR. International database for reference materials. BAM, Berlin. www.comar.bam.de (2006). Accessed 15 May 2015

[65] Eurachem Guide: The selection and use of reference materials (2002). www.eurachem.org

第5章
无机痕量元素分析中的样品分解技术

5.1 概述

许多现代仪器分析技术要求在分析前将固体或含固体的样品转化为溶液,"分解""破坏""消解""酸消解""溶解""灰化""湿灰化""氧化酸消解"和"矿化"等术语都是指这一过程。在本节中一般使用的表达是"分解",特指干法或湿法灰化。"矿化"仅指使分析物转化成无机化学形式的过程。可用于矿化的方法有很多,从将烧杯放置于电热板上进行常温常压湿法消解到采用专门仪器实现的高压微波加热。在经典的元素分析中分解样品需要将其矿化以除去有机成分,因此,将试样分解用于总元素测定似乎是所有情况下推荐使用的方法。

一般来说,分解过程需要将样品的原始化学状态转变成消解液,即分析物均匀分布的溶液。分解技术必须达到以下要求:

(1) 分解必须完全。无机材料应完全转化为可溶性化合物,有机材料必须完全矿化。

(2) 在检测中产生干扰的基体成分必须去除,残余物应定量溶于小体积的高纯酸中。

(3) 分解过程应尽可能简单,不需要复杂的设备。

(4) 分解应适应整个分析过程,能够调整分析物氧化状态从而与化学分解过程兼容。

(5) 应优先考虑一次实现分解和分离过程的分解方案。

(6) 为了减少分解过程中的由污染、元素损失以及分解不完全带来的系统误差,必须使用由惰性材料制成的清洁容器和最小量的高纯试剂,避免灰尘。反应室应尽可能小。应采取措施尽量减少分析物(元素)与容器材料吸附或化学反应以及挥发导致的损失。

(7) 分解操作不应对实验室人员造成危险或伤害。

(8) 分解步骤的产率应使用放射性示踪剂进行检验。

本章概述了不同材料的分解方法及其在近期的发展和应用。其他样品前处理方法,如化学萃取和浸出、碱溶、酶分解、热分解和阳极氧化,都超出了本书的范围而未加讨论。

5.2 参考书目

大量出版物提供了关于基体和分析物在各种可能的组合下进行分解的有效参考信息。一些综合性的书籍和综述文章包含与有机[1-4]基体或无机[5-10]基体以及两者兼有[11-21]的分解资料。

本章全面讨论分解技术是不可能的,更全面的信息许多综述和书籍是可用的。Šulcek 和 Povorda[8]、Bock[11] 以及 Krakovská 和 Kuus[17] 的书籍仅涉及分解方法,其他书籍专门涉及某种单一的技术如微波辅助样品前处理技术[22-23],该技术也在文献[24-34]进行过综述。不同基体的样品前处理分解方法推荐指南也可从分析化学百科全书中找到[35]。尽管无法引用这一领域发表的每篇论文,但本章的参考文献清单全面综述了这一主题的最新进展、潜在的应用、创新的发展以及分解技术的进展。

5.3 样品分解技术

表 5-1 概述了有机材料和无机材料的不同分解方法。其目的不是展示各种材料基体的分解过程细节,而是说明每种分解技术对某类样品的独特匹配特性。

表 5-1 不同分解方法列表

	分解技术	必需的试剂	应用
湿法化学分解	在开放系统内 — 酸分解(热对流湿法分解)	HNO_3, HCl, HF, H_2SO_4, $HClO_4$	无机/有机
	微波辅助湿法分解	HNO_3, HCl, HF, H_2SO_4, $HClO_4$, H_2O_2	无机/有机
	紫外分解(光解)	H_2O_2, $K_2S_2O_4$, $HClO_4$, HNO_3	水、悬浮液
	声辅助酸分解	H_2O_2, HNO_3	无机/有机
	密闭系统 — 使用传统方法加热(热对流高压消解)	HNO_3, HCl, HF, H_2O_2	无机/有机
	使用微波加热	HNO_3, HCl, HF, H_2O_2	无机/有机
	流动系统 — 使用传统方法加热	HNO_3, H_2SO_4, H_2O_2, HCl	无机/有机
	使用微波加热	HNO_3, H_2SO_4, H_2O_2, HCl	无机/有机
	紫外分解	H_2O_2, $K_2S_2O_4$, HNO_3	水、悬浮液

续表

分解技术			必需的试剂	应用
湿法化学分解	气相酸分解	使用传统方法加热	HNO_3, HCl, HF, H_2O_2	无机/有机
		使用微波加热	HNO_3, HCl, HF, H_2O_2	无机/有机
燃烧	开放系统	干灰化		无机/有机
		低温灰化(氧气流中的燃烧)		有机
		冷等离子体灰化(Wickbold 燃烧)		有机
	密闭系统	氧瓶燃烧(Schöniger 烧瓶)		有机
		氧弹燃烧		有机
		动态系统中的燃烧(Trace-O-Mat)		有机
熔融分解			熔剂	无机

5.3.1 湿法化学分解

开放系统中的湿法分解是古老但仍然常用的方法之一,湿法分解可用于封闭系统。样品湿法分解是将基体组分转化为简单的化学形式,这种分解是通过提供能量,如加热,或使用酸一类的化学试剂,或通过这两种方法的组合来实现。当使用试剂时,其性质根据基体性质来选择,所用试剂的量取决于样品量的大小,同时取决于测定方法的灵敏度。然而,将材料转化为溶液的过程往往是分析过程中的关键步骤,因为存在许多潜在的误差来源,包括分析物的部分分解、来自所使用容器的某种类型的污染等。讨论所有可能的系统误差超出了本书的范围,在文献[36]讨论了如何避免样品分解过程中的系统误差。

大多数湿法分解方法涉及使用氧化性酸,如 HNO_3、热浓 $HClO_4$、热浓 H_2SO_4 等及非氧化性酸(如 HCl、HF、H_3PO_4、稀 H_2SO_4、稀 $HClO_4$)和过氧化氢的一些组合。这些酸都是腐蚀性的,特别是在热和浓的情况下,应小心处理以免发生损伤和事故。具有高纯度的浓酸在商业上是可获得的,但它们也可以通过亚沸蒸馏进一步纯化[37]。

湿法消解的优点是对无机材料和有机材料都有效。它通常破坏或移除样品基体,从而有助于减少或消除某些类型的干扰。表 5-2 概述了用于样品前处理的常用矿物酸的物理性质。

表 5-2 用于湿法分解的常用矿物酸的物理性质

化合物	公式	分子量	含量 质量分数/%	含量 摩尔浓度/mol/L	密度/(kg/L)	沸点/℃	备注
硝酸(Ⅴ)	HNO_3	63.01	68	16	1.42	122	68% HNO_3,共沸混合物
盐酸	HCl	36.46	36	12	1.19	110	20.4% HCl,共沸混合物
氢氟酸	HF	20.01	48	29	1.16	112	38.3% HF,共沸混合物
高氯酸(Ⅶ)	$HClO_4$	100.46	70	12	1.67	203	72.4% $HClO_4$,共沸混合物
硫酸(Ⅵ)	H_2SO_4	98.08	98	18	1.84	338	98.3% H_2SO_4
磷酸	H_3PO_4	98.00	85	15	1.71	213	分解成 HPO_3
过氧化氢	H_2O_2	34.01	30	10	1.12	106	

从使用温度或试剂来说,大多数湿法分解过程的条件是极端的。因此,制造烧瓶、坩埚和其他工具所用的材料必须根据应用的具体程序选择。制作消解装置所用的材料通常是分析空白的来源,待测元素既可以从材料中溶解,也可以从消解容器表面解吸。因此材料的性质是非常重要的。评估消解容器材料的适宜性可从其耐热性和传导性、机械强度、耐酸性和耐碱性、表面性质、反应性和污染性等方面来考虑,也必须考虑有机材料和无机材料的特殊性质。表 5-3 列出了湿法分解容器的优选材料。在分解过程中与样品接触的容器材料也经常引起系统误差。元素可以从材料中溶解、解吸或吸附在容器表面上,其量值取决于材料、接触时间和温度。表 5-4 总结了各种容器材料可能遇到的无机杂质。硼硅酸盐玻璃含有几个浓度相对高的主要、次要和微量元素,通常不适合在极痕量范围内的元素测定。石英可以看作纯物质,并且很容易获得不同等级的纯度,高纯度石英是痕量金属分析中的大多数样品前处理过程的优选容器或工具材料。另外,高纯度合成聚合物可在许多分解过程中应用,如聚乙烯(PE)、聚丙烯(PP)、聚四氟乙烯(PTFE)和超高分子聚合物(如 PFA、FEP、TFM)。用于湿法分解过程的设备和容器必须严格清洗,并测试任何可能的污染。通常将烧杯在浓硝酸中煮沸,然后在使用前用超纯水冲洗几次。为了避免这种程序不充分,最有力的清洗程序之一是将容器置于装有硝酸

或盐酸的密封 TE 容器[38]用微波加热蒸煮。这一程序特别适用于石英、硼硅酸盐玻璃和聚四氟乙烯容器。

表 5-3 湿法分解容器的优选材料

材料	化学名称	工作温度/℃	热变形温度/℃	吸水率/%	备注
硼硅玻璃	SiO_2[①], B_2O_3[②]	<800[③]			常规实验室玻璃不适用于湿法分解程序
石英	SiO_2[④]	<1200			用于有机材料湿法分解的所有程序,石英是容器材料的最合适材料
玻璃碳	石墨	<500			玻璃碳以坩埚的形式和碱性熔融物的形式使用,作为湿分解过程的容器
PE	聚乙烯	<60			
PP	聚丙烯	<130	107	<0.02	
PTFE	聚四氟乙烯	<250	150	<0.03	PTFE 通常仅用作高压分解系统中的分解容器
PFA	全氟烷氧基	<240	166	<0.03	
FEP	四氟异戊二烯	<200	158	<0.01	
TFM	四氟甲氧西林			<0.01	

注:① SiO_2 含量为 81%~96%;
② B_2O_3 含量为 3%~5%;
③ 在 800℃温度时软化;
④ SiO_2 含量为 99.8%。

表 5-4 容器材料中的无机杂质

元素	硼硅酸盐玻璃	石英	聚乙烯	PTFE 特氟龙[①]	玻璃碳
Al	Main[②]	100~50000	100~3000	/	6000
As	500~22000	0.1~80.0	/	/	50
B	Main	10~100	/	/	100
Ca	10^6	100~3000	200~2000	/	80000
Cd	1000	0.4~10.0	/	/	10
Co	100	1	0.5	2	2
Cr	3000	3~5	20~300	30	80
Cu	1000	10~70	/	20	200
Fe	$2×10^5$	200~800	1000~6000	10~30	2000

续表

元素	硼硅酸盐玻璃	石英	聚乙烯	PTFE 特氟龙①	玻璃碳
Hg		1		10③	1
Mg	$6×10^8$	10	100~2000		100
Mn	6000	10			100
Na	Main	10~1000	200~10000	25000	350
Ni	2000				500
Pb	3000~50000		200		400
S	Main	Main			85000
Sb	8000	1~2		0.4	10
Ti	3000	100~800			12000
Zn	3000	50~100	100	10	300

注：数据单位为 ng/g；
① 特氟龙是 DuPont 的注册品牌；
② Main 代表重要，该元素的浓度很高，但没有特定的值；
③ 取决于存储条件。

硝酸是一种几乎通用的分解试剂，也是最广泛使用的分解有机物的初级氧化剂，因为它对大多数测定无干扰，并且在商业上具有足够的纯度。过氧化氢和盐酸可以有效地结合硝酸作为改善分解质量的一种手段。盐酸和硫酸会干扰稳定化合物的测定，盐酸通常用于主要含有无机基体的样品，与氢氟酸组合可用于分解不溶于其他酸的硅酸盐。使用高氯酸时，安全性尤其重要。

1. 开放式湿法分解

作为最古老的技术之一，开放式酸消解化学实验室中有机样品和无机样品分解或溶解最常用的方法。这种低成本的技术对于常规分析具有非常重要的价值，因为它可以很容易地实现自动化，所有时间、温度、分解试剂种类等相关参数都可以直接控制。

湿法分解与干法灰化相比，其主要优点是速度快。然而，该方法最大分解温度不能超过相应酸或酸混合物的大气压力下的沸点。例如，在硝酸溶液沸点 122℃ 温度下，硝酸对于许多基体的氧化能力是不充分的。解决办法之一是添加硫酸显著增加分解溶液的温度；但是否可行取决于基体和测定方法如高脂肪和高蛋白样品通常不会在大气压下完全分解。其他缺点主要是实验室环境污染、需加入大量试剂以及微量元素损失的风险，可以通过使用过量的酸与冷凝回流器，并通过优化温度和持续时间来降低损耗。然而，从工作场所的安全角度来看应优选在大气压下操作的系统。

1）常规加热湿法分解

常规加热湿法分解需要配备一个常规加热源,如本生灯、加热板、砂浴等,加热系统要么运行在一个固定的温度,要么响应于温度程序。酸分解通常可以在任何容器中完成,如玻璃或聚四氟乙烯材质的烧杯、锥形瓶中。然而,当样品在开放的湿法消解中消解时,需配备具有回流功能的冷凝器。Bethge[39]对其必要的设备进行了描述。在过去的几十年来,开放式加热分解系统在样品分析中很受欢迎,但一直受到易腐蚀和随之而来的污染风险等主要缺点的困扰(图5-1)。因此,开放式加热消解系统,即加热板技术不被认为是痕量和超痕量样品前处理的最先进的技术。石墨消解系统正变得越来越受重视,该系统克服了传统的由不锈钢或铝制装置的缺点,因为加热区是由石墨制造的,通常涂有一种含氟聚合物,以防止在处理样品过程中来自系统表面的金属污染。石墨消解系统目前改善了传统占主流的开放和密闭容器消化技术,它可以同时消解大量的样品,从而克服密闭容器的主要弱点。常用的分解试剂有硝酸、硫酸、氢氟酸、高氯酸和过氧化氢以及它们的各种组合。湿法分解的大多数应用涉及水或有机基体,如地表水、废水、生物和临床样品、食品样品,以及土壤、沉积物和污水污泥、煤、高纯材料等各种材料。最近,更多开放式消解系统已经研制发展起来,梯度分解通常用装有回流冷凝器的若干容器来消除一些分析物可能的挥发损失,并避免反应混合物的蒸发。这样的组合满足了消解大量样品的需要。现代商业上可用的 Hach Digesdahl 消解设备(HACH公司)被设计用于消解有机样品和矿物样品。

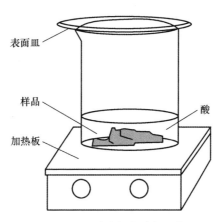

图5-1 开放式加热分解系统

2）微波加热湿法消解方法

湿法分解过程中最具创新性的能源是微波。由于加热分解发生在混合物内部,因此采用微波辅助分解比用常规加热方式分解更有效。对于某些难溶样品来说,使用微波不管是分解速度还是效率都有提高。此外,微波加热可以与一些仪器一起实现自动化。

自 Abu-Samra 等[40]报道了微波技术在生物样品湿法分解中的应用,微波辅助分解技术在元素分析方面发展迅速。文献[15-36]详细介绍了微波辅助分解在多种样品类型中的应用,如地质、生物、临床、植物学、食品、环境、污泥、煤和灰分、金属、合成物、混合物,以及将微波消解作为特定的实验条件来消化基体。最早的尝试是使用国产微波炉进行微波辅助敞口湿法消解(图5-2),当时没有可用的商业消解设备,考虑到安全性和功能性,家用微波炉不应在实验室中使用。在常压下的开放系统中,微波辅助分解通常只适用于简单基体,或经严格定义并遵守特定的分解参数时结果才是可复现的。White[24]和 Mermet[25]的综述聚焦微波辅助系统的性能和丰富的应用,Nóbrega 等[41]描述了微波辅助样品前处理高达 10g 的大量有机样品的实验方案,但其中的汞或有机金属化合物可能会有损失。使用常压设备时为了达到一个高的分解温度,添加硫酸是必不可少的,因为酸或酸混合物的沸点决定了最高分解温度。然而,硫酸盐对许多金属的测定都存在干扰。虽然非加压微波系统受到最高消解温度偏低的限制,但是由于不会发生超压的情况,它们依然是确保人员安全的最佳选择。此外,非加压微波消解适合于连续流动系统的在线分解。

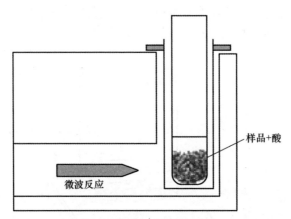

图 5-2 微波辅助敞口湿法消解

Matusiewicz[42]介绍了一种不同于其他文献研究中的用于固体和液体样品预处理的微波系统,它包括独特的仪器方法、各种商业上可用的系统及其操作参数和附件。

3) 紫外光分解方法

紫外(UV)光分解主要用于未污染天然水基体样品(如海水、淡水、河流、湖泊、地表、河口和海岸水等)或轻微污染的水样(如饮料、特殊工业废水、来自污水处理厂的水、土壤提取物等)在少量过氧化氢、酸(主要是HNO_3)或过硫酸盐的存在下,通过UV光辐射分解液体或悬浮液[43],溶解的有机物与待测物的络合物被分解并产生游离的金属离子。相应的分解容器应该放置在最接近光源UV低压或高压灯的位置,以确保高光通量(图5-3)。在光分解中,分解机理可以通过水和过氧化氢由UV光辐射引发而形成的OH^*和O_2^*自由基来表征[43]。这些反应性自由基能够氧化简单基体中存在的约含100mg/L碳的有机物质,生成二氧化碳和水。只有在简单的基体的光分解或光分解与其他分解技术相结合的情况下才有可能消除基体效应[44]。但是,该方法不能氧化所有可能存在于水中的有机成分,如氯化酚、硝基酚、六氯苯等类似化合物仅能部分氧化。因为高挥发性元素可能会发生损失,因此样品的有效冷却是必要的。为了产生澄清的样品溶液,过氧化氢的添加可能需要重复几次。现代有市售的UV分解系统(文献[43]和表5-1)。

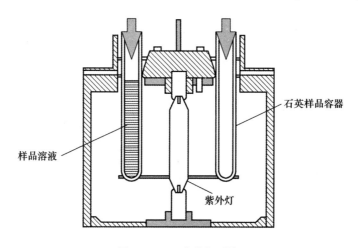

图5-3 UV光分解系统

4) 超声辅助酸消解方法

尽管人们对超声波的使用没有太多兴趣,但超声波能量通常超过其他常规辅助能量。因此,超声波对固体样品的预处理有很大帮助,能够促进和加速溶解、熔融和分解等。图5-4为超声辅助酸消解装置,其中包括水浴和超声波探头。该方法是采用最小频率16kHz的超声波通过超声效应产生快速的液体运动而形成大量微镜空腔,从而产生自由基分散化学层,并加速反应成分之间的接触。通常超声效应在非均相体系中比在均相体系中强烈得多,有利于两相体系产生乳化效应,并且促使两相体系中的质量传热增加,这些效应已应用于农业、生物和环境化学中的样品前处理[45-46]。

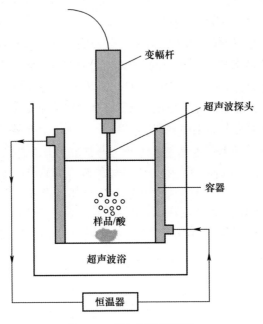

图 5-4 超声辅助酸消解

2. 封闭系统湿法分解

几十年来,用密闭容器进行湿法样品前处理的方法得到了广泛应用。封闭系统的优点是操作基本上与实验室大气隔离,从而使环境污染大大降低。样品的分解基本上是通过常规的湿法消解过程来完成,同时伴随高温高压的协同作用,在高于沸点的温度下发生分解。事实上,压力只不过是样品分解过程中不可避免的副产物,封闭消解技术通常比开放式系统中的常规湿法分解更有效,而且避免了挥发性元素的损失,也减少了对空白值的任何贡献,对分解难溶样品也更有效。几十年来,封闭消解方法的应用积累了大量相关经验,密闭系统分解特别适用于痕量和超痕量分析,尤其是当样品量受到限制时。

由于分解试剂的氧化能力明显依赖温度,低压分解和高压分解时的反应温度显著不同。低压分解(20bar, 1bar = 10^5 Pa)限制在约 180℃ 的温度下,而高压装置(70bar)的分解温度可以超过 300℃。

1) 常规热对流湿法压力分解

在密封管中进行无机和有机物质分解是 19 世纪末最早采用的压力消解方法,至今在某些领域仍在应用,而且不能用其他消解方法取代。密封玻璃管的使用可以追溯到 1980 年,由 Mitscherlich[47] 和 Carius[48] 首次提出,通常称为卡里乌斯(Carius)技术。Carius 在 250~300℃ 下用浓硝酸消解有机样品,样品和酸在厚壁石英安瓿瓶中混合、密封,安瓿瓶被转移到"弹筒"中,并在"弹炉"中加热几小时,然后冷却,打开,进行含量分析。Carius 管(图 5-5)在 240℃下分解产生超过 100bar 的内压。为了安全起见,使用足以容纳 Carius 管的不锈钢套筒作为外部压力容器[49],套筒中放置固体 CO_2 颗粒,加热时与管壁横面保持相等的压力。

随着 Carius 管的发展诞生了密闭容器分解技术。1894 年,Jannasch[50] 首先提出了将样品放入带有金属内反应容器的高压釜中进行分解,但由于金属容器的易被严重腐蚀等一系列缺点而未被广泛采用。

在分析过程中广泛使用压力分解技术始于 1960 年,这是因为当时在有机聚合物制造领域有了相当大的技术进步。对流加热压力容器系统被证明是保证固体样品完全或几乎完全消解的最有价值的系统,因为它们可以提供 200℃~230℃ 的较高消解温度[51]。尽管带有 PTFE 托架[57]的特殊石英容器或玻璃碳容器[58]适用于痕量分析,但大多数用于热对流压力消解的样品容器由 PTFE[52-54]、PFA[55] 或 PVDF[56]制成,并安装在不锈钢压力高压釜中,然后在实验室干燥箱、加热炉或加热块中加热到所需温度(图 5-6)。

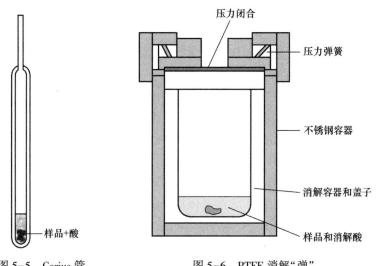

图 5-5　Carius 管　　　　图 5-6　PTFE 消解"弹"

为了满足大批量样品检测需要,已经开发出机械化多样品压力消解系统[59]。为了能将烘箱或加热块中消解罐里已处理好的样品溶液快速移出并检测[60],须将一个冷却回路插入金属外壳或护套中。通过混合反应物也可以加速溶解,最好使用覆盖 PTFE 搅拌棒[61]。Uchida 等[62]提出了另一种设计方案,将用于消解的小螺帽瓶放置在一个聚四氟乙烯(特氟龙)双消解容器内,设计了一种带有特氟龙内衬膜的压力计和热电偶的系统[63],改善了压力-温度和某些材料的碳平衡。最近,研究人员又提出了一种用于对流炉的消解容器[64],其独特的设计包括具有3嵌套结构的容器:最内层为 PTFE 的容量 30mL 的容器、容量为 100mL 的中间 PTFE 容器、外部不锈钢外壳。

高压热消解的缺点是预热和冷却消解液以及打开样品容器[65]均需要花费大量时间,样品大小也是有限的,并且不可能直观地观察消解的进展。Langmyhr 等为消解容器(或称为消解弹)的商业化做出了贡献,如今已经有广泛的应用市场,如流行的 Parr 酸消解弹(Parr 仪器公司,美国)、Uniseal 分解容器(Uniseal 分解容器,以色列)、特氟龙内衬的不锈钢压力容器(Berghof of Laborprodukte,德国)、压力分解系统 CAL 130FEP(Cal Laborgeräte,德国)和压力消解系统 PRAWOL(Fleischhacker KG,德国)。

为了避免在高温下失去机械稳定性,新研制的压力消解系统使用了石英制成的容器[66-67]。Knapp[68]介绍的高压灰化技术,不仅降低了有效消解时间,而且开发了碳、碳纤维、矿物油等耐高温材料的消解方法。由 Knapp 开发的高温高压(320℃,130 bar)湿法消解的优化系统,已经作为 HPA-S 高压灰化系统(Anton Paar,奥地利)进行商业应用。

最近,为了有效分解有机废料开发了一种使用红外加热的消化技术[69],其是 HPA(IR-HP-asher)装置的原型。物料在高压釜内的 6 个石英容器中进行高压分解,在 130 bar 的压力下,最高消解温度高达 300℃。这种技术的创新点在于红外加热的 HPA 系统设计。

由于金属高压釜价格昂贵,因此设计人员设计了一种没有金属外壳的压力容器。该容器可以用螺母[70]良好地密封。挥发性组分在加热过程中不损失,因此酸蒸气不会在实验室释放造成污染。全特氟龙厚壁 PTFE 容器已应用于难溶性海洋悬浮物消解,同时使用 HCl、HNO_3 和 HF[71]。半透明的 Nalelne 密封瓶已应用于 $HClO_4$ 和 HNO_3[72]的混合酸湿法消解鱼、鸟、植物组织等生物材料。另外,使用线性 PE 瓶的压力消解技术可用于天然基体样品[73],而且 PE 容器是透明的,可以观察整个消解过程。在完全由石英制成的封闭系统中,利用轻微超压(4 bar)对脂肪材料进行完全分解是可能的[74]。文献[75]设计了一种采用常规加热炉消解的 30mL 容量的封闭式聚四氟乙烯弹,由模压加工而成的带螺母和特氟龙-TFE 泄压棒的容器。

2)微波辅助加压湿法分解

封闭容器微波辅助分解技术是"清洁"化学应用的最佳消解方法之一,并且与其他封闭容器技术相比,它具有在相对较低的压力下就可以达到高温的独特优势,用于微波酸消解的容器是低压或高压弹。目前的微波封闭容器是两件式设计:由高纯度特氟龙或 PFA 制成的衬套和盖子,由聚醚酰亚胺和聚醚酮或另一种强微波透过复合材料制成的外壳(图 5-7)。最高工作温度为 260℃,即特氟龙的软化点,其压力极限为60~100bar。对于能在HNO_3和/或HCl中溶解的样品来说,密闭容器是理想的选择。

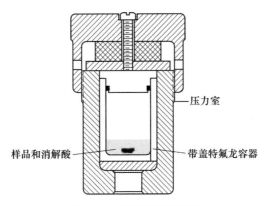

图5-7 微波酸消解"弹"

微波加热液体但蒸汽不吸收微波能量,因此,汽相的温度低于液相的温度,蒸汽在冷的容器壁上冷凝,实际蒸汽压低于预测蒸汽压力,这种持续的热非平衡动态方式是微波技术的一个关键优势,因为在相对低的压力下可以达到非常高的温度,所以可以缩短消化时间。

微波压力消解研究的最初灵感来自美国矿务局的报告[76],该报告描述了使用微波炉来加热聚碳酸酯瓶中的酸混合物样品,可以实现一些矿物样品的快速溶解。由于聚碳酸酯树脂具有优良的化学和力学性能,Smith等[77]用其代替了PFA。Buresch等[78]使用聚四氟乙烯或石英制成低压释放容器;Alvarado等[79]开发了带有PP螺母的改性厚壁Pyrex玻璃试管为加压容器;Kojima等[80]用PP夹套双层特氟龙容器对消解弹进行改进,使样品无泄漏、安全地分解;在文献[81]中也描述了一个密闭容器微波消解系统。在有机物酸分解过程中,对密闭PFA容器中升高的温度和压力的原位测量证实,通过监控温度和压力可以实现分解可控和分解机理的研究。

实验室用于低压或中压工作的全特氟龙弹,也适用于微波加热消解[82],特别适用于带有泄压孔、阀或防爆膜等的情况。

低容量微波辅助分解方法已开始进行小尺寸样品的应用研究,在大型消解设备中的样品损失是不可避免的。文献[83]中提到在一个改进后的特氟龙衬套的Parr微波酸消解弹中,用高纯硝酸分解的干重为5~100mg纸巾。文献[84]介绍了用7mL低容量特氟龙-PFA密闭容器消解100mg干质量的生物组织样品。

在对有机物含量较高的1g样品进行封闭微波酸分解的过程中,为了防止出现压力上升过快的情况,设计了一种带回流的开放容器预分解技术,释放二氧化碳等氧化产物,而不发生酸或分析物的蒸发损失。预分解之后,容器被封盖在微波压力下完成消解[85]。

为了缩短微波酸消解后安全打开特氟龙压力容器的时间,进而显著减少样品制备时间,须将消解用压力罐浸渍在液氮中[86]。还有研究是设计了一种特殊类型的特氟龙弹,可以通过冷却内部石英或特氟龙螺旋盖的方式使特氟龙弹中的蒸汽压力保持在中等水平(一般不超过 5bar),同时在消解过程中冷凝下来的酸和水蒸气的回流持续地更新样品的液相[87]。

Pougnet 等使用 500 W 或 1200W、2.45GHz 的基本模式微波导腔,开发了几种微波加热装置[88-89],目前该微波加热装置用在实验室样品分解和其他应用中。

Légère 和 Salin[90-91]详细综述了胶囊的概念。样品被包封在胶囊中直到在消解溶剂中消解。胶囊微波消解系统的操作按照几个步骤进行,过程中需监测温度和压力。

从前面的讨论中可以清楚地看出,微波酸消解适用于密闭容器消化,它的应用局限在于需要使用能够透过微波的非金属材料制成的封闭特氟龙衬里容器,操作时最大的上限安全压力在 60~100bar。为了突破这些限制,Matusiewicz[92-93]开发了一种聚焦微波加热弹,它能改善现有微波消解系统的消解能力,并允许构建一个集成的微波源/弹组合,能够进行水或液体的原位冷却。另一个替代容器结构集成了周围的微波室。这包括一个或多个微波可透过特氟龙、石英等材质的容器,再密封和封装在耐酸不锈钢腔室中[94],不锈钢腔室既是压力容器又是微波室。现代系统可以在 320℃和 130~200bar 的条件下处理酸分解。

最近,开发了一种新型微波辅助高温紫外消解系统,用于加速分解溶解的有机化合物或浆料[95-96]。该技术基于封闭的、加压的微波分解装置,装置中炉腔的微波场操作浸没式无极 Cd 放电灯(228nm)产生 UV 照射,使反应温度最高可达250~280℃,使矿化效率大大提高,有许多微波消解弹和系统可选用[42]。

3. 流动系统湿法分解

无论是在升高压力还是在大气压下的独立的容器消解系统,都需要大量的操作过程,即便在普通的炉子或微波炉中组装、关闭、打开和定位容器的过程是费时费力的。用通流管线圈来代替连续加热消解、紫外分解和微波消解系统中的容器的设计,克服了操作过程的一些局限性。样品被泵入含有消解装置的线圈,然后通过热量、UV 或微波等方式加热消解(图 5-8)。连续流动的载流溶剂通过时对系统进行了清洗,消除了需要清理容器的繁琐程序。这些系统可以控制温度和压力应对突然不稳定的发生或产生不稳定样品。目前已经有许多不同设计的流动消解系统,但很少能满足样品高性能分解的先决条件。

1) 常规加热分解

使用流动技术在封闭系统中自动进行样品前处理,可以克服样品消解的许多

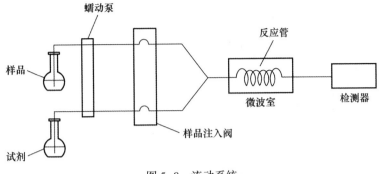

图 5-8 流动系统

缺点。

Gluodenis 和 Tyson[97]提出了一个成熟的消解系统,其中,PTFE 管被松散地置入电阻加热炉中。通过使用 PTFE 管,限制最大消解温度为 210℃。材料有限的机械强度仅允许 35bar 的最大工作压力,因此,通常的工作压力为 10~20bar。在停止流动、增压的条件下,以各种形式注入将 10% HNO_3 中形成的可可粉浆料进行消解,说明了该系统的潜力。

Berndt[98-100]开发了一种用于生物和环境样品连续分解的新型高温、高压流动系统,当电加热特氟龙内衬高效液相色谱管作为消解毛细管时,流动系统温度达 260℃,压力达 300bar,血液、肝脏、叶片等生物样品经消解后在流动系统的出口可以被收集。在随后的研究[97-98]中,电加热的 Pt/Ir 毛细管可在 320~360℃和 300 bar 的条件下用作能耐受浓酸的消解管,由于其不是玻璃材质,因此消解管可以通过加入氢氟酸来消解具有高硅酸盐含量的样品。

2) 紫外在线分解

紫外分解是一种清洁的样品前处理方法,它不需要使用大量的氧化剂。此外,紫外分解也是有效的,可以较容易地合并至流动注射模块中。样品在 H_2O_2、H_2SO_4 或 HNO_3 存在的情况下,流过盘绕在紫外线灯上的 PTFE、石英等消解管。文献[101]最近对这类流动系统进行了简要综述。市售的分析仪器由 Skalar Ana Analytical(荷兰)生产。

Fernandes[102]基于在线紫外/热诱导两阶段进行消解,开发了一种氧化反应模块。紫外消解装置是将一根长 4m 的 PTFE 管直接紧紧缠绕在 15W 的紫外线源上形成一个螺旋反应器。热消解容器是一根盘绕成螺旋物长 2m 的 PTFE 管,该螺旋管浸入在 90℃恒温浴中。

3) 微波加热流动消解系统

微波加热流动消解系统的许多不同设计已经发表[25, 32, 101],主要是为元素分析时全自动样品处理创造了新的可能性。

这一领域早期的工作是由 Burguera 等[103]报道的,他采用流动注射系统在线分解样品,并采用火焰原子吸收光谱(FAAS)法测定了 Cu、Fe、Zn 等金属元素。该方法是将实验试剂和样品在线混合,然后血清、血液或血浆等样品在位于微波炉内的 Pyrex 线圈中分解。这种方法本质上允许连续样品的分解,大大减少了样品处理时间,并且适用于需要温和分解条件的样品,尤其是液体样品。

根据系统中消解单元的位置,迄今为止的文献中描述了注射单元之前和之后两种类型。在前期制备模块中,样品以连续的流动[104]或中断流动模式[105]引入微波消解炉中,分解后注入的样品与待分解的试剂一起进入微波炉单元,然后在传送到检测器之前冷却和脱气[106]。在这两种情况下测量可以部分或全部离线或在线进行。

固体样品需要更复杂的流动系统,因为它们需要在高浓度酸的存在下消解,以达到迅速破坏有机基体的目的。1988 年报道了一个旨在简化消解操作的初步尝试[107],将冻干后的肝脏和肾脏样品研磨和称重后与无机酸一起放置在试管中,将其摇匀后再置于微波辐射,以避免剧烈的反应形成大量泡沫,试管被装入一个具盖 Pyrex 罐内并置于一个给定功率下特定时间内运行的家用微波炉内。Carbonell 等[104]建立了悬浮液方法与流动注射系统微波炉消解相结合的固体样品中金属元素的测定方法,用 FAAS 测定了铅。用同样的方法对洋蓟、巧克力、污水污泥、番茄叶等各种天然样品进行了测定,加入 HNO_3 和 H_2O_2 混合后进行磁力搅拌,然后用连续泵送至一个开放的再循环系统中,该循环系统的 120cm PTFE 管部分放置于家用微波炉内。

文献[108]设计了一种微波加热流动分解容器,即特氟龙缠绕管,其与商用聚焦微波系统(Prolabo A300)一起用于在线生物样品的前处理,包括牛奶、血液和尿液等。

硝酸完全地氧化有机样品组分温度需要超过 200℃。然而,所使用的 PTFE 管不能承受在 200℃ 或更高温度下分解样品混合物的蒸气压。因此,必须有替代品来克服这个限制。一种提高管的耐压性的方法是用高机械强度的塑料带包裹它们,文献[109]介绍了配备这种管的一种消解系统(CEM SpectroPrep)。在中等功率下使用 CEM SpectroPrep,对生物组织(0.5% m/v)和海洋沉积物(1% m/v)的悬浮液样品进行了在线分解,该系统的压力阈值接近 25bar。然而,当反应温度达到大约 250℃ 时,系统中的压力将高达 35bar 左右,最近新开发的装置能够通过一个的压力平衡系统耐受 40bar 的压力,可满足 250℃ 高温[110]。压力平衡系统可以保持 PTFE 或 PFA 消解管内外的压力平衡,即使对于非常快的氧化反应也是如此。但该系统仅能处理 1% m/v 悬浮液和较低浓度的生物材料悬浮液,只能使用最灵敏的元素检测装置如电感耦合等离子体质谱(ICP-MS)对处理液进行检测,因此该系统可分析的样品的类型有限。Mason 等[111]改进了 SpectroPrep 炉,设计了一种宽孔径连续微波消解系统,可用于王水提取的 Cd、Cr、Mn、Ni、Pb 等痕量金属元素的测定,该系统可不使用表面活性剂进行悬浮处理,可直接消解 250 μm 左右的较大粒径的实际土壤样品。

微波加热进行在线消解的优势导致了基于这种混合技术的商业仪器的不断问世[42]。

微波增强流动系统的优点包括样品前处理时间显著减少,能够避免在通常密闭容器中温度和压力的突然升高而产生的危险,以及具有处理瞬间分解或易分解的样品或中间体的能力。但是流动系统也有一定的局限性,所有的样品必须均匀且足够细小才能通过管路,因此大多数样品需要某种形式的预处理才可以进入管中。

4. 气相酸反应分解

另一种用于防止杂质污染的样品的酸消解的方法是利用气相反应。在过去的 40 年中,出现了用一个容器中产生的无机酸蒸气侵蚀和溶解另一个容器中的样品材料的新型样品消解技术。Matusiewicz[112]回顾总结了这种气相侵蚀溶解和分解方法无机物和有机物的痕量元素分析技术。因为在溶液中使用的消解试剂是气相试剂,所以目前这种方法只要适用,就可用于开放的、半封闭的、封闭的检测系统中。

在开放系统中,已证明光谱法测定痕量杂质时用氢氟酸和硝酸蒸气的组合作为消解剂对样品进行处理是有效的。ZilbershTein 等[113]使用这种方法溶解硅并在 PTFE 片上浓缩杂质,将残余物和 PTFE 片转移到石墨电极,随后作为直流电弧的一个电极用于光谱痕量分析。

关于半封闭系统,专门设计了产生氢氟酸蒸气的 PTFE 装置,以尽量减少在测定高纯硅、石英和玻璃中痕量元素时的污染[114]。样品安置在一个多孔 PTFE 板上的 PTFE 烧杯中,PTFE 板面保持在腔室中氢氟酸液面以上。Thomas 和 Smith[115]描述了一种简单的全玻璃装置,用于使用硝酸气相氧化高达 90% 的植物材料。加入高氯酸能快速和完全的氧化,并且在高氯酸氧化过程中由于硝酸的存在消除了爆炸危险。克里特尼克等采用同一技术,用简易 PTFE 消化池测定脑组织中锌的含量[116]。水合酸和硝酸蒸气作为消解剂的组合已被证明在制备用于开放系统的痕量杂质光谱测定样品时是有效的。Klitenick 等使用同样的技术,利用一个简易密封消解容器测定脑组织中的锌元素[116]。

有些物质不能在大气压力下通过酸消解法完全溶解。一个更有力的处理方法涉及弹型压力容器中的消解,该设计是将一个封闭的压力容器和气相消解技术结合在一个单元中(图 5-9)。这比文献[113-116]中描述的装置更容易制造,并且它只需要相当小的酸体积。可以在普通导电加热的烘箱或微波炉中完成加热。

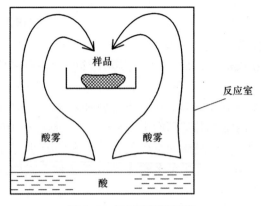

图 5-9　气相酸消解系统

Woolley 介绍了密闭容器气相样品消解这一概念的起源与发展[117],他描述了该装置的工作温度为 110℃ 低温和 250℃ 高温,该装置由一个密封的 PTFE 容器组成,它包含两个同心的腔室,即一个容纳样品杯的内腔和另一个容纳外腔,且均放置溶剂酸,如用 50∶50 浓 HNO_3 和 HF 的混合物来消解高纯玻璃。研究人员研制了一种用于消解更难消解的化合物,例如硅质材料[118]的具有温度梯度的、完全封闭的聚四氟乙烯弹或高压釜。Marinescu 将传统的单样品压力消解弹改进为多样品气相消化[119]装置,设计了一种直接放入消化弹的多位样品固定器,可用于消解有机和无机固体、半固体和液体样品。Kojima 等改进了一种密封的聚四氟乙烯弹,使用放置在 PTFE 外容器中的 PTFE 瓶,利用硝酸、盐酸和氢氟酸蒸气溶解高纯二氧化硅[120]。Matusiewicz 开发了一种实验室研制的带有 PTFE 微量进样装置的高压消解弹[121]。这种简单廉价的设备可以很方便地用于处理少量样品,并且可以易于通过修改可用的聚四氟乙烯弹来制造。值得注意的是,PTFE 微量进样装置可用于气相消解和原子光谱中的雾化进样技术。高纯材料密闭系统中的气相消解对痕量元素的光谱测定非常实用[122],该方法使用一个扩大空腔的石墨电极,不使用收集器。商业化的高压、高温消解装置(高压 Asher HPA;Anton Paar,奥地利)是在石英容器中产生蒸汽相酸的技术。文献[123]介绍了一种针对 50~165mg 少量生物样品在体积 3.1mL 的迷你石英样品容器中的消解技术,当生物标准物质在 230℃ 和 122bar 消解时,消解样品中的残余碳含量(RCC)小于 1.8%。

前面介绍的用于常规加热的封闭系统的成功的消解方法,但是使用微波能来进行气相消解的尝试很少见报道。尽管对消解容器设计的不断改进[125],但是这个低压微波装置的早期试验结果不令人满意[124]。由 Matusiewicz 先前报道所开发的方法是生物材料酸性气相热分解的延伸[126],微波辅助气相酸消解,采用一个适用于 250mg 子样本的特殊 PTFE 微量进样杯,分别使用硝酸和硝酸-氢氟酸用于消解海洋生物和沉积物标准物质,最大压力约为14bar[125]。最近发表的文献[127-131],研究讨论了商用压力微波消解系统或石英样品容器[127]、石英衬垫[128-129]和TFM内容器[128],在大气压下工作的聚焦微波炉和PTFE微量进样杯[130]的方式,并对Matusiewicz[125]开创性概念进一步应用和评价。

综上所述,使用酸性气相消解和侵蚀一些有机和无机基体是一种方便有效的样品前处理方法。密闭压力系统在避免元素挥发损失的同时,也能够通过试剂和工业级酸的等压蒸馏来保持极低空白值。

5. 湿法分解工艺的有效性

质量控制在分析化学中越来越重要。目前它主要针对仪器分析技术,不包括样品前处理。对于样品分解过程中的质量控制,有必要精确地测量和记录某些参数,以便后续对分解处理的过程进行控制。

仪器分析方法的精密测量要求样品必须完全分解,伏安和极谱测定法尤其是这种情况[132-135]。有机化合物不完全分解产生干扰的情况也会发生在使用原子光谱分析过程中,如AAS[136-137]、ICP-OES[138-139]和ICP-MS[140-141]。如前所述,硝酸是最常用的样品溶解介质,在200℃温度下,有机材料中所含的碳元素仅部分地被HNO_3转化为CO_2[14]。在这些情况下,延长消解时间和增加硝酸的量并不能改善分解程度。原则上,用RCC作为定量评估有效性,温度和消化分解时间最终决定消解效果[63,142],总之需要提高温度来实现完全分解的目的[143-144]。应该注意的是,分解的效果不应该从视觉的角度来判断,因为一种透明的、无色的与水不可分辨的溶液仍然可以含有大量的碳。在密闭系统中,压力不仅取决于温度,还取决于样品的类型和数量、容器的大小以及分解试剂的性质和总量。该压力不影响测定质量,但应当自动控制。Würfels等描述了用反相伏安法测定元素时残留有机化合物的显著干扰,证明了用纯硝酸加压分解样品所需的温度达到(300~320)℃时才能获得含碳量低于0.1%的溶液[145-148];否则,痕量元素不能用反相伏安法测定。这一点被Wasilewska等证明[149],他证明了用硝酸对有机化合物进行完全氧化时,分解温度应提高到300℃。如果消解时间长到足以达到静态温度,则消解设备无论是加热还是微波的影响是可以忽略不计的。220~250℃时用硝酸消解样品导致RCC在低百分比范围内。

消解容器的加热越来越频繁地使用微波技术,因此微波辅助湿法分解是有机材料中痕量元素测定常用的样品前处理技术。文献[124,150-154]开展了RCC作为分解效率的研究,采用气相色谱法,Stoeppler等[63]定量测定了常规加压分解的灰化能力。在原始样品中的碳含量与转化为CO_2的碳量之间的差异表明,所研究的生物和环境样品未被硝酸完全灰化。Würfels和Jackwerth测定了加压分解或用HNO_3蒸发的样品中残余的碳[155]。在大多数情况下,生物材料的微波分解是不完全的,随后对未消解的化合物进行了识别[145]。与Würfels和Jackwerth[155]

的研究相同,Pratt 等鉴定了牛肝脏硝酸消解的残留有机物种类[156]。Kingston 和 Jassie 评估了用 HNO_3 湿法分解的几种生物和植物样品的分解过程[157]。人尿样品的游离氨基酸浓度通常减少 10^5 倍,这说明蛋白质水解的相对效率不一定等于总碳氧化效率。Nakashima 等研究了 HNO_3 和 $HClO_4$ 混合酸的消解效率[152],测定了分解一些海洋生物标准物质(NRCC TROT-1)溶液中的总 RCC,并将其作为各种分解方案有效性的相对度量。实验表明,两阶段微波辅助分解优于单级分解,即使是两阶段程序消解,也仍然残留 24% 的碳。Hee 和 Boyle[153] 以及 Krushevska[154] 等描述了 ICP-OES 同时分析测定生物材料分解液中残留的碳。Krushevska 等比较了不同干法和湿法灰化工艺对牛乳样品的氧化效果[158],笔者认为,中压微波辅助分解得到的 RCC 在 5%~15% 之间变化,在 11bar 的中等压力下微波系统中使用 H_2O_2 或 H_2SO_4 与 HNO_3 的混合物进行氧化,并没有产生比纯硝酸更高的分解效率。然而,在高压/温度聚焦的微波加热 TFM-特氟龙弹装置上,有机材料仅单步过程就被硝酸完全氧化[92, 94]。Matusiewicz 和 Sturgeon[159] 评价了在线和高压/温密闭容器技术的分解效率,生物材料破坏的完整性用标准物质和有证标准物质消化后的溶液的 RCC 表征。在 TFM-特氟龙容器中的加压分解是最有效的方法,有机材料在一步过程中被硝酸完全氧化,而用 HNO_3、HNO_3-H_2O_2 和过二硫酸盐氧化,采用在线微波加热不能完全分解污水中的 56%~72% 的尿。最近,一个底部抗反射涂层(BARC)样品的剩余重量被成功地用作评价分解动力学的指示器[160]。在不同温度下,重量损失速率与样品质量无关,但强烈依赖消解酸体积和消解温度。以分解动力学为支撑,得到了预测 BARC 样品分解效率的数学模型。

过氧化氢是一种非常受欢迎的氧化剂,因为它在生物材料氧化过程中被转化为水和氧[124, 161-163]。然而,Matusiewicz 等采用 HNO_3 和 H_2O_2 的混合物进行的实验结果表明,所有 HNO_3 和 H_2O_2 的加压微波消解均未完全分解[122],添加 50% H_2O_2 的分解效率无明显改善。将此观测现象扩展到中压和高温微波加热,同样的观测结果得到了验证[164],即与仅使用 HNO_3 相比,添加 H_2O_2 的硝酸消解并没有提高分解效率。因此,需要另一种氧化剂能完全、安全地分解有机碳残留物。研究发现,臭氧在破坏天然有机化合物[165-167] 方面是非常有效的,并且有潜力被用作一种辅助分解和/或最终试剂[168]。

对一个复杂基体的完全分解来说,一个单一的消解程序往往是不够的,可使用两种或多种技术的组合。举两个例子说明这一观点[95, 133]。第一个例子是先采用压力消解,再 UV 光解。结果表明,使用伏安法分析橄榄叶中的重金属时,仅使用"加压消解"会导致结果变差。只有通过紫外照射补充消解确保了基体的充分分解,才能获得可靠的数据[133]。第二个例子是一种新的微波辅助高温紫外消解程序,用于在液体样品中痕量元素消解之前干扰溶解有机碳的加速分解。这种新技术显著地提高了 UV 氧化分解效率,由于其极低的污染风险,特别适用于超痕量分析[95, 169]。

为了研究无机材料溶解的彻底性,通常需要确定主要、次要和痕量元素测定的回收率和准确度。例如,硅酸盐通常是土壤、沉积物、淤泥、陶瓷和其他类似样品等基质中的主要无机成分,必须使用氢氟酸来实现完全溶解[170-171]。

6. 湿法分解技术的比较

对比应用几种消解技术是确保结果准确的唯一方法,尤其是在对特殊基体的消解几乎没有经验的情况下,或者在现有的报告矛盾的情况下,分析人员必须仔细选择样品前处理方法以确保该消解技术对于现有的分析技术是最佳的。然而,仍然没有通用的样品前处理方法。要尽可能避免挥发或残留造成的污染或损失,样品前处理的最佳选择似乎是对流加热或微波辅助湿法消解、石英衬里的高压湿法消解、紫外消解和气相酸消解,但这些技术都需要相当大的设备投资。在一个开放的容器中消解样品,尽管使用了回流冷凝器,但仍存在严重的分析物损失风险。如果考虑经济性,就意味着低采购消费、较短的消解时间以及高样品量,微波辅助湿法消解,特别是微波辅助加压在线消解似乎是最符合要求的。许多样品仅通过高压、高温特氟龙或石英内衬压力容器消解,或者通过密闭的湿式消解系统与 UV 照射相结合就可以实现完全消解。

表 5-5 总结了 5.3.1 节"湿法化学分解"中讨论的湿法分解技术在分析物损失、空白水平、污染问题、样品大小、消解时间、消解程度和经济性方面的优点和缺点。

表 5-5 湿法分解的优点和缺点

	分解技术	可能的损失途径	空白来源	样品质量/g		最大值		分解时间	分解程度	经济性方面
				有机	无机	温度/℃	压力/bar			
开放系统	传统加热	挥发	酸、容器、空气	<5	<10	<400		几小时	不完全	不贵,需要管理
	微波加热	挥发	酸、容器、空气	<5	<10	<400		1h	不完全	不贵,需要管理
	UV 分解	无		液体		<90		几小时	高	不贵,需要管理
	超声辅助分解	挥发	酸、容器、空气					几分钟	不完全	不贵,需要管理

续表

分解技术		可能的损失途径	空白来源	样品质量/g		最大值		分解时间	分解程度	经济性方面
				有机	无机	温度/℃	压力/bar			
密闭系统	传统加热	滞留	酸(低)	<0.5	<3	<320	<150	几小时	高	不需要管理
	微波加热	滞留	酸(低)	<0.5	<3	<300	<200	<1h	高	贵,不需要管理
流动系统	传统加热	不完全分解	酸(低)	<0.1(悬浮液)	<0.1(悬浮液)	<320	>300	几分钟	高	贵,不需要管理
	UV在线分解	不完全分解	无	液体		<90		几分钟	高	不贵,不需要管理
	微波加热	不完全分解	酸(低)	<0.1(悬浮液)	<0.3(悬浮液)	<250	<40	几分钟	高	贵,不需要管理
蒸汽相酸分解	传统加热	无	无	<0.1	<0.1	<200	<20	<1h	高	不需要管理
	微波加热	无	无	<0.1	<0.1	<200	<20	<20min	高	不需要管理

7. 分解系统

目前,仪器市场提供了许多自动化设备以使湿分解变得更有效和易于控制,但这些设备主要是通过微波能量实现的。

湿法分解无论是否带有回流装置,都在开放容器中进行。优化的消解时间和温度参数是至关重要的,消解自动化不仅有更高的样品处理效率,更少的人为干预,同时也避免了错误发生。最简单的自动化形式可以通过编程定时和自耦变压器控制的加热模块控温来实现,市场上有很多样式的加热模块。最大限度地自动化是在消解过程中将试剂流量控制进行混合。

这些自动化程序实现了成批操作。连续样品处理相比于不连续处理具有一些优点,前者通常能更好地匹配分析需求。自动湿法消解装置(VAO;anton paar,奥地利)就是这样一种连续操作的消解系统,是实验室需要大量处理相似样品的理想仪器。它可以执行所有的湿法化学分解方法[172],在微处理器的帮助下可以控制所有重要的消解参数,自动控制样品消解的时间-温度/压力程序,可以在最佳条件下处理不同的样品材料。高压消解罐完全自动化是不可能的,须手动完成装有样品材料高压消解罐的装载或通电。Berghof 压力消解系统[173]可在纯的、等静压 PTFE 或石英容器中,在 200~250℃温度和 100~200bar 压力下进行无机和有机基体样品的前处理。

目前,主要有大气压力、高压密闭容器和流动通过 3 种基本类型的微波消解系统。微波消解系统以多模式腔和聚焦型波导两种常见模式工作。在文献[42,174]中给出了商业上可获得的微波消解系统和加热系统概要,并对高压、大气压和流动通过 3 种类型的特征进行了说明。微波消解的简易性和有效性很容易实现自动化和智能化,已经开发出了能够进行称样、加酸、容器加盖和开盖、微波辅助消解、稀释消解液、转移容器,甚至清洗和再利用容器的系统。一旦使用这样的系统,就需要在系统能够识别的位置提供和放置代表性样品,然后启动控制程序。

8. 酸分解的安全性

在消解时所使用的试剂、仪器和操作具有潜在的危险,即使是按指导使用时也是如此,操作者必须用实验服、手套和安全眼镜,更重要的是面部防护来保护自己。HF、HNO_3、HCl 等高浓度发烟酸,只能在通风良好的通风橱中处理。HNO_3、$HClO_4$ 等氧化酸比 HCl、H_3PO_4、HF 等非氧化酸更危险,更容易发生爆炸,特别是在存在有机物等还原剂时。高氯酸只有在浓和热时才发生氧化,除非用硝酸稀释,否则不能与有机物质直接接触。

酸消解必须在配有清洗器的通风橱内操作。高氯酸的蒸发仅在合适的不锈钢、石器或 PP 罩中进行,并用洗涤设备来消除高氯酸盐沉积物。

"压力消解"法可实现快速、一步消化而不损失,但使用时应格外小心,因为压力消解容器(弹)含有酸性烟雾。一些反应,特别是自发反应,会产生超过容器安全极限的爆炸性气体。例如,HNO_3 和 H_2O_2 在密闭容器中分解有机物可能导致爆炸,这是由于容器内超极限的压力积聚。这些系统产生的高压尖峰可以通过减少样品量或逐级升温来避免。

微波辅助溶样有其自身的安全要求。由于直接能量吸收和快速加热的结果,微波技术可提供其他方法中没有的独特的安全优势。传统的实验室操作和微波实施方法之间的条件差异应该在用微波能量加热试剂或样品之前进行检查。文献[18-19,175]中对其进行了总结。

5.3.2 燃烧分解

1. 开放系统燃烧

1) 干灰化

"干灰化"意在包含基于气态的或固体灰化试剂的所有过程。相对于湿法分解过程,干灰化没有本质上的区别,但它确实提供了一些实用的优点。严格地说,干灰化是指处于空气中的物质在几百摄氏度的温度下氧化燃烧,其通常在马弗炉或类似的装置中完成。

对于含有大量有机物并分析非挥发性金属的样品,干灰化是一种相对简单的去除有机物的方法。它可以用于相对较大量的样品,一般为$(2\sim10)$g,并且不需要占用太多时间。经典的干灰化是有机样品在马弗炉或实验室火焰中,利用空气中的氧在$(400\sim600)$℃下热解和燃烧,以除去有机成分。有机质转化为CO_2和H_2O[176]。由此产生的无机"灰分"残渣一般溶于稀酸。用于灰化的坩埚通常由二氧化硅、石英、瓷器、铂、锆或派热克斯玻璃制成。

目前干灰化很少应用,在很大程度上已被湿法消解取代,因为它有一些缺点,如挥发造成的损失、一些材料的灰化效果不理想、灰化物质难以溶解及污染。该方法的优点是不用试剂,不需要过多牵扯操作者的精力。

最近已经开发了利用热和微波加热干燥灰化样品的分析仪器;如 APIONA 干式矿化仪(Tessek,捷克共和国)、MLS-1200P YRO 微波灰化炉(Milestone,美国)和 MAS 7000 微波灰化系统(CEM,美国)。

2) 低温灰化

测定有机物中的硒、砷、锑、镉、锌、铊等挥发性元素需要非常温和的处理。为此,在$1\sim5$Torr(1Torr=133.3Pa)的压力下激发氧一般在200℃低温灰化是合适的[177]。

氧等离子体可以通过射频电源或微波能量(300W,13.5MHz)产生。所产生的游离氧自由基和激发态氧等活性氧与有机样品表面有效地反应,形成有机灰渣,然后将吸附到冷却装置上的灰渣和元素用酸回流溶解。这种方法的优点是元素获得相对高的浓度,可用于木材、纸、煤、食品或聚合物等各种可燃固体的样品前处理。基于这种技术的典型商业仪器是冷等离子体管 CPA-4(Anton Paar,奥地利)(图 5-10)。

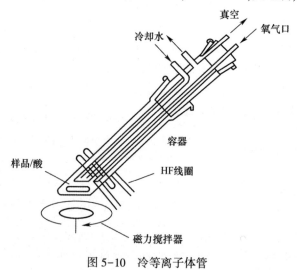

图 5-10 冷等离子体管

3) 冷等离子体灰化(Wickhold 燃烧)技术

冷等离子体灰化(wickbold)技术适用于石油产品等液体可燃性样品的处理,这些样品是其他技术难以分解的[178]。在 Wickbold 燃烧技术中,用氧-氢焰来实现在2000℃高温下的样品分解。液体样品直接引入火焰中,而固体样品需要在预燃烧单元中进行初步热解燃烧后,所得产物在石英表面上冷凝并被吸收在合适的溶液中。已经有商业化的 Wickbold 燃烧装置(V5;Heraeous,德国)。

2. 密闭系统中的燃烧

1) 氧瓶燃烧(Schöniger)

氧瓶燃烧技术的氧瓶燃烧过程,适用于卤族元素、硒、硫、磷、硼、汞、砷、锑等易挥发元素测定。在密封容器中用氧气进行燃烧,反应产物被吸收在合适的溶剂中,然后打开反应容器。

Schöniger[179]提出了一种简单的氧化装置(图 5-11)。它由一个容量为500~1000mL 烧瓶和一个磨砂玻璃塞组成,塞子与一个铂金网相连,铂金网上可以装载2~200mg 的样品。如果样品是固体,则用低灰分含量的滤纸包裹。液体样品可以在称入明胶胶囊后用滤纸包裹。在烧瓶底部放置一小部分吸收剂溶液,在燃烧过程中,为了防止挥发性氧化产物的逸出,将"烧瓶"倒置。完全燃烧后打开容器,将含有分析物的所得溶液转移并稀释以进行分析。氧瓶分解所需的时间通常小于10min,材料和设备也相对便宜。然而,该过程需要分析人员手工操作并持续关注燃烧过程,并且通常一次只能处理一个样品。

Schöniger 系统的烧瓶型燃烧装置已商业化(Mikro K;Heraeus,德国)。

2) 氧弹燃烧

氧燃烧弹是成功应用于几种材料和分析物的经典技术[180](图 5-12)。在这种处理技术中,样品被放置在带有两个铂丝的燃烧杯中,两个铂丝与两个电极连接。在容器底部添加约 10mL 的吸收溶液,容器由不锈钢制成或覆盖有铂。系统用氧气加压至 20~30bar,然后使用电流进行点火,样品燃烧后所产生的气体被吸收,待冷却后打开系统,转移吸收溶液。燃烧弹具有较高的分解效力,可以燃烧相对大量的样品,质量可达 0.5 g,整个过程可以在 1h 内完成。

已经商业化的系统如 PARR 氧燃烧"弹"(Paar 仪器公司,美国)和 Bioklav 压力分解装置(西门子,德国)可以用于样品前处理。

动态系统中的燃烧(Trace-O-Mat)是在 20 世纪 80 年代由 Knapp 等开发的,实现样品燃烧的污染最小[181]。该系统不仅能在封闭系统中燃烧,而且可以进一步处理。图 5-13 展示了 Trace-O-Mat 燃烧单元 VAE-Ⅱ(Kürner,德国)。

样品在由石英制成的系统中的纯氧气流中完全燃烧。燃烧室上方的冷却系统的体积仅为 75mL 并装有液氮,将所有挥发性的痕量物质以及 CO_2 和 H_2O 等燃烧产物一起冷凝。燃烧后氮被蒸发,剩余的灰分和冷凝的挥发性元素在(1~2)mL 高

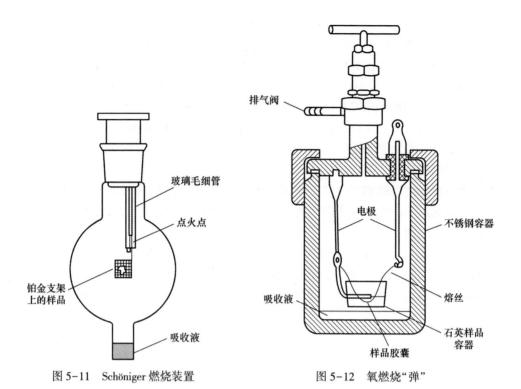

图 5-11 Schöniger 燃烧装置

图 5-12 氧燃烧"弹"

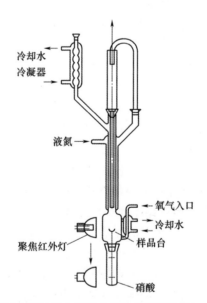

图 5-13 用于在氧气流中燃烧有机物的石英装置(Trace-O-Mat)

纯度 HCl 或 HNO$_3$ 中被回流溶解,然后收集到下面的放置试剂容器中。固体样品的最大样品量为 0.7~0.8g,可压制成球团;液体有机样品可以用最小 1mL 的酸制成高元素浓度的溶液,需用一个特殊的样品固定器。

5.3.3 熔融分解

熔融,特别是碱熔融,对痕量元素含量水平相对较高的有机基体以及具有高硅和高氧化铝含量的基体来说是一种强有力的技术。一般来说,盐熔融是通过将样品与盐混合,加热熔化混合物,待冷却后溶解固化的熔融物来完成的。根据路易斯酸-碱的定义,酸可以接受电子对,碱可以提供电子对,熔融特性从酸性到碱性也可以定义为氧化或还原。

虽然酸侵蚀是溶解硅酸盐样品的经典方法,但偏硼酸锂使用最早的是 Ingamells[182] 的创新成果,它可以简便且快速地制备一种硅酸盐透明水溶液。熔融分解大多数物质时,需要在(500~1000)℃高温下进行,为了使试剂与样品充分接触也需要使用高浓度的试剂。样品以非常细的粉末形式与 10 倍过量的熔剂在石墨或铂坩埚(有时是镍或锆)中混合。然后将坩埚置于马弗炉中在(500~1000)℃加热几分钟至几小时以得到一个"熔融物",冷却后溶解熔融物。

不同的样品可以使用不同的熔剂,用于熔融酸性物质的碱性熔剂包括碳酸盐、氢氧化物、过氧化物和硼酸盐等,硫代硫酸盐可作为酸性熔剂。如果需要氧化性熔剂,可以使用过氧化钠,也可以将少量的硝酸盐或氯酸盐与碳酸钠混合替代过氧化钠。碱性和酸性熔剂分别溶于酸或碱性介质中。

熔融需要的熔剂量较大,增加了熔剂空白值升高的风险,同时从熔融中得到的水溶液具有高的盐浓度,这可能会给随后的分析步骤带来困难。熔融所需的高温增加了挥发损失的风险。这些不足使得熔融成为一种不太理想的极端痕量元素测定技术。然而,在对粉煤灰、硅酸盐、炉渣和灰尘等基体中主要成分、次要成分甚至一些痕量元素的测定可以得到良好的结果。

5.4 结论与发展趋势

5.3 节对有机样品和无机样品主要的分解方法进行了评价。这些技术在各种样品基体中应用的简要总结列于表 5-6 中。目前,固体和液体样品分解的方法很多,可以根据分析物的基体和类型,选择最合适的方法,兼顾考虑完成分析过程而使用的后续步骤。尽管如此,样品分解也不应被视为一个孤立的步骤,而是整个分析过程中的其中一个步骤。

表 5-6 湿法分解在材料分析(元素测定)中的应用综述

材料/基体/样品		需要的酸/s①	分解技术(模式)②
水(s)		H_2O_2, HNO_3	UV 照射
环境样品	煤	HNO_3, HCl, HF	开放或密闭系统
	粉煤灰	王水③+HF④	开放或密闭系统
	灰尘	王水+HF	开放或密闭系统
	催化剂	王水	开放系统
废弃物	污水污泥	HNO_3, HCl	开放或密闭或流动系统
	废水	HNO_3	流动系统
植物学体系	植物学	HNO_3+HCl+HF	开放或密闭系统
	植物	HNO_3+HCl+HF	开放或密闭系统
	临床	HNO_3	开放或密闭系统
	海洋	HNO_3	开放或密闭系统
法庭样品		HNO_3	开放或密闭系统
食品(s)		HNO_3	开放或密闭系统
饮料		HNO_3, H_2O_2	开放或密闭或流动系统
硅酸盐	土壤	王水+HF	开放和/或密闭系统
	沉积物	王水+HF	开放和/或密闭系统
	玻璃	HF	开放系统
地质样品	岩石	王水+HF⑤	开放和/或密闭系统
	矿石	王水+HF	开放和/或密闭系统
	矿物质	HF+H_2SO_4, HCl	开放系统
石油产品	燃料	HNO_3+HCl	开放或密闭系统
	油液	HNO_3+HCl	开放或密闭系统
药物和样品		HCl, HNO_3	开放系统
金属	铁类	HNO_3+(HF 或 HNO_3 或 H_2SO_4)	开放系统
	非铁类	HCl 或 HNO_3 或 HF	开放系统
	合金	王水+HF	开放系统
	钢铁	HCl+ HNO_3, $HClO_4$⑥	开放系统
化工产品		HCl, HNO_3, HF, H_2SO_4	开放或密闭系统
聚合物		HCl, HNO_3, HF, H_2SO_4	开放或密闭系统

续表

	材料/基体/样品	需要的酸/s[①]	分解技术(模式)[②]
难溶化合物[⑦]	陶瓷	HNO_3, HCl, HF, H_2SO_4, H_2O_2	开放或密闭系统
	复合材料	HNO_3, HCl, HF, H_2SO_4, H_2O_2	开放或密闭系统
	核材料	HNO_3 或 HCl, H_3PO_4, $HClO_4$	开放或密闭系统

注：①通常使用浓酸,H_2O_2浓度为30%,大多数情况下,根据分析者的要求选择分解方式；

②传统或微波；

③不稳定；

④仅使用特氟龙容器,加入氢氟酸是为了得到 Cr 的定量回收率；

⑤加入 H_3BO_3 通过生成 BF_4 来中和 HF；

⑥有爆炸风险；

⑦某些难溶材料不会被分解,需要通过熔融进行溶解。

要关注高温和高压的分解,高压分解是目前应用最普遍的分解体系,其分解温度范围较大,是适用于绝大多数无机材料和有机材料技术。随着联用分解技术的发展,进一步提高样品制备效率的新方法应运而生。一种新型的微波辅助高温紫外光消解用于加速分解有机化合物或浆料[95],具有较低的空白值和酸浓度,这种新技术是理想的极端痕量分析前处理方法,也可用于离子色谱法测定非金属元素。另外,在特氟龙内衬消解容器的极限范围内,可以通过添加一定浓度的强氧化剂如臭氧或氧气来实现分解效率的提高,这是处理生物样品有效的分解方法,具有不增加分析空白的优点。气相酸消化也是一种可选用的样品处理方案,它可以降低消解液中酸的浓度,可以使用分析纯等普通纯度级别的酸而不会影响分析空白。另一个获得分解和溶解改进的例子是使用微波和超声波能量相结合的反应器[42],这两种方法有望开辟一个新的研究领域——"联合消解技术"。

可以肯定地说,未来大多数消解将通过微波辅助进行。过去几年,微波消解在减少系统误差、提高检测限、操作自动化方面有了很大的进步。加压密闭容器系统是一个明显的趋势,它可以实现高温分解与痕量分析。一些研究人员主张在100bar的高压和(250~300)℃温度下消解样品以破坏难降解化合物的干扰,而制造商正在设计能够承受这些样品消解条件的容器。

近年来,全自动在线分析技术逐渐成为研究热点。微波辅助高压流动消解系统使用 PTFE 管或 PFA 管,消解温度高达250℃,为全自动样品前处理开辟了新的可能性[42]。另外,已经开发了新的高温/高压消解系统,包括用于连续消解各种样品的电阻加热毛细管,该毛细管与原子光谱仪器连接[98-100]。流动系统被预测将成为液体样品和浆料的主流分析技术,仅使用微克级样品或微升级试剂并通过将样品前处理与分析相结合,扩大仪器分析法的分析能力。这些研究的最终目标是建立标准化、批量化在线消解系统,并直接与分析仪结合。

很明显,湿法分解仍然是一个充满发展活力的领域,需要设计新的消解技术来解决仪器的局限性并最大限度地发挥其潜力。常规和微波仪器的发展趋势将集中在样品消解能力、容器性能指标的增强、新材料的使用、原位容器的温度和压力控制、入射和反射的微波功率的进一步优化,以及由计算机控制的具有自动化能力的样品消解器。

最后,人们无法干预自动化的固体样品湿法分解,其开发只能通过使用机器人工作站来实现[183]。然而,一些辅助能量和市售模块可以帮助和/或加快分析程序,减少以溶液的形式从固体样品中获得分析物的样品处理程序所耗费的时间和精力。

参考文献

[1] Gorsuch, T.T.: The Destruction of Organic Matter. Pergamon, Oxford (1970)

[2] Sansoni, B., Panday, V.K.: Ashing in trace element analysis of biological material. In: Fachetti, S. (ed.) Analytical Techniques for Heavy Metals in Biological Fluids, pp. 91–131. Elsevier, Amsterdam (1983)

[3] Subramanian, K.S.: Determination of metals in biofluids and tissue: sample preparation methods for atomic spectroscopic technique. Spectrochim. Acta Part B 51, 291–319 (1996)

[4] Iyengar, G.V., Subramanian, K.S., Woittiez, J.R.W.: Sample decomposition. In: Iyengar, G. V., Subramanian, K.S., Woittiez, J.R.W. (eds.) Element Analysis of Biological Samples. Principles and Practice, pp. 103–134. CRC, Boca Raton (1998)

[5] Matusiewicz, H.: Review of sample preparation methodology for elemental analysis of cellulose type materials. ICP Inf. Newsl. 25, 510–514 (1999)

[6] Matusiewicz, H.: Solubilization: trends of development in analytical atomic spectrometry for elemental food analysis. In: Caroli, S. (ed.) The Determination of Chemical Elements in Food. Applications for Atomic and Mass Spectrometry, pp. 19–50. Wiley, Chichester (2007)

[7] Šulcek, Z., Povondra, P., Doležal, J.: Decomposition procedures in inorganic analysis. CRC Crit. Rev. Anal. Chem. 6, 255–323 (1977)

[8] Šulcek, Z., Povondra, P.: Methods of Decomposition in Inorganic Analysis. CRC, Boca Raton (1989)

[9] Povondra, P., Šulcek, Z.: Modern methods of decomposition of inorganic substances. In: Zýka, J. (ed.) Instrumentation in Analytical Chemistry, pp. 188–245. Ellis Horwood, New York (1991).

[10] Chao, T.T., Sanzolone, R.F.: Decomposition techniques. J. Geochem. Explor. 44, 65–106 (1992)

[11] Bock, R.: A Handbook of Decomposition Methods in Analytical Chemistry (translated and revised

by Marr, I.L.). International Textbook Company, Glasgow (1979)

[12] Bajo, S.: Dissolution of matrices. In: Alfassi, Z.B., Wai, C.M. (eds.) Preconcentration Techniques for Trace Elements, pp. 3–29. CRC, Boca Raton (1992)

[13] Vandecasteele, C., Block, C.B.: Sample preparation. In: Modern Methods for Trace Element Determination, pp. 9–52. Wiley, Chichester (1993)

[14] Stoeppler, M. (ed.): Sampling and Sample Preparation. Springer, Berlin (1997)

[15] Hoenig, M., de Kersabiec, A.-M.: Sample preparation steps for analysis by atomic spectroscopy methods: present status. Spectrochim. Acta Part B 51, 1297–1307 (1996)

[16] Hoenig, M.: Preparation steps in environmental trace element analysis: facts and traps. Talanta 54, 1021–1038 (2001)

[17] Krakovská, E., Kuss, H.-M.: Rozklady v Analitickej Chémii. VIENALA, Košice (2001) (In Slovak)

[18] Matusiewicz, H.: Wet digestion methods. In: Mester, Z., Sturgeon R. (eds.) Sample Preparation for Trace Element Analysis, pp. 193–233. Elsevier, Amsterdam (2003)

[19] Arruda, M.A.Z. (ed.): Trends in Sample Preparation. Nova Science, New York (2007)

[20] de Oliveira, E.: Sample preparation for atomic spectroscopy: evolution and future trends. J. Braz. Chem. Soc. 14, 174–182 (2003)

[21] Korn, M.G.A.: Sample preparation for the determination of metals in food samples using spectroanalytical methods. A review. Appl. Spectrosc. Rev. 43, 67–92 (2008)

[22] Kingston, H.M., Jassie, L.B. (eds.): Introduction to Microwave Sample Preparation. Theory and Practice. ACS, Washington, DC (1988)

[23] Kingston, H.M., Haswell, S.J. (eds.): Microwave-Enhanced Chemistry. Fundamentals, Sample Preparation, and Applications. ACS, Washington, DC (1997)

[24] White Jr., R.T.: Open reflux vessels for microwave digestion: botanical, biological, and food samples for elemental analysis. In: Kingston, H.M., Jassie, L.B. (eds.) Introduction to Microwave Sample Preparation. Theory and Practice. ACS, Washington, DC (1988)

[25] Mermet, J.M.: Focused-microwave-assisted reactions: atmospheric-pressure acid digestion, on-line pretreatment and acid digestion, volatile species production, and extraction. In: Kingston, H.M., Haswell, S.J. (eds.) Microwave-Enhanced Chemistry. Fundamentals, Sample Preparation, and Applications. ACS, Washington, DC (1997)

[26] Matusiewicz, H., Sturgeon, R.E.: Present status of microwave sample dissolution an decomposition for elemental analysis. Prog. Analyt. Spectrosc. 12, 21–39 (1989)

[27] Kuss, H.M.: Applications of microwave digestion technique for elemental analyses. Fresenius J. Anal. Chem. 343, 788–793 (1992)

[28] Zlotorzynski, A.: The application of microwave radiation to analytical and environmental chemistry. Crit. Rev. Anal. Chem. 25, 43–76 (1995)

[29] Smith, F.E., Arsenault, E.A.: Microwave-assisted sample preparation in analytical chemistry. Talanta 43, 1207–1268 (1996)

[30] Burguera, M., Burguera, J.L.: Microwave sample pretreatment in analytical systems. A review. Quim. Anal. 15, 112–122 (1996)

[31] Mamble, K.J., Hill, S.J.: Microwave digestion procedures for environmental matrices. Analyst 123, 103R–133R (1998)

[32] Burguera, M., Burguera, J.L.: Microwave-assisted sample decomposition in flow analysis. Anal. Chim. Acta 366, 63–80 (1998)

[33] Chakraborty, R., Das, A.K., Cervera, M.L., de la Guardia, M.: Literature study of microwave-assisted digestion using electrothermal atomic absorption spectrometry. Fresenius J. Anal. Chem. 355, 99–111 (1996)

[34] Kingston, H.M., Walter, P.J.: The art and science of microwave sample preparations for trace and ultratrace elemental analysis. In: Montaser, A. (ed.) Inductively Coupled Plasma Mass Spectrometry, pp. 33–81. Wiley-VCH, New York (1998)

[35] Meyers, R.A.: Encyclopedia of Analytical Chemistry. Wiley, Chichester (2000)

[36] Mester, Z., Sturgeon, R. (eds.): Sample Preparation for Trace Element Analysis, pp. 193–233. Elsevier, Amsterdam (2003)

[37] Kuehner, E.C., Alvarez, R., Paulsen, P.J., Murphy, T.J.: Production and analysis of special high-purity acids purified by sub-boiling distillation. Anal. Chem. 44, 2050–2056 (1972)

[38] Barnes, R.M., Quináia, S.P., Nóbrega, J.A., Blanco, T.: A fast microwave-assisted acid-vapor, steam-cleaning procedure for autosampler cups. Spectrochim. Acta Part B 53, 769–771 (1998)

[39] Bethge, P.O.: An apparatus for the wet ashing of organic matter. Anal. Chim. Acta 10, 317–320 (1954)

[40] Abu-Samra, A., Morris, J.S., Koirtyohann, S.R.: Wet ashing of some biological samples in a microwave oven. Anal. Chem. 47, 1475–1477 (1975)

[41] Nóbrega, J.A., Trevizan, L.C., Araújo, G.C.L., Nogueira, A.R.A.: Focused – microwaveassisted strategies for sample preparation. Spectrochim. Acta Part B 57, 1855–1876 (2002)

[42] Matusiewicz, H.: Systems for microwave-assisted wet digestion. In: de Moraes Flores, É.M. (ed.) Microwave-Assisted Sample Preparation for Trace Element Determination, pp. 77–98. Elsevier, Amsterdam (2014)

[43] Golimowski, J., Golimowska, K.: UV-photooxidation as pretreatment step in inorganic analysis of environmental samples. Anal. Chim. Acta 325, 111–133 (1996)

[44] Trapido, M., Hirvonen, A., Veressinina, Y., Hentunen, J., Munter, R.: Ozonation, ozone/UV and UV/H_2O_2 degradation of chlorophenols. Ozone Sci. Eng. 19, 75–96 (1997)

[45] Priego-Capote, F., Luque de Castro, M.D.: Analytical uses of ultrasound I. Sample preparation. Trends Anal. Chem. 23, 644–653 (2004)

[46] Seidi, S., Yamini, Y.: Analytical sonochemistry; developments, applications, and hyphenations of ultrasound in sample preparation and analytical techniques. Cent. Eur. J. Chem. 10, 938–976

(2012)

[47] Mitscherlich, A.: Beitrage zur analytischen chemie. J. Prakt. Chem. 81, 108-117 (1860)

[48] Carius, G.L.: Über elementar analyse. Ber. Dtsch. Chem. Ges. 3, 697-702 (1870)

[49] Long, S.E., Kelly, W.R.: Determination of mercury in coal by isotope dilution cold-vapor generation inductively coupled plasma mass spectrometry. Anal. Chem. 74, 1477-1483(2002)

[50] Jannasch, P.: Über die aufschliessung von silikaten unter druck durch konzentrierte salzsäure. Z. Anorg. Allgem. Chem. 6, 72-73 (1894)

[51] Jackwerth, E., Gomišček, S.: Acid pressure decomposition in trace element analysis. Pure Appl. Chem. 56, 479-489 (1984)

[52] Ito, J.: A new method of decomposition for refractory minerals and its application to the determination of ferrous iron and alkalis. Bull. Chem. Soc. Jpn. 35, 225-228 (1962)

[53] Langmyhr, F.J., Sveen, S.: Decomposability in hydrofluoric acid of the main and some minor and trace minerals of silicate rocks. Anal. Chim. Acta 32, 1-7 (1965)

[54] Bernas, B.: A new method for decomposition and comprehensive analysis of silicates by atomic absorption spectrometry. Anal. Chem. 40, 1682-1686 (1968)

[55] Okamoto, K., Fuwa, K.: Low-contamination digestion bomb method using a Teflon double vessel for biological materials. Anal. Chem. 56, 1758-1760 (1984)

[56] Ravey, M., Farberman, B., Hendel, I., Epstein, S., Shemer, R.: A vessel for low-pressure acid dissolution of mineral and inorganic samples. Anal. Chem. 67, 2296-2298 (1995)

[57] Uhrberg, R.: Acid digestion bomb for biological samples. Anal. Chem. 54, 1906-1908 (1982)

[58] Kotz, L., Henze, G., Kaiser, G., Pahlke, S., Veber, M., Tölg, G.: Wet mineralization of organic matrices in glassy carbon vessels in a pressure-bomb system for trace element analysis. Talanta 26, 681-691 (1979)

[59] Stoeppler, M., Backhaus, F.: Pretreatment studies with biological and environmental materials I. Systems for pressurized multisample decomposition. Fresenius' Z. Anal. Chem. 291,116-120 (1978)

[60] Kotz, L., Kaiser, G., Tschöpel, P., Tölg, G.: Decomposition of biological materials for the determination of extremely low contents of trace elements in limited amounts with nitric acid under pressure in a Teflon tube. Fresenius' Z. Anal. Chem. 260, 207-209 (1972)

[61] Tomljanovic, M., Grobenski, Z.: The analysis of iron ores by atomic absorption spectrometry after pressure decomposition with hydrofluoric acid in a PTFE autoclave. At Absorption Newsl. 14, 52-56 (1975)

[62] Uchida, T., Kojima, I., Iida, C.: Determination of metals in small samples by atomic absorption and emission spectrometry with discrete nebulization. Anal. Chim. Acta 116, 205 - 210 (1980)

[63] Stoeppler, M., Müller, K.P., Backhaus, F.: Pretreatment studies with biological and environmental materials III. Pressure evaluation and carbon balance in pressurized decomposition with nitric acid. Fresenius' Z. Anal. Chem. 297, 107-112 (1979)

[64] Takenaka, M., Kozuka, S., Hayashi, M., Endo, H.: Determination of ultratrace amounts of metallic and chloride ion impurities in organic materials for microelectronics devices after a microwave digestion method. Analyst 122, 129–132 (1997)

[65] Lechler, P.J., Desilets, M.O., Cherne, F.J.: Effect of heating appliance type on Parr bomb temperature response performance. Analyst 113, 201–202 (1988)

[66] Knapp, G.: Development of mechanized sample preparation for plasma emission spectrometry. ICP Inf. Newsl. 10, 91–104 (1984)

[67] Knapp, G., Grillo, A.: A high pressure asher for trace analysis. Am. Lab. 3, 1–4 (1986)

[68] Knapp, G.: Sample preparation techniques: an important part in trace element analysis for environmental research and monitoring. Int. J. Environ. Anal. Chem. 22, 71–83 (1985)

[69] Strenger, S., Hirner, A.V.: Digestion of organic components in waste materials by high pressure ashing with infrared heating. Fresenius J. Anal. Chem. 371, 831–837 (2001)

[70] Hale, M., Thompson, M., Lovell, J.: Large batch sealed tube decomposition of geochemical samples by means of a layered heating block. Analyst 110, 225–228 (1985)

[71] Eggimann, D.W., Betzer, P.R.: Decomposition and analysis of refractory oceanic suspended materials. Anal. Chem. 48, 886–890 (1976)

[72] Adrian, W.J.: A new wet digestion method for biological material utilizing pressure. At Absorption Newsl. 10, 96 (1971)

[73] Kuennen, R.W., Wolnik, K.A., Fricke, F.L., Caruso, J.A.: Pressure dissolution and real sample matrix calibration for multielement analysis of raw agriculture crops by inductively coupled plasma atomic emission spectrometry. Anal. Chem. 54, 2146–2150 (1982)

[74] May, K., Stoeppler, M.: Pretreatment studies with biological and environmental materials IV. Complete wet digestion in partly and completely closed quartz vessels for subsequent trace and ultratrace mercury determination. Fresenius' Z. Anal. Chem. 317, 248–251 (1984)

[75] Matusiewicz, H.: Modification of construction of Teflon digestion bomb. Chem. Anal.(Warsaw) 28, 439–452 (1983)

[76] Matthes, S.A., Farrell, R.F., Mackie, A.J.: A microwave system for the acid dissolution of metal and mineral samples. Tech. Prog. Rep. U.S. Bur. Mines 120, 1–9 (1983)

[77] Smith, F., Cousins, B., Bozic, J., Flora, W.: The acid dissolution of sulfide mineral sample under pressure in a microwave oven. Anal Chim. Acta 177, 243–245 (1985)

[78] Buresch, O., Hönle, W., Haid, U., Schnering, H.G.: Pressure decomposition in the microwave oven. Fresenius Z. Anal. Chem. 328, 82–84 (1987)

[79] Alvarado, J., León, L.E., López, F., Lima, C.: Comparison of conventional and microwave wet acid digestion procedures for the determination of iron, nickel and vanadium in coal by electrothermal atomization atomic absorption spectrometry. J. Anal. At. Spectrom. 3, 135–138 (1988)

[80] Kojima, I., Uchida, T., Iida, C.: Pressurized microwave digestion of biological samples for metal determination. Anal. Sci. 4, 211–214 (1988)

[81] Levine, K.E., Batchelor, J.D., Rhoades, C.B., Jones, B.T.: Evaluation of a high-pressure, high-temperature microwave digestion system. J. Anal. At. Spectrom. 14, 49–59 (1999)

[82] Xu, L.Q., Shen, W.X.: Study on the PTFE closed-vessel microwave digestion method in food elemental analysis. Fresenius' Z. Anal. Chem. 332, 45–47 (1988)

[83] Nicholson, J.R.P., Savory, M.G., Savory, J., Wills, M.R.: Micro-quantity tissue digestion for metal measurements by use of a microwave acid-digestion bomb. Clin. Chem. 35, 488–490 (1989)

[84] Baldwin, S., Deaker, M., Maher, W.: Low-volume microwave digestion of marine biological tissues for the measurement of trace elements. Analyst 119, 1701–1704 (1994)

[85] Reid, H.J., Greenfield, S., Edmonds, T.E., Kapdi, R.M.: Reflux pre-digestion in microwave sample preparation. Analyst 118, 1299–1302 (1993)

[86] Reid, H.J., Greenfield, S., Edmonds, T.E.: Liquid nitrogen cooling in microwave digestion. Analyst 118, 443–445 (1993)

[87] Heltai, G., Percsich, K.: Moderated pressure microwave digestion system for preparation of biological samples. Talanta 41, 1067–1072 (1994)

[88] Pougnet, M., Michelson, S., Downing, B.: Microwave digestion procedure for ICP-AES analysis of biological samples. J. Microw. Power Electromagn. Energy 26, 140–145 (1991)

[89] Pougnet, M., Downing, B., Michelson, S.: Microwave irradiation systems for laboratory pressure vessels. J. Microw. Power Electromagn. Energy 28, 18–24 (1993)

[90] Légère, G., Salin, E.D.: Capsule-based microwave digestion system. Appl. Spectrosc. 49, 14A–20A (1995)

[91] Légère, G., Salin, E.D.: Design and operation of a capsule-based microwave digestion system. Anal. Chem. 70, 5029–5036 (1998)

[92] Matusiewicz, H.: Development of a high pressure/temperature focused microwave heated Teflon bomb for sample preparation. Anal. Chem. 66, 751–755 (1994)

[93] Matusiewicz, H.: New technology for in situ visualization, monitoring and controlling microwave chemical reaction progress using a focused microwave high pressure-temperature closed-vessel digestion system. Analyst 134, 1490–1497 (2009)

[94] Matusiewicz, H.: Development of a high-pressure asher focused microwave system for sample preparation. Anal. Chem. 71, 3145–3149 (1999)

[95] Florian, D., Knapp, G.: High-temperature, microwave-assisted UV digestion: a promising sample preparation technique for trace element analysis. Anal. Chem. 73, 1515–1520 (2001)

[96] Matusiewicz, H., Stanisz, E.: Characteristics of a novel UV-TiO_2-microwave integrated irradiation device in decomposition processes. Microchem. J. 86, 9–16 (2007)

[97] Gluodenis, T.J., Tyson, J.F.: Flow injection systems for directly coupling on-line digestion with analytical atomic spectrometry. Part 1. Dissolution of cocoa under stopped-flow, high-pressure conditions. J. Anal. At. Spectrom. 7, 301–306 (1992)

[98] Gräber, C., Berndt, H.: Development of a new high temperature/high pressure flow systems for

the continuous digestion of biological samples. J. Anal. At. Spectrom. 14, 683-691 (1999)

[99] Haiber, S., Berndt, H.: A novel high-temperature (360℃) high-pressure (30MPa) flow system for online sample digestion applied to ICP spectrometry. Fresenius J. Anal. Chem.368, 52-58 (2000).

[100] Jacob, P., Berndt, H.: Online element determination in biological and environmental samples by flame AAS coupled with a high-temperature/high-pressure flow digestion system. J. Anal.At. Spectrom. 17, 1615-1620 (2002).

[101] Burguera, J.L., Burguera, M.: Flow injection systems for on-line sample dissolution/decomposition.In: Sanz-Medel, A. (ed.) Flow Analysis with Atomic Spectrometric Detectors,pp. 135-149. Elsevier, Amsterdam (1999)

[102] Fernandes, S.M.V., Lima, J.L.F.C., Rangel, A.O.S.S.: Spectrophotometric flow injection determination of total phosphorus in beer using on-line UV/thermal induced digestion.Fresenius J. Anal. Chem. 366, 112-115 (2000)

[103] Burguera, M., Burguera, J., Alarcón, O.M.: Flow injection and microwave-oven sample decomposition for determination of copper, zinc and iron in whole blood by atomic absorption spectrometry. Anal. Chim. Acta 179, 351-357 (1986)

[104] Carbonell, V., de la Guardia, M., Salvador, A., Burguera, J.L., Burguera, M.: On-line microwave oven digestion flame atomic absorption analysis of solid samples. Anal. Chim.Acta 238, 417-421 (1990)

[105] Karanassios, V., Li, F.H., Liu, B., Salin, E.D.: Rapid stopped-flow microwave digestion system. J. Anal. At. Spectrom. 6, 457-463 (1991)

[106] Haswell, S.J., Barclay, D.: On-line microwave digestion of slurry samples with direct flame atomic absorption spectrometric elemental detection. Analyst 117, 117-120 (1992)

[107] Burguera, M., Burguera, J., Alarcón, O.M.: Determination of zinc and cadmium in small amounts of biological tissues by microwave-assisted digestion and flow injection atomic absorption spectrometry. Anal. Chim. Acta 214, 421-427 (1988)

[108] Martines Stewart, L.J., Barnes, R.M.: Flow-through, microwave-heated digestion chamber for automated sample preparation prior to inductively coupled plasma spectrochemical analysis. Analyst 119, 1003-1010 (1994)

[109] Sturgeon, R.E., Willie, S.N., Methven, B.A., Lam, W.H., Matusiewicz, H.: Continuous-flow microwave-assisted digestion of environmental samples using atomic spectrometric detection.J. Anal. At. Spectrom. 10, 981-986 (1995)

[110] Pichler, U., Haase, A., Knapp, G., Michaelis, M.: Microwave-enhanced flow system for high-temperature digestion of resistant organic materials. Anal. Chem. 71, 4050-4055 (1999)

[111] Mason, C.J., Coe, G., Edwards, M., Riby, P.: Development of a wide bore flow through microwave digestion device for the determination of trace metals in soil. Analyst 125, 1875-1883 (2000)

[112] Matusiewicz, H.: A review of acid vapour-phase sample digestion of inorganic and organic ma-

trices for elemental analysis. Spectroscopy 6, 38-46 (1991)
[113] Zilbershtein Kh, I., Piriutko, M.M., Jevtushenko, T.P., Sacharnova, I.L., Nikitina, O.N.: Spectrochemical analysis of high purity silicon. Zavod. Lab. 25, 1474-1476 (1959)
[114] Mitchell, J.W., Nash, D.L.: Teflon apparatus for vapor phase destruction of silicate materials. Anal. Chem. 46, 326-328 (1974)
[115] Thomas, A.D., Smythe, L.E.: Rapid destruction of plant material with concentrated nitric acid vapour (vapour phase oxidation). Talanta 20, 469-475 (1973)
[116] Klitenick, M.A., Frederickson, C.J., Manton, W.I.: Acid-vapor decomposition for determination of zinc in brain tissue by isotope dilution mass spectrometry. Anal. Chem. 55, 921-923 (1983)
[117] Wooliey, J.F.: PTFE apparatus for vapour-phase decomposition of high-purity materials. Analyst 100, 896-898 (1975)
[118] Karpov, Y.A., Orlova, V.A.: Analytical autoclaves. Vysokochist. Veshchestva 2, 40-56 (1990)
[119] Marinescu, D.M.: Multisample pressure decomposition in the vapor phase. Analusis 13, 469-470 (1985)
[120] Kojima, I., Jinno, F., Noda, Y., Iida, C.: Vapour-phase acid decomposition of highly pure silicas in a sealed PTFE bomb and determination of impurities by "one-drop" atomic spectrometry. Anal. Chim. Acta 245, 35-41 (1991)
[121] Matusiewicz, H.: PTFE microsampling device for acid-vapour-phase decomposition in Teflon digestion bomb. Chem. Anal. (Warsaw) 33, 173-175 (1988)
[122] Pimenov, V.G., Pronchatov, A.N., Maksimov, G.A., Shishov, V.N., Shcheplyagin, E.M., Krasnova, S.G.: Atomic emission spectrochemical analysis of high purity germanium with preconcentration of impurities by vapour phase autoclave sample decomposition in the electrode. Zh. Anal. Khim. 39, 1636-1640 (1984)
[123] Amaarsiriwardena, D., Krushevska, A., Argentine, M., Barnes, R.M.: Vapour-phase acid digestion of micro samples of biological material in a high-temperature, high-pressure asher for inductively coupled plasma atomic emission spectrometry. Analyst 119, 1017-1021 (1994)
[124] Matusiewicz, H., Sturgeon, R.E., Berman, S.S.: Trace element analysis of biological material following pressure digestion with nitric acid-hydrogen peroxide and microwave heating. J. Anal. At. Spectrom. 4, 323-327 (1989)
[125] Matusiewicz, H., Sturgeon, R.E., Berman, S.S.: Vapour-phase acid digestion of inorganic and organic matrices for trace element analysis using a microwave heated bomb. J. Anal. At. Spectrom. 6, 283-287 (1989)
[126] Matusiewicz, H.: Acid vapour-phase pressure decomposition for the determination of elements in biological materials by flame atomic emission spectrometry. J. Anal. At. Spectrom. 4, 265-269 (1989)
[127] Amaarsiriwardena, D., Krushevska, A., Barnes, R.M.: Microwave-assisted vapor-phase nitric

acid digestion of small biological samples for inductively coupled plasma spectrometry. Appl. Spectrosc. 52, 900–907 (1998)

[128] Czégény, Z., Berente, B., Óvári, M., Garcia Tapia, M., Záray, G.: Microwave-assisted vaporphase acid digestion of cellulose nitrate filters for elemental analysis of airborne dust samples. Microchem. J. 59, 100–106 (1998)

[129] Eilola, K., Perämäki, P.: Microwave heated vapor-phase digestion method for biological sample materials. Fresenius J. Anal. Chem. 369, 107–112 (2001)

[130] Han, Y., Kingston, H.M., Richter, R.C., Pirola, C.: Dual-vessel integrated microwave sample decomposition and digest evaporation for trace element analysis of silicon material by ICPMS: design and application. Anal. Chem. 73, 1106–1111 (2001)

[131] Araújo, G.C.L., Nogueira, A.R.A., Nóbrega, J.A.: Single vessel procedure for acid-vapour partial digestion in a focused microwave: Fe and Co determination in biological samples by ETAAS. Analyst 125, 1861–1864 (2000)

[132] Würfels, M., Jackwerth, E., Stoeppler, M.: About the problem of disturbances of inverse voltammetric analysis after pressure decomposition of biological samples. Fresenius' Z. Anal. Chem. 329, 459–461 (1987)

[133] Hertz, J., Pani, R.: Investigation of the completeness of digestion procedures prior to voltammetric trace metal analysis of olive leaves. Fresenius' Z. Anal. Chem. 328, 487–491 (1987)

[134] Schramel, P., Hasse, S.: Destruction of organic materials by pressurized microwave digestion. Fresenius J. Anal. Chem. 346, 794–799 (1993)

[135] Reid, H.J., Greenfield, S., Edmonds, T.E.: Investigation of decomposition products of microwave digestion of food samples. Analyst 120, 1543–1548 (1995)

[136] Nève, J., Hanocq, M., Molle, L., Lefebvre, G.: Study of some systematic errors during the determination of the total selenium and some of its ionic species in biological materials. Analyst 107, 934–941 (1982)

[137] Welz, B., Melcher, M., Nève, J.: Determination of selenium in human body fluids by hydride generation atomic absorption spectrometry: optimization of sample decomposition. Anal. Chim. Acta 165, 131–140 (1984)

[138] Knapp, G., Maichin, B., Baumgartner, U.: Interferences in ICP-OES by organic residue after microwave-assisted sample digestion. At. Spectrosc. 19, 220–222 (1998)

[139] Machat, J., Otruba, V., Kanicky, V.: Spectral and non-spectral interferences in the determination of selenium by inductively coupled plasma atomic emission spectrometry. J. Anal. At. Spectrom. 17, 1096–1102 (2002)

[140] Ashley, D.: Polyatomic interferences due to the presence of inorganic carbon in environmental samples in the determination of chromium at mass 52 by ICP-MS. At. Spectrosc. 13, 169–173 (1992)

[141] Begerov, J., Turfeld, M., Duneman, L.: Determination of physiological palladium, platinum, iridium and gold levels in human blood using double focusing magnetic sector field inductively

coupled plasma mass spectrometry. J. Anal. At. Spectrom. 12, 1095–1098 (1997)

[142] Matusiewicz, H., Golik, B., Suszka, A.: Determination of the residual carbon content in biological and environmental samples by microwave-induced-plasma atomic emission spectrometry. Chem. Anal. (Warsaw) 44, 559–566 (1999)

[143] Carrilho, E.N.V.M., Nogueira, A.R.A., Nóbrega, J.A., de Souza, G.B., Cruz, G.M.: An attempt to correlate fat and protein content of biological samples with residual carbon after microwave-assisted digestion. Fresenius J. Anal. Chem. 371, 536–540 (2001)

[144] Costa, L.M., Silva, F.V., Gouveia, S.T., Nogueira, A.R.A., Nóbrega, J.A.: Focused microwave-assisted acid digestion of oils: an evaluation of the residual carbon content. Spectrochim. Acta Part B 56, 1981–1985 (2001)

[145] Würfels, M., Jackwerth, E., Stoeppler, M.: Residues from biological materials after pressure decomposition with nitric acid. Part 1. Carbon conversion during sample decomposition. Anal. Chim. Acta 226, 1–16 (1989)

[146] Würfels, M., Jackwerth, E., Stoeppler, M.: Residues from biological materials after pressure decomposition with nitric acid. Part 2. Identification of the reaction products. Anal. Chim. Acta 226, 17–30 (1989)

[147] Würfels, M., Jackwerth, E., Stoeppler, M.: Residues from biological materials after pressure decomposition with nitric acid. Part 3. Influence of reaction products on inverse voltammetric element determination. Anal. Chim. Acta 226, 31–41 (1989)

[148] Würfels, M.: Voltammetric determination of metal traces in marine samples after nitric acid decomposition. Marine Chem. 28, 259–264 (1989)

[149] Wasilewska, M., Goessler, W., Zischka, M., Maichin, B., Knapp, G.: Efficiency of oxidation in wet digestion procedures and influence from the residual organic carbon content on selected technique for determination of trace elements. J. Anal. At. Spectrom. 17, 1121–1125 (2002)

[150] Matusiewicz, H., Suszka, A., Ciszewski, A.: Efficiency of wet oxidation with pressurized sample digestion for trace analysis of human hair material. Acta Chim. Hung. 128, 849-859 (1991)

[151] Krushevska, A., Barnes, R.M., Amarasiriwaradena, C.J., Foner, H., Martines, L.: Comparison of sample decomposition procedures for the determination of zinc in milk by inductively coupled plasma atomic emission spectrometry. J. Anal. At. Spectrom. 7, 851–858 (1992)

[152] Nakashima, S., Sturgeon, R.E., Willie, S.N., Berman, S.S.: Acid digestion of marine samples for trace element analysis using microwave heating. Analyst 113, 159–163 (1988)

[153] Hee, S.S.Q., Boyle, J.R.: Simultaneous multielement analysis of some environmental and biological samples by inductively coupled plasma atomic emission spectrometry. Anal. Chem. 60, 1033–1042 (1988)

[154] Krushevska, A., Barnes, R.M., Amarasiriwaradena, C.J.: Decomposition of biological samples for inductively coupled plasma atomic emission spectrometry using an open focused mi-

crowave digestion system. Analyst 118, 1175–1181 (1993)

[155] Würfels, M., Jackwerth, E.: Investigations on the carbon balance in decomposition of biological materials with nitric acid. Fresenius' Z. Anal. Chem. 322, 354–358 (1987)

[156] Pratt, K.W., Kingston, H.M., MacCrehan, W.A., Koch, W.F.: Voltammetric and liquid chromatographic identification of organic products of microwave-assisted wet ashing of biological samples. Anal. Chem. 60, 2024–2027 (1988)

[157] Kingston, H.M., Jassie, L.B.: Microwave energy for acid decomposition at elevated temperatures and pressures using biological and botanical samples. Anal. Chem. 58, 2534–2541 (1986)

[158] Krushevska, A., Barnes, R.M., Amarasiriwaradena, C.J., Foner, H., Martines, L.: Determination of the residual carbon content by inductively coupled plasma atomic emission spectrometry after decomposition of biological samples. J. Anal. At. Spectrom. 7, 845–850 (1992)

[159] Matusiewicz, H., Sturgeon, R.E.: Comparison of the efficiencies of on-line and high-pressure closed vessel approaches to microwave heated sample decomposition. Fresenius J. Anal.Chem. 349, 428–433 (1994)

[160] Ko, F.-H., Chen, H.-L.: Study of microwave digestion kinetics and establishment of a model for digestion efficiency prediction. J. Anal. At. Spectrom. 16, 1337–1340 (2001)

[161] Matusiewicz, H., Barnes, R.M.: Tree ring wood analysis after hydrogen peroxide pressure decomposition with inductively coupled plasma atomic emission spectrometry and electrothermal vaporization. Anal. Chem. 57, 406–411 (1985)

[162] Sah, R.S., Miller, R.O.: Spontaneous reaction for acid dissolution of biological tissues in closed vessels. Anal. Chem. 64, 230–233 (1992)

[163] Veschetti, E., Maresca, D., Cutilli, D., Santarsiero, A., Ottaviani, M.: Optimization of H_2O_2 action in sewage-sludge miocrowave digestion using pressure vs. temperature and pressure vs. time graphs. Microchem. J. 67, 171–179 (2000)

[164] Matusiewicz, H.: Critical evaluation of the effectiveness of nitric acid oxidizing systems: pressurized microwave-assisted digestion procedures. Chem. Anal. (Warsaw) 46, 897–905(2001)

[165] Clem, R.G., Hodgson, A.T.: Ozone oxidation of organic sequestering agents in water prior to the determination of trace metals by anodic stripping voltammetry. Anal. Chem. 50, 102–110 (1978)

[166] Filipović-Kovačević,Ž., Sipos, L.: Voltammetric determination of copper in water samples digested by ozone. Talanta 45, 843–850 (1998)

[167] Sasaki, K., Pacey, G.E.: The use of ozone as the primary digestion reagent for the cold vapor mercury procedure. Talanta 50, 175–181 (1999)

[168] Jiang, W., Chalk, S.J., Kingston, H.M.: Ozone degradation of residual carbon in biological samples using microwave irradiation. Analyst 122, 211–216 (1997)

[169] Matusiewicz, H., Stanisz, E.: Evaluation of high pressure oxygen microwave-assisted wet decomposition for the determination of mercury by CVAAS utilizing UV-induced reduction.Micro-

chem. J. 95, 268–273 (2010)

[170] Schramel, P., Lill, G., Seif, R.: A complete HF digestion method for element and trace element determination in soils, sediments, sludges and other similar samples using a closed system. Fresenius' Z. Anal. Chem. 326, 135–138 (1987)

[171] Matusiewicz, H.: High-pressure microwave dissolution of ceramics prior to trace metal determinations by microwave induced plasma atomic emission spectrometry. Mikrochim.Acta 111, 71–82 (1993)

[172] Budna, K.W., Knapp, G.: Continuous decomposition of organic materials with H_2O_2–H_2SO_4. Fresenius' Z. Anal. Chem. 294, 122–124 (1979)

[173] Stainless steel pressure vessels with Teflon inserts, Berghof Laborprodukte GmbH, Eningen, Germany

[174] Matusiewicz, H.: Development of high-pressure closed-vessel systems for microwaveassisted sample digestion in microwave-enhanced chemistry. In: Kingston, H.M., Haswell, S.J. (eds.) Microwave-Enhanced Chemistry. Fundamentals, Sample Preparation, and Applications, pp. 353–369. ACS, Washington, DC (1997)

[175] de Moraes Flores, É.M. (ed.): Microwave-Assisted Sample Preparation for Trace Element Determination. Elsevier, Amsterdam (2014)

[176] Jorhem, L.: Dry ashing, sources of error, and performance evaluation in AAS. Mikrochim.Acta 119, 211–218 (1995)

[177] Gleit, C.E., Holland, W.D.: Use of electrically excited oxygen for the low temperature decomposition of organic substances. Anal. Chem. 34, 1454–1457 (1962)

[178] Wickbold, R.: Neue Schnellmethode zur Halogenbestimmung in Organischen Substanzen. Angew. Chem. 64, 133–135 (1952)

[179] Schöniger, W.: Eine Mikroanalytische Schnellbestimmung von Halogen in Organischen Substanzen. Mikrochim. Acta 1, 123–129 (1955)

[180] Souza, G.B., Carrilho, E.N.V.M., Oliveira, C.V., Nogueira, A.R.A., Nobrega, J.A.: Oxygen bomb combustion of biological samples for inductively coupled plasma optical emission spectrometry. Spectrochim. Acta Part B 57, 2195–2201 (2002)

[181] Knapp, G., Raptis, S.E., Kaiser, G., Tölg, G., Schramel, P., Schreiber, B.: A partially mechanized system for the combustion of organic samples in a stream of oxygen with quantitatively recovery of the trace elements. Fresenius' Z. Anal. Chem. 308, 97–103 (1981)

[182] Ingamells, C.O.: Absorptiometric methods in rapid silicate analysis. Anal. Chem. 38, 1228–1234 (1966)

[183] Luque de Castro, M.D., Luque Garcia, J.L.: Automation of sample preparation. In: Mester, Z., Sturgeon, R. (eds.) Sample Preparation for Trace Element Analysis, pp. 649–680.Elsevier, Amsterdam (2003)

第6章
痕量分析中的萃取方法

6.1 概述

在自然界和环境中经常发生萃取现象,萃取技术极其重要并具有巨大的发展潜力。因此,萃取技术在许多领域得到了广泛应用,在化学及相关技术以及化学分析中都具有特殊用途。

萃取技术是通用的、标准化的和常规的样品制备方法[1-4]。20年来,出现了多种萃取技术,这些技术应用于分析包括DNA和RNA在内的各种有机化合物和元素及其形态和形式,这些化合物和元素以痕量或微量形式存在于不同的样品基体中。

在色谱法、光谱法、电分析法或电泳法等分析方法应用之前,萃取可看作一种有效和选择性的样品制备方法[5-10]。国际标准化组织、美国食品药品监督管理局和美国国家环境保护局的国际规范建议,在食品和环境及药物样品的分析中应用萃取方法。关于萃取的新思想和新观点引起了许多关于术语的争议,随后萃取和其他分析样品处理技术之间的分界线面临重新划定,并开始变得模糊。

萃取是多相和多组分体系中的质量交换过程,它涉及将一种或多种组分转移和分配成两个不混溶相,因此,必须考虑热力学和动力学因素,或至少考虑其中之一。萃取过程可用适当的热力学和动力学方程描述,也可用热力学函数描述。液-液萃取是动态平衡状态,在固体样品的浸出或固相萃取(SPE)等过程是不平衡处理状态。

在有机化合物的提取中,通常进行衍生化反应,因为它们既保证了某些有机物的高效分离,又可以在仪器测量过程中通过改变选定的官能团来提高检测灵敏度。一些新发展的更加复杂的萃取方法大大降低了能源和化学品的消耗,并且允许所使用萃取系统的循环回收,满足了绿色化学的要求。

本书中只提出与萃取有关的技术和针对痕量分析的选择方法,当然,每种选择以及选择的依据都是主观的。

6.2 萃取方法的分类及分析性能

已经开发和报道了许多提取方法和技术,最常见和最简单的分类方法是类似于色谱法中基于分析物被转移到不同相态进行分类。可以按照萃取物状态将其分为液体、固体、气体和超临界流体相萃取物。更精确的是用分析物分布的两个相来描述,如液-液或固-液(浸出)萃取,后一种方法也称溶剂萃取。

可依据萃取过程中使用的方法和技术进行分类,结果如下:
(1)周期性的和连续性的萃取;
(2)根据有和没有任何额外能量供给进行萃取系统分类,能量供给通常是微波或超声波能量;
(3)单一或多重萃取;
(4)对流和并流萃取;
(5)根据在开放还是密闭系统中,采用的是低压、大气压或高压装置进行萃取技术分类;
(6)柱萃取;
(7)有或无加速溶剂萃取。

萃取提取方法也可以根据在分离过程中所涉及现象的机理进行分类。有些人会重点考虑分析物和/或样品的类型,另一些人会考虑色谱和仪器方面的适用性。根据上述类别,可以对20年来报道的各种新提取方法进行分类。例如,中空纤维液相微萃取、加速溶剂萃取、超分子溶剂萃取和分散液-液微萃取都属于溶剂萃取类。将吸附剂的微萃取、搅拌棒吸附萃取等基于固相萃取(SPE)的微萃取和基于溶剂萃取的微萃取进行区分分类是比较合理的。

如果在痕量分析中必须通过萃取分离样品组分,那么需要选择适当的萃取方法和建立适当的操作参数以确保萃取的高效率,最终获得高质量的分析性能。提取条件的选择对于分析物的定量回收至关重要,或者至少使分析结果具有足够的有效性至关重要。如果水溶液是萃取相之一,那么诸如络合、水解和溶剂化之类的问题会起关键作用。从水相提取元素到有机相通常需要选择合适的配体和控制pH值。

在溶剂萃取的过程中,选择的洗脱剂应具有以下特征:
(1)对分析物有非常好的溶解性;
(2)受体与供体相不混溶性;
(3)无毒性;
(4)难燃性;
(5)易于获得,价格低廉;

(6) 具有稳定性与不活跃性；

(7) 沸点与主溶剂有很大不同。

另外,溶剂萃取必须考虑与所使用的痕量测量方法相匹配。例如,气相色谱与质谱联用(GC-MS)需要挥发性溶剂,而高效液相色谱(HPLC)使用非挥发性溶剂。低表面张力的溶剂溶液生成乳液并导致较长的相分离时间。

SPE过程的第一阶段是将分析物固定在吸附剂上,再用适当的溶剂选择性洗脱目标组分。选择固体吸附相时应选择具有高亲和力的固体吸附相,其对分析物的吸附力应明显高于分析物和供体相之间的吸附力。洗脱液的选择是影响SPE效率的另一个因素。

6.3 提取方法的选择

在本章介绍的提取方法的选择中,主要考虑方法的认可度、成熟度、操作复杂性和成本以及可改进性,在众多的综述文献中可以发现许多对传统萃取方法的改进。本章省略了在本书应用部分中介绍的一些其他方法,如索氏方法。

6.3.1 固相萃取

"固相萃取"(SPE)一词是由J. T. Baker公司在1982年首先提出的,该方法是将液体或气体样品保留到固体固定相,然后使用适当的洗脱剂去除固定相中的分析物。方法的主要目的是分离和富集感兴趣的化合物或样品,以便清除和简化基体。此样品制备技术的应用也可以提取分馏,由于有机溶剂用量明显减少、回收率高以及实现过程自动化的可能性,SPE是常规液-液萃取的良好替代方法。根据固定相对化合物的吸附性质,分为以下三类:

(1) 非极性化合物:用于极性样品基质中的非极性和中等极性化合物。

(2) 极性化合物:从非极性溶液中萃取极性或中等极性分析物。

(3) 离子化合物:用于分离阴离子或阳离子化合物。

许多新型吸附剂被发明并用于简化和改进萃取程序[11-13]。

用碳链改性的二氧化硅凝胶,包括最普遍应用的十八烷基硅(-C18),其烷基有18个碳原子,通常用于反相分离。不同的碱性或芳基基团被用于二氧化硅的改性阳离子,如—C_2,—C_4,—C_8,—CN和—NH_2。这些官能团都是疏水性的,因此分析物的保留是非极性-非极性吸引力或分散力的结果,碳链改性的硅胶吸附剂在较窄的pH值范围内稳定。

聚合物吸附剂如苯乙烯-二乙烯苯共聚物,或石墨化炭黑,通常用作保留分析物的反相物质,这些吸附剂在较宽的 pH 值范围内稳定并且具有良好的表面张力,因此大部分溶解于水的分析物被很好地保留。分析物在这种吸附剂上的保留主要基于分析物的大小和形状。近年来,具有乙酰基、羟甲基、苯甲酰或羧苄基等官能团的化学改性聚合物吸附剂已经上市,通过将不同官能团引入聚合物树脂中而与水样提供更好的表面接触,提高了 SPE 的性能。化学改性聚合物吸附剂具有高疏水性,其回收率优于未改性吸附剂。具有明确孔径大小的聚合物凝胶允许分子尺寸级混合物组分的分离,例如用树脂吸附剂分离低分子量和高分子量的分子,这个过程称为凝胶过滤。

分析物在离子交换吸附剂上的吸附是基于分析化合物中的带电官能团与固定相上二氧化硅表面上的带正电荷或负电荷基团之间的静电式相互作用。为了提高吸附剂的选择性和分离效率,开发了具有两个官能团的多功能吸附剂。这类吸附剂可以提供混合保留机制,包括疏水、离子和/或 PI-阳离子相互作用。不同的相互作用机制赋予了具有不同极性和酸性化合物的有效保留,这种吸附剂应用于一些药物、食品样品、生物制剂、动物组织和废物的分析。

最具特异性和选择性的吸附剂是免疫吸附剂和分子印迹聚合物(MIP)等新型固定相。由于免疫吸附剂有一种特异性抗体附着在二氧化硅表面,因此具有高度分子选择性,用于去除溶液中相应的抗原[14]。选择性增强是特异性抗原-免疫抗体相互作用的结果。MIP 是一个由目标分析物模板与功能单体之间的混合物获得的三维聚合物网络[15]。在从聚合物中除去模板后形成模板识别腔,其形状、尺寸和化学功能与模板分子互补。分析物结合是通过分子间相互作用力,如氢键、偶极-偶极相互作用以及模板分子与聚合物基体官能团之间的离子相互作用实现的。

通常,吸附剂放置在萃取柱或萃取管中,现在可以获得许多尺寸和类型的商业化固定相萃取盒或萃取盘。

从固定相中分离分析物通常借助选定的洗脱溶剂,有时也可以通过解吸实现。在两种或多种洗脱溶剂的帮助下进行的顺序洗脱,用于选择性分离两种或多种感兴趣的化合物或相似的分析物组。选择洗脱溶剂时应考虑感兴趣的分析物的种类、基体组成、固定相的类型和检测方法等多种因素。对于极性固定相,洗脱强度由溶剂极性和极化率决定。对于非极性吸附剂,洗脱强度取决于范德华力。选择洗脱溶剂时还必须考虑溶剂黏度、水溶性、折射率、紫外截止波长、专用检测器的适用性和环境可比性等条件,例如,洗脱系列可通过它们从某一吸附剂中除去分析物的能力来标记溶剂。SPE 在药物产品及生物和环境样品痕量分析中的应用实例见表 6-1[16]。

使用 SPE 萃取的有机物种类型和分析样品的种类显著增加,主要是新的智能材料用作吸附剂材料,如免疫吸附剂、分子印刷聚合物、碳纳米材料等。本节将讨论与这一重要技术相关的基本信息。关于各种特殊设计用作固体吸附剂的现代材料的更多细节和它们的未来前景参见文献[17-19]。

表6-1 SPE在药物产品及生物和环境样品痕量分析中的应用实例

分析物	样品	吸附剂类型	洗脱溶剂	测定方法
β-内酰胺抗生素(如青霉素G、氨苄西林、阿莫西林)	废水	Oasis Max（聚合物）	四正丁基铵甲醇中的硫酸氢盐	HPLC-UV
四环素、磺胺类、大环内酯类	河水	Oasis HLB（聚合物）	甲醇	LC-MS
磺胺类药物、四环素	牛奶	Oasis HLB（聚合物）	甲醇	LC-ESI-MS
四环素、大环内酯类、磺胺类药物	土壤,PLE后的液体样品	Oasis HLB（聚合物）	甲醇	LC-ESI-MS
四环素、大环内酯类化合物、磺胺类药物	沉积物,LLE后的液体样品	Oasis HLB（聚合物）	甲醇	LC-MS
四环素	组织,LLE后的液体样品	MAA作为功能单体 EGDMA作为交联剂	KOH甲醇溶液	HPLC-UV

注：PLE—加压液体萃取；LLE—液萃取；MAA—甲基丙烯酸；EGDMA—乙二醇二甲基丙烯酸酯；HPLC—高效液相色谱；LC—液相色谱；ESI—电喷雾电离；MS—质谱。

1. 固相微萃取

固相微萃取(SPME)是一种快速的无溶剂改性固相萃取技术[20]。SPME使用有涂层的纤维吸附剂,如熔融石英纤维或聚合物包覆熔融纤维。该方法适用于从液体或气体等各种介质中提取不同种类的挥发性和非挥发性分析物。在直接萃取模式下,将带涂层的纤维浸泡在样品中直到样品基体与固相达到平衡,分析物通过扩散到萃取相,然后对分析物进行解吸和测定。对于挥发性化合物可以将其解吸并在高温条件下直接注入气相色谱仪系统,或引入载气流中进行测定[21-22]。

目前,聚二甲基硅氧烷、聚丙烯酸酯和石蜡-二乙烯基苯或聚二甲基硅氧烷-二乙烯基苯是常用的固定相。非极性分析物采用聚二甲基硅氧烷,极性化合物适于采用聚丙烯酸酯。

SPME技术的发展必须满足工业、实验室和教学实验中绿色化学理论的要求[23]。SPME技术的改进以减少溶剂使用、减小分析样品体积以及分析物直接测定为目的。微型化技术的典型例子是搅拌棒吸附萃取(SBSE)[24-26]和填充吸附剂微萃取(MEPS)[27-29]。在SBSE中,分析物被提取到磁性搅拌棒上的聚合物涂层,大多数SBSE的应用需要进行热脱附。MEPS一般是自动化技术,聚硅氧烷基吸附剂集成放置于微升注射器药筒中,其主要优点是分析样品体积小、洗脱溶剂用量少而且样品制备和注射所需时间短。MEPS在环境、食品和生物医学样品痕量分析中的应用实例见表 6-2[30-31]。

表 6-2 MEPS 在环境、食品和生物医学样品痕量分析中的应用实例

被分析物	样品	吸附剂	测定方法
安非他命	尿、头发	聚二甲基硅氧烷纤维	HPLC
利多卡因	血清、尿液	聚二甲基硅氧烷纤维	GC-FID，HPLC-UV
杂环芳香胺	食品	聚乙二醇，聚乙二醇-二乙烯基苯，聚二甲基硅氧烷二乙烯苯聚丙烯酸酯	HPLC-UV
农药/有机磷	蜂蜜	聚二甲基硅氧烷	LC-MS
Cr(Ⅲ)	水	聚合物涂层石英光纤	GC-FPD

注：GC-FID—气相色谱离子化检测；HPLC-UV—高效液相色谱紫外检测；LC-MS—液相色谱-质谱联用；GC-FPD—气相色谱法火焰光度检测。

2. 顶空技术

"顶空"是指气体、液体或固体样品上方的气体层[32]。顶空萃取是将挥发性或半挥发性有机成分从样品基质中分离到顶空气体中。在此萃取过程中，挥发性样品成分从基体扩散到气相中，从而形成顶空气体。经典的静态顶空技术即气相萃取，是指被测化合物从密封小瓶中的样品里热蒸发，在一定的温度下达到扩散平衡，并在该温度下对顶空蒸气进行采样，将样品上方的气体采集并导入 GC 装置的注入口。静态顶空提取更有效的形式是多次顶空提取，其中对同一样品瓶重复提取相同样品的顶空气体，这可以测量样品中存在的分析物总量。

动态顶空萃取即吹扫和捕集萃取模式，是载气持续吹扫加热的样品，将挥发物吹扫到吸附剂捕集器并解吸到气相色谱仪中。

顶空固相微萃取(HS-SPME)是 SPME 技术的一种快速无溶剂的改进，其中在顶空气体中插入带有聚合物涂层的熔融二氧化硅纤维，以提取有机化合物，并将它们直接转移到气相色谱仪的注射器中用于热脱附和分析。在该技术中，萃取到纤维上的化合物的量取决于固定相的极性和厚度、萃取时间和样品中挥发物的浓度。

顶空萃取的典型应用是食品中的香味剂、土壤中的挥发性有机化合物和药物产品中的残留溶剂[33-34]。顶空萃取的主要优点是样品前处理少，可以将顶空气体直接引入气相色谱仪。

本节仅介绍了顶空技术的基本信息和基本原理、理论，关于气体到气相色谱的转移、应用和设备的更多细节可以在许多指南和文章[35-37]中找到。

6.3.2 膜萃取

膜萃取起源于 20 世纪 60 年代，由于其高选择性、高富集因子和可自动化而被

广泛使用。膜是一种选择性屏障,溶质分子由于浓度梯度从作为供体相的样品传递到接收相。在某些情况下,被测化合物通过气态、液体或固相从接收相提取。通过膜实现挥发性化合物从气态样品到气态受体的提取称为渗透。

膜面积越大,厚度越小,分子的运输越快。同时,温度和溶质分配系数影响扩散,扩散系数随温度升高而升高。膜萃取的有效性与扩散和分配系数有关[38-40]。

膜萃取技术可根据膜结构或参与萃取过程的相数进行分类[41-42]。过滤、透析和电渗析是基于多孔膜的分离技术。在透析中,供体和接收溶液均通过膜孔隙持续接触,孔隙尺寸允许小分子的扩散。扩散是膜的供体和接收相之间存在浓度梯度的结果,为了提高处理效率,可对接收相进行处理。有时透析会导致大分子如蛋白质和腐殖质化合物溶液的显著提纯,但实现分析物的预浓缩通常很困难,因此透析和 SPE 的耦合是有必要的。

还有一些技术涉及使用非多孔、固体或液体膜,通过明显的相边界将供体相与接收相分离。最常用的是包括供体相、膜和接收相的三相系统或两相系统,其中相边界的作用与膜相同。固体膜一般为化学性质稳定的疏水性聚合物如 PTFE、PVDF、PS、PP、硅酸盐等,或金属如 Pd 合金,或陶瓷材料。膜组件的通道体积为 $10\sim1000\mu L$,根据几何形状可以分为平面的或者是纤维状的。可以使用两种模式建立膜系统:膜可以浸没在样品中,即样品中的膜(MIS);将样品引入膜中,即膜中的样品(SIM)。在这两种模式中,只有少量的样品与膜直接接触,因为膜表面积与样品体积的比例很小。

液膜以宽层、乳液、固定相、杂化或聚合物的形式存在。宽层膜系统最简单但仅偶尔地用于分析,其中分离两个水相的有机相厚为几厘米到十几厘米,并且通常包含支持分析物传输的物质。液体乳化膜(LEM)是由不混溶两相乳化形成的,液滴分散在第三相。乳化膜的主要优点是膜相面积大、接受相体积小、萃取效率高、分析物富集率高。在分析物乳液从供体相中分离(通常依靠重力)之后,再使用电学、化学或热方法破坏乳液。在分析实验室中广泛应用的一组重要液膜是支撑液膜(SLM),这种膜是有机溶剂通过毛细管力保持在多孔膜支撑片或多孔中空纤维的孔中,膜厚为 $100\sim200\mu m$,有机相的体积为 μL 级,可以使用脂肪族醇、烃类或醚(如二己基醚、磷酸二辛酯或煤油)。SLM 萃取过程的驱动力可以是供体和受体相之间的 pH 差、浓度差或电位差,从供体相输送的溶液溶解在有机膜中,随后由于供体和接收相之间的浓度梯度而转移到第二水相。提取和再萃取过程取决于水相和膜之间的分析物分离系数。如果被测物转化成不溶于有机膜的形式,则可以依靠扩散流动进行选择性传输,也可以通过存于液膜中能够与分析物形成复合物的化合物来促进传输。SLM 萃取主要用于提取极性组分,如有机酸、无机酸和碱或金属离子。比 SLM 更稳定的是聚合物包含膜(PIM),其通过注入含有增塑剂的聚合物溶液和某类分析物载体形成。在这种类型的膜中,有机相充满聚合物膜中的所有孔体积,最常用的是硅橡胶、聚氯乙烯或三醋酸纤维素。分析物运输由载体、络合剂或离子交换器提供[41]。

聚合物膜在液-液萃取过程中分离水相和有机相的应用称为微孔膜液-液萃取(MMLLE)。在疏水膜孔中的有机接收溶剂与膜表面附近的水相保持直接接触,在膜表面发生传质。这种萃取类似于SLME,但在两相系统中进行,并且由于没有载体试剂,因此速度比较慢且选择性较低。由于聚合物膜是不溶性的,因此可以任意组合水相和有机相,萃取效率主要取决于分配系数。

膜萃取的一个显著优点是通过在流动系统中接入分析仪器,样品处理过程可以实现自动化[42]。质谱仪、色谱仪、原子光谱仪是最常用的金属分析仪器。可将膜置于真空系统内,将分析物直接引入电离源与质谱分析相结合。用于气体或挥发性化合物渗透的膜通常是疏水性非多孔硅膜,因此膜萃取和气相色谱的耦合更加复杂。表6-3列出了近年来膜萃取在痕量分析中的一些应用[38-42]。

表6-3 膜萃取在痕量分析中的应用

分析物	样品/基体	膜萃取技术
药物	水溶液、生物液体	SLM, MMLLE
羧基酸	水、空气、土壤	SLM
除草剂、杀菌剂	天然水、食品	MMLLE, PME
氨基酸	合成水溶液	ELM, SLM
金属及其形态形式	水、尿液、废水、海水	SLM, ELM, PIM
碘	水溶液	PIM
酚类化合物	天然水、燃料	SLM, PME, MMLLE

注:SLM—支撑液膜;MMLLE—微孔膜液液萃取;ELM—乳状液膜;PIM—聚合物夹杂物膜。

6.3.3 微波萃取

在微波萃取(MAE)过程中,微波能量用于加热溶剂或/和样品,以加速被测化合物从样品到溶剂的转移[43-44]。微波是一种波长1m~1mm的电磁辐射,频率为1~100GHz。

微波辐射可以由不同的化合物通过偶极极化或离子传导来传递。偶极极化决定由分子偶极如水、乙酸、氯仿组成的极性化合物的加热。在这种情况下,微波能量激发应用区域偶极的旋转和重组。在离子化合物中,微波能与离子场引起离子运动,离子的迁移和溶液间流动的阻力导致溶液的加热。1986年,Ganzler等首次从食品中提取粗脂肪和抗营养剂,并从土壤中提取农药[45],微波能量在液-液萃取中的应用导致萃取时间显著缩短。与传统加热不同,微波被整个样品吸收而不受容器材料影响,因为微波能量通过分子相互作用传递到材料中,并且电磁能量被转换成热能。偶极分子和离子吸收微波能量,但非极性溶剂如正己烷不会因微波作用而变热。溶剂介电常数越大,释放出的热能越多,萃取溶剂或样品基体达到所需温度的速度就越快。表 6-4 列出了在 MAE 中用作萃取剂的选定溶剂的物理参数。

表 6-4 选定溶剂的物理常数

溶剂	介电常数	偶极矩(20℃)	沸点/℃ ($p=101.3kPa$)
丙酮	20.7	2.69	56.3
乙腈	35.9	3.44	81.6
乙醇	24.6	1.69	78.3
正己烷	1.89	0.08	68.7
甲醇	32.7	2.87	64.6
2-丙醇	19.9	1.66	82.3
水	78.9	1.87	100.0

MAE 方法的使用基本原则如下:

(1) 样品是用吸收微波能的一种溶剂或溶液混合物提取的;

(2) 萃取剂是高介电常数和低介电常数溶剂的混合物;

(3) 用不吸收微波能量的试剂浸出高介电常数样品。

通常,在实验室中应用两套装置进行 MAE:一种为允许调节和控制温度、压力的密闭容器系统;另一种为常压下运行的开放容器系统。在开放的容器系统中,最大温度是由萃取剂沸点决定的,微波辐射的吸收发生在整个样品中,因此加热是有效和均匀的。开放容器系统的主要缺点是挥发性化合物可能会损失,可以通过将回流系统装配在提取容器顶部来减少损失。

MAE 工艺的优化包括选择合适的溶剂或溶剂混合物,以及优化其他影响因素,如溶剂-进料比、提取时间和温度、微波功率和基体特性(包括含水量)。

在大多数公开文献中,萃取剂通常具有高的介电常数并极易吸收微波能量。常见的单一萃取剂应用是:用甲醇或乙醇萃取紫杉烷,用四氢呋喃浸出低分子量的低聚物。采用两种极性溶剂乙腈和甲醇的混合物提取药物中的非洛地平。使用二氯甲烷和甲醇提取土壤中的莠去津残留及其衍生物[46]。

通常用于提取多环芳烃(PAHs)或多氯联苯(PCB)的极性和非极性溶剂混合物是己烷和丙酮的组合物,可以用乙酸乙酯和环己烷萃取氯化有机化合物[46]。

对于高介电常数的样品,应使用不吸收微波能量的溶剂。从高湿度样品中提取热不稳定的分析物就是一个典型例子[46-47]。微波加热基体中的水分,将其蒸发并在电解池内产生内部压力,导致细胞破碎并释放溶质,从而提高提取效率。该方法不同于经典的溶剂萃取——溶剂扩散到基体中并根据其溶解度释放分析物。

不同于常规萃取,萃取剂体积的增加不能提高萃取效率。在某些情况下,非常小的溶剂体积足以提供定量提取,例如,10mL的提取剂就可用于从5g土壤样品中浸出酚类化合物[46]。

无论使用哪种技术,温度都是影响萃取效率的一个非常重要的因素。MAE在密闭容器中进行时,温度可以超过溶剂沸点。温度升高导致萃取效率提高的原因是,分析物从基体中解吸效率更高、分析物增溶作用增强以及溶剂能更好地渗透到样品中。从土壤和沉积物中提取有机污染物如多环芳烃或石油衍生碳氢化合物的最常用温度为115~120℃。农药提取过程中可以使用温度80~110℃。为了浸出热稳定性差的有机成分如芳香胺或氨基酸,需要选择较低的工艺温度,防止分析物发生热分解。

萃取效率的其他影响参数是时间、压力、微波辐射功率、pH值以及样品质量。所有设置的参数都取决于样品的种类和水分含量。在许多应用中,不同类型样品在相同条件下进行萃取的效果明显不同。因此,样品处理程序的个体适应性是分析新材料的必要前提。

商用微波辅助系统可以在密闭容器中同时处理多个样品,并能精确控制温度和压力。交付的软件允许控制并设定适用于独立分析条件。

关于MAE在不同材料分析中的应用参见文献[48-52]。在痕量分析中使用MAE的例子见表6-5。

表6-5 MAE在痕量分析中的应用实例

分析物	样品/基体	提取剂	提取时间/min
PAH	灰尘、沉积物、空气、土壤	己烷-丙酮、丙酮、甲苯-水	5~20
PCB	沉积物、土壤	正己烷-丙酮、甲苯-水	5~10
PCDD/PCDF	沉积物	甲苯	20
烃类	土壤、沉积物	丙酮	15
农药	沉积物、土壤、鱼类组织	异辛烷、己烷-丙酮、四氢呋喃、乙基辛烷-环己烷	20~30
咪唑	土壤	氨缓冲液、pH10	3
噻嗪类	土壤	DCM-甲醇(9:1)	20
药物	药物、血清、生理液、土壤、沉积物	甲醇、乙腈、氯仿、己烷-异戊醇	1~15
有机汞	土壤	甲苯+HCl	10
砷化合物	鱼类组织	甲醇-水(80:20)	4
金属	沉积物、植物和动物样品、药物、土壤		10~20

6.3.4 超声萃取

声波通过振动机械能分散于固体、液体或气体等所有弹性传输介质中传播。超声波是频率大于20kHz的振荡声压波,与其他声波类似,它们通过粒子或分子振动能量的传播而传播。液体中,压缩和膨胀运动的能量产生机械纵波,膨胀导致低压,因此高强度超声波产生小真空泡,在高压循环中,当气泡不能吸收更多的能量时会剧烈地崩裂,这种现象称为空化现象。气体和蒸汽的突然压缩会导致温度和压力的局部升高,空化气泡的侵入产生液体射流,膨胀气泡的势能转化为液体射流的动能。由于空化气泡的尺寸与溶液的总体积相比非常小,因此温度和压力的局部变化对溶液参数没有影响。液体射流对固体物质的影响是极其严重的,它甚至能引发超声波分解,即细胞壁破裂和细胞内物质的提取。超声波是样品制备过程中的重要辅助手段,可促进和加速有机化合物和无机化合物的提取、浆料分散、样品溶解、乳化、均质、雾化、清洗、衍生和样品脱气。

超声辅助提取是从不同样品中提取多种分析物的有效方法[52-55]。与索氏装置中溶剂萃取等传统萃取技术相比,该方法简单、快速、经济。超声辅助固-液萃取是一种有效、省时的提取方法,超声处理加速了两相的传质过程。使用超声波使操作温度降低,允许提取温度敏感成分。与其他新的提取技术如 MAE 相比,超声设备更经济且易于操作。

通常在实验室中使用超声波浴和超声波探头两种类型的超声波装置。然而,由于超声能量在整个溶液中分布的不均匀性和功率随时间的衰减,超声波浴实验条件的重复性和再现性往往不能令人满意。利用超声波探头将能量聚焦在小样本区域上,这显著地提高了空化效率,从而提高了提取效率[56]。

有许多综述文章涉及微波提取在食品技术中的应用[57]以及从草药和其他植物材料中分离生物活性物质[58],也应用于从环境和工业样品中浸出重金属[59]。在连续提取微量元素的过程中,超声的应用明显地缩短了提取过程,但是为了获得满意的效果,必须提高温度和改进基体。

超声波应用的另一个例子是从不同种类的样品中浸出有机杂质,主要分析物是多环芳烃,广泛存在于土壤、沉积物、灰尘和颗粒样品中[55]。推荐使用超声提取作为快速、有效、直接的环境样品前处理方法,用于测定多氯联苯、硝基酚、农药或聚合物添加剂。利用超声波能量,可以从动植物组织中提取微量有机金属和生物活性化合物(如维生素 A、维生素 D 和维生素 E)[59]。表6-6列出了超声萃取在生物和环境样品痕量分析中的一些典型的应用实例[60]。

表 6-6 超声萃取在痕量分析中的应用实例

分析物	样品	提取剂	提取时间/min
酚类酸	草本(唇形科)	乙醇	30
丁基苯基锡	贻贝	甲醇/己烷/甲苯,络合剂	15~30
Ca、Mg、Mn、Zn	蔬菜	硝酸(V)	30、10
Ag、As、Cd、Cu、Pb	土壤	王水	9
汞化合物	尿液	环己烷中的双硫腙	3
杀菌剂	土壤	乙基辛烷	20
雌激素、孕酮	土壤	乙醇+丙酮(1+1)	15
伊斯维安	大豆	乙醇(50%)	20
PCB	大蓝鹭蛋	硫酸钠(VI)/己烷	15
PAH	土壤	己烷/丙酮,0.5mol/L 十二烷基钠磺胺嘧啶-X-100	60
抗生素	动物饲料	甲醇、乙酸	15
抗氧化剂	中药	丁酮、乙醇、乙基辛烷	15、30、45

6.3.5 顺序分离

顺序分离过程包括使用一系列提取剂顺序处理样品,这些提取剂是根据其与基体中不同分析物的溶解能力而选择的。使用模拟自然现象(如酸雨)的顺序提取程序可以提供关于微量元素的起源、发生方式、生理作用和痕量元素转移等详细信息,还可以评估以不同形式存在的金属和存在这些金属不同阶段上的毒理学风险。越来越具有强烈破坏性试剂的连续作用,使顺序分离出具有类似化学性质的含有待测金属组分。顺序提取程序常用于环境样品,如土壤、沉积物和废料。典型的顺序提取过程基于 Tessier 等[61]提出的五步程序,在这个过程中获得以下组分:

(1) 离子交换,采用 $MgCl_2$ 溶液提取;
(2) 与碳酸盐结合,采用醋酸钠溶液提取;
(3) 与铁和锰氧化物络合后,被还原的部分金属,采用盐酸羟胺提取;
(4) 金属与有机物结合后氧化的部分成分,用硝酸和过氧化氢浸出,然后用乙酸铵提取;
(5) 在晶体结构中包覆金属中的矿物残留物,用 HF 和 $HClO_4$ 的混合物消解。

在这种预处理方法中,各阶段处理的样品通常在 1000r/min 下进行 30min 的离心分离,用吸管除去残渣上面的溶液。残留物用去离子水洗涤,随后离心后丢弃。用光谱法对所获得的液体组分中的金属进行定量。

另一个常用的顺序分离程序是由欧洲共同体委员会共同体参考局提出的协议方法,被称为BCR协议。为了使土壤和沉积物样品分析的条件一致,在不同实验室共同研究的基础上提出了这种方法。1992年的研究数据显示,使用乙二胺四乙酸(EDTA)或乙酸(CH_3COOH)溶液从土壤样品中消除有机金属化合物是适当且有效的[62]。对于其他样品,在应用具有下列提取剂的3个阶段程序之后获得最佳结果:

(1) 乙酸溶液,提取生物金属物和有机金属化合物;
(2) 盐酸羟胺溶液,提取与铁和锰氧化物络合的金属;
(3) 过氧化氢和乙酸铵的混合物,提取与有机物和硫化物结合的金属。

有时去离子水作为评价水溶性元素形态的首选提取剂。直接影响萃取效率和程序重复性的因素包括提取剂的化学性质、选择性、加入顺序、提取时间、样品转移到提取剂中的质量体积浓度以及再吸附过程。另外,在优化提取过程中还应考虑其他参数,如pH值、溶液浓度、温度和相分离条件。

用于土壤、沉积物和废料的典型顺序提取程序基于Tessier等提出的五步程序[61]。通常对植物材料(如水生苔藓[3]和菠菜[4])进行水、EDTA、石油醚、乙酸乙酯、丁醇、甲醇和盐酸的提取顺序,以分离与痕量元素结合的有机化合物。采用超声波或微波能量代替传统的顺序提取法,如将样品置于水浴中摇动和加热会明显地缩短实验时间。表6-7列出了常规提取与超声和微波能量辅助处理的实验条件比较[63-64]。

表6-7 常规提取与超声和微波能量辅助处理的实验条件比较

步骤	分馏/提取	浸出过程条件		
		传统提取	超声辅助提取	微波辅助提取
1	可交换/($1mol/dm^3$) $MgCl_2$, $8cm^3$, pH=7	1h, 25℃	3min, 50%振幅	30s, 90W
2	碳酸盐结合/($1mol/dm^3$) CH_3COONa, $8cm^3$, pH=5	5h, 25℃	1min, 50%振幅	30s, 90W
3	可还原的/0.04(mol/dm^3) $NH_2OH \cdot HCl$/25%(质量分数) CH_3COOH, $20cm^3$	6h, 96℃	7min, 50%振幅	30s, 90W
4	可氧化的/0.02(mol/dm^3) HNO_3/30% H_2O_2 + 30% H_2O_2 + 3.2mol/dm^3 CH_3COONa, $3cm^3/5cm^3$ + $3cm^3 + 5cm^3$	2h, 85℃; 3h, 85℃; 30min, 25℃	7min, 50%振幅; 2min, 50%振幅	30s, 270W; 10s, 270W

模拟实际环境条件开发了几种对不同材料的处理程序。在燃料油燃烧过程中产生的粉煤灰中元素分馏提取的实例详见表6-8[65-66]。

表6-8 粉煤灰中元素分馏提取的实例(修改的BCR协议)

元素	灰化材料	水溶性馏分/%	馏分Ⅰ/%	馏分Ⅱ/%	馏分Ⅲ/%	馏分Ⅳ/%
Zn	石油	30.8	1.1	0.1	0.1	60.6
	重油	19.2	0.7	0.3	0.7	79.2
P	石油	26.8	0.3	42.5	9.2	21.2
	重油	24.7	0.3	<0.1	1.3	73.7
Cd	石油	15.3	0.3	2.7	0.1	81.7
	重油	41.1	0.5	1.3	<0.1	57.2
Fe	石油	10.8	0.5	4.0	<0.1	84.7
	重油	43.4	0.6	1.6	<0.1	54.4
Co	石油	31.8	0.8	0.1	0.5	66.9
	重油	51.3	0.8	0.3	0.8	46.7
Ni	石油	20.6	0.6	<0.1	0.2	78.6
	重油	34.3	1.2	0.2	1.7	62.6

6.3.6 生物酶反应

酶法提取过程用于评估固体和液体食品中的微量元素、其他必需营养素以及活性化合物的生物可给性[67-70],该方法也适用于土壤样品,以评估土壤进入人体后消化系统释放化学成分的数量。应用体外酶提取程序可测定样品基体在人体胃肠道消化过程中释放化合物的数量水平。使用含有消化酶的溶液来模拟胃和肠液进行的提取提供了关于食物成分的信息,并可以评估营养物的生物可用性。人类消化道的第一部分是口腔,口腔分泌的唾液含有可以分解淀粉和糖原的淀粉酶以及消化脂肪的唾液脂肪酶。在大多数常用的体外酶提取程序中,由于食物停留在口腔中的时间很短,因此这一阶段被省略。下一个消化阶段发生在胃里,胃消化是通过使用与胃液成分相似的溶液来模拟的,该溶液包含胃蛋白酶以及作为胃蛋白酶原激活剂的水解蛋白和盐酸。胃蛋白酶的最佳pH值为1~2。通常,模拟胃消化的过程持续2~4h。在实验室中,肠消化用含有胰腺酶的溶液来模拟,通常选择淀粉酶、脂肪酶、胰蛋白酶和胰凝乳蛋白酶或含有前面所列酶混合物的胰酶,其中淀粉酶催化淀粉二糖和三糖的水解;脂肪酶对膳食脂肪起作用,将甘油三酯转化为单甘酯和脂肪酸;胰蛋白酶和糜蛋白酶是水解肽键,将蛋白质分解成小肽的蛋白酶,即氨基酸。类似肠液的萃取剂也含有胆盐、黏液和中和来自胃中盐酸的物质。胰腺酶具有最大活性的最佳pH值为7~8。样品在肠液替代物中培养(1~6)h,最常用的是2h。

基于人体生理学的消化过程可以在实验室采用体外消化过程来模拟,通过一步或一系列阶段模拟发生在口腔、胃和小肠中的食物消化过程,最常见的是模拟胃消化和肠消化两个阶段。消化过程的模拟常常在简单的实验室玻璃器皿中进行,并构建无人操作模型来模拟咀嚼、蠕动和营养吸收,并使用包含消化道各部分特异性酶的萃取剂溶液,添加酸或碱(通常是HCl和$NaHCO_3$)调节酶活性的最佳酸度。

所有酶提取程序均在37℃下进行,模拟正常体温。通过摇动、机械或磁力搅拌或旋转样品(通常在水浴中)来模拟再现酶水解食物时的胃肠运动和混合。使用超声波能量酶水解所需时间减少至30min[71]。体外酶法提取效果直接受温度、提取剂酸度、酶浓度和水解/提取时间的影响,其中酸度影响酶活性。

使用体外胃肠道模型进行酶提取方法经济快捷,而且重复性好,并且在伦理上比选择用人和动物来研究更容易接受。研究营养物生物利用率、痕量元素对营养学家、药剂师和毒理学家来说很重要,应用顺序提取程序可以实现分析物分馏,这些分析物通常是金属,酶只选择性地浸出所测定元素的某些物种形式。表6-9列出了痕量元素分析和形态分析中酶提取过程的应用实例[71-72]。

表6-9 痕量元素分析和形态分析中酶提取的应用实例

分析物	样品	酶及分解形式	pH值
Ag、As、Cd、Cu、Fe、Mg、Pb、Zn	贻贝,角鲨肌肉,角鲨肝脏	链霉蛋白酶/脂肪酶;肠消化	7
As形态形式	儿童补充剂	胰蛋白酶/胃蛋白酶;肠消化	7
Se形态形式	富硒补充剂	胃蛋白酶;胃消化	2
Se	酵母,牡蛎,贻贝	胃蛋白酶;肠消化	7
As形态形式	鱼,角鲨	胰蛋白酶;肠消化	7
Al、As、Cd、Cr、Cu、Mn、Ni、Pb、Zn	贝类	胃蛋白酶;胃消化	2
		胰酶/淀粉酶	9 (As, Cd, Cu, Fe, Ni)
		胰酶/脂肪酶;肠消化	6 (Al, Cr, Mn, Pb, Zn)
Cu、Pb	红酒	胃蛋白酶;胃消化	3.5
		胰酶+淀粉酶;肠消化	7.4
As	土壤	胃蛋白酶;胃消化	2.5
		胰酶;肠消化	7.0

6.3.7 浊点提取

浊点提取（CPE）也称为胶束介导提取，是基于一些表面活性剂的水溶液所表现出的相分离行为，所用表面活性剂通常是非离子、阴离子或两性离子。CPE 由于具有成本低、环境友好、分析时间短、样品量大同时预处理、回收率高等优点，已被公认为常规液-液萃取的一种替代方法。当表面活性剂水溶液浓度高于临界胶束浓度（CMC）时，温度或压力的改变，或引入合适的添加剂，水溶液就形成分子聚集体，即胶束而出现浊点现象。根据表面活性剂的特性、浓度和温度，还可以形成一系列结构的胶束。胶束内部有规律性的非极性核和外部延伸的极性层，即为疏水尾和亲水头。关于表面活性剂使用的原理和理论背景以及分析方面都有详细记载[73-75]。

在水溶液中，可溶解的或不溶于水的化学物质可与胶束缔合和结合，从而溶解在溶液体系中。表面活性剂与分析物之间的相互作用可以是静电的、疏水的或两者的组合[76]，增溶点位随溶解物种和表面活性剂的性质而变化[77]。非离子表面活性剂胶束表现出对多种化合物的最大增溶能力，例如，可以在水溶液中溶解烃类或金属络合物，或在非极性有机溶液中溶解极性化合物。随着非离子表面活性剂水溶液的温度升高，溶液变得浑浊，发生相分离，从而产生小体积的富表面活性剂相（SRP），在 SRP 中含有捕集在胶束结构中的分析物，相当于富集了原水相中稀释浓度的分析物。发生相分离的温度称为浊点。CMC 和浊点都取决于表面活性剂的结构和添加剂的种类。表 6-10 列出了主要的非离子表面活性剂的 CMC 和浊点温度，这些表面活性剂是 CPE 过程中最常用的表面活性剂。

表 6-10 某些非离子表面活性剂的 CMC 和浊点温度

表面活性剂	CMC/(mmol/L)	浊点/℃
Triton X-100	0.17~0.30	64~65
Triton X-114	0.20~0.35	23~25
PONPE 7.5	0.085	5~20
Brij-30（$C_{12}E_4$）	0.02~0.06	2~7
Brij-56（$C_{16}E_{10}$）	0.0006	64~69
Genapol X80（$C_{12}E_8$）	0.05	75

在非离子表面活性剂的情况下，温度的升高促进聚氧乙烯链的脱水和聚集体的生长。相分离通常在一个很窄的温度范围内发生并且可逆。有机物与水相的体积比可以很小 0.007~0.04，因此包含在胶束内或与表面活性剂聚集体连接的分析

物可在有机相中富集、分离和预浓缩。这种分离分析物的方法有许多优点:实验简单、经济,所用的试剂无毒,并且比有机溶剂的危害小。CPE 是公认的符合绿色化学原理的方法。

CPE 的最早使用方法是由 Watanabe 等[78]提出的用于分离和富集金属离子的方法,该方法在测定各种样品中的不同金属离子以及提取有机分析物方面越来越受到人们重视。CPE 方案[73-82]:第一阶段,向分析物溶液中添加高于 CMC 浓度的表面活性剂;第二阶段,在水、超声波浴或微波炉中将溶液加热到浊点以上,直到相分离,当温度比浊点高 15~20℃时,出现最佳平衡温度;第三阶段,在自发重力沉降或离心后,分析物完全留在有机相中。如果需要,可以通过在水相中加入表面活性剂来重复上述过程[83]。为了增加 SRP 的黏度,并确保更好地附着在容器壁,混合物在用氯化钠和丙酮的冰浴中冷却。其可以通过滗析、移液管或注射泵等进行水相的去除,也可以通过在氩气、氦气或氮气流中的挥发去除水。如果是热稳定化合物(如金属配合物)的情况,那么可以在空气中加热至高于 100℃。将有机相体积减少到初始体积决定了预浓缩因子大小,该因子定义为 SRP 中的分析物浓度与提取前的原始水溶液中分析物浓度的比例。为了便于最终溶液的仪器分析,有必要通过添加适当的稀释剂来降低 SRP 的黏度。根据所使用的检测系统,可以使用少量的水、甲醇、乙醇、乙腈或无机酸。

将 CPE 与流动注射分析(FIA)[73,84]进行耦合,使用 FIA 将 SRP 引入各种分析设备,有助于 SRP 以小体积溶解以提高预浓缩因子,从而改善了重现性问题[73]。

达到定量提取效果的表面活性剂的最佳浓度范围很窄,应针对每个工艺单独确定。通常,CPE 在高温下进行,这对动力学有益,并引起胶束的部分脱水,从而增加相体积比(V_{aq}/V_{org})。

为了达到最大效率和预浓缩,必须考虑和优化表面活性剂的种类和浓度、添加剂的离子强度、pH 值、平衡温度和时间以及离心条件。离子强度影响相分离和浊点的设计和建立。

对于有机分子,pH 值是调节目标分析物在胶束相中分配的最关键的因素之一[73,76,79]。特别是对于可电离物种,如酚和胺,在目标分析物以不带电的形式大量存在的某个 pH 值下以达到最大的提取效率。在元素分析中,金属可以是离子形式或在适当条件下反应后产生的疏水螯合物,此时 pH 值是定量络合物形成的关键参数,它在不添加螯合剂的情况下对提高金属的提取效率起着重要作用,因为它影响分析物的总体电荷以及金属与表面活性剂聚氧乙烯基团之间形成络合物[77,80-82]。还应优化其他实验参数,如配体浓度、表面活性剂类型和浓度、离子强度、溶剂类型和体积。对于疏水螯合物形式的金属离子的分离,常用的是氨基甲酸酯、吡啶偶氮、喹啉和萘酚衍生物、双硫腙、8-羟基喹啉和 O,O-二乙基二硫代磷酸盐[73,80,85]。表6-11列出了CPE在有机物和无机物痕量分析中的应用实例。

表6-11　CPE在有机物和无机物痕量分析中的应用实例

分析物	样品	检测方法
有机化合物(多环芳烃、苯酚和衍生物、多氯联苯、二苯并呋喃和二苯并二噁英、农药、腐殖化合物、邻苯二甲酸酯类、胺类、药物类)	天然水、废物、血清、生物液体、植物和动物样品、沉积物	LC、HPLC
金属离子 (Ag, Al, As, Au, Ba, Be, Bi, Cd, Co, Cr, Cu, Dy, Er, Fe, Ga, Gd, Hg, In, La, Mn, Mo, Ni, Pb, Pd, Pt, Rh, Ru, Se, Tl, U, V, Zn)	天然水、生物液体、海水、废物、肥料、头发	AAS、ICP-OES、ICP-MS、CE、分光光度法、荧光光度法
形态分析(Cr, Cu, Fe, Hg, Mn, Sb, Se, Sn)	合成样品、水、红酒	AAS、ICP-OES、ICP-MS、分光光度法

注：AAS—原子吸收光谱仪，ICP-OES—电感耦合等离子体发射光谱仪，ICP-MS—电感耦合等离子体质谱仪，CE—毛细管电泳分析仪。

6.3.8　超临界流体萃取

超临界流体萃取(SFE)利用超临界流体的特性从固体样品中提取分析物。超临界流体(SCF)是介于典型气、液两相之间超过临界温度和压力的物质。低黏度、近零表面张力和汽化热使SCF比液体溶剂更快地渗入固体中，这导致更有利的传质。SCF的密度接近液体密度，而其黏度接近气体黏度。SCF可以像气体一样在固体中扩散并像液体一样溶解物质。此外，接近临界点，压力和/或温度的微小变化会导致密度大的变化。这使许多SCF特性被精确地调整设计并且实现显著的可选择性。

表6-12中列出了用作溶剂或SCF的一些物质的临界参数。因为CO_2具有临界压力和温度优选的低值，所以大多数分析用SFE使用超临界CO_2作为萃取剂[86-87]。此外，CO_2无毒、无色、无臭、不易燃、价廉、易得。由于其非极性性质，CO_2不能用于溶解极性分子。对于极性化合物的提取，N_2O或$CHClF_2$更适用于超临界状态，但在常规分析中，它们相对很少使用[88]。极性化合物在超临界CO_2中的溶解度和工艺选择性可以通过添加少量(1%~10%)的其他极性溶剂(称为修饰剂或共溶剂)来提高。所有的SCF都是完全互溶的，所以如果超过混合物的临

界点,就可以保证混合物为单相。提取含氧化合物的常用改性剂有乙醇、甲醇、丙酮、四氢呋喃和2-丙醇。添加乙腈通常用于提取氮化合物,而甲酸则用于提取酸。对于含硫化合物,CS_2、SO_2或SF_6可作为改性剂[1]。甲醇、乙酸和苯胺作为酸性/碱性改性剂大大提高了多氯联苯的萃取率。甲苯、二乙胺和CH_2Cl等是高分子量PAH的SFE最佳改性剂[89-90]。由于金属离子电中和和溶质-溶剂相互作用较弱,超临界流体直接萃取金属离子是无效的。然而,如果金属离子转化为中性金属螯合物,则它们在超临界CO_2中的溶解度将增加。这是通过添加络合剂进行SCF改性实现的[90-91]。

表6-12 超临界流体萃取中使用的选定物质的临界参数

物质	临界温度/℃	临界压力/atm	临界密度/(kg/dm³)
CO_2	31.3	72.9	0.47
N_2O	36.5	72.5	0.45
SF_6	46.5	37.1	0.74
NH_3	132.5	112.5	0.24
H_2O	374	227	0.34
$n-C_4H_{10}$	152	37.5	0.23
CHF_3	25.9	46.9	0.52

SFE最适合的样品是具有良好渗透性的粉末状固体,如土壤、沉积物、干燥的植物和生物材料。湿的或液体样品和溶液的提取是比较困难的[56,90]。SFE的基本组件包括CO_2储罐、高压泵、萃取池、加热炉、控制SCF压力的流量节流器和收集容器。改性剂由一个单独的递送系统提供[1]。将样品放入提取容器中加热。

在填充有固体吸附剂的柱上,在含有适当溶剂的容器中,在连接到色谱仪的收集装置中,或者在组合的固相-溶剂捕集器上,通过减压收集提取物[92]。对于挥发性化合物的提取,使用丙酮、CH_2Cl_2、甲醇或液氮等溶剂。硅胶柱是最常用的捕集固体的方法。在这种情况下,通过吸附剂选择性地洗脱[88,92],可以提高工艺的选择性。SFE可以在静态模式下进行,其中样品和溶剂混合,并在用户指定的时间内保持恒定的压力和温度,或者在动态模式下,溶剂以连续方式通过样品[56]。所提取的分析物可以收集到离线装置或转移到在线色谱系统用于直接分析。

SFE是一种高效、快速的提取技术,符合绿色化学策略。SFE在食品、环境和药物分析、工业和生物医学实验室,以及形态分析中有广泛的应用[86,88,93-97]。表6-13列出了超临界流体萃取在痕量分析中的应用实例。

表 6-13　超临界流体萃取在痕量分析中的应用实例

分析物	基体/样品	提取物
二噁英类	土壤、海洋沉积物、植物、污泥、城市和工业废物	CO_2+甲苯
PAH		CO_2+甲苯、CO_2+CH_3OH
PCB	土壤	CO_2+CH_2Cl_2
中药,杀虫剂	土壤、水果和蔬菜,以及他们的蜜饯、动物组织、食品	CO_2+甲醇、丙酮、乙醇、水
PAH	土壤、沉积物	
药物,麻醉剂	血液、尿液、软组织、头发	CO_2、CO_2+CH_3OH、NH_3
金属-有机化合物,形态形式(As, Hg, Sn, Pb)	溶液、沉积物、食品、贝类、鱼、沙子、灰尘	CO_2+HCOOH、CO_2+CH_3OH
重金属	水、土壤、粉尘、木材、沉积物、沙子、灰尘	CO_2+络合剂

6.3.9　QuEChERS 技术

QuEChERS(quick, easy, cheap, effective, rugged and safe)是样品前处理过程名称的英文首字母缩写,通常基于从样品中溶剂提取分析物和为清洗萃取剂用 SPE 分散的组合。

QuEChERS 程序由许多简单的分析操作组成,不易出错。一般可以分为两个步骤:第一步是获得粗提取物,用有机溶剂(通常是乙腈)从均匀化样品中提取分析物,然后加入盐(通常是 $MgSO_4$ 和 NaCl),为了从萃取液中分离出水(相分离),通常须用缓冲液调节 pH 值。第二步通过使用加入吸附剂和 $MgSO_4$ 的清洗程序获得萃取液。通过 GC 和/或 LC 方法分析所述提取物,并且优选以直接方式进行。

QuEChERS 程序开发和广泛商业化的通用试剂盒,提供了极高的回收率和极好的再现性。完整的商业 QuEChERS 试剂盒在技术标准化和确保高质量分析成果方面具有很大的优势,与美国分析化学家协会(AOAC)和欧洲标准化委员会(CEN)当前推荐的方法兼容。

QuEChERS 方法是在 2003 由 Anastassiades 等发明的,它是一种快速、简单、廉价、方便地用于果蔬样品中农药多残留分析的前处理方法[98]。目前,该方法用于测定农药、农药残留物和其他与环境有关的化合物(如苯酚衍生物、过氧基化合物和氯化烃),食品和农产品中的药物化合物,环境样品(如土壤、沉积物和水)等(文献[99-102])。

6.4 小结

分析萃取方法及其应用趋势是与现代分析化学,特别是痕量分析领域面临的挑战和要求相适应的。因此,萃取分析技术的趋势和发展必须考虑:有效降低测定分析物的定量限,通过验证和可追溯性确保测量的高质量和可靠性,减少能源和材料(特别是有机溶剂)消耗以符合生态学和经济学要求,使用由精密电子系统操作和控制的小型分析设备。许多不同的微萃取技术顺应了这些趋势和要求,利用微波和超声波能量、高压和溶剂加速辅助萃取过程已成为广泛和有前途的方法,可提高效率并缩短程序时间。

适用于分析物组的 SPE 技术的改进和无溶剂技术的开发是有前景和重要的,通过成功开发和实施具有特殊、可预测和编程性质的新型吸附材料可达到此目的。还有注意到不同的分析方法和技术的联用,以及这些方法和技术与色谱方法(GC、HPLC、nanoHPLC)的结合是典型的技术发展趋势。新的提取技术可以通过利用由复杂的电子系统操作和控制的小型分析仪器来实施。

现代技术方案实现了提取过程及其现场分析的小型化和自动化,在萃取反应器的小型化和自动化方面令人瞩目的进展是先进的原位分析和可移动仪器的商业化。

利用萃取技术进行元素分馏和形态分析的现状仍然没有达到人们的需求和期望,这是由于可用标准物质数量不足和缺乏多样性。

参考文献

[1] Mitra, S. (ed.): Sample Preparation Techniques in Analytical Chemistry. Wiley, Hoboken (2003)

[2] Boyd, R.K., Basic, C., Bethem, R.A.: Trace Quantitative Analysis by Mass Spectrometry. Wiley, Chichester (2008)

[3] Pawliszyn, J., Lord, H.L. (eds.): Handbook of Sample Preparation. Wiley, Hoboken (2010)

[4] Pawliszyn, J. (ed.): Sampling and Sample Preparation in Field and Laboratory. Fundamentals and New Directions in Sample Preparation (Comprehensive Analytical Chemistry), vol. 37. Elsevier, Amsterdam (2002)

[5] Piao, C., Chen, L., Wang, Y.: A review of the extraction and chromatographic determination methods for the analysis of parabens. J. Chromatogr. B 969, 139–148 (2014)

[6] Teo, C.C., Chong, W.P.K., Tan, E., Basri, N.B., Low, Z.J., Ho, Y.S.: Advances in sample

preparation and analytical techniques for lipidomics study of clinical samples. Trends Anal. Chem. 65, 1–18 (2015)

[7] Nearing, M.M., Koch, I., Reimer, K.J.: Complementary arsenic speciation methods: a review. Spectrochim. Acta B 99, 150–162 (2014)

[8] Soodan, R.K., Pakade, Y.B., Nagpal, A., Katnoria, J.K.: Analytical techniques for estimation of heavy metals in soil ecosystem: a tabulated review. Talanta 125, 405–410 (2014)

[9] Lum, T.-S., Tsoia, Y.-K., Leung, K.S.-Y.: Current developments in clinical sample preconcentration prior to elemental analysis by atomic spectrometry: a comprehensive literature review. J. Anal. At. Spectrom. 29, 234–241 (2014)

[10] Yadava, S.K., Chandrab, P., Goyala, R.N., Shim, Y.-B.: A review on determination of steroids in biological samples exploiting nanobio-electroanalytical methods. Anal. Chim. Acta 762, 14–24 (2013)

[11] Poole, C.F.: New trends in solid-phase extraction. Trends Anal. Chem. 22, 362–373 (2003)

[12] Fontanals, N., Marcé, R.M., Borrull, F.: New hydrophilic materials for solid-phase extraction. Trends Anal. Chem. 24, 394–406 (2005)

[13] Yea, N., Shi, P.: Applications of graphene-based materials in solid-phase extraction and solid-phase microextraction. Sep. Purif. Rev. 44(3), 183–198 (2015)

[14] Pichon, V., Bouzige, M., Miège, C., Hennion, M.-C.: Immunosorbents: natural molecular recognition materials for sample preparation of complex environmental matrices. Trends Anal. Chem. 18(3), 219–235 (1999)

[15] Stevenson, D.: Molecular imprinted polymers for solid-phase extraction. Trends Anal. Chem. 18(3), 154–158 (1999)

[16] Mutavdžić Pavlović, D., Babić, S., Horvat, A.J.M., Kapelan-Macan, M.: Sample preparation in analysis of pharmaceuticals. Trends Anal. Chem. 26, 1062–1075 (2007)

[17] Buszewski, B., Szultka, M.: Past, present, and future of solid phase extraction. A review. Crit. Rev. Anal. Chem. 42, 198–213 (2012)

[18] Tobiasz, A., Walas, S.: Solid-phase-extraction procedures for atomic spectrometry determination of copper. Trends Anal. Chem. 62, 106–122 (2014)

[19] Chen, L., Wang, H., Zeng, Q., Xu, Y., Sun, L., Xu, H., Ding, L.: On-line coupling of solid-phase extraction to liquid chromatography-a review. J. Chromatogr. Sci. 47(8), 614–623 (2009)

[20] Pawliszyn, J.: Theory of solid-phase microextraction. J. Chromatogr. Sci. 38, 270–278 (2000)

[21] Pawliszyn, J.: New directions in sample preparation for analysis of organic compounds. Trends Anal. Chem. 14, 113–122 (1995)

[22] Arthur, C.L., Pawliszyn, J.: Solid phase microextraction with thermal desorption using fused silica optical fibers. Anal. Chem. 62, 2145–2148 (1990)

[23] Duan, C., Shen, Z., Wu, D., Guan, Y.: Recent developments in solid-phase microextraction for on-site sampling and sample preparation. Review article. Trends Anal. Chem. 10, 1568–1574

(2011)

[24] David, F., Sandra, P.: Stir bar sorptive extraction for trace analysis. J. Chromatogr. A 1152, 54–69 (2007)

[25] Baltussen, E., Sandra, P., David, F., Cramers, C.A.: Stir bar sorptive extraction (SBSE) a novel extraction technique for aqueous samples: theory and principles. J. Microcolumn Sep. 11, 737–747 (1999)

[26] Camino-Sánchez, F.J., Rodríguez-Gómez, R., Zafra-Gómez, A., Santos-Fandila, A., Vílchez, J.L.: Stir bar sorptive extraction: recent applications, limitations and future trends. Talanta 130, 388–399 (2014)

[27] Abdel-Rehim, M.: Microextraction by packed sorbent (MEPS): a tutorial. Anal. Chim. Acta 701, 119–128 (2011)

[28] Altun, Z., Abdel-Rehim, M., Blomberg, L.G.: New trends in sample preparation: On-line microextraction in packed syringe (MEPS) for LC and GC applications. Part III Determination and validation of local anaesthetics in human plasma samples using a cation-exchange sorbent and MEPS-LC-MS-MS. J. Chromatogr. B 813, 129–135 (2004)

[29] Rani, S., Malik, A.K., Singh, B.: Novel micro-extraction by packed sorbent procedure for the liquid chromatographic analysis of antiepileptic drugs in human plasma and urine. J. Sep. Sci. 35, 359–366 (2012)

[30] Jinno, K., Ogawa, M., Ueta, I., Saito, Y.: Miniaturized sample preparation using fiber-pacing capillary as the medium. Trends Anal. Chem. 26, 27–35 (2007)

[31] Spieteluna, A., Pilarczyka, M., Kloskowskia, A., Namieśnik, J.: Polyethylene glycol-coated solid-phase microextraction fibres for the extraction of polar analytes—a review. Talanta 87, 1–7 (2011)

[32] Snow, N.H., Slack, G.C.: Head-space analysis in modern gas chromatography. Trends Anal. Chem. 21, 608–617 (2002)

[33] Möller, J., Strömberg, E., Karlsson, S.: Comparison of extraction methods for sampling of low molecular compounds in polymers degraded during recycling. Eur Polym J 44, 1583–1593 (2008)

[34] Pecoraino, G., Scalici, L., Avellone, G., Ceraulo, L., Favara, R., Candela, E.G., Provenzano, M.C., Scaletta, C.: Distribution of volatile organic compounds in Sicilian groundwaters analysed by head space-solid phase micro extraction coupled with gas chromatography mass spectrometry (SPME/GC/MS). Water Res. 42(14), 3563–3577 (2008)

[35] Tipler, A.: An introduction to headspace sampling in gas chromatography. Fundamentals and theory.www.perkinelmer.com

[36] Jerkovic, I., Marijanovic, Z.: A short review of headspace extraction and ultrasonic solvent extraction for honey volatiles fingerprinting. Croat. J. Food Sci. Technol. 1(2), 28–34 (2009)

[37] Hakkarainen, M.: Developments in multiple headspace extraction. J. Biochem. Biophys. Methods 70(2), 229–233 (2007)

[38] Jakubowska, N., Polkowska, Ż., Namieśnik, J., Przyjazny, A.: Analytical applications of membrane extraction for biomedical and environmental liquid sample preparation. Crit. Rev. Anal. Chem. 35, 217–235 (2005)

[39] Hylton, K., Mitra, S.: Automated, on-line membrane extraction. J. Chromatogr. A 1152, 199–214 (2007)

[40] Jönsson, J., Mathiasson, L.: Membrane extraction in analytical chemistry. J. Sep. Sci. 24, 495–507 (2001)

[41] Nghiem, L.D., Mornane, P., Potter, I.D., Perera, J.M., Cattrall, R.W., Kolev, S.D.: Extraction and transport of metal ions and small organic compounds using polymer inclusion membranes (PIMs). J. Membr. Sci. 281, 7–41 (2006)

[42] Barri, T., Jönsson, J.: Advances and developments in membrane extraction in gas chromatography. Techniques and applications. J. Chromatogr. A 1186, 16–38 (2008)

[43] Chemat, F., Cravotto, G. (eds.): Microwave-Assisted Extraction for Bioactive Compounds. Theory and Practice. Springer, New York (2013)

[44] Teo, C.C., Pooi, W., Chong, K., Ho, Y.S.: Development and application of microwave-assisted extraction technique in biological sample preparation for small molecule analysis. Metabolomics 9, 1109–1128 (2013)

[45] Ganzler, K., Salgò, A., Valkò, K.: Microwave extraction. A novel sample preparation method for chromatography. J. Chromatogr. 371, 299–306 (1986)

[46] Eskilsson, C.S., Björklund, E.: Analytical-scale microwave-assisted extraction. J. Chromatogr. A 902, 227–250 (2000)

[47] Nobrega, J.A., Trevizan, L.C., Araújo, G.C.L., Nogueira, A.R.A.: Focused-microwave-assisted strategies for sample preparation. Spectrochim. Acta B 57, 1855–1876 (2002)

[48] Camel, V.: Microwave-assisted solvent extraction of environmental samples. Trends Anal. Chem. 19, 229–248 (2000)

[49] Srogi, K.: A review: application of microwave techniques for environmental analytical chemistry. Anal. Lett. 39, 1261–1288 (2006)

[50] Madej, K.: Microwave-assisted and cloud point extraction in determination of drugs and other bioactive compounds. Trends Anal. Chem. 28, 436–446 (2009)

[51] Routray, W., Orsat, V.: Microwave-assisted extraction of flavonoids: a review. Food Bioprocess Technol. 5, 409–424 (2012)

[52] Azmir, J., Zaidul, I.S.M., Rahman, M.M., Sharif, K.M., Mohamed, A., Sahena, F., Jahurul, M.H.A., Ghafoor, K., Norulaini, N.A.N., Omar, A.K.M.: Techniques for extraction of bioactive compounds from plant materials: a review. J. Food Eng. 117, 426–436 (2013)

[53] Shirsatha, S.R., Sonawanea, S.H., Gogate, P.R.: Intensification of extraction of natural products using ultrasonic irradiations-a review of current status. Chem. Eng. Process. 53, 10–23 (2012)

[54] Luque de Castro, M.D., Delgado-Povedano, M.M.: Ultrasound: a subexploited tool for sample

preparation in metabolomics. Anal. Chim. Acta 806, 74–84 (2014)

[55] Picó, Y.: Ultrasound-assisted extraction for food and environmental samples. Trends Anal. Chem. 43, 84–99 (2013)

[56] Chen, Y., Guo, Z., Wang, X., Qiu, C.: Sample preparation. Review. J. Chromatogr. A 1184, 191–219 (2008)

[57] Mason, T.J., Paniwynk, L., Lorimer, J.P.: The uses of ultrasounds in food technology. Ultrason. Sonochem. 3, 253–260 (1996)

[58] Vinatoru, M.: An overview of the ultrasonically assisted extraction of bioactive principles from herbs. Ultrason. Sonochem. 8, 303–313 (2001)

[59] Luque-García, J.L., Luque de Castro, M.D.: Ultrasound: a powerful tool for leaching. Trends Anal. Chem. 22, 41–47 (2003)

[60] Priego-Capote, F., Luque de Castro, M.D.: Analytical uses of ultrasound. Sample preparation. Trends Anal. Chem. 23, 644–653 (2004)

[61] Tessier, A., Campbell, P.G.C., Bisson, M.: Sequential extraction procedure for speciation of particulate trace metals. Anal. Chem. 51, 844–851 (1979)

[62] Ure, A.M., Quevauviller, P., Muntau, H., Gripink, B.: Speciation of heavy metals in soils and sediments. An account of the improvement and harmonization of extraction techniques undertaken under auspicies of the BCR of the Commission of the European Communities. Int. J. Environ. Anal. Chem. 51, 135–151 (1993)

[63] Filgueiras, A.V., Lavilla, I., Bendicho, C.: Chemical sequential extraction for metal partitioning in environmental solid samples. J. Environ. Monitoring 4, 823–857 (2002)

[64] Gleyzes, C., Tellier, S., Astruc, M.: Fractionation studies of trace elements in contaminated soils and sediments: a review of sequential extraction procedures. Trends Anal. Chem. 21, 451–467 (2002)

[65] Pardo, P., López-Sánchez, J.F., Rauert, G.: Characterisation, validation and comparison of three methods for the extraction of phosphate from sediments. Anal. Chim. Acta 376, 183–195 (1998)

[66] Pérez-Cid, B., Lavilla, I., Bendicho, C.: Comparison between conventional and ultrasound accelerated Tessier sequential extraction schemes for metal fractionationin sewage sludge. Fresenius J. Anal. Chem. 363, 667–672 (1999)

[67] Basua, A., Nguyena, A., Bettsa, N.M., Lyons, T.J.: Strawberry as a functional food: an evidence-based review. Crit. Rev. Food Sci. Nutr. 54(6), 790–806 (2014)

[68] Kadam, S.U., Tiwari, B.K., O'Donnell, C.P.: Application of novel extraction technologies for bioactives from marine algae. Agric. Food Chem. 61(20), 4667–4675 (2013)

[69] Wijesinghe, W.A.J.P., Jeon, Y.-J.: Enzyme-assistant extraction (EAE) of bioactive components: a useful approach for recovery of industrially important metabolites from seaweeds: a review. Fitoterapia 83, 6–12 (2012)

[70] Puri, M., Sharma, D., Barrow, C.J.: Enzyme-assisted extraction of bioactives from plants.

Trends Biotechnol. 30(1), 37-44 (2012)

[71] Bermejo, P., Capelo, J.L., Mota, A., Madrid, Y., Camara, C.: Enzymatic digestion and ultrasonication: a powerful combination in analytical chemistry. Trends Anal. Chem. 23, 654-653 (2004)

[72] Intawongse, M., Dean, J.R.: In-vitro testing for assessing oral bioscessibility of trace metals in soil and food samples. Trends Anal. Chem. 25, 876-886 (2006)

[73] Paleologos, E.K., Giokas, L., Karayannis, M.I.: Micelle-mediated separation and cloud-point extraction. Trends Anal. Chem. 24, 426-436 (2005)

[74] Burguera, J.L., Burguera, M.: Analytical applications of organized assemblies for on-line spectrometric determinations: present and future. Talanta 64, 1099-1108 (2004)

[75] Hinze, W.L., Pramauro, E.: A critical review of surfactant-mediated phase separations (cloud point extractions): theory and applications. Crit. Rev. Anal. Chem. 24, 133-177 (1993)

[76] Xie, S., Paau, M.C., Li, C.F., Xiao, D., Choi, M.M.: Review. Separation and preconcentration of persistent organic pollutants by cloud point extraction. J. Chromatogr. A 1217, 2306-231 (2010)

[77] Bezerra, M.A., Arruda, M.A.Z., Ferreira, S.L.C.: Cloud point extraction as a procedure of separation and pre-concentration for metal determination using spectroanalytical techniques: a review. Appl. Spectrosc. Rev. 40, 269-299 (2005)

[78] Watanabe, H., Tanaka, H.: A non-ionic surfactant as a new solvent for liquid-liquid extraction of zinc (II) with 1-(2-pyridylazo)-2-naphtol. Talanta 25, 585-589 (1978)

[79] Fererra, Z.S., Sanz, C.P., Santana, C.M., Santana Rodriquez, J.J.: The use of micellar systems in the extraction and pre-concentration of organic pollutants in environmental samples. Trends Anal. Chem. 23, 469-479 (2004)

[80] Silva, M.F., Cerutti, E.S., Martinez, L.D.: Coupling cloud point extraction to instrumental detection systems for metal analysis. Microchim. Acta 155, 349-364 (2006)

[81] Bosch, O.C., Sánchez, R.F.: Separation and preconcentration by a cloud point extraction procedure for determination of metals: an overview. Anal. Bioanal. Chem. 394, 759-782 (2009)

[82] Samaddar, P., Sen, K.: Review. Cloud point extraction: a sustainable method of elemental preconcentration and speciation. J. Ind. Eng. Chem. 20, 1209-1219 (2014)

[83] Wen, Y., Li, J., Liu, J., Wenhui, L., Ma, J., Chen, L.: Dual cloud point extraction coupled with hydrodynamicelectrokinetic two-step injection followed by micellar electrokinetic chromatography for simultaneous determination of trace phenolic estrogens in water samples. Anal. Bioanal. Chem. 405, 5843-5852 (2013)

[84] Frizzarin, R.M., Rocha, F.R.P.: An improved approach for flow-based cloud point extraction. Anal. Chim. Acta 820, 69-75 (2014)

[85] Pytlakowska, K., Kozik, V., Dabioch, M.: Complex-forming organic ligands in cloud-point extraction of metal ions: a review. Talanta 110, 202-228 (2013)

[86] Machado, B.A.S., Pereira, C.G., Nunes, S.B., Padilha, F.F., Umsza-Guez, M.A.: Supercriti-

cal fluid extraction using CO_2: main applications and future perspectives. Sep. Sci. Technol. 48, 2741-2760 (2013)

[87] Zougagh, M., Valcárcel, M., Rios, A.: Supercritical fluid extraction: a critical review of its analytical usefulness. Trends Anal. Chem. 23, 399-405 (2004)

[88] Lang, Q., Wai, C.M.: Supercritical fluid extraction in herbal and natural product studies - a practical review. Talanta 53, 771-782 (2001)

[89] Anitescu, G., Tavlarides, L.L.: Supercritical extraction of contaminants from soil and sediments. J. Supercrit. Fluids 38, 167-180 (2006)

[90] Sunarso, J., Ismadij, S.: Decontamination of hazardous substances from solid matrices and liquids using supercritical fluids extraction: a review. J. Hazard. Mater. 161, 1-20 (2009)

[91] Quach, D.L., Mincher, B.J., Wai, C.M.: Supercritical fluid extraction and separation of uranium from other actinides. J. Hazard. Mater. 274, 360-366 (2014)

[92] Turner, C., Eskilsson, C.S., Björklund, E.: Collection in analytical-scale supercritical fluid extraction. J. Chromatogr. A 947, 1-22 (2002)

[93] Reverchon, E., DeMarco, I.: Supercritical fluid extraction and fractionation of natural matter. J. Supercrit. Fluids 38, 146-166 (2006)

[94] de Melo, M.M.R., Silvestre, A.J.D., Silva, C.M.: Supercritical fluid extraction of vegetable matrices: applications, trends and future perspectives of a convincing green technology. J. Supercrit. Fluids 92, 115-176 (2014)

[95] Bielska, L., Smidova, K., Hofman, J.: Supercritical fluid extraction of persistent organic pollutants from natural and artificial soil and comparison with bioaccumulation in earth-worms. Environ. Pollut. 176, 48-54 (2013)

[96] Bayona, J.M.: Supercritical fluid extraction in speciation studies. Trends Anal. Chem. 19, 107-112 (2000)

[97] Herrero, M., Mendiola, J.A., Cifuentes, A., Ibanez, E.: Supercritical fluid extraction: recent advances and applications. J. Chromatogr. A 1217, 2495-2511 (2010)

[98] Anastassiades, M., Lehotay, S.J., Stajnbaher, D., Schenck, F.J.: Fast and easy multiresidue method employing acetonitrile extraction/partitioning and "dispersive solid-phase extraction" for the determination of pesticide residues in produce. J. AOAC Int. 86(2), 412-431 (2003)

[99] Wilkowska, A., Biziuk, M.: Determination of pesticide residues in food matrices using the QuEChERS methodology. Food Chem. 125, 803-812 (2011)

[100] Golge, O., Kabak, B.: Evaluation of QuEChERS sample preparation and liquid chromatography-triple-quadrupole mass spectrometry method for the determination of 109 pesticide residues in tomatoes. Food Chem. 176, 319-332 (2015)

[101] Bruzzoniti, M.C., Checchini, L., De Carlo, R.M., Orlandini, S., Rivoira, L., Del Bubba, M.: QuEChERS sample preparation for the determination of pesticides and other organic residues in environmental matrices: a critical review. Anal. Bioanal. Chem. 406, 4089-4116 (2014)

[102] Ribeiro, C., Ribeiro, A.R., Maia, A.S., Gonçalves, V.M.F., Tiritan, M.E.: New trends in sample preparation techniques for environmental analysis. Crit. Rev. Anal. Chem. 44(2), 142-185 (2014)

第二部分
痕量分析的应用

第7章
选择性有机化合物的痕量分析

7.1 概述

有机化合物的痕量分析主要用于检测和测定自然产生的有害组分(如菌毒素),以及由人类活动产生的有不可预期影响的有害成分,尤其是在工农业方面。后者既可能来自有意的合成(杀虫剂、阻燃剂、化学武器等),也有可能来自工艺过程中以非可控方式所释放出的不需要的杂质,或是来源于燃料和废弃物的不当燃烧[1-2],因此更加引人关注。

有机痕量分析的主要应用包括以下几方面:
(1) 环境化学(自然物质循环的分析);
(2) 生物化学(基因分析、蛋白质组学、代谢组学等);
(3) 生态毒理学(如微量元素的生物积累研究);
(4) 医疗诊断;
(5) 食品分析(对成分和杂质的分析);
(6) 材料工程(特别是在晶体学和半导体工业中使用的高纯度材料领域)。

即使是被公认为有害成分的总量为痕量,也会对活体器官的总体健康产生很大的影响,因此,生物化学和医疗诊断中的痕量分析成为人们首要关心的问题。检测和识别致癌物质和雌激素物质(癌症的主要原因)是非常重要的。根据 Kundson 介绍的分类[3],癌症发病的主要原因如下:
(1) 遗传素质;
(2) 环境因素;
(3) 不确定因素。

应该注意的是,自然环境不仅包括水、空气、土壤,还应包括食物、生活方式、职业暴露、药物,以及人与环境接触的所有因素。假设可以避免接触潜在的致癌物质和改变生活方式,则绝大多数癌症是可以预防的。

7.2 致癌物

致癌物是一种可以引起原始基因信息改变(突变)并诱发癌症的化合物。

已知的具有致癌作用的物质属于不同的化学基团,而且无法鉴别出所有致癌物共有的结构和属性。目前已证实可致癌的基础化合物包括以下几类：

(1) 无机化合物,如砷、铬和镍的盐类(这些不在本书的讨论范围)；
(2) 有机化合物,如苯,2-萘胺、氯乙烯和多环芳烃；
(3) 复杂物质,如煤烟、焦油和矿物油等；
(4) 天然物质,如黄曲霉毒素、丝霉素 C、佛波酯以及亚硝胺。

食物中的致癌物质有黄曲霉毒素,二噁英,真菌毒素,N-亚硝胺和其他有机化合物。

黄曲霉毒素是水溶性的,很容易穿过动植物的细胞膜、组织以及动物皮肤。这些化合物在体内的积累会导致功能紊乱、发病甚至死亡。

二噁英构成一组氯化芳香族化合物,该分子具有极高的耐热性、化学抗氧化性和生物降解过程。这些化合物的全称是多氯二苯并对二噁英。这些物质属于一组人工合成的高毒活性化合物。二噁英由两个氧原子连接两个苯环组成,并且有 1~8 个氯原子附着在这两个苯环上。俗称的"二噁英"是指所有可能的氯化硫代蒽(二苯并-1,4-二噁英)。由于自然界中也存在类似毒性的物质,因此,二噁英组还包括多氯二苯并呋喃和多氯联苯(PCB)。

霉菌毒素是真菌(霉菌)各种丝状物的二次代谢产物,这些产物是作为新陈代谢的副产物或防御目的产生的。霉菌毒素具有很强的毒性、诱变性或致畸性,在不同环境条件下的许多农产品中均有发现。

7.3 类雌激素活性化合物

7.3.1 雌激素

内分泌干扰物(EDCs)包含大量的化合物,它们通过模仿、改变甚至破坏生命体内的自然活性来阻断激素的活动,从而潜在地干扰生物的自然功能。这些化合物包括有机氯化物农药、有机磷农药、软化剂(增塑剂)、邻苯二甲酸盐、PCB、多溴联苯醚(PBDE)、溴系阻燃剂(BFR)、全氟碳化合物、抗生素、非甾体类抗炎药、心血管药物、激素制剂、表面活性化合物及其代谢物(酚类)。它们大部分存在于环境中,并且非常持久,在食物链中进行累积,并储存在脂肪组织中被缓慢地代谢和分泌[4]。

EDC主要来源于自然界化合物,然而,类似的合成物质越来越多地由人类生产,并以各种方式释放到环境中。这些物质称为外源性雌激素或异种雌激素。活性雌激素化合物是一类具有特殊致癌作用的化合物[5],它们可以通过与受体结合来模仿天然雌激素化合物。人类产生的许多化合物表现出雌激素活性,这些人造雌激素化合物包括一些有机氯杀虫剂、邻苯二甲酸盐和壬基酚。尽管EDC的功能已被人们熟知多年,但最近这些化合物才引起人们的兴趣,并成为环境污染最具争议的议题之一。

无论来源于何处的外源性雌激素,都有能力与内分泌系统相互作用,并以雌激素特有的方式干扰人体内分泌系统的正常活动。雌激素类药物(避孕药、激素替代疗法)的开发和传播已导致这些物质大量地释放到环境中,人们怀疑这些物质导致哺乳动物的生育能力下降、缺少阳刚之气等一系列副作用,这些现象在海洋和内陆水域的野生动物中越来越普遍。同时饮用水中存在的外源性雌激素也导致男性生育问题和人类胎儿性别失调的问题。异种雌激素还存在于塑料瓶、儿童玩具、化妆品、食品包装、城市废水污染的天然水体、游泳池的水和加工食品(肉类、豆制品)中。

考虑到内分泌系统的复杂性,外源性化学物质引起功能障碍的作用机制极其复杂,尚不完全清楚。由于婴儿和儿童缺乏成熟的机制和反馈来调节激素的运作、合成和排泄,因此外源性化学物质的影响尤其危险,这种变化可能是不可逆的,会在很长一段时间后才显现出来。一些文献广泛讨论了接触EDC和生物体健康之间的关系。似乎成年人接触EDC会导致暂时性或永久性的激素紊乱,进而对胎儿造成永久性的损害。由于胎儿体积小,生长过程动力高,对有害物质的解毒能力低,对EDC特别敏感[6]。大多数EDC会通过胎盘进入胎儿的循环系统。所以几十年来,人们对EDC的透过率进行了许多研究,并对其在脐血血清和母体脂肪组织中的含量进行了研究[7-20]。在怀孕期间和哺乳期间,脂肪储备的增加意味着胎儿以及后来的婴儿会明显地接触到高含量的EDC[21]。而且研究发现,产前接触EDC会对胎儿造成出生体重过轻、早产、心理运动迟缓和认知功能改变的问题[22-27]。

食物是接触外源性雌激素的主要来源。此外,在较小程度上它们可通过呼吸系统或皮肤进入人体。当外源性雌激素到达血浆后,它们有可能会与蛋白质结合,也可能保持自身未结合状态。有些通过代谢转化为非活性物质或通过尿液或其他途径排泄;有些被运送到目标组织和器官,通过与受体结合或通过其他机制产生特殊效果;有些化合物只有在体内转化为活性代谢物后才具有激素活性,这也可能导致化合物活性的改变。如果外源化合物是持久性化合物,那么它可以在组织和器官(如在脂肪组织中)积累,然后缓慢地释放到体内。在很多情况下,外源化合物持续作用的时间可以是整个生物体的生命周期。

7.3.2 双酚 A

近几年,与人们接触的双酚 A(4′-二羟基- 2,2-二苯基,BPA;图 7-1)因其雌激素特性受到了人们关注。BPA[28]通常用作工业增塑剂,在油漆、阻燃剂、不饱和聚酯树脂、塑料食品包装、水容器、婴儿奶瓶和储藏食品用箔层中都有发现。一些研究已经证实,在许多食品容器中也检测到了 BPA[29-31]。对于 BPA 潜在危险性的看法各不相同。基于 BPA 的聚合物和非聚合物单体中酯键的水解是该组化合物广泛污染的原因。然而,在文献中也可以找到关于 BPA 无害的信息。根据这些报告,BPA 不能被视为重要的生物污染,因为它可以相对较快地代谢并从体内排出。另外,文献中也有报告表明 BPA 的传输是通过胎盘进行的,胎盘中只有一部分被代谢和排泄,其余的留在妇女体内,使母亲和胎儿遭受长时间接触 BPA 的风险并产生负面影响[32-34]。有必要认识到,即使在非常低的浓度下,BPA 也显示出潜在的雌激素性质[35-36]。大量的生化和毒理学研究证实双酚 A 对雌激素受体具有雌激素性质和兴奋剂活性。最近的研究表明,将 BPA 归为破坏动物和人类内分泌系统平衡的外在物质(激素活性物质)是完全有根据的。

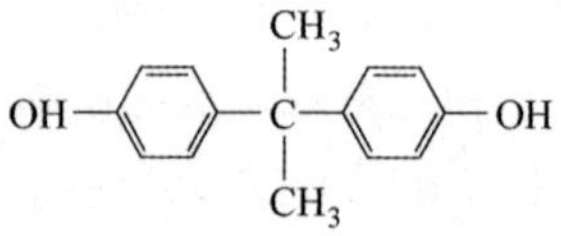

图 7-1 2,2-双-(4-羟苯基)-丙烷

2001 年,世界上 BPA(1957 年在美国开始生产)的产量估计约为 250 万 t。

含有 BPA 的产品在生活的许多领域都有应用,例如,在金属盒、罐子(内部)和其他用来储存水、食品和药品的容器涂上的清漆中。此外,聚碳酸酯塑料因其具有重量轻、耐久性好、抗拉强度高、模量高、熔点高、玻璃化转变温度高、吸水少、耐光性强、耐高温、电气绝缘性能优良等特点,广泛应用于生产、医疗设备(如用于透析和血液氧合)、婴儿奶瓶、餐具等。

多年来,BPA 一直被认为是一种无毒的物质。20 世纪 90 年代初,环境中以及饮用水和食品包装中 BPA 的检出引起了人们的注意。1996 年,欧洲委员会将 BPA 列为对身体和后代健康有害的外在物质。

7.3.3 双酚 A 衍生物

除纯双酚 A 外,双酚 A 二缩水甘油醚(BADGE)和双酚 F 二缩水甘油醚(BFDGE)都是双酚 A 在工业上非常重要的衍生物,图 7-2 显示了这两种化合物的结构式。

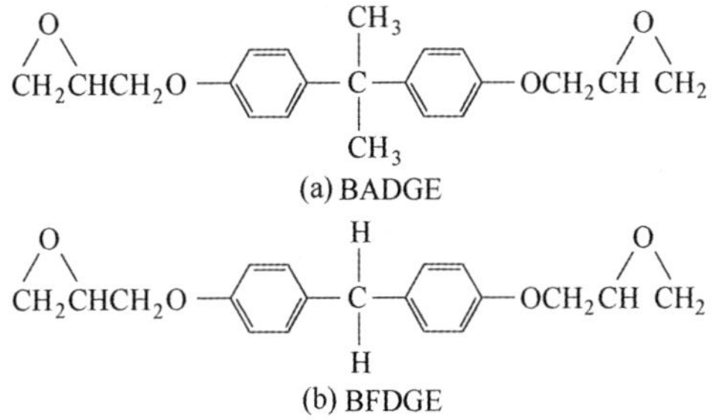

$$CH_2CHCH_2O-\!\!\bigcirc\!\!-\underset{CH_3}{\overset{CH_3}{C}}-\!\!\bigcirc\!\!-OCH_2CHCH_2$$

(a) BADGE

$$CH_2CHCH_2O-\!\!\bigcirc\!\!-\underset{H}{\overset{H}{C}}-\!\!\bigcirc\!\!-OCH_2CHCH_2$$

(b) BFDGE

图 7-2 BADGE 和 BFDGE 的结构式

BADGE 是双酚 A 和环氧氯丙烷反应的产物,是工业中用于生产环氧树脂的主要化合物,广泛用作食品容器的涂料。据估计,目前使用的 75%环氧树脂来自 BADGE。此外,这种化合物被用作罐头和其他用于储存食品容器的涂层。BADGE 被归类为诱导有机体突变的化合物(引起 DNA 变化的基因突变物质)。突变的物质有致癌概率,主要导致接触人的后代中遗传性疾病(所有致癌物的原理都是基因突变,但这些突变并不总是遗传给后代),因为这些物质作用于生殖细胞的 DNA,如果损害的数量超过了睾丸或卵巢细胞矫正机制的能力,生殖细胞就会发生遗传记录的改变,会导致后代身上许多疾病和缺陷。

BFDGE 用作生产环氧树脂的原料和覆盖食品罐头内部的涂层。

从专门致力于双酚 A 及其衍生物测定的文献可以看出,关于这一主题的出版物每年都在增加。这些化合物大部分是通过气相色谱(GC)和高效液相色谱(HPLC)结合质谱(MS)技术来测定的。大多数工作论述了地表水、污水、河水、地下水,以及储存在聚碳酸酯罐头和包装袋中的食品样品中双酚 A 的测定[37-38]。研究人员还研究了与婴儿用聚碳酸酯瓶接触的自来水和人的血液中是否存在这种化合物。表 7-1 列出了特定产品中的 BPA 含量。

表 7-1 特定产品中的 BPA 含量

食物	0.07~0.42ng/g
婴儿食物	0.1~13.2ng/g
奶粉	约 45ng/g
水	0.016~0.5μg/L
血清	0.46~0.56μg/L
脐血	0.46~0.62μg/L

相关文献还描述了 BPA、BADGE 和 BFDGE 从聚合物迁移到食品的能力,特别是在高温和错误使用聚合物容器和包装的情况下。在过去的几年里,从许多环境

保护组织和全球工业组织已经研究了低水平的 BPA 从聚碳酸酯产品到食品和饮料中迁移的情况。这些研究一致表明，双酚 A 向食物的迁移率较低，一般低于 5 份/百万份(小于 5ppm)，与制造商提供的关于聚碳酸酯制成的产品说明书一致。根据研究结果估计，从食物中迁移出来的聚碳酸酯 BPA 的接收量不到每天 0.0000125mg/kg 体重。最高限量水平 3mg/kg 比这一水平大 24 万倍。因此，迁移研究表明进入食物的聚碳酸酯 BPA 含量很低。

食品科学委员会(SCF)是欧盟委员会的一个独立咨询委员会，它估算了食品中 BPA 的安全水平。SCF 将双酚 A 的可耐受日摄入量(TDI)确定为每天 0.01mg/kg 体重，并对涵盖毒性各个方面的所有科学事实进行了全面和详尽的分析。欧盟委员会对 BPA 的食物迁移限制量是 3mg/kg 体重，而对 BADGE 和 BFDGE 的食物迁移限制量是 1mg/kg 体重。

7.3.4 烷基酚

在属于 EDC 组的化合物中有单独的一类，通常被认为会引起明显的雌激素效应，这类物质与类固醇和烷基酚乙氧基酯(APEL)有关。烷基酚于 20 世纪 40 年代投入使用。它们用作油漆、除草剂、杀虫剂、某些非离子型洗涤剂、化妆品、塑料容器、复合材料的原料，以及纺织品、皮革等的脱脂和表面处理的介质等方面。

4-壬基酚(4NP)和 4-辛基酚(4OP)的结构式如图 7-3 和图 7-4 所示，是微生物降解乙氧基化壬基酚的代谢产物，在水环境和污水处理厂中均有发现[3]。化合物 4-叔-壬基酚(4tNP)(图 7-3)在工业上仍广泛使用。这种化合物除了具有雌激素活性外，对包括人类在内的生物也有毒性。

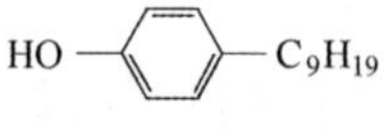

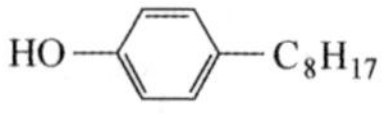

HO—⟨ ⟩—C_9H_{19} HO—⟨ ⟩—C_8H_{17}

图 7-3 4-壬基酚结构式(4NP)   图 7-4 4-辛基酚结构式(4OP)

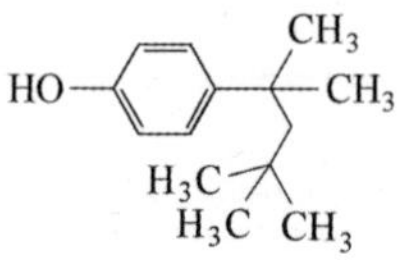

图 7-5 4-叔-壬基酚结构式(4tOP)

这些化合物从塑料材料中释放进入地表水并积累。在地表水中这些外来雌激素的含量达到了每立方米中毫克甚至是微克的水平。在英格兰地区壬基酚含量最高达到 180mg/m³，在威尔士地区辛基酚含量最高达到 13mg/m³。表 7-2 总结了在环境样品(表层水体)中测定的外源性雌激素的浓度。

表 7-2　外源性雌激素在环境样品中的含量[39]　　　mg/m³

国家	4-叔-壬基酚	4-壬基酚
丹麦	<0.1	0.1~0.29
德国	0.0004~0.036	0.001~0.221
	<0.189<0.01~0.189	0.01~0.485
荷兰	0.05~6.3	0.11~4.1
英格兰	<1	0.2~180
威尔士	<0.01~13	0.03~5.2
美国	—	0.11~0.64

已证实壬基酚能引起鱼类内分泌紊乱。壬基酚以及其他类似物质会干扰内分泌活动(或具有雌激素活性),倾向于在生物组织的脂肪层中积累。这些来自工农业的垃圾填埋场和废水排放的化合物通过雨水和大气中气溶胶粒子沉淀进入地表水和地下水。壬基酚的生物积累浓度因子(BCF,有机物中污染物浓度与周围水体中污染物浓度的比值)约为300[40]。

7.3.5　金属雌激素

近年来的研究表明,一些金属及其化合物是一种新型的干扰雌激素活性的化合物。目前,镉、铜、钴、镍、铅、汞、锡、铬、钒阴离子和砷酸盐已被认定是能与雌激素受体结合的金属雌激素。这些金属是被称为"微量元素"的化学元素的一部分,它们在地壳和生物环境中都以微量的形式存在。正如人们理解的那样,特别是在涉及环境问题时,其中一些金属也被称为"重金属"[41]。这一组中各元素的化学性质有很大的差异,因此无法明确其特性。根据含量和性质,它们对生物体产生不利的影响,也可以在生物体构造和发育过程中构成其组成部分。据估计,这些化合物中大约有18种是人体正常运作所必需的。

在环境中积累比例最高(10~600)的元素中,常见的是镉、铅、锌和铜,而汞和铬很少被发现[42]。如果这些金属含量超过食物链下层活体生物的生物界限水平[43],这些元素就会对人体造成极大的危害。

金属雌激素主要存在于水(铬、汞、铜,1~800μg/L)和土壤(铬、汞、铜,40~459mg/L)中。这些金属雌激素也存在于鱼类(Ni、Cr、Hg、Pb、Cu,81~328mg/L)和谷类(Cu,1~14g/L)、空气和香烟烟雾(不吸烟者肾脏中 Cd,15~20mg/L、吸烟者肾脏中 Cd,30~40mg/L)中。

如前所述,一些金属代表一种能够干扰雌激素活性的新化合物。雌激素受体是具有DNA结构域的蛋白质,它能与锌结合。雌激素受体上的锌又能与DNA半胱氨酸残基相互作用形成锌指结构。先前的研究表明,这种金属可被另一个金属(如镍、铜)取代,进而阻断它们与DNA结合域及17β-雌二醇的结合[44]。

某些金属结合α-雌激素受体的能力表明,这些金属会增加内分泌系统紊乱的风险(荷尔蒙系统、内分泌)。长期接触铅和汞会导致不孕、流产和早产等问题。

根据最近的研究发现,Ni、Co、Hg、Pb、Cr^{6+}等金属会导致乳腺癌,影响肾、肺、肝和胰腺功能[42]。砷还引发癌变(身体细胞的变化导致肿瘤),诱导姊妹染色单体交换(染色体复制产生的遗传物质之间的交换),染色体转换和基因扩增,导致不同类型的突变。由于As^{3+}与巯基的相互作用,砷还会改变DNA修复酶并干扰双链DNA的复制[45]。

如前所述,会破坏内分泌系统的金属包括镉、铅、汞、钴、铜、镍、锡、铬以及钒阴离子和砷[45-46]。

7.3.6 苯甲酮

紫外防晒剂是水生环境中存在的另一组有机化合物。它们是新兴的污染物,正在引起一定的关注。每年,消费者市场上都有新的配方,旨在为皮肤提供适当的紫外线防护。人们首次对海中的紫外防晒剂进行检测,是从发现澳大利亚沿海珊瑚礁的大规模灭绝与拥挤的旅游景点相关后开始的。类似的污染问题不仅局限于海洋,也涉及大多数内陆水域。河流、灌溉运河、湖泊,以及人工湖泊也越来越多地受到污染。

紫外防晒剂中最有名和最常用的成分是二苯甲酮类化合物(BP),主要是2,4-二羟基二苯甲酮(BP-1)和2-羟基,4-甲氧基二苯甲酮(BP-3),其结构如图7-6和图7-7所示。

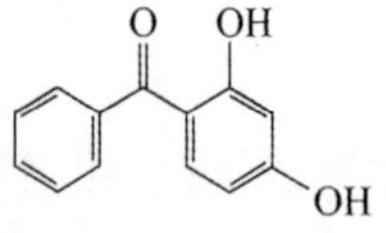

图7-6 2,4-二羟基二苯甲酮(BP-1) 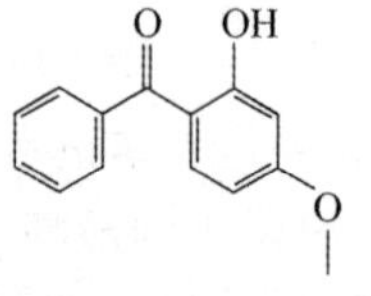　　图7-7 2-羟基,4-甲氧基二苯甲酮(BP-1)

这些化合物起化学过滤的作用,以防止紫外线辐射的不良影响。这种辐射会破坏胶原纤维,进而导致免疫反应减弱。此外,紫外线辐射会产生破坏蛋白质结构的自由基,导致肿瘤形成[47]。根据已发表的数据,这些化合物不仅存在于废水中,也存在于海水、湖泊、河流、污水污泥和土壤中[48-49]。

近年来,人们针对紫外防晒剂的雌激素特性进行了大量的研究[50-52]。结果表明,即使在低浓度下紫外防晒剂也会干扰软体动物的内分泌系统,而在较高浓度下对许多水生生物是有毒的[53]。对于鱼类(斑马鱼),固定剂量的紫外线防晒剂(如乙基己基甲氧基肉桂酸酯,EHMC),即使在低浓度(2.2mg/dm^3)下,也可以在基因水平上引起显著变化。分析显示,EHMC 对参与体内激素代谢的基因转录有影响。

紫外防晒剂也会对组织的构建和重建以及免疫系统功能产生负面影响,并会引起炎症和 DNA 损伤[54]。

紫外防晒剂对内分泌系统的负面影响已在生物体内得到证实。大鼠出现子宫增生,而呆鲦鱼的糖蛋白卵黄蛋白原(一种卵黄前体蛋白)水平增加了 800 倍。但是,实验中使用的紫外防晒剂浓度始终高于环境样品中的浓度[53-57]。可以推断,这些物质不应对生物构成任何威胁。然而,人们不应该低估紫外防晒剂化合物对环境的影响。考虑到紫外防晒剂亲脂性的相关风险,紫外线防晒剂可积聚在活的生物体内,特别是长寿物种的脂肪组织中。BP 的高亲脂性使其能够迅速穿过真皮组织,在人体中产生生物积累。敷用几个小时后,这些 BP 紫外防晒剂可以在血浆、胆汁和尿中检测到[58]。此外,在人乳中也检测到一些紫外防晒剂[59]。

不应该忽视这些化合物相互强化的可能性,因为存在一些其他物质表现出潜在的协同特性[47]。此外,一个重要方面是这些化合物通过紫外线辐射产生代谢产物,这些代谢产物可能会对环境产生类似甚至更强的负面影响[60]。

上述研究表明,这些化合物的浓度取决于采样地点和时间。此外,季节甚至采样的日期都很重要。在夏季,特别是在阳光明媚的日子里,海水中二苯基酮衍生物的浓度显著增加而且在下午达到最高。表 7-3 总结了环境样品中最常用的 3 种苯甲酮的浓度,表 7-4 列出了环境和生物群中紫外防晒剂的浓度[58]。

表 7-3 环境样品中苯甲酮的浓度

组分	样品类型	制样时间	浓度/($\mu g/m^3$)	参考文献
苯甲酮-4	未处理过的污水	夏天	1481	文献[61]
	河水		849	
	海水		138	
苯甲酮-3	未处理过的污水	夏天	1195	文献[62]
苯甲酮-4			4150	
苯甲酮-3	湖水	寒冬	35	文献[48]
		夏天	125	

续表

组分	样品类型	制样时间	浓度/(μg/m³)	参考文献
苯甲酮-3	工业废水	夏天高含量,其他季节低含量	6~697	文献[63]
	生活污水	夏天高含量,其他季节低含量	720~7800	
苯甲酮-3	泉水	春天	300	文献[64]
苯甲酮-1			1000	
苯甲酮-1与苯甲酮-3的和	未处理过的污水		2700~4800	

表7-4 环境与一个地区动植物中紫外防晒剂的含量

环境样品	紫外防晒剂	含量/(ng/L, mg/kg dw)	位置	参考文献
湖水	4-MBC	80	瑞典	文献[52]
	BP-3	125		
	EHMC	19		
	OC	27		
	BMDBM	24		
	BP-3	85	斯洛文尼亚	文献[65]
	EHMC	92		
	Et-PABA	34		
	OC	31		
	BH	85	朝鲜	文献[66]
河水	HMS	345	斯洛文尼亚	文献[65]
	BP-3	114		
	EHMC	88		
	OC	34		
	DHB	47	朝鲜	文献[66]
海水(海滩)	4-MBC	488	挪威	文献[67]
	BP-3	269		
	EHMC	238		
	OC	4461		

续表

环境样品	紫外防晒剂	含量 /(ng/L,mg/kg dw)	位置	参考文献
生饮用水	EHMC	5610	加利福尼亚	文献[68]
未处理的废水	4-MBC BP-3 EHMC OC	6500 7800 19000 12000	瑞典	文献[69]
	BP-3 EHMC	6240 400	加利福尼亚	文献[68]
处理过的废水	4-MBC BP-3 EHMC OC	2700 700 100 270	瑞典	文献[69]
泳池水	4-MBC BP-3	330 400	斯洛文尼亚	文献[65]
鱼(湖)	4-MBC HMS EHMC BP-3	3.80mg/kg(lw) 3.10mg/kg(lw) 0.31mg/kg(lw) 0.30mg/kg(lw)	德国	文献[70]
	4-MBC BP-3 EHMC OC	0.17mg/kg(lw) 0.12mg/kg(lw) 0.07mg/kg(lw) 0.02mg/kg(lw)	瑞典	文献[69]
鱼(河)	4-MBC OC	0.42mg/kg(lw)[①] 0.63mg/kg(lw)[①]	瑞典	文献[71]
淤泥废物	4-MBC EHMC OC OMC	1.78mg/kg(dm)[①] 0.11mg/kg(dm)[①] 4.84mg/kg(dm)[①] 5.51mg/kg(dm)[①]	瑞典	文献[72]

注:dw:干重,lw:脂类质量,dm:干物质,EHMC:乙基己基甲氧基肉桂酸酯,4-MBC:3′-(4′-甲基亚苄基)樟脑,BP-3:二苯甲酮-3,OC:奥克立林,Et-PABA:2-4-二甲基氨基苯甲酸乙基酯,BM-DBM:丁基甲氧基二苯甲酰甲烷,HMS:3,3,5-三甲基环己基水杨酸酯,BH:二苯甲醇,DHB:2,4-二羟基二苯甲酮,OMC:辛基-4-甲氧基肉桂酸酯上标[a]:平均浓度。

BP不仅添加在专门的化妆品中起到保护皮肤免受辐射致癌的作用,它们还被添加到药品和许多日常产品,如沐浴露、洗发水、泡沫浴和发胶中。这些化合物也用于合成材料的包装,以延长被阳光照射产品的使用寿命。其他应用还有轮胎添加剂、铸件、涂料、颜料、纺织品和其他产品,这都是为了增强抗紫外线能力[49,64,73-74]。

由于BP能够吸收有害的UV-B辐射,因此BP是全世界最常用的UV防晒剂。然而,欧洲联盟要求对化妆品中这些物质的含量进行系统的监测,因为它们可能导致皮肤过敏,而且BP-3及其代谢物BP-1被怀疑有雌激素活性[49]。

文献[48,61]数据表明,BP通常是通过生物工艺从废水中高效去除的。各污水处理厂对BP-3的去除率在(28~30)%到(68~96)%之间。缺少生物废水处理过程明显降低了它们在处理装置中的去除率[61]。

7.4 亚硝胺

亚硝胺(图7-8)是最危险的EDC,具有多种诱变和致癌特性[75]。

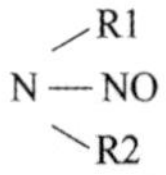

$$N-NO\begin{matrix}R_1\\R_2\end{matrix}$$

图7-8 亚硝胺的通式

根据波兰卫生和社会保障部的指令以及欧盟指令93/11/EEC[76-77],N-亚硝胺被评为致癌物或可能是致癌物[78]。

每年全世界有超过5000万吨含有硝基的废物产生。硝基化合物不仅通过胺的硝化而形成,还通过酰胺、尿素、胍、氨基甲酸酯、氰化物和磺酰胺的硝化而形成。由于它们的化学性质不稳定,易分解成可水解的硝基亚胺和相对稳定的N-亚硝胺。亚硝胺是稳定的化合物,当暴露在光线下或酸性水溶液中时会慢慢分解。

大量的亚硝胺从制药、食品工业、塑料工业、纺织工业、废物运输(机动车辆)、工业废水(染料、润滑剂、橡胶)和溶剂的生产中释放到环境中。燃料制造厂和炼油厂,以及垃圾填埋场和化石燃料燃烧过程(产生热量和动力)都是亚硝胺的重要排放源。这些化合物还通过动物粪便自然渗透到环境中。

亚硝胺是在高温条件下通过各种微生物的作用由仲胺和叔胺形成的。科技的进程增加了它们形成的风险。它们也会在恶劣的食物储存期间形成,或在胃液的影响下形成于胃肠中。最常见的亚硝胺如下:

(1) 亚硝基二甲胺(NDMA),由甘氨酸和缬氨酸产生,在啤酒中它的前驱体可能是肌氨酸(N-甲基甘氨酸);

(2) 亚硝基二乙胺(NDEA);

(3) 由丙氨酸产生的亚硝基甲基乙基胺(NMEA);

(4) 亚硝基二丙胺(DPNA);

(5) 亚硝基哌啶(NPIP);

(6) 亚硝基吡咯烷(NPYP),来自脯氨酸。

N-亚硝基化合物的生物活性不同。与 N-亚硝酰胺形成鲜明对照的是,N-亚硝胺显示出诱变和致畸作用。N-亚硝基化合物从胃肠道被迅速吸收,其生物半衰期小于24h。基于涉及 300 种 N-亚硝基化合物的实验研究结果显示,超过80%的 N-亚硝基化合物是致癌的。质量浓度为 $20\mu g/g$ 的 N-亚硝基二甲胺(NDMA)烷基化 DNA 链可导致肝癌。

在空气、水和食物中,特别是添加了硝酸盐(Ⅲ)和某些鱼产品的肉制品中都检测到低浓度的 N-亚硝基化合物。在城市空气样本中检测到 NDMA。在纯净水(供水系统)和河流中也发现了 N-亚硝基化合物的存在。通过实验研究发现,在活生物体中检测到硝酸盐(Ⅲ)和(Ⅴ)与胺和酰胺反应形成 N-亚硝胺,就像食品中存在的季铵碱反应。

氨基甲酸酯类农药残留物和某些药物导致 N-亚硝胺的形成。某些化合物(氯化物、碘化物、溴化物、三氰化物)催化仲胺和硝酸盐(Ⅲ)形成 N-亚硝基化合物;其他组分(抗坏血酸、没食子酸、亚硫酸钠)抑制这个过程。N-亚硝胺可以在食品加工和储存时形成,如同处在动物的肠胃和人的胃液中。许多硝化反应表明,反应不仅取决于前驱体的存在,还取决于它们的数量、温度和 pH 值。对于弱碱性的胺,反应更快,产率更高。然而,由强碱性胺(二甲基和二乙基)形成的亚硝胺的数量更多。人体和动物胃肠道中 N-亚硝胺的形成也与其他各种非胺化食品成分的存在有关。酚类化合物也会被硝化,这种反应通过催化完成,例如,硫氰酸酯、硫氰酸酯是唾液的组成成分。对于吸烟者来说,发现的反应物是尼古丁和氮氧化物。

有证据表明,工业和家庭上的食品加工过程影响亚硝胺的形成。例如,在鲜鱼中 N-亚硝胺的含量大于 $5g/kg$,而在加工过的鱼(储藏)中 N-亚硝胺的含量达到 $13g/kg$。这些化合物在熏制品中含量更高。对 300 种加工猪肉的研究显示,NDMA 平均含量为 $3g/kg$,最高含量为 $50g/kg$。

奶酪在高温作用下会产生一些致癌的亚硝胺。氨基酸是蛋白质的基本组成部分,氨基酸的降解导致奶酪中含有胺。将火腿和奶酪这样的产品混合在一起,会导致体内 N-硝基或尼古丁(NNA)浓度过高。此外,这些产品在高温处理过程中(如披萨或砂锅菜)会导致含量超出限制。虽然硝酸盐(Ⅲ)和硝酸盐(Ⅴ)被广泛应用于某些类型的生产,但关于奶酪中亚硝胺的形成还没有确切的数据。据推测腌制肉类中的亚硝胺可能是硝酸盐(Ⅲ)与黑胡椒、辣椒等香料相互作用的结果。

啤酒中的亚硝胺也能达到有害的限量,这迫使生产商改变生产工艺。在啤酒中检测到高浓度的亚硝胺,含量达到3mg/L。在西班牙,多达52%的啤酒产品被亚硝胺污染,主要是NDMA和N-亚硝基二乙胺(NDEA)。这些化合物是由烤箱中大麦麦芽直接干燥形成的,其氮氧化物反应是在烤箱中(1500~1800)℃条件下进行的。可以通过改变啤酒生产工艺的热交换介质使空气温度降低到100℃,进而显著降低亚硝胺的产量。亚硝胺在不同啤酒中的浓度是相似的,这主要取决于添加的麦芽大麦的数量。啤酒的类型也会影响NNA的检测量。含酒精的啤酒比不含酒精的饮料含有更少的有害化合物,淡啤酒比黑啤酒更有害。除此之外,在其他一些酒精饮料(白兰地、朗姆酒)和药物(氨基苯酚、土霉素)中也检测到NNA。

从NDMA的角度来看,N-亚硝胺在人类饮食中最重要的来源是炸培根、熏肉、熏鱼0.4~440mg/kg、葡萄酒10~21mg/kg和啤酒0.1~0.5mg/kg[79]。

大气、工作场所、食品和常用物品中存在的N-亚硝胺都应该被检测并消除掉。对于直接接触食物和人体的产品进行亚硝胺的检测尤其重要。这尤其适用于婴儿和儿童专用的橡胶制品(如奶嘴和玩具)和食品工业中使用的产品(如软管、皮带、机械零件)。尽管橡胶制品中的亚硝胺浓度很低(百万分之一或更低),但这些化合物仍会威胁健康。

橡胶制品中的N-亚硝胺是橡胶混合物硫化过程中发生反应的副产品[80],它们是由仲胺(促进剂、抗氧化剂)化合物通过周围空气中的氮氧化物进行亚硝化形成的,这些反应既可以发生在橡胶制品内部也可以发生在表面。从橡胶内部产生的N-亚硝胺可以扩散到物品的表面,然后扩散到使用该产品的环境或介质中。在制备过程中,由于原料的污染,N-亚硝胺也可以混入橡胶混合物中。

基于动物研究,N-亚硝胺是一种已被证实具有致癌作用的化合物。寻找它们的来源和减少人类接触它们是当今最重要的研究难题之一。目前,一个主要问题是在食品中存在的N-亚硝胺及其前身。食品中亚硝胺的测定依赖于在含有液体石蜡的碱性介质中,通过真空蒸馏分离这些化合物,然后用二氯甲烷萃取,预浓缩样品,并使用GC分析。与气相色谱联用的热分析仪可作为这些化合物的有效检测器。

水中富含氮的无机化合物肥料有助于亚硝胺的形成。氮肥被雨水从草地上冲走,进入湖泊或河流等地表水。即使是超过200天,这些化合物也聚集在土壤中。在pH值为7~8时,产生这些化合物的风险最小。pH值超过10时,亚硝胺的生成显著增加。NNA也是水与其他有害物质分解物通过氯化过程形成的。

出于制药目的,必须测定三乙醇胺中的N-亚硝基二乙醇胺。该化合物在实际样品中具有高挥发性和低浓度的特点。因此,要测定痕量的这种化合物,需要预浓缩。真空蒸馏是可采用的方法之一[81]。为了定量测定亚硝胺,常用的方法有GC和HPLC[82-83]。根据《欧洲药典》第五版,N-亚硝基二乙醇胺的最大含量不应超过25ng/g。

7.5 阻燃剂

阻燃剂（FR）是一种有机物质和无机物质的混合物，可以添加到材料中，以降低其可燃性，减少火灾后的火势蔓延。

阻燃性可以通过以下方式实现：
(1) 通过减少易燃、挥发性低分子量物质含量来改变热解过程；
(2) 加速破坏聚合物链的过程；
(3) 在聚合物和热源之间形成较少易燃、烧焦的外壳屏障；
(4) 切断空气供应；
(5) 通过自由基失活来限制火势蔓延；
(6) 通过脱水和脱羧等吸热反应减少烟雾产生过程中散发的热量[84-85]。

使用阻燃剂可以限制物质的可燃性，降低火灾的蔓延程度，甚至避免火灾。它还能延长火灾期间离开房间和建筑物所需的时间，并增加消防队到达时灭火的可能性。

阻燃剂可以分为以下几种[86]：
(1) 无机化合物，如氢氧化铝和氢氧化镁；
(2) 有机化合物的卤化衍生物，如氯化和溴化阻燃剂（BFR）；
(3) 有机磷化合物（OPFR）；
(4) 其他化合物，如三聚氰胺、三聚氰胺氰酸盐和红磷。

有机化合物的卤化衍生物（主要是含有氯和溴的物质）可作为 FR。这些化合物非常有效，并且经济实惠。然而，这些化合物对人类和环境都是高毒性的，并且涉及 Sb_2O_3 的使用。溴化 FR 的效率通常是氯化化合物的 2 倍，而且分解温度更高。这些材料燃烧的产物往往显酸性，因此它们不用于电缆的生产。这组 FR 的作用是清除燃烧过程中形成的高能自由基和羟基氢[87-89]。

最重要的 OPFR 是亚磷酸盐、磷酸盐和磷酸盐酯（V）。OPFR 中磷的比例会影响其延迟材料燃烧的能力。因此，取代基为脂肪链的化合物比芳香族化合物更受欢迎，因为取代基中磷的比例更大。脂肪族化合物的使用受限于对含芳香族取代基化合物水解分离具有较高的敏感性。为了提高磷的比例和降低水解的不稳定性，合成了含有芳香族取代基和环取代基成分的化合物。通过在其结构中加入氯、溴或氮原子，可进一步提高 OPFR 的阻燃效果。卤原子主要加在烷基链上，氮原子主要加在氨基上。磷与引入的氯、溴、氮之间的协同作用可增强阻燃性能[84,90-91]。

阻燃剂主要用作塑料的添加剂，特别是用于电子电气设备的制造。FR 可添加到电路板、电缆、连接器、插头和室内组件设备中。这些化合物可应用于工程塑料、热塑性塑料和弹性体元件以及绝缘材料。阻燃剂也可以用在家具、床垫、地毯、窗帘、服装(主要是儿童防护或运动衫),以及生产汽车、公共汽车、飞机和军事装备的聚合材料中。应用于任何材料中的 FR 必须满足易燃性安全标准[87]。

最常用的 BFR 是多溴二苯醚(PBDE)、四溴双酚 A(TBBPA)和六溴环十二烷(HBCD)。近年来,FR 的使用迅速增长。BFR 是最常用的 FR,它们的市场仍在扩展。然而,据估计西欧的 OPFR 年度使用量几乎是所有 BFR 总量的 2 倍。许多 FR 因其潜在的毒性且在环境存在和人体组织中的积累而被禁止使用。从市场上撤出的 FR 可能会被其他的 FR 取代。尽管欧洲已经引入了 REACH(化学品的注册、评估、授权和限制)监管体系以保护人类健康和环境,但仍需要监控环境样品中的 FR[84,88]。

尽管 FR 在不同的行业中大规模使用,但在许多国家它们的使用量正在系统地减少。这与这些化合物的高毒性(特别是有机化合物的卤素衍生物)直接相关。纺织工业中使用的 FR 会刺激皮肤和黏膜。类似于其他化合的燃烧物,FR 也会产生许多有毒物质(CO、CO_2、PO_x、NO_x、HBr、HCl 和 HCN),这些物质会引起过敏症,呼吸道过敏或严重中毒。FR 的一个非常危险的特征是慢性毒性,这与卤素 FR 分解过程中二噁英的形成有关。许多有机磷化合物具有神经毒性作用,有些还具有致癌和致突变作用[84,88]。

由于许多卤化 FR 具有耐久性、生物积累以及对环境和生物毒性的特点,因此相关部门对其增加了控制,甚至完全禁止了部分产品的生产。这些限制使人们提高了有机磷阻燃剂重视程度和总体使用量,2006 年欧盟有机磷阻燃剂占总使用量的 20%[84,92]。空气中 FR 的主要来源是油漆、电气设备、建筑和装修材料、家具、家庭资源以及其他物质的释放。因为这些化合物经久耐用,所以它们包含的材料具有稳定的成分。这种稳定性化合物的缺点是,即使在使用对象或材料废弃之后,其 FR 也可以通过降雨、城市和工业污水以及许多其他生产活动和日常人类活动渗透到环境中[88,92]。长时间接触 FR,生物累积会对健康造成不利影响。这些化合物可以影响自然激素的新陈代谢,或者可以激活或阻断受体的活动,甚至改变它们的数量。接触 FR 最常见的影响是免疫系统弱化,生育能力下降,神经发育缺陷和癌症。此外,FR 可以在脂肪组织、血液和身体中积累。胎儿在发育过程中长期接触 FR 可能会出现生长障碍、智力或身体发育迟缓,或骨骼、神经和内分泌系统发育受损[84,88,92]。

因此,环境和生物不同基质中的 FR 必须被检测,现有的文献主要集中于对已经存在多年甚至长期被禁止的化合物(如 PCB)的测定。由于 FR 高耐久性和在禁止使用之前所生产的目前仍在市场上销售的产品释放这些化合物到环境中的结果,这就是研究之所以重要的原因。因为随着新 FR 的发展,这些研究可以为各种基体的测定提供分析程序,特别是那些我们最容易接触到的材料。相关文献主要论述了地表水[93-94]、土壤[95-96]、沉积物[97-98]和尘埃[99-100]样品中 FR 的测定。此外,由于这些含有 FR 的环境基体很容易与动物和人类接触,因此监测这些基体中 FR 的浓度是非常重要的。

由于环境基质中含有大量的各种干扰素,且 FR 浓度相对较低,因此需要使用各种提取技术将这些化合物从环境样品中分离出来,并对样品进行富集纯化。根据基体的类型,可以应用不同的技术。对于水样,首先需要进行预处理,最常用的技术是固相萃取(SPE),根据需要选用不同的 SPE 柱(C18、ENV、HLB、PAD3)和不同的洗脱液。为了从水样中富集 FR,需要使用 500mL 甚至 1000mL 的取样量[93-94,101]。这些研究中最重要的发现是一些化合物在地表水中含量相当高。例如,磷酸三(2-氯-1-甲基乙基)酯(TCPP)被定量测定含量高达 26μg/L[94];然而,这些化合物大多数被发现在纳克每升级别,不超过 500ng/L。

在分析土壤、沉淀物和灰尘样品之前以类似的方式制备样品。在预处理步骤和均匀化之后,使用不同的固液萃取技术从样品中提取 FR。最常用的技术是加速溶剂提取(ASE)法,它可以使用不同的溶剂(如己烷和二氯甲烷[98-100])快速提取样品。针对这些样品常用的其他技术还有使快速提取成为可能的超级微波辅助提取(UAE)法[97],以及更耗时但非常有效的索氏提取技术[96]。文献[95]还论述了一些不太常见的技术,如微型 SPE 法。有资料显示,许多不再生产的 FR(主要是多氯联苯和多溴二苯醚)以极高的量存在于粉尘、土壤和沉积物中,甚至达到 390μg/g[98]。

FR 在环境样品中的广泛存在,可能对动物和鱼类以及人类造成直接威胁。这就是鱼和鸟类组织[102]以及人类血液和血清中特定 FR 的测定原因[103],在这些工作中准备并优化了应对此类威胁发生时的检测方法。为了实现 FR 的预浓缩,针对动物组织提出了微波辅助提取法,针对人体血清和血液提出了 SPE 方法。

可以使用不同的方法来测定被检物中 FR 的最终浓度。前面提到的技术制备的提取物主要用于色谱技术分析中,特别适用于测定各种有机化合物的浓度。最常用的技术是 GC 耦合 MS[9,93,96,98,102]或串联 MS/MS[98,100,103]。文献[95]论述了不同检测器结合 GC 系统(μECD)的使用。一些研究人员不仅使用了液相色谱法,还使用了 HPLC 与串联质谱/质谱联用技术[99],以及超高效液相色谱(UHPLC)与串联质谱/质谱检测器[100]或紫外检测器[101]结合的使用技术。使用 UHPLC 技术可以将分析时间缩短 10 倍。

7.6 小结

本章介绍了与选择性化合物的痕量分析有关的基本问题。尤其重要的是发现和确定致癌物和雌激素化合物,它们是引发癌症的主要原因之一。出于这个原因,人们的注意力集中在选定具有这种特性的有机化合物上。这些物质包括二苯甲酮、N-亚硝胺、雌激素活性化合物、邻苯二甲酸盐和壬基酚等致癌物质。尤其是具有雌激素特性的双酚 A 及其衍生物引起了研究者的关注。此外,基于阻燃剂的普遍使用和这些化合物在自然环境中的广泛存在,本章还论述了与阻燃剂分析相关的一些问题。

参考文献

[1] Manahan, S. E.: Toxicological Chemistry and Biochemistry. CRC Press, Boca Raton, FL(2003). in Polish

[2] Zakrzewski, S. F.: Fundamentals of environmental toxicology. Wyd. Nauk. PWN, Warsaw(1997), in Polish

[3] Seńczuk, W.: Toxicology. PZWL, Warsaw (2004), in Polish

[4] Daughton, C. G.: "Emerging" pollutants, and communicating the science of environmental Spectrom. 12, 1067–1076(2001)

[5] Siemiński, M.: Environmental Health Risks. Wyd. Nauk. PWN Warsaw (2007), in Polish

[6] Rauch, S. A., Braun, J. M., Boyd Barr, D., Calafat, A. M., Khoury, J., Montesano, M. A., Yolton, K., Lanphear, B. P.: Associations of prenatal exposure to organophosphate pesticide metabolites with gestational age and birth weight. Environ. Health Perspect. 120, 1055-1060(2012)

[7] Bergonzi, R., Specchia, C., Dinolfo, M., Tommasi, C., de Palma, G., Frusca, T., Apostoli, P.: Distribution of persistent organochlorine pollutants in maternal and foetal tissues: data from an Italian polluted urban area. Chemosphere. 76, 747–754 (2009)

[8] Bergonzi, R., Specchia, C., Dinolfo, M., Tommasi, C., de Palma, G., Frusca, T., Apostoli, P.: Persistent organochlorine compounds in fetal and maternal tissues: evaluation of their potential influence on several indicators of fetal growth and health. Total. Environ. 409, 2888–2893 (2011)

[9] Shen, H., Main, K. M., Virtanen, H. E., Damggard, I. N., Haavisto, A. M., Kaleva, M., Boisen, K. A., Schmidt, I. M., Chellakooty, M., Skakkebaek, N. E., Toppari, J., Scrhamm, K. W.: From mother to child: investigation of prenatal and postnatal exposure to persistent bioaccumulating toxicants using breast milk and placenta biomonitoring. Chemosphere 67, S256-S262 (2007)

[10] Pulkrabovà, J., Hràdkovà, P., Hajslova, P., Poustka, J.: Brominated flame retardants and other organochlorine pollutants in human adipose tissue samples from the Czech Republic. J. Environ. Int. 35, 63–68 (2009)

[11] Jimenez-Diaz, I., Zafra-Gòmez, A., Ballesteros, O., Navea, N., Navalòn, A., Fernandez, M. F.: Determination of Bisphenol A and its chlorinated derivatives in placental tissue samples by liquid chromatography-tandem mass spectrometry. J. Chromatogr. B 878, 3363–3369(2010)

[12] Pathak, R., Suke, S. G., Ahmed, R. S., Tripathi, A. K., Guleria, K., Sharma, C. S., Makhijani,

S. D. , Mishra, M. , Banerjee, B. D. : Endosulfan and other organochlorine pesticide residues in maternal and cord blood in North Indian population. Bull. Environ. Toxicol. 81,216-219(2008)

[13] Jimenez-Torres, M. , Campoy, F. C. , Canabatr, R. F. , Rivas, V. A. , Cerrillo, G. I. , Mariscal, A. M. , Olea-Serrano, F. : Organochlorine pesticides in serum and adipose tissue of pregnant women in Southern Spain giving birth by cesarean section. Sci. Total Environ. 372,32-38 (2006)

[14] Fukata, H. , Omori, M. , Osada, H. , Todaka, E. , Mori, C. : Necessity to measure PCBs and organochlorine pesticide concentrations in human umbilical cords for fetal exposure assessment. Environ. Health Perspect. 113,297-303 (2005)

[15] Mustafa, M. D. , Pathak, R. , Tripathi, A. K. , Ahmed, R. S. , Guleria, K. , Banerjee, B. D. : Maternal and cord blood levels of aldrin and dieldrin in Delhi population. Environ. Monit. Assess. 171,633-638 (2010)

[16] Daglioglu, N. , Gulmen, M. K. , Akcan, R. , Efeoglu, P. , Yener, F. , Unal, I. : Determination of organochlorine pesticides residues in human adipose tissue, data from Cukurova, Turkey. Bull. Environ. Contam. Toxicol. 85,97-102 (2010)

[17] Myllynen, P. , Pasanen, M. , Pelkonen, O. : Human placenta: a human organ for developmental toxicology research and biomonitoring. Placenta 26,361-371 (2005)

[18] Schonfelder, G. , Wittfoht, W. , Hopp, H. , Talsness, C. E. , Paul, M. , Chahoud, I. : Parent bisphenol A accumulation in the human maternal-fetal-placental unit. Environ. Health Perspect. 110, A703-A707 (2002)

[19] Yamada, H. , Furuta, I. , Kato, E. H. , Kataota, S. , Usuki, Y. , Kobashi, G. : Maternal serum and amniotic fluid bisphenol A concentrations in the early second trimester. Reprod. Toxicol. 16,735-858 (2002)

[20] Padmanabhan, V. , Siefert, K. , Ranson, S. , Johnson, T. , Pinkerton, J. , Anderson, L. : Maternal bisphenol-A levels at delivery: a looming problem? J. Perinatol. 28,258-263 (2008)

[21] Stefanidou, M. , Maravelias, C. , Spiliopoulou, C. : Human exposure to endocrine disruptorsand breast milk. Curr. Drug Targets 9,269-273 (2009)

[22] Rylander, L. , Stromberg, U. , Hagmar, L. : Lowered birth weight among infants born to women with a high intake of fish contaminated with persistent organochlorine compounds. Chemosphere 40,1255-1562 (2000)

[23] Ezkenasi, B. , Rosas, L. G. , Marks, A. R. , Bradman, A. , Harley, K. , Holland, N. , Johnson, C. , Fenster, L. , Barr, D. B. : Pesticide toxicity and the developing brain. Basic Clin. Pharmacol. 102,228-236 (2008)

[24] Siddiqui, M. K. J. , Srivastava, S. , Srivastava, S. P. , Mehrota, P. K. , Mathur, N. , Tandon, I. : Persistent chlorinated pesticides and intra-uterine foetal growth retardation: a possible association. Int. Arch. Occup. Environ. Health 76,75-80 (2003)

[25] Ranjit, N. , Siefert, K. , Padmanabhan, V. : Bisphenol-A and disparities in birth outcomes: a review and directions for future research. J. Perinatol. 30,2-9 (2010)

[26] Yolton, K. , Xu, Y. , Strauss, D. , Altaye, M. , Calafat, A. M. , Khoury, J. : Prenatal exposure to bi-

sphenol A and phthalates and infant neurobehavior. Teratoxicol. Neurol. 33,558-564(2011)

[27] Perera, F. P. , Rauh, V. , Tsai, W. Y. , Kinney, P. , Camann, D. , Barr, D. , Bernert, T. , Garfinkel, R. , Tu, Y. H. , Diaz, D. , Dietrich, J. , Whyatt, R. M. : Effects of transplacental exposure to environmental pollutants on birth outcomes in a multiethnic population. Environ. Health Perspect. 111, 201-205 (2003)

[28] Lapworth, D. , Baran, N. , Stuart, M. , Ward, R. : Emerging organic contaminants in groundwater: a review of sources, fate and ccurrence. Environ. Pollut. 163, 287-303 (2012)

[29] Vanderberg, L. N. , Maffini, M. V. , Sonnenschein, C. , Rubin, B. S. , Soto, A. M. : Bisphenol-A and the great divide: a review of controversies in the field of endocrine disruption. Endrocr. Rev. 30, 75-95 (2009)

[30] Kuo, H. W. , Ding, W. H. : Trace determination of bisphenol A and phytoestrogens in infant formula powders by gas chromatography-mass spectrometry. J. Chromatogr. A 1027, 67-74(2004)

[31] Le, H. H. , Carlson, E. M. , Chua, J. P. , Belcher, S. M. : Bisphenol A is released from polycarbonate drinking bottles and mimics the neurotoxic actions of estrogen in developing cerebellar neurons. Toxicol. Lett. 176, 149-156 (2008)

[32] Matsumoto, A. , Kunugita, N. , Kitagawa, K. , Isse, T. , Oyama, T. , Foureman, G. : Bisphenol A levels in human urine. Environ. Health Perspect. 111, 101 (2003)

[33] Brock, J. W. , Yoshimura, Y. , Barr, J. R. , Maggio, V. L. , Graiser, S. R. , Nazakawa, H. : Measurement of bisphenol A levels in human urine. J. Expo. Anal. Environ. Epidemiol. 11, 323-328 (2001)

[34] Arakawa, C. , Fujimaki, K. , Yoshinaga, J. , Imai, H. , Serizawa, S. , Shiraishi, H. : Daily urinary excretion of bisphenol A. Environ. Health Prev. Med. 9, 22-26 (2004)

[35] Vom Saal, F. S. , Huges, C. : An extensive new literature concerning low-dose effects of bisphenol A shows the need for a new risk assessment. Environ. Health Perspect. 113, 926-933 (2005)

[36] Welshons, W. V. , Nagel, S. C. , vom Saal, F. S. : Large effects from small exposures. III Endocrine mechanisms mediating effects of bisphenol A at levels of human exposure Endocrinology. 147, S56-S69 (2006)

[37] Rykowska, I. , Wasiak, W. : Advances in stir bar sorptive extraction coating: a review. Acta Chromatogr 25(1), 27-46 (2013)

[38] Rykowska, I. , Wasiak, W. : Novel stir-bar sorptive extraction coating based on chemical bonded silica for the analysis of polar organic compounds and heavy metal ions. Mendeleev Commun. 23 (2), 88-89 (2013)

[39] Dudziak, M. , Bodzek, M. : Determination of contents of xenoestrogens in water by means of sorption extraction method. Environ. Prot. 31, 11-14 (2009). in Polish

[40] Biłyk, A. , Nowak-Piechota, G. : Environmental pollution substances that cause endocrine disruption of the body's functions. Environ. Prot. 26, 29-35 (2004). in Polish

[41] Kabata-Pendias, A. , Pendias, H. : Biogeochemistry of trace elements. PWN, Warsaw (1999) in Polish

[42] Martin, M., Reiter, R., Pham, T.: Estrogen-like activity of metals in MCF-7 breast cancer cells. Endocrinology 144, 2425–2436 (2003)

[43] Sohoni, P., Sumpter, J.: Several environmental oestrogens are also anti-androgens J. Endocrinol. 158, 327–339 (1998)

[44] Świtalska, M., Strza dal a, L.: Non-genomic action of estrogens. Postepy Hig. Med. Dosw. 61, 541–546 (2007). in Polish

[45] Darbre, P. D.: Metalloestrogens: an emerging class of inorganic xenoestrogens with potential to add to the oestrogenic burden of the human breast. J. Appl. Toxicol. 26(3), 191 (2006)

[46] Woźniak, M., Mrias, M.: Xenoestrogens: endocrine disrupting compounds. Ginekol. Pol. 79, 785–790 (2008). in Polish

[47] Gackowska, A., Gaca, J., Załoga, J.: Determination of selected UV filters in water samples Chemist 66, 615–620 (2012). in Polish

[48] Giokal, D. L., Salvador, A., Cisvent, A.: UV filters: from sunscreens to human body and the environment. Trends Anal. Chem. 26(5), 360–374 (2007)

[49] Zenker, A., Schutz, H., Fent, K.: Simultaneous trace determination of nine organic UV-absorbing compounds (UV filters) in environmental samples. J. Chromatogr. A1202, 64–74 (2008)

[50] Morohoshi, K., Yamamoto, H., Kamata, R., Shiraishi, F., Koda, T., Morita, M.: Estrogenic activity of 37 components of commercial sunscreen lotions evaluated by in vitro assays Toxicol. In Vitro 19, 457–469 (2005)

[51] Kunz, P. Y., Galicia, H., Fent, K.: Comparison of in vitro and in vivo estrogenic activity of UV filters in fish. Toxicol. Sci. 90, 349–361 (2006)

[52] Poiger, T., Buser, H. R., Balmer, M. E., Per-Anders, B., Müller, M. D.: Occurrence of UV filter compounds from sunscreens in surface waters: regional mass balance in two Swiss lakes Chemosphere 55, 951–967 (2004)

[53] Kerdivel, G., Le Guevel, R., Habauzit, D., Brion, F., Ait-Aissa, S.: Estrogenic potency of benzophenone UV filters in breast cancer cells: proliferative and transcriptional activity substantiated by docking analysis. PLoS One 8(4), e60567 (2013)

[54] Zucchi, S., Oggier, D. M., Fent, K.: Global gene expression profile induced by the UV-filter 2-ethyl-hexyl-4-trimethoxycinnamate (EHMC) in zebrafish (Danio rerio). Environ. Pollut 159, 3086–3096 (2011)

[55] Fent, K., Zenker, A., Rapp, M.: Widespread occurrence of estrogenic UV-filters in aquatic ecosystems in Switzerland. Environ. Pollut. 158, 1817–1824 (2010)

[56] Jeon, H. K., Chung, Y., Ryu, J. C.: Simultaneous determination of benzophenone-type UV filters in water and soil by gas chromatography-mass spectrometry. J. Chromatogr. A 1131, 192–202 (2006)

[57] Schmitt, C., Oetken, M., Dittberner, O., Wagner, M., Oehlmann, J.: Endocrine modulation and toxic effects of two commonly used UV screens on the aquatic invertebrates Potamopyrgus antipo-

darum and Lumbriculus variegatus. Environ. Pollut. 152,322-329(2008)

[58] Fent,K. , Kunzac, P. Y. , Gomezd, E. : UV filters in the aquatic environment induce hormonal effects and affect fertility and reproduction in fish. Chimia. 62,68-74 (2008)

[59] Schlumpf, M. , Durrer, S. , Faass, O. , Ehnes, C. , Fuetsch, M. : Developmental toxicity of UV filters and environmental exposure: a review. Int. J. Androl. 31,144-151 (2008)

[60] Próba,M. : Chosen anthropogenic factors of surface water pollution analysis of the phenomenon. Eng. Environ. Prot. 16(1),113-119 (2013). in Polish

[61] Rodil,D. , Quintana, J. B. , Lope-Mahia, P. , Muniategui-Lorenzo, S. , Prada-Rodriguez, D. : Multiclass determination of sunscreen chemicals in water samples by liquid chromatography-tandem mass spectrometry. Anal. Chem. 80(4),1307-1315 (2008)

[62] Kasprzyk-Hordern,B. , Dinsdale, R. M. , Guwy, A. J. : The removal of pharmaceuticals, personal care products,endocrine disruptors and illicit drugs during wastewater treatment and its impact on the quality of receiving waters. Water Res. 43,363-380 (2009)

[63] Bursey,J. T. , Pellizzari, E. D. : USEPA, analysis of industrial wastewater for organic pollutants in consent decree survey, Contract Number 68-03-2867, Athens, GA, USA, (USEPA Environ. Res. 13 Lab.) (1982)

[64] Dabrowska,A. , Binkiewicz, K. , Nawrocki, J. : Determination of contents of benzophenone in surface water by means of GC-ECD. Environ. Prot. 31(3),57-63 (2009). in Polish

[65] Cuderman, P. , Health, E. : Determination of UV filters and antimicrobial agents in environmental water samples. Anal. Bioanal. Chem. 387,1343-1350 (2007)

[66] Jeon,H. K. , Chung, Y. , Ryu, J. C. : Simultaneous determination of benzophenone-type UV filters in water and soil by gas chromatography-mass spectrometry. J. Chromatogr. A 113, 192 - 202 (2006)

[67] Langford, K. , Thomas, K. V. : Inputs of chemicals from recreational activities to the Norwegian coastal zone. In: SETAC Europe 17th Annual Meeting,Porto, Portugal (2007)

[68] Loraine, G. A. , Pettigrove, M. E. : Seasonal variations in concentrations of pharmaceuticals and personal care products in drinking water and reclaimed wastewater in Southern California. Environ. Sci. Technol. 40,687-695 (2006)

[69] Balmer, M. E. , Buser, H. R. , Müller, M. D. , Poiger, T. : Occurrence of some organic UV filters in wastewater,in surface waters, and in fish from Swiss Lakes. Environ. Sci. Technol. 39,953-962 (2005)

[70] Nagtegaal,M. , Ternes,T. A. , Baumann, W. , Nagel, R. : UV-Filtersubstanzen in Wasser und Fischen UWSF-Z. Umweltchem. Ökotoxikol. 9,79-81 (1997)

[71] Buser,H. R. , Balmer, M. E. , Schmid, P. , Kohler, M. : Occurrence of UV filters 4-methylbenzylidene camphor and octocrylene in fish from various Swiss rivers with inputs from wastewater treatment plants. Environ. Sci. Technol. 40,1427-1431 (2006)

[72] Plagellat,C. , Kupper,T. , Furrer, R. , de Alencastro, L. F. , Grandjean, D. , Tarrradellas, J. : Concentrations and specific loads of UV filters in sewage sludge originating from a monitoring network

in Switzerland. Chemosphere 62,915-925 (2006)

[73] Diaz-Cruz, S. M., Llorca, M., Barcelo, D.: Organic UV filters and their photodegradates, metabolites and disinfection by-products in the aquatic environment. Trends Anal. Chem. 10, 873-887 (2008)

[74] Giokas, D. L., Sakkas, V. A., Albanis, T. A.: Determination of residues of UV filters in natural waters by solid-phase extraction coupled to liquid chromatography-photodiode array detection and gas chromatography-mass spectrometry. J. Chromatogr. A 1026,89-293 (2004)

[75] Seńczuk, E.: Modern toxicology. PZWL, Warsaw (2005), in Polish

[76] Directive of Ministry of Heath, 2 Sept. 2003, Law gazette No 199, 24 Nov. 2003

[77] Standard PN-EN 12868:2001, in Polish

[78] Davarani, S. S. H., Masoomi, L., Banitaba, M. H., Zhad, H. R. L. Z., Sadeghi, O., Samiei, A.: Determination of N-nitrosodiethanolamine in cosmetic products by headspace solid phase microextraction using a novel aluminum hydroxide grafted fused silica fiber followed by gaschromatography-mass spectrometry analysis. Talanta 105, 347-353 (2013)

[79] Domański, W.: N-Nitrosodiethylamine and N-nitrosodimethylamine-determination method. Principles and Methods of Working Environment Assessment 4/34, 115-120 (2002), in Polish

[80] Parasiewicz, W.: Analysis of chemical hazards in the rubber manufacturing process. Elastomers 5 (1), 17-28 (2001). in Polish

[81] Kupruszewski, G., Sobocińska, M., Walczyna, R.: Basic preparation of organic compounds, Gdańsk Publishing 44, (1998), in Polish

[82] Byun, M. W., Ahn, H. J., Kim, J. H., Lee, J. W., Yook, H. S., Han, S. B.: Determination of volatile N-nitrosamines in irradiated fermented sausage by gas chromatography coupled to a thermal energy analyzer. J. Chromatogr. A 1054, 403-407 (2004)

[83] Sauches, F. P., Rios, A., Valcárcel, M., Zanin, K. D., Caramão, E. B.: Determination of nitrosamines in preserved sausages by solid-phase extraction-micellar electrokinetic chromatography. J. Chromatogr. A 985, 503-512 (2003)

[84] Van der Veen, I., De Boer, J.: Phosporous flame retardants: properties, production, environmental occurrence, toxicity and analysis. Chemosphere 88, 1119 (2012)

[85] Laoutid, F., Bonnaud, L., Alexandre, M., Lopez-Cuesta, J., Dubois, P.: New prospects in flame retardant polymer materials: from fundamentals to nanocomposites. Mater. Sci. Eng. R63 (2009)

[86] Weil, E. D.: Flame retardancy. In: Encyclopedia of Polymer Science and Technology, 3rd edn. Wiley-Interscience, New York (2004)

[87] Sutker, B. J.: Flame retardants. In: Ullmann's Encyclopedia of Industrial Chemistry, 7th edn. Wiley-Interscience, Weinheim (2011)

[88] Al-Odaini, N. A., Yim, U. H., Kim, N. S., Shim, W. J., Hong, S. H.: Isotopic dilution determination of emerging flame retardants in marine sediments by HPLC-APCI-MS/MS. Anal. Methods 5, 1771 (2013)

[89] Dufton, P. W.: Flame retardants for plastics: market report. Rapa Technol., Shawbury (2003)

[90] Hoang, D., Kim, J., Jang, B. N.: Synthesis and performance of cyclic phosphorus-containing flame retardants. Polym. Degrad. Stab. 93, 2042 (2008)

[91] Vothi, H., Nguyen, C., Lee, K., Kim, J.: Thermal stability and flame retardancy of novel phloroglucinol based organo phoshorous compound. Polym. Degrad. Stab. 95, 1092 (2010)

[92] Covaci, A., Sakai, S., Van den Eede, N.: A novel abbreviation standard for organobromine, organochlorine and organophosphorus flame retardants and some characteristics of the chemicals. Environ. Int. 49, 57 (2012)

[93] Bollmann, U. E., Moller, A., Xie, Z.: Occurrence and fate of organophosphorus flame retardants and plasticizers in coastal and marine surface waters. Water Res. 46, 531 (2012)

[94] Cristale, J., Katsoyiannis, A., Sweetman, A. J.: Occurrence and risk assessment of organophosphorus and brominated flame retardants in the River Aire (UK). Environ. Pollut. 179, 194 (2013)

[95] Zhou, Y. Y.: The use of cooper(II) isonicotinate-based micro-solid-phase extraction for the analysis of polybrominated diphenyl ethers in soils. Anal. Chim. Acta 747, 36 (2012)

[96] Liu, M.: Heavy metals and organic compounds contamination in soil from an e-waste region in South China. Environ. Sci. 15, 919 (2013)

[97] Cristale, J., Lacorte, S.: Development and validation of a multiresidue method for the analysis of polybrominated diphenyl ethers, new brominated and organophosphorus flame retardants in sediment, sludge and dust. J. Chromatogr. A 1305(267) (2013)

[98] Labunska, I.: Levels and distribution of polybrominated diphenyl ethers in soil, sediment and dust samples collected from various electronic waste recycling sites within Guiyu town, southern China. Environ. Sci. 15, 503 (2013)

[99] Fang, M.: Investigating a novel flame retardants known as V6: measurements in baby products, house dust, and car dust. Environ. Sci. Technol. 47, 4449 (2013)

[100] Kim, J. W., Isobe, T., Sudaryanto, A.: Organophosphorus flame retardants in house dust from the Philippines: occurrence and assessment of human exposure. Environ. Sci. Pollut. Res. Int. 20(2), 812 (2013)

[101] Kowalski, B., Mazur, M.: The simultaneous determination of six flame retardants in water samples using SPE pre-concentration and UHPLC-UV method. Water Air Soil Pollut. 225, 1866 (2014)

[102] Ma, Y., Cui, K., Zeng, F.: Microwave-assisted extraction combined with gel permeation chromatography and silica gel cleanup followed by gas chromatography-mass spectrometry for the determination of organophosphorus flame retardants and plasticizers in biological samples. Anal. Chim. Acta 786, 47 (2013)

[103] Lin, Y. P., Pessah, I. N., Puschner, B.: Simultaneous determination of polybrominated diphenyl ethers and polychlorinated biphenyls by gas chromatography-tandem mass spectrometry in human serum and plasma. Talanta 113, 41 (2013)

第8章
药物杂质分析

8.1 概述

尽管从药物学的观点来看痕量分析的发展始于相关痕量元素的测定研究,但是这一科学领域的起始应该追溯到药物分析,尤其是不同类型的药物汤液、酊剂和浸泡液中活性组分的研究。从罂粟汁中提取鸦片和后来的吗啡,或者从金鸡纳树皮中提取奎宁,都只不过是进行痕量分析前样品的预处理和浓缩[1-3]。

本章的目的是讨论药物痕量分析中最重要的问题之一——药物杂质分析。"药物杂质"与"药物纯度"这两个术语可以互换,它们通常代表着化学纯度、色谱纯度,或者药物纯度。根据美国药学[4]和欧盟药学[5],每种药物在美国或欧盟国家获得市场授权时都要满足适当的纯度,包括化学纯度和微生物纯度。一种药品满足药物标准,只能说明它符合了药典一系列标准中的化学和微生物部分的要求。

在开始论述药品中化学纯度和药物中化学杂质的痕量分析研究之前,需要给出杂质的定义。目前,最通行的杂质定义指与药物活性成分在分子结构与组成上有任何不同的每一种化学成分,水除外(因为水被认为不是杂质)"[5]。

根据这一定义,杂质也包括所有让药物变得稠密、稳定、可溶、有味道以及其他特点的配制物辅料。这显然与目前化学杂质的定义相矛盾,因为在药品分析和药物分析中所有的辅料都被认为是有效组分。这意味着必然遇到与"药物纯度"(不能含有含量多于活性物质的杂质)术语相同的标准,以及其他涉及化学结构和性能的标准,以此类推。

实际上,获得100%纯度的药物是不可能的,这就使得核实药物杂质类型很有必要。研究包括杂质的离析、鉴别、定量,以及在许多情况下明确它们的生物学活性和毒性。实施这些研究是药物发展过程中很重要的一部分,也是临床治疗学安全性的一部分,目的是防止类似1938年悲剧的发生,那年105人在注射了2-乙氧基乙醚污染的长生不老药磺酸之后死亡[6]。

有效组分(有效的药物组分)以及最终产品中(药物产品)污染物的测定是一个宽泛并且耗时的过程,目的是保证药品的质量。这是药物治疗安全性最重要的影响因素之一。

根据人用药品注册技术要求国际协调会议(ICH)的定义,杂质是有效组分和最终药物产品中一个单独的化学极小量,这个极小量不是有效组分或者药物辅料。它包含所有的光学同分异构体、降解产物,以及多晶态。杂质的药理性能和生物学活性可能类似或异于有效组分,因此,杂质能以药物产品相同或相异的方式导致药物活性的增减,并对人体产生影响[7]。

8.2 药品杂质控制条例

药物药典专著描述了为保证产品的质量,需要满足的基本要求。大多数专著表明药物纯度在98%~99%,这意味着1%~2%的杂质是允许的。但在使用高剂量药物的情况下,如维生素或抗生素,它们每天的剂量从几百毫克到几克,那么1%~2%的杂质会对病人产生严重的后果。

ICH 文件 Q3A、Q3B 和 Q3C [7-9]中提供了有关药品纯度的更多限制性说明:

Q3A 文件描述了药物材料中的杂质。根据此规定,每种新药物必须评估以下杂质水平:

(1) 有机杂质:如果杂质浓度大于0.1%,那么需要评估所有已知或未知结构的杂质;如果杂质浓度均小于0.1%,那么需要评估所有有害性质;评估杂质总量。

(2) 无机杂质。

(3) 溶剂残留水平。

Q3A 条例不包括生物或生物技术来源物质、蛋白质和放射性药物[7]。

Q3B 文件同样描述了药物制剂中的杂质。该条例主要关注降解产物的含量水平和活性物质与辅料、包装组分的反应产物。该文件还规定了必须识别和适当限定的杂质量级。基于这些规定,根据每日摄入活性物质的总剂量将药物产品分组:小于1mg,1~10mg,10~100mg,100mg~2g,大于2g。例如,ICH 指南要求,日剂量在(1~10)mg 范围内且杂质量高于0.5%的药物,需要确定杂质的结构和生物活性;如果杂质低于0.5%,则必须计算患者接受药物治疗的每日量,如果高于20μg/剂,则必须确定杂质的结构,如果该量高于50μg/剂,则必须确定杂质的生物活性(表8-1)[8]。

表 8-1 ICH 药物和药品杂质报告与识别和量值极限指南[7-8]

每日最大剂量		限值		
		报告	记录	限量
药物	≤2g	0.05%	每天 0.01mg 或者 1.0mg	每天 0.15mg 或 1.0mg
	>2g	0.03%	0.05%	0.05%
药品	≤1g	0.1%		
	>1g	0.05%		
	<1mg		每天 1.0μg 或 5μg	每天 1.0μg 或 50μg
	1~10mg		每天 0.5μg 或 20μg	每天 1.0μg 或 50μg
	10~100mg		每天 0.2mg 或 2mg	每天 0.5μg 或 200μg
	100mg~2g		每天 0.2mg 或 2mg	每天 0.2μg 或 2mg
	>2g		0.1%	0.1%

Q3C 文件重点介绍药物中溶剂的残留,它描述了用于合成活性物质和辅料的有机和无机溶剂。根据 ICH 指南,溶剂可分为三类:

Ⅰ类:具有已证实或潜在致癌性质的溶剂,以及对环境有害的溶剂,如苯、1,2-二氯乙烷、1,1-二氯乙烷和四氯化碳。这些溶剂不应用于制药行业。如果必须使用它们,则必须检测它们在药品中的含量,并且它们不能超过 ICH 指南里的要求(表 8-2)[9]。

表 8-2 根据 ICH 指南 Q3C 的有机溶剂类别[9]

Ⅰ类		Ⅱ类		Ⅲ类	
溶剂	含量限度/ppm	溶剂	含量限度/ppm	溶剂	含量限度/ppm
苯	2	2-甲氧基乙醇	50	1-丁醇	0.5
四氯化碳	4	甲基异丁基酮	50	1-戊醇	0.5
1,2-二氯乙烷	5	硝基甲烷	50	1-丙醇	0.5
1,1-二氯乙烷	6	氯仿	60	2-丁醇	0.5
1,1,1-三氯乙烷	1500	1,1,2-三氯乙烷	80	2-丁醇	0.5
—	—	1,2-二甲氧基乙烷	100	2-丙醇	0.5
—	—	四氢化萘	100	3-甲基-1-丁醇	0.5
—	—	乙二醇单乙醚	160	乙酸	0.5
—	—	环丁砜	160	丙酮	0.5
—	—	吡啶	200	苯甲醚	0.5
—	—	甲酰胺	220	乙酸丁酯	0.5

续表

Ⅰ类溶剂	含量限度/ppm	Ⅱ类溶剂	含量限度/ppm	Ⅲ类溶剂	含量限度/ppm
—	—	己烷	290	二甲亚砜	0.5
—	—	氯苯	360	乙酸乙酯	0.5
—	—	1,4-二恶烷	380	乙醚	0.5
—	—	乙腈	410	甲酸乙酯	0.5
—	—	二氯甲烷	600	甲酸	0.5
—	—	乙二醇	620	庚烷	0.5
—	—	四氢呋喃	720	乙酸异丁酯	0.5
—	—	N,N-二甲基甲酰胺	880	乙酸异丙酯	0.5
—	—	甲苯	890	乙酸甲酯	0.5
—	—	二甲基乙酰胺	1090	甲基乙基酮	0.5
—	—	甲基环己烷	1180	甲基异丁基甲酮	0.5
—	—	1,2-二氯乙烯	1870	戊烷	0.5
—	—	二甲苯	2170	乙酸丙酯	0.5
—	—	甲醇	3000	甲基叔丁基醚	0.5
—	—	环己烷	3880		
—	—	N-甲基吡咯烷酮	4840		

Ⅱ类:具有可能的有毒活性或具有可逆毒性的溶剂。该类包括25种溶剂,如乙腈、甲醇、环己烷和二氯甲烷。对于每种溶剂,都有允许的每日接触剂量(PDE)和最终产品中可接受的浓度水平(表8-2)。

Ⅲ类:人体中毒性低的溶剂,如丙酮、乙醇和乙烯醚。该类的每日可接受水平为50mg,但在某些情况下可以超过此水平。该类含有27种溶剂(表8-2)。

溶剂残留物是在工艺过程中难以完全去除的杂质,这使确定其在最终产品中的含量水平很有必要。根据ICH标准,Ⅰ类残留物是必须识别和量化的。如果Ⅱ类和Ⅲ类的残留浓度水平超过可接受的标准,则需要采取类似的措施[9]。

8.3 药物杂质特征

药物杂质根据其化学性质可以分为有机成分和无机成分。根据它们的来源,它们可以分为以下三类:

(1) 由活性物质或辅料的合成过程产生的杂质;

(2) 由活性物质或辅料降解过程产生的杂质；

(3) 意外杂质。

其中,最大的一类杂质是由活性物质合成过程带来的。根据 Sandor Görög 等[10-12,14,16]的研究发现,这些物质可以分为以下七类：

(1) 合成物的基体和半成品:预测(可能的)杂质结构可以通过合成物的反应过程来确定。

(2) 竞争反应产物:由与主要合成反应平行的反应产生的可预测杂质。

(3) 副反应产物:结构难以预测的杂质。

(4) 源自原材料基体杂质:原材料基体的杂质会留在最终产物中,因为它们经历相同的反应过程,往往会产生同分异构杂质。

(5) 源自溶剂的杂质:溶剂和溶剂杂质也可能遗留在反应的最终产品中。

(6) 源自反应催化剂的杂质:合成反应的催化剂有可能成为反应的基体,进而导致最终产品污染。

(7) 无机杂质:缓冲剂、无机试剂、催化剂、重金属和重金属离子(主要来自设备),还原和氧化因素。

另一类杂质是降解产物。这些杂质可以与合成过程的杂质相同,也可能不同。降解产物在溶液和固相中形成,主要来自羟基化、氧化、还原、异构化和环氧化的反应。降解产物可具有自身的药理活性,有时它们甚至显示出毒性作用。药物的稳定性测试可以明确药物降解的机理和动力学,确定不同因素对药物稳定性的影响,并明确哪种储存条件使产物降解程度最小化。

还有一类杂质是意外杂质[13-15]。这些化学物质与药物的合成或降解无任何关联。包括：

(1) 金属:通常来自用于合成的设备,目前通过使用更先进的技术和材料来减少。

(2) 消毒剂:用于生产线的准备。

(3) 除草剂、杀虫剂、重金属:植物来源的药物。

有机杂质可分为手性杂质和遗传毒性杂质。药物的化学结构和物理化学性质与它们的药理活性紧密相关,因此微小的变化就可以改变分子的生物活性,例如不对称碳的不同构型。对于手性药物,可以认为光学同分异构体是杂质,例如,青霉胺的S-对映异构体可用于类风湿性关节炎的治疗,但 R-对映异构体显示出毒性迹象[13-15]。

遗传毒性杂质由于其生物学特性,是一类单独分类的有机杂质。这些化合物可引起基因突变、染色体变化和诱发癌症。最终的药物制剂中即使存在很少痕量的这种杂质也会导致极糟的后果。该类物质包括芳香胺、环氧化物、过氧化物和杂环氮化合物等化合物[17]。

综上所述,活性物质和最终药物制剂中的杂质可能与药物分子的结构和化学特性相似,也可能完全不同,它们既可能表现出无生物活性,也可能表现出明显的药理活性(甚至毒性),并可能导致副作用的增强(图8-1)。

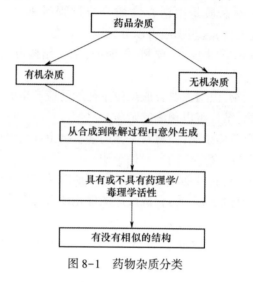

图8-1 药物杂质分类

8.4 药物杂质的仪器分析方法

确定某种活性物质或最终药物产品的杂质大体情况,需要使用多种分析方法来全面且精确地识别所有杂质。

分析药物杂质的方法,如紫外法(UV)、傅里叶变换红外(FT-IR)、核磁共振(NMR)、质谱(MS),用于分离、识别和定量杂质,也可以确定结构。目前最有效的方法是联用技术,如 GC-MS、LC-MS、液相色谱-二极管阵列检测-质谱(LC-DAD-MS)、LC-NMR、LC-DAD-NMR-MS 等[18-20]。

图8-2 显示了确定活性物质杂质概况的实验方法。首先,在至少三个分组系统中使用色谱或电泳方法分离杂质。将保留时间与标准品的保留时间进行比较,并对洗脱液进行定量分析。接下来,使用光谱法(UV、FT-IR、MS)结合色谱/电泳方法来确定未知杂质的结构。如果这还不够,则应进行制备液相色谱(薄层色谱(TLC)、高效液相色谱(HPLC)),并使用 NMR 光谱法或使用 HPLC-NMR-MS 组合检查所得结果。在下一阶段,合成鉴定的杂质,通过色谱分析,比较合成标准物的保留时间与杂质的保留时间。使用该方案确定药物的杂质概况可确保杂质得到充分鉴定[19]。

基于集体的研究[18-20]和出版物的论述[21-24],我们提出了用于分析药物杂质的最有用的仪器方法。

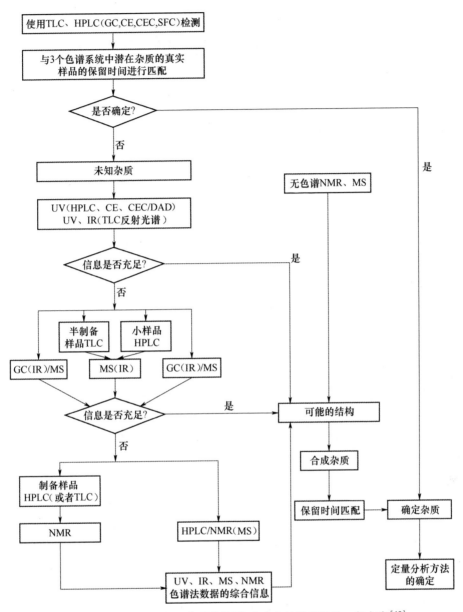

图 8-2 药物中相关有机杂质的检测、鉴定和结构确认的一般方案[19]

8.4.1 分离方法

杂质分析的第一阶段是将杂质与活性物质和辅料分离。为此,须使用以下方法:

(1) 色谱:TLC,HPLC,GC,超临界流体色谱(SFC)。

(2) 电动方法:毛细管电泳(CE),胶束电动色谱(MEKC),微乳液电动色谱(MEEKC),毛细管电色谱(CEC)。

上述方法还可以鉴别和定量杂质[19]。

根据Niessen[25]和Olsen等的研究,目前药物分析中最常用的方法是HPLC,它也是杂质分析中最常用的方法[26]。HPLC的用途很广泛,包括以下例子:

(1) 对乙酰氨基酚中7种杂质的鉴定和定量[27];

(2) 鉴定氯替泼诺依碳酸酯的几种杂质和降解产物[28];

(3) 使用整体填充柱从鸟嘌呤[29]中分离阿昔洛韦和从4种杂质中分离利福平[30];

(4) 使用HPLC和油/水乳液引入流动相,如微乳液色谱(MELC)用于定量辛伐他汀[31]和5种扑热息痛杂质[32]。

在上述实例中需要使用紫外灯进行检测。对于不吸收紫外线的药物可以使用电分析检测器,例如用于分析硫酸新霉素[33]和阿米卡星[34]及其杂质的电流检测器。

除HPLC外,TLC也常用于药物分析。虽然TLC比其他色谱方法选择性更低,精度更低,但它的多功能性、快速性和低成本使其成为一种常用的方法。传统的TLC方法逐渐被HPTLC取代,HPTLC使用新的固定相和自动光密度检测。目前,HPTLC与HPLC在药物杂质的分析方法上形成了竞争[19-20]。

TLC已被用于环丙沙星、双氯芬酸、氯丙嗪、三氟丙嗪、多西平、阿普唑仑、奥美拉唑、泮托拉唑、四环素和金霉素的杂质分析[35-41],其他例子可以在Ferenczi-Fodor等[42]的著作中找到。

作为色谱方法的替代方法,电泳方法可用于确定药物的杂质概况。该方法包括CE、MEKC和MEEKC。最有用的方法之一是CEC,该技术将CE的高效率与LC的能力相结合。在药物杂质分析中使用电泳方法的实例列于表8-3[43-48]。

表8-3 在药物杂质分析中使用电泳方法

活性药物成分	分析方法	参考文献
阿莫西林	使用MEKC分离和测定	[43]
二乙酰吗啡	使用MEKC测定杂质分布	[44]
氢氯噻嗪,氯噻嗪	使用MEKC分离和测定	[45]
溴泮	用CE测定杂质分布	[46]
四环素盐酸盐	用CE测定杂质分布	[47]
布洛芬	用CE测定杂质分布	[48]

注:MEKC—胶束电动色谱;CE—毛细管电泳。

8.4.2 光谱方法

在没有使用色谱进行预分离的情况下使用UV-VIS和FT-IR光谱这两种方法的选择性都很低,用它们来测定药物杂质是有局限性的。但在某些情况下,当杂质具有与活性物质非常不同的光谱特性时,直接测量吸光度也可以提供关于杂质结构的一些有用数据。然而,在许多情况下,杂质具有与活性物质相似的结构,因此,它们的紫外和红外吸收光谱相似,并且相互重叠。这意味着在药物纯度分析中直接使用光谱方法是有局限性的[49]。

使用光谱导数法可以提高UV方法的选择性[50]。导数分光光度法是一种分析前对经典的紫外光谱(零阶光谱)进行波长区分的化学计量方法,它比传统的紫外光谱更具选择性和精确性[50],在药物纯度分析中使用光谱导数方法的实例见表8-4。

表8-4 在药物纯度分析中使用光谱导数方法的实例[50,53-61]

活性药物成分	分析方法	参考文献
阿昔洛韦,糠酸二甲氧脲	用二阶导数分光光度法测定阿昔洛韦和鸟嘌呤(杂质) 用三阶导数法测定糠酸二甲氧脲及其降解产物	[53]
醋氯芬酸	用三阶导数法测定醋氯芬酸和双氯芬酸(主要降解产物)	[54]
氨氯地平	用三阶导数测定氨氯地平和光降解产物	[55]
头孢他啶(Ⅰ) 头孢呋辛(Ⅱ) 头孢噻肟(Ⅲ)	使用一阶导数测定Ⅰ、Ⅱ和Ⅲ及其降解产物	[56]
氢氯噻嗪	用一阶和二阶导数测定氢氯噻嗪及其降解产物(甲氧基氢氯噻嗪,羟基氢氯噻嗪和5-氯-2,4-二磺酰氨基苯胺)	[57]
利奈唑胺	使用一阶导数测定利奈唑胺及其降解产物	[58]
美洛昔康	用一阶导数法测定美洛昔康及其降解产物	[59]
奥美拉唑	使用一阶、二阶和三阶导数测定奥美拉唑及其降解产物(次磺酰胺和苯并咪唑硫醚)	[60]
塞尼达唑	用一阶导数法测定塞尼达唑及其降解产物	[61]

药物分析中最常用的光谱分析方法是质谱法。它可以(取决于电离技术)在纳克级或皮克量级测量杂质,如果使用离子轰击技术,甚至可以达到亚皮克量

级[51]。如果分析物之前没有进行色谱分离,则可使用串联质谱(MS/MS),因为它使用了与色谱分离不同的电离分离技术。使用 MS 分析螺内酯纯度的方法如图 8-3 所示[52]。

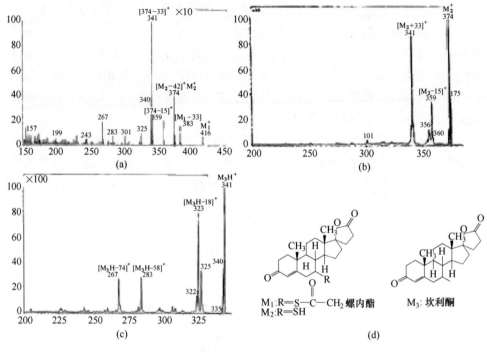

图 8-3 螺内酯及其杂质的 MS 谱[52]

8.4.3 串联方法

有机药物杂质分析的最佳结果是通过使用串联(联用)方法获得的[62-80],该方法将色谱/电泳方法的分离性质与光谱方法的定性和定量能力相结合。在实践中最常见的组合包括使用具有 UV、NMR 或 MS 检测的 LC 或 GC,或电泳方法与 MS 检测的组合(CE-MS、MEKC-MS 和 CEC-MS)。

利用 UV 光谱,主要使用光电二极管阵列,可以定量分析停留时间与活性物质不同的药物杂质。Görög 通过使用 HPLC-UV 测定哌库溴铵的杂质类型[22],其原理是该药物缺乏显色团,不能吸收紫外范围内的电磁辐射,但其杂质(烯醇衍生物)可以吸收紫外辐射(最大吸收波长为 236nm)。

Zhou 等广泛描述了在药物纯度分析中 UV 检测的使用[62],Görög[19]也做了这些工作。

目前,常用的药物杂质测定方法是 HPLC-MS。这种分析需要样品的适当制备、分离,适当调整分离参数和选用合适的电离方法。使用软电离使杂质产生分子离子,然后确定其分子量。使用串联 MS/MS 质谱测定法,可以通过标记碎片来确定杂质的化学结构。在多项综述文章中描述了采用液相色谱结合 MS 检测的方法对扎来普隆[63]、依托考昔[64]、乙二醇、二乙酸[65]、双氯[66]、喹那普利[67]等进行的纯度研究[68-70]。

GC-MS 在药物分析中的应用仅限于具有适当热稳定性和挥发性的药物。尽管 HPLC-MS 的使用已影响到 GC-MS 在药物分析中的重要性,但 GC-MS 分析过程的改进(例如引入高选择性 MS 检测器)意味着 GC-MS 仍然是药物杂质分析的重要方法[71-74]。

串联电泳方法,如 CE-MS、MEKC-MS、CEC-MS,可以替代 LC 方法来分析药物杂质:

(1) CE-ESI-MS/MS 法分析盐酸加兰他敏杂质的方法[75];

(2) CE-ESI-MS 法分析 16 种潜在的药物杂质[76];

(3) CE-MS 法鉴定和定量酮康唑中的顺式酮康唑杂质[77];

(4) MEKC-MS 法分析布洛芬和磷酸可待因的方法[78];

(5) MEKC-APPI-MS 法分析美贝维林中的杂质[79];

(6) MEKC-ESI-MS 法分析氢溴酸加兰他明纯度及氢溴酸异丙托铵降解产物的方法[80]。

综上所述,不同的高效检测器的联用是目前药物杂质分析的最佳选择。它们可以识别和定量分析药物中存在的杂质,即使是在百万分之一或百万分之一的水平。

8.4.4 无机杂质分析方法

无机杂质最大的一类是金属,通常使用原子光谱分析技术(如原子吸收光谱(AAS)法、电感耦合等离子-原子发射光谱(ICP-AES)法或质谱分析(ICP-MS)[19]进行分析。

最常用的是 AAS(火焰和非火焰),它可以识别和定量固体、液体和气体样品中的金属[81]。ICP-AES 可分析痕量水平的杂质(纳克级和皮克级痕量水平)[81]。ICP-MS 可检测和分析元素周期表的大部分元素,以及定量分析不同基质(体液、水、污水等)中元素的不同同位素浓度。许多元素的检出限为 $10^{-13} \sim 10^{-10}$ mg/L [81]。

ICP-MS 是用于药物分析的最新仪器方法之一,已用于以下方面:

(1) 药物代谢研究:大鼠尿液中 4-溴苯胺代谢物的定量[82]。

(2) 活性物质的定量:维生素 B_{12}[83],西咪替丁[84],含杂原子(S、P、Cl、Br、I)的药物或金属[85],如顺-二氯二氨络铂[86]。

(3) 放射性药物分析:^{99}Tc 至 ^{99m}Tc 比的定量,定量限为 50pg/mL [87]。

(4) 生物医学分析:DNA、蛋白质分析,阿尔茨海默病和帕金森病 P、Zn、Cu、Fe 的定量分析,这些都是毒性元素[88-89]。

表 8-5 列出了使用 GF-AAS、ICP-OES 和 ICP-MS 方法鉴定药物中 16 种元素的结果。表 8-6 列出了使用 ICP-MS 分析药物中存在的元素的例子。

表 8-5 药品中金属的限值[90]

元素		口服日剂量/(PDE)(μg/天)	成分限量/(μg/g)		仪器检出限/(μg/g)		
			口服	注射	ICP-OES	GFAAS	ICP-MS
高毒性	砷(无机)	15	1.5	0.15	3.5	0.1	0.01
	镉	25	2.5	0.25	0.06	0.0008	0.002
	铅	10	1.0	0.10	2.0	0.04	0.003
	汞(Hg^{+2})	15	1.5	0.15	3.0	0.6	0.001
中度毒性	铬Ⅲ	250	25	2.5	0.3	0.05	0.02
	钼	250	25	2.5	0.12	0.006	0.002
	镍	250	25	2.5	0.6	0.05	0.02
	钯	100	10	1.0	4.0	0.05	
	铂	100	10	1.0	2.0	0.02	0.003
	锇	100	10	1.0	2.0		
	铑				2.0	0.01	
	钌				5.0	1.0	
	铱				2.0	0.05	
	钒	250	25	2.5	0.78	0.1	0.004
低毒性	铜	2500	250	25	0.2	0.001	0.01
	锰	2500	250	25	0.05	0.005	0.02

注:PDE—每日允许最大接触量。

表 8-6 无机杂质分析中的检测器和定量限的例子[91-95]

药品	元素	定量限	检测器	参考文献
双环胺,乙胺丁醇,吡嗪酰胺,呋喃唑酮	Ti,Cr,Mn,Co,Ni,Cu,Zn,Cd,Hg,Pb	—	Varian Ultra Mass 700	[91]
硅酸铝镁	Al,Mg	40μg/g 和 60μg/g	CETAC LSX-100,Perkin-Elmer SCIEX,ELAN 6000 ICP-MS	[92]

续表

药品	元素	定量限	检测器	参考文献
维生素E	Ti,V,Cr,Mn,Fe,Co,Cu,Mo,Ag,Cd,Ni,Sn,Pb	0.01~3.02 ppb（乳胶），0.01~1.26 ppb（15% HNO3）	Agilent7500 ICP-MS,1100W,Meinhard 雾化器	[93]
盐酸甲基苯丙胺	Na,Br,Pd,Ba,I	—	Seiko ICP-MS SPQ-6100,1.35 kW 四极杆型	[94]
福辛普利钠	Pd	0.1μg/g	四级杆等离子体 PQ 11 涡轮增压 ICP-MS Jacketed Scott 喷雾器	[95]

8.4.5 挥发性溶剂残留物的分析方法

药物中的挥发性溶剂残留物是用于合成活性物质或最终制剂的有机化合物的剩余物。有机溶剂在制药工业中起着重要作用，但是，其中的许多有机溶剂对人类和环境都有毒。从最终产品中完全去除这些溶剂比较困难，所以需要建立它们的档案[90]。

热重分析(TGA)可用于分析地毯中的溶剂残留物。TGA 通过测量样品质量随时间和温度的变化识别和定量溶剂，定量限为 100 ppm[96]。

IR 方法可用于定量四氢呋喃、二氯苯和二氯甲烷[96]。

GC 是最有用的溶剂残留分析方法，可以通过直接注射技术，或通过顶空、固相微萃取(SPME)或单滴微萃取(SDME)技术[96]进行分析。GC 选择性高、特异性好、操作简单、样品制备简单。现代毛细管气相色谱可以分离许多化合物，并对其进行鉴定和定量[96]。GC 可以使用不同的检测器系统，具体如表 8-7 所列。

表 8-7 残留溶剂分析常用的气相色谱检测器总结[97]

检测器	类型	检出限/g
热导检测器(TCD)	通用	4×10^{-10}
火焰离子化检测器(FID)	通用(有机/碳化合物)	2×10^{-12}
光化电离检测器(PID)	通用	2×10^{-13}
^{63}Ni 电子俘获探测器(ECD)	选择性(卤素和其他吸电子基团)	5×10^{-15}
质量光谱(MS)	通用的或选择性的	达到 25×10^{-15}

8.5 小结

在现代社会,痕量分析在生态学、化学工程、食品加工、生物医学分析和药物化学等领域发挥着重要作用。先进的、灵敏的和精确的分析方法可以鉴别和定量不同的化合物,包括有机和无机化合物,以及大分子配合物。这些方法具有非常好的检测和定量分析水平,可达到每万亿分之一或每千万亿分之一。

痕量分析对于分析药物中活性物质和最终药物制剂中存在的杂质非常重要。药物杂质的鉴定和表征受国际标准管制,用于确保药品的质量。

本章根据 ICH 标准[7-9]介绍了杂质类别、不同杂质的要求以及用于表征药物杂质概况的分析方法;清楚地阐述了用于药物质量保证的各种方法。这些方法的不断发展确保了药物的质量不断提高,从而提高了药物疗法的安全性。

参考文献

[1] Lahiri, S.: Advanced Trace Analysis. Narosa Publishing House, New Delhi (2009)

[2] Bartley, E. H.: Text-Book of Medical and Pharmaceutical Chemistry. Nabu Press, London (2010)

[3] Gunzler, H., Williams, A.: Handbook of Analytical Techniques. Wiley, Weinheim (2001)

[4] United States Pharmacopeia 38th ed. Rockville, Maryland (2014)

[5] The European Pharmacopoeia, 8th ed. Council of Europe, European Committee on Pharmaceuticals and Pharmaceutical Care (2014)

[6] Geiling, E. M. K., Cannon, P. R.: Pathogenic effects of elixir of sulfanilamide (diethylene glycol) poisoning. JAMA 111, 919-926 (1938)

[7] International Conference on Harmonisation of Technical Requirements for Registration of Pharmaceuticals for Human Use, ICH Q3A(R3): Impurities in New Drug Substances (2002)

[8] International Conference on Harmonisation of Technical Requirements for Registration of Pharmaceuticals for Human Use, ICH Q3B(R3): Impurities in New Drug Products (2003)

[9] International Conference on Harmonisation of Technical Requirements for Registration of Pharmaceuticals for Human Use, ICH Q3C(R3): Impurities: Residual Solvents (2005)

[10] Gorog, S., Babjak, M., Balogh, G., Brlik, J., Csehi, A., Dravecz, F., Gazdag, M., Horvath, P., Lauko, A., Varga, K.: Drug impurity profiling strategies. Talanta 44, 1517-1526 (1997)

[11] Gorog, S.: Chemical and analytical characterization of related organic impurities in drugs. Anal. Bioanal. Chem. 377, 852-862 (2003)

[12] Gorog, S.: The importance and the challenges of impurity profiling in modern pharmaceutical analysis. Trends Anal. Chem. 25, 755-757 (2006)

[13] Ng, L., Lunn, G., Faustyno, P.: Organic impurities in drug substance: origin, control and measurement. In: Smith, R. J., Webb, M. L. (eds.) Analysis of Drug Impurities. Blackwell, Oxford (2007)

[14] Gorog, S.: The nature and origin of the impurities in drug substances. In: Gorog, S. (ed.) Identification and Determination of Impurities in Drugs. Elsevier, Amsterdam (2000)

[15] Ahuja, S.: Overview: isolation and characterization of impurities. In: Ahuja, S., Alsante, K. M. (eds.) Handbook of Isolation and Characterization of Impurities in Pharmaceuticals. Academic, San Diego (2003)

[16] Gorog, S.: Drug safety, drug quality, drug analysis. J. Pharm. Biomed. Anal. 48, 247-253 (2008)

[17] McGovern, T., Jacobson-Kram, D.: Regulation of genotoxic and carcinogenic impurities in drug substances and products. Trends Anal. Chem. 25, 790-795 (2006)

[18] Smith, R. J., Webb, M. L. (eds.): Analysis of Drug Impurities. Blackwell, Oxford (2007)

[19] Gorog, S. (ed.): Identification and Determination of Impurities in Drugs. Elsevier, Amsterdam (2000)

[20] Ahuja, S., Alsante, K. M. (eds.): Handbook of Isolation and Characterization of Impurities in-Pharmaceuticals. San Diego, Academic Press (2003)

[21] Bartos, D., Gorog, S.: Recent advances in the impurity profiling of drugs. Curr. Pharm. Anal. 4, 215-230 (2008)

[22] Gorog, S.: New safe medicines faster: the role of analytical chemistry. Trends Anal. Chem. 22, 407-415 (2003)

[23] Rahman, N., Najmul, S., Azmi, H., Wu, H. F.: The importance of impurity analysis in pharmaceutical products: an integrated approach. Accred. Qual. Assur. 11, 69-74 (2006)

[24] Baertschi, S. W.: Analytical methodologies for discovering and profiling degradation-related impurities. Trends Anal. Chem. 25, 758-767 (2006)

[25] Niessen, W. M. A.: LC MS and CE MS strategies in impurity profiling. Chimia 53, 478-483 (1999)

[26] Olsen, B. A., Castle, B. C., Myers, D. P.: Advances in HPLC technology for the determination of drug impurities. Trends Anal. Chem. 25, 796-805 (2006)

[27] Kamberi, M., Riley, C. M., Ma, X., Huang, C. W. J.: A validated, sensitive HPLC method for the determination of trace impurities in acetaminophen drug substance. J. Pharm. Biomed. Anal. 34, 123-128 (2004)

[28] Yasueda, S. I., Higashiyama, M., Shirasaki, Y., Inada, K., Ohtori, A. J.: An HPLC method to evaluate purity of a steroidal drug, loteprednol etabonate. J. Pharm. Biomed. Anal. 36, 309-316 (2004)

[29] Tzanavaras, P. D., Themelis, D. G. J.: High-throughput HPLC assay of acyclovir and its major impurity guanine using a monolithic column and a flow gradient approach. J. Pharm. Biomed. Anal. 43, 1526-1530 (2007)

[30] Liu, J., Sun, J., Zhang, W., Gao, K., He, Z. J.: HPLC determination of rifampicin and related

compounds in pharmaceuticals using monolithic column. J. Pharm. Biomed. Anal. 46, 405 – 409 (2008)

[31] Malenovic, A. , Medenica, M. , Ivanovic, D. , Jancic, B. : Monitoring of simvastatin impurities by HPLC with microemulsion eluents. Chromatographia 63, 95–100 (2006)

[32] McEvoy, E. , Donegan, S. , Power, J. , Altria, K. D. : Optimisation and validation of a rapid and efficient microemulsion liquid chromatographic (MELC) method for the determination of paracetamol (acetaminophen) content in a suppository formulation. J. Pharm. Biomed. Anal. 44, 137– 143 (2007)

[33] Hanko, V. P. , Rohrer, J. S. J. : Determination of neomycin sulfate and impurities using high-performance anion – exchange chromatography with integrated pulsed amperometric detection. J. Pharm. Biomed. Anal. 43, 131–141 (2007)

[34] Zawilla, N. H. , Li, B. , Hoogmartens, J. , Adams, E. : Improved reversed-phase liquid chromatographic method combined with pulsed electrochemical detection for the analysis of amikacin. J. Pharm. Biomed. Anal. 43, 168–173 (2007)

[35] Council of Europe: European Pharmacopoeia, vol. 2, 5th edn, p. 968. Council of Europe, Strasbourg (2005)

[36] Berezkin, V. : The discovery of thin layer chromatography. J. Planar Chromatogr. 8, 401 – 405 (1995)

[37] Krzek, J. , Hubicka, U. , Szczenpanczyk, J. : High performance thin layer chromatography with densitometry for the determination of ciprofloxacin and impurities in drugs. J. AOAC Int. 88, 1530 –1536 (2005)

[38] Maslanka, A. , Krzek, J. : Use of TLC with densitometric detection for determination of impurities in chlorpromazine hydrochloride, trifluoperazine dihydrochloride, promazinehydrochloride, and doxepin hydrochloride. J. Planar Chromatogr. 20, 463–475 (2007)

[39] Anjaneyulu, Y. , Marayya, R. , Linga, R. D. , Krishna, R. P. : High performance thin layer chromatographic determination of the related substances in alprazolam drug. Asian J. Chem. 19, 3375– 3381 (2007)

[40] Agbaba, D. , Novovic, D. , Karljikovic-Rajic, K. , Marinkovic, V. : Densitometric determination of omeprazole, pantoprazole, and their impurities in pharmaceuticals. J. Planar Chromatogr. 17, 169– 172 (2004)

[41] Naidong, W. , Hua, S. , Roets, E. , Hoogmartens, J. : Assay and purity control of tetracycline, chlortetracycline and oxytetracycline in animal feeds and premixes by TLC densitometry with fluorescence detection. J. Pharm. Biomed. Anal. 33, 85–93 (2003)

[42] Ferenczi-Fodor, K. , Végh, Z. , Renger, B. : Thin-layer chromatography in testing the purity of pharmaceuticals. Trends Anal. Chem. 25, 778–789 (2006)

[43] Okafo, G. N. , Camilleri, P. : Micellar electrokinetic capillary chromatography of amoxicillin and related molecules. Analyst 117, 1421–1424 (1992)

[44] Lurie, I. S. , Hays, P. A. , Garcia, A. E. , Panicker, S. : Use of dynamically coated capillaries for

the determination of heroin, basic impurities and adulterants with capillary electrophoresis. J. Chromatogr. A 1034, 227–235 (2004)

[45] Thomas, B. R., Fang, X. G., Chen, X., Tyrrell, R. J., Ghodbane, S.: Validated micellar electrokinetic capillary chromatography method for quality control of the drug substances hydrochlorothiazide and chlorothiazide. J. Chromatogr. B Biomed. Sci. Appl. 657, 383–394 (1994)

[46] Hansen, S. H., Sheribah, Z. A.: Comparison of CZE, MEKC, MEEKC and non-aqueous capillary electrophoresis for the determination of impurities in bromazepam. J. Pharm. Biomed. Anal. 39, 322–327 (2005)

[47] Tjornelund, J., Hansen, S. H.: Determination of impurities in tetracycline hydrochloride by non-aqueous capillary electrophoresis. J. Chromatogr. A 737, 291–300 (1996)

[48] Stubberud, K. P., Astrom, O.: Separation of ibuprofen, codeine phosphate, their degradation products and impurities by capillary electrophoresis: II. Validation. J. Chromatogr. A 826, 95–102 (1998)

[49] Clarke, H. J., Norris, K. J.: Sample selection for analytical method development. In: Ahuja, S., Alsante, K. M. (eds.) Handbook of Isolation and Characterization of Impurities in Pharmaceuticals. Academic, San Diego (2003)

[50] Karpińska, J.: Derivative spectrophotometry – recent applications and directions of developments. Talanta 64, 801–822 (2004)

[51] Lang, J. K. (ed.): Handbook on Mass Spectrometry: Instrumentation, Data and Analysis and Applications. Nova Science, New York (2010)

[52] Mak, M., Czira, G., Brlik, J.: Mass spectrometry in impurity profiling. In: Görög, S. (ed.) Identification and Determination of Impurities in Drugs. Elsevier, Amsterdam (2000)

[53] Daabees, H. G.: The use of derivative spectrophotometry for the determination of acyclovir and diloxanide furoate in presence of impurity or degradation product. Anal. Lett. 31, 1509–1522 (1998)

[54] Hasan, N. Y., Abdel-Elkawy, M., Elzany, B. E., Wagieh, N. E.: Stability indicating methods for the determination of aceclofenac. Farmaco 58, 91–99 (2003)

[55] Ragno, G., Garofalo, A., Vetuschi, C.: Photodegradation monitoring of amlodipine by derivative spectrophotometry. J. Pharm. Biomed. Anal. 27, 19–24 (2002)

[56] Kelani, K., Bebavy, L. I., Abdel, F. L.: Stability – indicating spectrophotometric and densitometric methods for determination of some cephalosporins. J. AOAC Int. 81, 386–393 (1998)

[57] Bebawy, L. I., El Kousy, N.: Stability-indicating method for the determination of hydrochlorothiazide, benzydamine hydrochloride and clonazepam in the presence of their degradation products. Anal. Lett. 30, 1379–1397 (1997)

[58] Hassan, E. M.: Determination of ipratropium bromide in vials using kinetic and first-derivative spectrophotometric methods. J. Pharm. Biomed. Anal. 21, 1183–1189 (2000)

[59] Bebawy, L. I.: Stability-indicating method for the determination of meloxicam and tetracaine

hydrochloride in the presence of their degradation products. Spectrosc. Lett. 31,797-820(1998)

[60] El Kousy,N. M. ,Bebawy,L. I. : Stability-indicating methods for determining omeprazole and octylonium bromide in the presence of their degradation products. J. AOAC Int. 82, 599 – 606 (1999)

[61] Moustafa,A. A. ,Bibawy,L. I. : Stability-indicating assay of secnidazole in the presence of its degradation products. Spectrosc. Lett. 32,1073-1098 (1999)

[62] Zhou,L. ,Mao,B. ,Reamer,R. ,Novak,T. ,Ge,Z. : Impurity profile tracking for active pharmaceutical ingredients: Case reports. J. Pharm. Biomed. Anal. 44,421-429 (2007)

[63] Bharathi,C. ,Prabahar,K. J. ,Prasad,C. S. ,Kumar,M. S. ,Magesh,S. ,Handa,V. K. ,Dandala, R. ,Naidu,A. : Impurity profile study of zaleplon. J. Pharm. Biomed. Anal. 44,101-109 (2007)

[64] Hartman, R. , Abrahim, A. , Clausen, A. , Mao, B. , Crocker, L. S. , Ge, Z. : J. Liq. Chromatogr. Relat. Technol. 26,2551-2566 (2003)

[65] Babják,M. ,Balogh,G. ,Gazdag,M. ,Gorog,S. : Estimation of impurity profiles of drugs and related materials: part XXI. HPLC/UV/MS study of the impurity profile of ethynodiol diacetate. J. Pharm. Biomed. Anal. 29,1153-1157 (2002)

[66] Raj,T. J. S. ,Bharati,C. H. ,Rao,K. R. ,Rao,P. S. ,Narayan,G. K. ,Parikh,K. : Identification and characterization of degradation products of dicloxacillin in bulk drug and pharmaceutical dosage forms. J. Pharm. Biomed. Anal. 43,1470-1475 (2007)

[67] Shinde,V. ,Trivedi,A. ,Upadhayay,P. R. ,Gupta,N. L. ,Kanase,D. G. ,Chikate,R. C. : Degradation mechanism for a trace impurity in quinapril drug bytandem mass and precursor ions studies. Rapid Commun. Mass Spectrom. 21,3156-3160 (2007)

[68] Tollsten,L. : HPLC/MS for drug impurity identification. In: Görög,S. (ed.) Identification and Determination of Impurities in Drugs. Elsevier,Amsterdam (2000)

[69] Smyth,W. F. : Recent studies on the electrospray ionisation mass spectrometric behaviour of selected nitrogen-containing drug molecules and its application to drug analysis using liquid chromatography-electrospray ionisation mass spectrometry. J. Chromotogr. B 824,1-20(2005)

[70] Qiu,F. ,Norwood,D. L. :Identification of pharmaceutical impurities. J. Liq. Chromatogr. Relat. Technol. 30,877-935 (2007)

[71] Laniewski,K. ,Wännman,T. ,Hagman,G. :Gas chromatography with mass spectrometric,atomic emission andFourier transform infrared spectroscopic detection as complementary analytical techniques for the identification of unknown impurities in pharmaceutical analysis. J. Chromatogr. A 985,275-282 (2003)

[72] Laniewski,K. ,Vager € o,M. ,Forsberg,E. ,Forngren,T. ,Hagman,G. :Complementary use of gas chromatography – mass spectrometry, gas chromatography-atomic emission detection and nuclear magnetic resonance for identification of pharmaceutically related impurities of unknown structures. J. Chromatogr. A 1027,93-102 (2004)

[73] Pan,C. ,Liu,F. ,Ji,Q. ,Wang,W. ,Drinkwater,D. ,Vivilecchia,R. :The use of LC/MS,GC/MS, and LC/NMR hyphenated techniques to identify a drug degradation product in

pharmaceutical development. J. Pharm. Biomed. Anal. 40,581-590 (2006)

[74] Dayrit, F. M. , Dumlao, M. C. : Impurity profiling of methamphetamine hydrochloride drugs seized in the Philippines. Forensic Sci. Int. 144,29-36 (2004)

[75] Visky, D. , Jimidar, I. , Van Ael, W. , Vennekens, T. , Redlich, D. , deSmet, M. : Capillary electrophoresis-mass spectrometry in impurity profiling of pharmaceutical products. Electrophoresis26, 1541-1549 (2005)

[76] Vassort, A. , Barrett, D. A. , Shaw, P. N. , Ferguson, P. D. , Szucs, R. : A generic approach to the impurity profiling of drugs using standardised and independent capillary zone electrophoresis methods coupled to electrospray ionisation mass spectrometry. Electrophoresis 26, 1712-1723 (2005)

[77] Castro-Puyana, M. , García-Ruiz, C. , Cifuentes, A. , Crego, A. L. , Marina, M. L. J. : Identification and quantitation of cis-ketoconazole impurity by capillary zone electrophoresis-mass spectrometry. J. Chromatogr. A 1114,170-177 (2006)

[78] Stubberud, K. , Callmer, K. , Westerlund, D. : Partial filling - micellar electrokinetic chromatography optimization studies of ibuprofen, codeine and degradation products, and coupling to mass spectrometry: Part II. Electrophoresis 24,1008-1015 (2003)

[79] Mol, R. , De Jong, G. J. , Somsen, G. W. : Atmospheric pressure photoionization for enhanced compatibility in online micellar electrokinetic chromatography mass spectrometry. Anal. Chem. 77, 5277-5282 (2005)

[80] Mol, R. , Kragt, E. , Jimidar, I. , De Jong, G. J. , Somsen, G. W. : Micellar electrokinetic chromatography - electrospray ionization mass spectrometry for the identification of drug impurities. J. Chromatogr. B 843,283-288 (2006)

[81] Kellner, R. , Otto, M. , Widmer, H. M. , Mermet, J. M. (eds.) : Analytical Chemistry. Wiley, New York (1998)

[82] Nicholson, J. K. , Lindon, J. C. , Scarfe, G. , Wilson, I. D. , Abou-Shakra, F. , Castro-Perez, J. , Eaton, A. , Preece, S. : High-performance liquid chromatography and inductively coupled plasma mass spectrometry (HPLC ICP MS) for the analysis of xenobiotic metabolites in rat urine: application to the metabolites of 4-bromoaniline. Analyst 125,235-236 (2000)

[83] Baker, S. A. , Miller-Ihli, N. J. : Determination of cobalamins using capillary electrophoresis inductively coupled plasma mass spectrometry. Spectrochim. Acta B 55,1823-1832 (2000)

[84] Evans, E. H. , Wolff, J. C. , Eckers, C. : Sulfur-specific detection of impurities in cimetidine drug substance using liquid chromatography coupled to high resolution inductively coupled plasma mass spectrometry and electrospray mass spectrometry. Anal. Chem. 73,4722-4728 (2001)

[85] Wind, M. , Edler, M. , Jakubowski, N. , Linscheid, M. , Wesch, H. , Lehmann, W. D. : Analysis of protein phosphorylation by capillary liquid chromatography coupled to element mass spectrometry with 31P detection and to electrospray mass spectrometry. Anal. Chem. 73,29-35 (2001)

[86] Falter, R. , Wilken, R. D. : Determination of carboplatinum and cisplatinum by interfacing HPLC with ICP MS using ultrasonic nebulisation. Sci. Total Environ. 225,167-176 (1999)

[87] Hill, D. M., Barnes, R. K., Wong, H. K., Zawadzki, A. W.: The quantification of technetium in generator-derivedpertechnetate using ICP-MS. Appl. Radiat. Isot. 53, 415-419 (2000)

[88] Edler, M., Jakubowski, N., Linscheid, M.: Styrene oxide DNA adducts: quantitative determination using 31P monitoring. Anal. Bioanal. Chem. 381, 205-211 (2005)

[89] Beauchemin, D., Kisilevsky, R.: Ionization within a cylindrical capacitor: electrospray without an externally applied high voltage. Anal. Chem. 70, 201-212 (1998)

[90] Abernethy, D. R., DeStefano, A. J., Cecil, T. L., Zaidi, K., Williams, R. L.: Metal impurities in food and drugs. Pharm. Res. 27, 750-755 (2010)

[91] Murty, A. S. R. K., Kulshresta, U. C., Rao, T. N., Talluri, M. V.: Determination of heavy metals in selected drug substances by inductively coupled plasma - mass spectrometry. Indian J. Chem. Technol. 12, 229-231 (2005)

[92] Lam, R., Salin, E. D.: Analysis of pharmaceutical tablets by laser ablation inductively coupled plasma atomic emission spectrometry and mass spectrometry (LA ICP AES and LA ICP MS). J. Anal. Atom. Spectrom. 19, 938-940 (2004)

[93] Ponce de Leon, C. A., Montes-Bayan, M., Caruso, J. A.: Trace element determination in vitamin E using ICP MS. Anal. Bioanal. Chem. 374, 230-234 (2002)

[94] Kishi, T., Res, J.: Analysis of trace elements in methamphetamine hydrochloride by inductively coupled plasma-mass spectrometry. J. Res. Nat. Bur. Stand. 93, 469-471 (1988)

[95] Lewen, N., Schenkenberger, M., Larkin, T., Conder, S., Brittain, H.: The determination of palladium in fosinopril sodium (monopril) by ICP MS. J. Pharm. Biomed. Anal. 13, 879-883 (1995)

[96] Camarasu, C., Madichie, C., Williams, R.: Recent progress in the determination of volatile impurities in pharmaceuticals. Trends Anal. Chem. 25, 768-777 (2006)

[97] B'Hymer, C.: Residual solvent testing: a review of gas-chromatographic and alternative techniques. Pharm. Res. 20, 337-344 (2003)

第9章
食品研究中的痕量元素分析

9.1 概述

矿物质是人体不可或缺的成分,在细胞代谢的调节中起到了至关重要的作用。矿物质通常存在于组织中,或是以离子的形式存在于体液中。它们也参与代谢过程,如电解质和荷尔蒙循环、血液生成以及神经和骨骼系统的发育[1-2]。

矿物成分的失衡会导致细胞功能紊乱和疾病症状的出现。自然环境的人为改变通常会造成食物质量下降并对人们的健康产生负面影响。这种不利的变化对某些维持人体正常机的矿物元素的存在和稳定性影响很大。饮食和食物中的化学元素会在不同程度上导致疾病的发生,不适当的饮食是循环系统疾病和癌症的起因[3],而癌症又是过早死亡的主要原因,这已经成为一个严重的社会和经济问题。

由于人类造成的环境污染[4],大多数食品中都被检测出微量的有毒金属。现在,随着人们健康意识的增强,对食物质量的要求也越来越高。因此,食品质量的监测已成为政府、组织和食品生产商的一项基本任务。为确保食品的质量,必须不断开发新技术和测试方法。通过计算元素成分与每日消耗量可以测量饮食中摄入的金属的总量,从而确定食品中有毒元素的危害性。对食品中元素含量的分析也能表明该产品是否已受到环境或工艺污染物的污染。食品中通常分析的元素有Ca、Mg、Na、K、P、Zn、Cu、Mn、Fe、Cr、Co、Ni、Se、Sn、Mo、As、Cd、Pb和Hg。

保持食品质量比保证非食用产品的质量要复杂得多。分析方法是一种控制健康所需物质必不可少的监控工具,也可用于监测对健康有害的有毒物质(包括重金属)的含量。通常检测食品中元素的技术有火焰原子吸收光谱(FAAS)法、石墨炉原子吸收光谱(GF-AAS)法、氢化物原子吸收光谱(HG-AAS)法、冷蒸气原子吸收光谱(CV-AAS)法、电感耦合等离子体原子发射光谱(ICP-AES)法、电感耦合等离子质谱(ICP-MS)法和中子活化分析(NAA)法。

9.2 分析样品的前处理

样品的正确前处理[5]是每种分析方法的关键步骤。测定样品中的目标分析物,首先要确定合适的取样量。这个量的大小取决于可用食品的数量,也取决于目标分析物的估计含量[6]。其次样品必须具有代表性,也就是说,样品经过采集、制备和储存,其化学成分与被分析的食品的化学成分尽可能接近[6]。

在对样品进行称重前,要对材料进行均匀化。如果样品含有水或其他杂质污染,它们就无法真实反映所检测物质的含量[7]。样品的均匀化过程取决于食品的物理化学性质。必须经过切割、混合、切碎或粉碎才能得到均匀的混合物,再从中提取分析样本[8]。Hoenig[9]详细介绍了用于痕量分析的样品制备方法。被均匀化的食品可能会被用于均匀化的设备所污染。Gouveia 等[10]和 Cubadda 等[11]列举了此类污染物的例子。后来有人发现了一些易污染样品的元素,如 Al、Cd、Co、Cr、Cu、Fe、Mn、Mo、Ni、Pb。

称重后,分析样品必须通过消解完全溶解。该过程旨在将样品中的有机化合物通过分解和氧化转化成无机衍生物。消解主要有干型和湿型。

干型消解技术包括焚化(灰化)、氧等离子体低温消解、有氧消解和熔化。

焚烧是食品分析中的常用方法,即将有机物在 400~600℃ 的电炉中分解。这是一个相当长的过程,产生的灰分主要由碳酸盐和氧化物组成,可以溶解在合适的酸或酸的混合物中[12-16]。

Mader 等[17]研究了用于干型消解的程序,即添加硝酸镁后在 450~500℃ 时进行灼烧。Vassileva 等[18]发现在消解混合物中加入 $MgNO_3$ 可以减少 As 和 Se 的损失。Fecher 和 Ruhnke[19]证明,在 450℃ 时没有任何添加剂情况下,干型消解过程中可能存在 Cd 和 Pb 的交叉污染。

可以使用干型消解解决的食品产品包括咖啡[13]、糖果[20-21]、调味品[22]、肉制品、水果和蔬菜、谷物制品[12,20]。

湿型消解技术可根据热能来源如进行分类,传统的热源、超声波、紫外线辐射和微波也可以根据它们的使用方式开放系统、(高压)密闭系统、固相系统、液相系统和气相系统来分类。

通常采用一种或多种强无机酸对基质进行分解,在加入或不加入其他氧化剂的情况下都可以进行。这些氧化剂通常是浓酸(HNO_3,H_2SO_4,$HClO_4$)和 H_2O_2[6,16,22]。这种技术已用于溶解鱼类和海产品[23-25]、蘑菇[26]、谷物产品、乳制品、肉制品、水果、蔬菜[27-28]和调味品[22]等食品中。

表 9-1 给出了湿消解过程中使用的消解混合物和加热系统的例子,具体取决

于目标元素和分析的材料。

表 9-1 湿型消解技术

消解技术		材料类型	需要的试剂	参考文献
开放系统	传统加热	植物材料和菌类	HNO_3,H_2SO_4,$HClO_4$,H_2O_2	[29]
	微波加热	磷虾	HNO_3,H_2O_2	[30]
	紫外辐射消解	海水	H_2O_2	[31]
密闭系统	传统加热	鱼类	HNO_3	[32]
	微波加热	海鲜	HNO_3	[25]
液相系统	传统加热	血清	HNO_3,H_2SO_4,$HClO_4$	[33]
	紫外辐射消解	水、牡蛎	HNO_3,H_2O_2	[34]
	微波加热	人血	HNO_3	[35]
	酸气相消解	生物材料	HNO_3	[36]

对流加热系统:这一过程是基于用对流加热和氧化酸分解样品。该系统的缺点是不能处理具有挥发性的样品;优点是价格便宜,而且消解发生得很快。Tinggi 等[37]比较了混合酸 HNO_3-$HClO_4$-H_2SO_4、HNO_3-H_2SO_4、HNO_3-H_2SO_4-HF 的消解过程,结果表明,使用 GF-AAS 测定各种各样的食物中的 Mn 和 Cr 时,添加 HF 并不能获得准确的测量结果。Muñoz 等[38]用这种方法来测定了圣地亚哥(智利)居民日常食物中的 As。

微波加热:溶解样品的标准技术是利用微波能量进行高压消解。该方法是在聚四氟乙烯罐内的高温下使样品与氧化酸反应。气体释放产生的压力使酸的沸点达到比开放系统中更高的温度,从而缩短了反应时间[5]。

微波消解的优点[8,39]:

(1) 良好的重复性;
(2) 操作简单;
(3) 反应时间短;
(4) 试剂消耗最少;
(5) 在广泛的浓度范围内测定大量元素;
(6) 样品损耗小;
(7) 样本污染最小;
(8) 对有机物和无机物都能有效分解。

微波消解的缺点[8,39]：

(1) 仪器价格昂贵；

(2) 只能使用少量的样品。

目前，微波消解是用于消解食品样品的最常用的技术之一，但想要充分取代传统技术还远远不够。Lamble 和 Hill[40]述评了这种技术所使用的程序。孙等[41]采用三种湿法消解程序(HNO_3、HNO_3-H_2O_2 和 HNO_3-H_2O_2-HF 的混合物)制备了基于食品的标准物质样品，以确定 13 种元素(Al、B、Ca、Cu、Fe、K、Mg、Mn、Na、P、S、Sr、Zn)的含量。用 HNO_3-H_2O_2-HF 进行的消解对除 3 种材料外，所有材料都获得了令人满意的结果，其他方法仅对 Al 的结果是准确的。Gawalko 等[42]在开放和封闭系统中应用微波加热来溶解谷物产品样品，获得的 Cd、Cu、Pb 和 Se 测定结果是令人满意的。Dolan 和 Capar[43]根据目标食品的能量值，根据要分析的食品样本的质量 0.41~9.5g，利用微波加热开发了湿消解程序。Krzysik 等[44]使用微波消解来确定快餐食品和方便食品中的 Cr 含量，而 Kwoczek 等[25]使用相同的技术测定了海鲜样品中的必需元素和有毒金属。

9.3 食品中元素含量的测定

火焰原子吸收光谱法(FAAS)是最古老的分析技术之一，并沿用于食品分析中。此方法分析通常在空气-乙炔或氧化亚氮-乙炔火焰中进行。该技术是测量高温((1000~4000)K)下产生的自由原子对电磁辐射的吸收[6]。

FAAS 可用于测定食品中许多常见元素和有毒金属。它速度快，具有令人满意的选择性、准确性和灵敏度水平。根据 AOAC 的规定，基于该技术的标准方法有测定食品中的 Cd[45]和 Pb[46]，食品中的 Cu、Fe 和 Zn[47-48]，罐装产品中的 Sn[49]，食品中的 Zn[50]，水果和蔬菜中的 Zn[51]、Cu[52]和 Fe[53]，牛奶中的 Zn[54]，奶酪中的钙和镁[55]，以及专业食品中的 Ca、Cu、Fe、Mg、Mg、K、Na 和 Zn[56]。表 9-2 列出了 FAAS 在各种食物中元素组分定量分析中的应用。Grembecka 等[13]使用 FAAS 来测定了咖啡和糖果产品中某些基本元素和有毒金属的含量。Milacic 和 Kralj[68]也用它来测定了斯洛文尼亚食品中 Zn、Cu、Cd、Pd、Ni 和 Cr 的含量。FAAS 的另一个应用例子是 Khajeh[69]在样品提取后对 Zn 和 Cd 的分析。Santelli 等[64]评估了食品样品消解的有效程度。Miller-Ihli[70]采用 FAAS 评估了食品标注成分中 Ca、Cu、Cr、Fe、Mg 和 Zn 等元素是否符合管理的规定。在 FAAS 和用于计算数据的统计方法的帮助下，实现了有机咖啡和传统咖啡的区分[71]。

表 9-2　火焰原子吸收在食物中元素组分定量分析中的应用

食物	元素	参考文献
膳食补充剂	Cd,Fe,Pb,Zn	[57]
总膳食	Ca,Cu,Mg,Na,P,Zn	[58]
咖啡	Cd,Pb,Co,Cr,Cu,Mn,Fe,Ni,Zn,Ca,Mg,Na,K	[13]
海鲜、肉、奶酪、谷物制品、豆类、干果	Ni	[59]
蔬菜	K,Na,Ca,Mg,Fe,Mn,Cu,Zn	[60]
蔬菜	Cd,Cu,Ni,Pb,Zn	[61]
罐装食物	Cu,Zn,Mn,Fe	[62]
婴儿食物	Cu,Zn,Mn,Fe	[63]
谷物制品	Fe,Zn,Mn	[64]
水果、蔬菜、豆类	Ca,Cu,K,Mg,Mn,Zn	[65]
红茶	Cu,Ni,Cr	[15]
啤酒	Fe,Mn,Zn,Cu,Mg,Ca,Al	[66]
生加工食物	Zn,Fe,Ca	[67]
海鲜	Co,Cr,Cu,Mn,Fe,Ni,Zn,Ca,Mg,Na,K	[25]

石墨炉原子吸收光谱法(GF-AAS)是基于自由原子能够吸收由辐射源发射的特定元素的共振线[6,72]的原理来实现特定元素的测定。

GF-AAS 具有高的灵敏度和选择性,样品无须事先预处理即可进行分析。可以分析微量的样品,检测限为 μg/L。使用这种技术,食品中的常规元素和有毒元素都可以检测到,而且检测限比 FAAS 更低。它的缺点是重复性差,仅测定单个元素,基体效应会明显影响分析结果的准确性[72]。

采用该途径的标准方法包括食品中 Cd 和 Pb 的测定[47-48],水果和蔬菜中的 Cd[73] 和 Pb 的测定[74],罐装炼乳中的 Sn 的测定[75],油脂中的 Cd 的测定[76],动植物脂肪中的 Pb[77]、P[78]、Cu、Fe 和 Ni 的测定[79],淀粉产品中的 Cd[80] 和 Pb 的测定[81],糖和糖浆中 Pb[82] 的测定。

表 9-3 列出了 GF-AAS 在食品分析中的应用。Julshamn 等[95]开发了一种在封闭体系中微波消解的方法,用于 GF-AAS 法测定海产品中 As 的含量。他们发现这种方法的灵敏度高,可以在 2.5mg/kg(干质量)水平下进行测定。相反地,Fedorov 等[96]通过比较 GF-AAS 与 HG-AAS 在测定 As 方面的差异,发现后者的准确度和精密度更高。Veillon 和 Patterson[97]认为使用 GF-AAS 确定 Cr 时会比较困难。

表9-3 GFAAS在食物分析中的应用

食物	元素	参考文献
总膳食	Cd,Pb,Ni	[83]
补充膳食、果汁	Al	[84]
食物、方便食品	Cd,Cr,Ni,Pb	[85]
肉、牛奶、蔬菜、谷物	As,Cd,Pb	[86]
谷类、蔬菜、水果、肉、鱼、蛋、饮料	Cd,Pb,Ni,Se	[87]
肉、海鲜、乳制品和谷物制品、蔬菜、水果、油脂、坚果	Cr	[88]
海鲜、蔬菜、橄榄油、奶制品、能力饮料、果汁、软饮料	Al	[89-90]
豆类、坚果	Cu,Cr,Fe,Zn,Al,Ni,Pb,Cd	[91]
油	Al,Cu,Co,Cr,K,Ni,Mn,Pb	[92]
罐装食品	Se,Al,Cr,Ni,Co	[62]
海鲜、谷物制品和奶制品、蔬菜、橄榄油	Cr	[93]
调味品、坚果、谷类制品、咖啡、婴儿食品、蔬菜和干果	Pb,Cd	[94]

氢化物发生原子吸收光谱法(HG-AAS)是利用As、Bi、Ge、Pb、Sb、Se、Sn、Te等元素与氢形成挥发性化合物的方法。

HG-AAS通常用于测定食品中的As和Se[98-101]。用于分析的样品制备通常涉及湿消解过程,即在$HNO_3-HClO_4-H_2SO_4$的混合物中,或将其与硝酸镁和氧化镁一起焚烧。Kabengera等[102]开发了一种相当简单的HG-AAS技术,可用于测定$HNO_3-H_2O_2$消解混合物中的As。为了使用HG-AAS测定肉制品中的硒,Tinggi[103]建议先在浓$HNO_3-HClO_4-H_2SO_4$混合物中进行消解。

HG-AAS用于测定大部分食物[83,104-196]、肉类[107]、精选食物[108]、海鲜[109]、豆浆和婴儿食品[110]、谷类[111]、咖啡、鱼和牡蛎[112]中As和Se的含量。冷蒸气原子吸收光谱法(CV-AAS)是测定极低汞含量的最佳、最灵敏的方法;它利用了该元素的原子可以在室温下以气态形式存在的特点。

基于CV-AAS的标准方法有测定果蔬[113]和鱼类[114]中Hg等元素。CV-AAS也被用于检测日常食物[15,83,106]、食品[86-87]、蘑菇[116]和海鲜[115,117-121]中的Hg含量。

Dabeka等[122]开发了一种利用低温消解法测定食品中低含量汞的方法。Larsen等[87]采用高压焚烧法和CV-AAS法来测定各种食品中的Hg。Zenebon等[123]用H_2O_2和H_2SO_4混合物在水浴加热至80℃的开放容器中消解食品样品来进行Hg的测定。这些方法的优点是简单、快速(消解在一夜之间完成)、灵敏和低成本[8]。

电感耦合等离子体质谱(ICP-MS)分析测量精度高,检测限低[124]。ICP-MS的优点如下[16,124-125]:

(1) 可同时分析一个或多个元素以及高激发电位元素,如 W、Cl、Br、I、S、U;
(2) 测量线性范围宽,对应于 4~5 个数量级的浓度差异;
(3) 可以确定同一样品中的主成分和痕量成分;
(4) 无污染,因为电极不用于激发;
(5) 检测下限低,0.01~10μg/L;
(6) 出色的灵敏度和选择性;
(7) 测试快速、直接;
(8) 基质效应的干扰很小。

ICP-MS[16,124]的缺点如下:

(1) 该方法不适用于大量稀释的情况,因为较低浓度水平的测定不够准确;
(2) 在循环测量期间必须移除干扰物;
(3) 必须减少或消除基体效应;
(4) 必须使用纯度很高的酸;
(5) 分析成本很高;
(6) 氩气和其他试剂必须具有非常高的纯度;
(7) 必须溶解具有复杂基体的样品;
(8) 样品必须稀释。

表 9-4 列出了 ICP-MS 在食物分析中的应用。Zbinden 和 Andrey[126]在高压灰化后,采用 ICP-MS 建立了测定食品中 Al、As、Cd、Hg、Pb 和 Se 的常规方法。他们通过在样品溶液中加入异丙醇来处理干扰物。Bhandari 和 Amarasiriwardena[127]描述了在微波系统中湿消解枫糖浆样品后使用 ICP-MS 测定 Pb 和其他 7 种元素的方法。ICP-MS 也被用于检测婴儿食品[128]和饮食产品[129]中的 I。Whyte 等[131]利用 ICP-MS 评估贻贝的营养质量和 Hg、Sn、As、Cd 和 Pb 等元素对新西兰人的毒性作用。鉴于美国人不断接触铅的危险,Bagga 和 Jarret[132]试图确定美国温室蔬菜中这种元素的含量。他们发现,大量的 Pb 积累在白菜、莴苣和芜菁中,这些蔬菜是从接触了中等污染环境中的土地上收获的。Chung[133]等研究了中国香港学生体内可测到的铅、锑和汞等元素的含量。Jorhem 等[134]对瑞典的一系列食品进行了 As(ICP-MS)、Pb 及 Cd (GF-AAS)含量测定。另外,Leblond 等[135]测试了海洋来源的膳食补充剂中 Cd、Co、Cr、Cu、Hg、Ni、Pb 和 Zn 的含量。除 Hg(汞分析仪)、Cr(ICP-AES)外,其余元素均采用了 ICP-MS 法进行测定。

表9-4 ICP-MS在食物分析中的应用

食物	元素	参考文献
海鲜	CH_3Hg,Zn,As,Se,Cd,Hg,Pb	[136]
总膳食	As,Cd,Pb Br,I	[137-138]
补充膳食	Al,As,Ca,Cd,Co,Cr,Cu,Fe,Hg,K,Li,Mg,Mn,Mo,Na,Ni,Pb,Sb,Se,Sn,Zn	[139]
洋葱、豌豆	Ag,Al,Au,B,Be,Bi,Ca,Cd,Ce,Co,Cr,Cs,Cu,Dy,Er,Eu,Fe,Ga, Gd,Ge,Hf,Ho,In,Ir,K,La,Li,Lu, Mn,Mo,Na,Nb,Nd,P,Pb,Pr,Pt, Rb,Re,Ru,S,Sb,Sc,Se,Si,Sm,Sn, Sr,Tb,Te,Th,Ti,Tl,Tm,U,V,W, Y,Yb,Zn,Zr	[140]
橘汁	Ba,Co,Cu,Li,Lu,Mn,Mo,Ni,Rb,Sn,V,Zn	[141]
野生食用菌	Ag,Rb,Cd,Hg,Pb,Cs,Sr,Tl,In,Bi,Th,U,Ce,Pr,Nd,Sm,Eu,Gd, Tb,Dy,Ho,Er,Tm,Yb,La,Lu,Zn,Fe,Cu,Mn,Ba	[142]
大米	Al,Ca,Mg,P,Na,K,Mn,Fe,Co,Ni,Cu,Zn, Sr,Mo,Cd,Pb,B,Se,Rb,Gd,Ho,W	[143]
黑橄榄和绿橄榄	Mg,Cr,Co,Ni,Fe,Cu,Zn,Sn,Cd,Pb	[144]
罐装蔬菜	Mg,Al,Cr,Mn,Fe,Ni,Zn,Cu,Mo,Cd,Sn,Sb,Ba,Pb,Bi	[145]
白酒	Li,Be,Al,V,Cr,Mn,Fe,Co,Ni,Cu,Zn,Ga,As,Se, Rb,Sr,Ag,Cd,In,Cs,Ba,Hg,Tl,Pb,Bi,U	[146]

其他分析技术,如电感耦合等离子体发射光谱(ICP-OES)或ICP-AES可用于同时测定多种基质中70种元素的平均值。这种技术的优点如下[16]:

(1) 可以分析一种或多种元素,对于具有高激发电位的元素(如W、Cl、Br、I、S、U)适用;

(2) 测量范围宽,涵盖4~5个数量级的浓度;

(3) 可以在同一样品中确定主要成分和痕量成分;

(4) 无污染(电极不用于激发);

(5) 检测下限低,0.01~10μg/L;

(6) 使用多色仪可在几分钟内测定约60种元素;

(7) 没有自吸现象;

(8) 不存在氧气干扰。

ICP-OES的缺点如下:

(1) 分析成本高;

(2) 需要使用氩气和其他纯度很高的试剂;

(3) 特定元素的检测限差异有几个数量级,这也阻碍了多元素分析;
(4) 富含发射线元素的光谱干扰,如 U、W、Co、Fe;
(5) 日常操作和设备处理对人员技术要求高;
(6) 相似波长的元素之间发生干扰;
(7) 需要监控环境温度和湿度;
(8) 需要非常好的单色器;
(9) 难以确定第Ⅰ族金属。

由于多元素分析的可能性,ICP-AES 被广泛用于食品检测,例如,婴幼儿食品中 Ca、Cu、Fe、Mg、Mn、P、K、Na、Zn[147];植物物质中的 B、Ca、Cu、K、Mg、Mn、P、Zn[148];动植物脂肪中的 P[149]。Miller-Ihli[150]验证了 ICP-AES 在干型和湿型消解后的食品样品测试中,存在 Ca、Co、Cu、Cr、Fe、Mg、Mn、Ni、P、V 和 Zn 元素情况下的应用。Dolan 和 Capar[43]使用 ICP-AES 测定了食品中的 20 种元素。Lomer 等[151]开发了一种利用 ICP-AES 检测食品中 Ti 的方法,目的是测定添加二氧化钛的含量,该方法需要使用 H_2SO_4 加热至 250℃来消解样品。Rao 等[152]利用 ICP-AES 测定食用紫菜湿消解后 Na、K、Ca、Mg、B、Pb、Cr、Co、Fe、Zn、Mn、Hg、Cu、As、Ni、Cd 和 Mo 的含量。这些分析结果证实了膳食补充剂的效用是因为其有益的矿物质成分。应用 ICP-AES、四极杆(Q)ICP-MS 和高分辨率(HR)ICP-MS 结合湿微波消解,可以测定水稻样品中 As、Cd、Co、Cr、Cu、Fe、Mn、Pb、V 和 Zn 的污染水平[153]。表 9-5 列出了 ICP-AES 在食物分析中的应用。

表 9-5　ICP-AES 在食物分析中的应用

食物	元素	参考文献
总膳食	Ca,Cu,Fe,Mg,Mn,K,P,Na,Zn	[83]
野生蘑菇	Al,Ca,K,Mg,Na,P,Si	[142]
橘汁	Al,B,Ca,Fe,Mg,P,K,Si,Na,Sr,Ti	[141]
茶	Al,Ba,Ca,Cu,Fe,K,Mg,Mn,Na,Sr,Zn	[155]
速溶咖啡	Na,K,Mg,Al,P,S,Ca,Mn,Fe,Ni,Cu,Zn,Cd,Sb,Pb,Cr,Sn	[156]
油	Ca,Fe,Mg,Na,Zn	[92]
调味品	Al,B,Ba,Bi,Ca,Cd,Cr,Cu,Fe,K,Li,Mg,Mn,N,Na,P,Pb,S,Se,Se,V,Zn	[157]
即食汤和调味混合物	Na,K,Ca,Mg,P,Cd,Cr,Cu,Fe,Mn,Ni,Pb,Zn	[158]
茴香饮料	Zn,B,Fe,Mg,Ca,Na,Si	[159]
蜂蜜、甜食	Pb,Cd,Cu,Cr,Co,Ni,Mn,Zn	[160]

中子活化分析这种技术可对元素进行定量和定性鉴定。它的原理是稳定的原子核被中子轰击后变成放射性原子核,然后测量放射性原子核发射的辐射[154]。NAA 的优点如下[16,154]:

(1) 不灭性;
(2) 出色的灵敏度;
(3) 可以同时确定 50~65 个元素;
(4) 检测限低,达 μg/kg 级;
(5) 样品需要前处理。
NAA 的缺点如下[16,153]。
(1) 操作复杂;
(2) 耗时长;
(3) 确定的元素总体水平无法区分其化学形式或物理状态;
(4) 一些元素(如 Pb)无法确定;
(5) 需要引入核反应;
(6) 成本高。

Kucera 等[161]对食品碘的三种测定方法进行了比较分析。基于食品标准物质对其准确度和精密度进行检验。检测下限是 1μg/kg。在葡萄牙,人们用仪器 NAA(INAA)和重复样品 INAA(RSINAA)测试了各种食品的总硒含量[162]。INAA 和放射化学 NAA(RNAA)也被用于评估希腊瓶装水的质量[163]。在水浴蒸发后的干残余物中测定了 U、Ba、La、Sb、Ca、Cr、Zn、K、As、Br、Se、Co 等元素[163]。Islam 等[164]使用 NAA 测定水、土壤和植物物质中 As 的含量。表 9-6 列出了 NAA 在食品分析中的应用。

表 9-6　NAA 在食物元素分析中的应用

食物	元　素	参考文献
总膳食	Ca,Co,Cr,Cs,Fe,I,K,Se,Sr,Th,Zn Al,As,Br,Ca,Cl,Co,Cr,Cs,Fe,I,K,Mg, Mn,Na,Rb,Sc,Se,Sm,Sr,Th,U,Zn	[165-166]
补充膳食	Br,Ca,Cl,Co,Cr,Cs,K,Fe,Mn,Mg,Mo,Na,Rb,Se,Zn	[167]
食物	Ca,Cl,Co,Cr,Cu,Fe,K,Mg,Mn,Mo,Na,Se,Sn,Zn	[168]
传统墨西哥食物	Fe,Al	[169]
谷类制品	Hg,Se	[170]
谷类制品、蔬菜、调味品	Na,K,P,Cl,Br,Co,Cr,Cs,Cu,Fe,Hg,Mn,Mo,Rb,Sb,Sc,Se,Sr,Th,Zn	[171]

9.4　元素的形态分析

形态是指样品中元素的不同物理和化学形式。形态分析涉及识别元素的物理和化学形式以及它们在样品中的测定。

全世界现行的大多数法律规定都是基于食品中元素的总体水平,只有少数几个元素与特定形态有关。环境保护署(EPA)、世界卫生组织(WHO)、欧洲环境委员会(EEC)和美国食品药品监督管理局(USFDA)等组织已经认识到形态分析的重要性,并建议限制各种形态的元素(如 As、Hg 和 Sn)的污染[8,172]。

发生在不同氧化态中元素(如 Cr、Fe、As 和 Se)形态的测定,以及各种形式的有机金属化合物(如三丁基锡、甲基汞、砷甜菜碱),对分析人员来说是一个严峻的挑战,这已经成为全球性的研究课题。由于元素形态的形式通常是不稳定的,并且在各种基质中含量很低(如 μg/kg),因此需要灵敏度高和选择性好的技术来确定它们[173]。形态分析中使用的标准分析技术是色谱方法(GC、HPLC)配合各种检测器,包括 AAS、火焰光度检测器(FPD)、ICP-AES、ICP-MS 和原子荧光光谱(AFS)。ICP-MS 结合不同的分离技术,如 LC、GC、超临界流体色谱(SFC)和毛细管电泳(CE),也常用于形态分析[136,174-178]。Vieira 等[179]论述了形态分析中除色谱方法外的其他技术。他们探索了一系列提取程序,包括单滴微萃取(SDME)以及衍生化技术。饮用水和食品中元素形态分析技术应具有合适的选择性和灵敏度[180-185]。

在环境样品和食品测定中,常常应用到很多先进的分析技术,尤其是基本元素与有毒元素,如 Fe、Zn、Cu、Co、V、Pt、Ag、Ca[186-192], As[185,193-202], Sb[203-205], Mn[206], Cr[185,187,207-211], Hg[120,185,212], Se[185,213-214], Cd[187-188,215-217], Pb[187,218-225], Ni[187-188,226], Tl[227], Sn[172,228-229], Te[230], S、N、P、Cl、I[231-232]和 Al[233]形态的定量分析中。

形态分析的步骤是样品采集,样品稳定和储存,杂质的去除,提取/消解,衍生化,分离,检测,校准,结果计算和结果评估[212,234-235]。

Huo 等描述并讨论了人工合成同位素稀释质谱(SID-MS)在形态分析中的应用[236]。溶液的同位素分析可用于评估食品和药物中各种元素形态分析的应用情况[237]。

从分析的角度来看,有证标准物质在食品监测中发挥着至关重要的作用,因为它们可以评估质量和测量精度。Quevauviller[182]和 Emons[238]证明了应用这些标准物质监控元素形态分析测量质量方面的必要性。

1. 砷

砷是一种可以有机和无机形式存在的类金属,它们的毒性明显不同。主要存在于海产品中的砷的有机化合物(如砷胆碱和砷甜菜碱)被认为是无毒的[172,189,239],但这种元素的无机形式具有强毒性并且可能致癌[240-246]。所以砷在海洋生物中经历的新陈代谢变化仍然是许多研究的主题。例如,鱼中砷含量高,但大多数是有机形式,因此食用大量鱼类不会有中毒的风险。由于没有关于砷糖或该元素的其他有机化合物毒性的数据,无法评估其潜在风险。因此,如果要对鱼及其产品的质量进行适当评估,简单、有效的分析方法以及基于形态分析的相关法规是必不可少的[247]。

相关文献报道了许多用于多种基体(包括食物)中的 As 的形态分析技术。常用的方法是 GC 或 LC 结合光谱或电化学检测[185,248],标准检测器为 AAS、ICP-AES 和 ICP-MS[192,246,249]。As 的形态分析中最受欢迎的技术之一是 HG-AAS 结合 LC[172]。牡蛎和海草[250-252]、啤酒[253]、水和其他食品[185]、大米[178,254-256]、鱼油[257,258]和婴儿食品[259]中的 AS 元素的形态已被确定。

As(Ⅲ)、As(Ⅴ)、一甲基胂酸(MMA)、二甲基胂酸(DMA)、砷甜菜碱(AB)、三甲基胂氧化物(TMAO)、砷胆碱(AC)和四甲基胂离子(TMI)等的形态已经使用阳离子和阴离子交换色谱与 ICP-MS 结合进行了测定。Wrobel 等[248]使用 HPLC + ICP-MS 测定了鱼组织中各种形态的砷。这种快速和灵敏的技术用于贻贝可食用组织中 As 的形态分析,并用于监控其质量[260]。利用阴离子交换梯度层析结合 ICP-MS 的技术检测了龙虾组织中 As 的形态[261]。

离子交换色谱与 HG-AAS 结合用于测定市售海产品中 As 的形态[172]。Chatterjee[250]使用微波湿法提取法和 HPLC + ICP MS 测定了牡蛎可食用组织中各种形态的砷。使用 ICP-MS 和电喷雾电离串联质谱(ESI MS/MS)[251-252]以及多维液相色谱法测定了海藻和牡蛎中各种形态 As 的比例。Coelho 等[253]使用 HPLC、HG-AFS 直接测定啤酒中的 As(Ⅲ)、DMA、MMA 和 As(Ⅴ);使用 HPLC 微波(MW)HG-AAS 测定鱼中的 As(Ⅲ)、DMA、MMA、As(Ⅴ)、AB 和 AC[262]。龚等[263]认为在生物样品(包括食品)中对于 As 的形态分析最有用的技术是色谱分离技术,结合 ICP-MS、HG-AAS 和 ES-MS 来检测。

2. 汞

大多数的汞是通过人类活动进入环境的。随着对该元素的痕量分析技术的研究,可以确定汞的来源,例如,工业过程和农业活动,以及天然火山和地下水排放。汞的形态不同其毒性也不同,这取决于人体的哪一部分接触它们。通过消化道吸收的有机汞化合物和通过呼吸道进入人体的金属汞是对人类健康的最大威胁[264-266]。与金属汞接触相当罕见,但与有机汞化合物的接触常见。接触有机汞的主要来源是海鲜和鱼类[265]。

许多人类悲剧是食物中的有机汞化合物引起的。20 世纪,汞化合物通常是毒性最强的一种,被用作种子的杀菌剂。由于一个令人震惊的错误,1971—1972 年在伊拉克发生了大规模的中毒事件,当时被 Hg 化合物污染的种子被出售供食用。此后,Hg 化合物在农业实践中的应用逐渐减少,但仍有相当数量的 Hg 元素留在土壤中。

1953—1970 年,涉及汞的最大悲剧发生在日本水俣湾。当时含汞化合物的工业废水未经处理就排入海湾,汞在鱼体内富集,然后被该海岸的居民捕获和食用。

这些人很快就出现了中毒的症状,最严重的出现了瘫痪、耳聋、失明甚至死亡[265,267]。

美国食品药品监督管理局规定,海产品受甲基汞污染的最高限量为 1mg/kg[185]。因此,在进入消费者市场之前必须在实验室中对食品的质量进行常规形态分析。通常用于监测食品质量的技术是 LC 结合 CV-AAS 或 CV-ICP-MS[172]。从 Van Dael 的综述[185]中得出的推论是:一些先进的技术,如 CV-AAS、CV-AFS、FT-IR 和微波诱导等离子体 AES(MIP-AES),以及不太复杂的方法,可以提供相似的选择性和灵敏度来确定食品中汞的形态,以满足食品质量监测要求。

Taylor 等[136]利用湿微波消解作用结合 GC-ICP-MS,对来自海洋的生物样品中 Hg 进行了形态分析,该方法特别适用于少量的样品。Magalhaes 等[268]使用 CV-AAS 法检测到鱼类中甲基汞和总汞的生物累积的差异。Vereda Alonso 等[120]通过先使用选择性固相萃取,然后流动注射(FI)CV-AAS 和 CV-GF-AAS 确定了海产品中汞的形态。该方法的优点是不需要色谱方法,却保留了高选择性和灵敏度。Carro 和 Mejuto[269]论述了食品样品中 Hg 有机衍生物的形态鉴定技术。用于检测甲基汞的方法有 GC-ICP-MS[270]、GC 电捕获检测(ECD)[271]、HSA-GC-MIP-AES[272]、顶空分析(HSA)GC-AFS[273] 和 GC-MIP-AES[274]。Pereiro 和 Diaz[222]论述了 GC-MIP 原子发射检测(AED)法在 Hg、Sn 和 Pb 定量形态分析中的应用。多毛细管气相色谱联用 ICP-MS 也已用于形态分析[275]。Grinberg 等[276]使用固相微萃取 GC 炉雾化等离子体发射光谱(SPME-GC-FAPES)法检测鱼中的甲基汞和无机汞,分别达到 $1.5\mu g/kg$ 和 $0.3\mu g/kg$ 的检测限。电热蒸发(ETV)ICP-MS 能够实现 Hg 的定量分析,无须事先进行样品预处理或将 CH_3Hg 和 $Hg(Ⅱ)$ 转化为金属汞[277]。这种转化在采用 GC-ICP-MS 的测试中被检测到。适用于鱼类中汞的形态分析的技术是 ETV-ICP-MS,它的 CH_3Hg 和 $Hg(Ⅱ)$ 的检测限分别为 $2\mu g/kg$ 和 $6\mu g/kg$[8]。Liang 等[278]建议使用 HPLC CV-AFS 测定 4 种形态的汞,即氯化汞、氯化甲基汞、氯化乙基汞和氯化苯基汞。该方法成功用于测定海产品中甲基氯化汞的含量。

3. 硒

硒的毒性与其形态密切相关。因此,研究者已研究出可以在生物和食品样品中测定 Se 化合物的分析程序。该元素安全的食用范围很窄:Se 过量和 Se 缺乏均会导致严重疾病,其生物利用度与其化学形态有关。已经确定,大量的 Se 是从有机化合物中积累的。鉴于此,确定食物中各种形态的 Se 是非常重要的[8]。

Van Dael[185]论述了 Se 的形态及其在大豆、小麦、浓缩大蒜、洋葱、西兰花、富硒酵母、熟鳕鱼、乳品(人、牛、羊)等食品中的分布。许多用于测定 Se 水平的方法都基于 GC 与 AAS、AFS、AES 和 ICP 的结合。Moreno 等[279]利用渗透蒸发(aporation,PV) GC-AFS 分析大蒜中挥发性硒化合物,实现了 0.45~1mg/kg 的定量范围。通过使用酶消化蛋白酶(ⅩⅣ)、脂肪酶(ⅤⅡ)、蛋白酶(ⅤⅧ)和 HPLC 微波辅助消化(MAD)HG-AFS[197]来进行 Se 的 5 种形态的提取。这些酶可以从样品中分离出各种形态的硒,但必须事先纯化样品,以便使用 HPLC-AFS 进行分析。然而,通过这种方法获得的这些不同形态的产率不尽如人意,这可能是由于它们从贻贝组织中没有完全提取[197]。酶提取法也应用于煮熟的鳕鱼片,利用阴离子交换色谱法结合 ICP-MS 测定其 Se 的种类及含量[172]。Kannamkumarath 等[280]在使用蛋白酶 K 进行预先酶促提取后,使用 HPLC-ICP-MS 测定了各种坚果中 Se 的分布和形态。根据 Uden 的论述[281],利用反相离子对色谱和离子色谱与 ICP-MS 结合,实现了富硒酵母、大蒜、蘑菇和海藻提取物中硒形态的分离和定量。反相离子对色谱与 ICP-MS 结合也用于生物样品(包括酵母)中 SeCys、Se-M 和 Me_3Se^+ 的形态分析[172]。另外,还使用反相手性色谱与 MW HG-ICP-MS[172]分析了酵母。为了确定 Se 的无机形态(如 Se(Ⅵ)和 Se(Ⅳ)),开发了使用流动注射(FI)HG-AAS 的方法[282]。Dernovics 等[283]提出了一种使用酶促法从蘑菇中提取硒的改进方法。根据 Stefánek 等的说法[284],可以使用 HPLC 结合液压高压雾化(HHPN)分析经酶处理的蘑菇样品 Se 的形态。硒化合物的分析已经在各种物质中开展,例如煮熟的鳕鱼片[172],各种坚果[280],酵母和坚果[285],大蒜、酵母、蘑菇和海藻提取物[281],选定的食品和水[286],婴儿食品和膳食补充剂[287,288],调味品[22]。Guerin 等[289]论述了用于分析食物基质,例如、鱼、贻贝、蔬菜、蘑菇和矿泉水中的 Se 和 As 的常用形态分析技术。最近,使用毛细管电泳对膳食补充剂和人乳中硒的形态分析取得了实质性进展[290]。Pedrero 和 Madrid 广泛论述了 Se 的作用以及在 Se 形态分析中用到的技术,如 ICP-MS 与 ESI-MS 联用技术,ESI-MS/MS 和基质辅助激光解吸/电离飞行时间(MALDI-TOF)联用技术[175]。Gosetti 等[287]使用 HPLC-MS/MS 测定了膳食补充剂中的 Se。Infante 等[291]论述了质谱分析应用于食品中硒形态的分析方法。

4. 锡

锡是广为人知的用于涂覆食品罐头的金属,钢罐上薄薄的一层锡就可防止钢材腐蚀,从而防止食品变质。锡无毒,不腐蚀,易在其他金属上形成涂层,是一种理想的材料。除了少数酸性值的罐装产品[145,292-295],它对人类没有任何危害。锡的形态在有机化合物中起着重要作用。

三丁基锡(TBT)是一种防污化合物,可用于防止微生物和真菌的生长,以及防止藤壶和其他生物附着在船体上。TBT 对水生生物包括鱼类和海洋哺乳动物有毒,并可在它们的体内迅速富集[296]。

由于 TBT 毒性大,人们对其进行了深入研究[222,241,297-300]。根据 Belfroid 等[301]的研究,在 22 个国家中至少有 9 个国家和地区(加拿大、法国、意大利、日本、韩国、波兰、中国、泰国和美国)的海产品中 TBT 残留量超标。采用气相色谱

(GC-FPD)法测定了海洋软体动物[302-303]、鱼类[302,304]和食鱼鸟类[304]中TBT、二丁基锡(DBT)和单丁基锡(MBT)等有机锡衍生物的含量。采用气相色谱-质谱联用(SIM)技术测定了贻贝和水中的有机锡化合物含量[181]。除了MBT[305]外，还有利用离子阱检测器(ITD)结合GC测定锡的有机衍生物含量。采用高效液相色谱(HPLC)法测定了海洋软体动物体内有机锡化合物的含量[172]。Marcic 等[306]采用高压萃取,气相色谱脉冲火焰光检测器(PFPD)测定蔬菜中有机锡化合物的含量。Dong 等[307]论述了在制备生物样品进行特定同位素稀释(SSID)、分析时苯基蛋白的转化过程。Üveges 等[308]使用GC-ICP-MS测定了贻贝和沉积物中丁基锡化合物的含量。HS搅拌棒吸附萃取(SBSE)和热消解(TD)GC-MS是水样中丁基锡和甲基汞形态分析的首选技术[309]。Pereiro 和 Diaz[222]利用GC-MIP-AED进行了Hg、Sn和Pb化合物的形态分析。

5. 其他元素

元素的物理化学形态对其生物利用度和在人体中的作用有重要影响。例如，铬有生物活性[Cr(Ⅱ)]和毒性[Cr(Ⅵ)]两种形态,因此从食品安全的角度来看，了解这两种形态的Cr至关重要[208]。Kotas和Stasicka[209]详细论述了Cr的形态分析方法,以及样品的制备和储存方法。采用的技术包括流动注射分析(FIA)、HP-LC(包括离子色谱)、离子对色谱和离子对反相色谱,并与UV-VIS、GF-AAS、ECD、X射线荧光(XRF)和ICP-AES等检测器相结合。Ambushe 等[174]应用动态反应细胞(DRC)ICP-MS评估了牛奶冻干样品中Cr的形态及含量。

铅的毒性因其形态而异,Pb的有机化合物比无机化合物毒性大得多[310]。Barałkiewicz等[311]采用HP-LC-ICP-MS和HP-LC-ESI-MS确定了豌豆中Cd和Pb的形态及含量。

利用阴离子交换色谱结合ICP-MS对鱼提取物中的As、Se、Sb、Te化合物同时进行形态分析[230]。尺寸排阻色谱(SEC)与特定检测器相结合经常用于分析富含蛋白质的材料(如肉和植物组织的提取物)中的痕量元素种类。例如,用ICP-MS联用SEC对豆科植物样品中的Cu和Zn进行形态分析[191]。同样的技术可用来分析鱼类中Cu、Cd、Zn、Se、As和Ca的形态[220],以及贻贝中Fe、Zn、Cu、Ag、Cd、Sn和Pb的形态[189]。SEC高效液相色谱法与GF-AAS联用对测定婴儿食品中Fe种类非常有效[312]。使用凝胶渗透色谱(GPC)GF-AAS,在两种受Cd元素污染的蔬菜中测定了其形态[216]。

9.5 评估痕量分析结果的化学计量技术

化学计量是科学技术的一个分支,使用统计学和数学从多维测量数据中提取有用信息。它应用于许多科学学科,包括食物分析[313-315]。多维分析常用技术包括主成分分析(PCA)、因子分析(FA)、线性判别分析(LDA)、典型判别函数分析(DA)、聚类分析(CA)和人工神经网络(ANN)。

多维技术经常用于与动植物食品质量控制中化学元素水平相关的测量数据的分析评估。已得到肉类和肉类产品[316-318]、鱼类[319-321]、海鲜[25,322-328]、牛奶和乳制品[329-332]以及蜂蜜[333-339]动物产品的化学计量解释。多维技术对大米[143]、谷物[340]、蔬菜[140,341-346]、水果和蜜饯[347]、茶[155,348-350]、咖啡[13,155,351-352]、蘑菇[26]、果汁[141]、糖果[21,353]、坚果[354]、葡萄酒[355-358]、啤酒[66,359]和其他酒精饮料[159,360-361]植物产品也有类似的解释。

掌握了动、植物食品中矿物成分的知识,人们可以确定它们的地理出处、生物属性和种类,以及加工程度方面的差异。

参考文献

[1] Mineral Components in Foods, pp. 363–388. CRC Press/Taylor Francis Group, London/New York (2007)

[2] Ziemlański, Ś.: Normy Żywienia Człowieka. Fizjologiczne Podstawy, pp. 1–532. WydawnictwoLekarskie PZWL, Warszawa (2001)

[3] Baer-Dubkowska, W.: Chemoprewencyjne i kancerogenne składnikiżywnośći. NowinyLekarskie 74, 505–509 (2005)

[4] Nasreddine, L., Parent-Massin, D.: Food contamination by metals and pesticides in the European Union. Should we worry? Toxicol. Lett. 127, 29–41 (2002)

[5] Matusiewicz, H.: Wet digestion methods. In: NamieŚnik, J., Chrzanowski, W., Żmijewska, P. (eds.) New Horizons and Challenges in Environmental Analysis and Monitoring, pp. 224–259. CEEAM, Gdańsk (2003)

[6] Kocjan, R.: Chemia analityczna T. 2. Analiza instrumentalna. Wydawnictwo Lekarskie PZWL, Warszawa (2000)

[7] Lichon, M. J.: Sample preparation. In: Nollet, L. M. (ed.) Residues and Other Food Component Analysis. Handbook of Food Analysis, vol. 2, 2nd edn, pp. 1485–1512. CRC Press, London (2004)

[8] Capar, S. G., Szefer, P.: Determination and speciation of trace elements in foods. In: Otles, S. (ed.) Methods of Analysis of Food Components and Additives, pp. 111–158. CRC Press, Boca Raton, FL (2005)

[9] Hoenig, M.: Preparationsteps in environmental trace element analysis—facts and traps. Talanta 54, 1021–1038 (2001)

[10] Gouveia, S. T., Lopes, G. S., Fatibello-Filho, O., Nogueira, A. R. A., Nóbrega, J. A.: Homogenization of breakfast cereals using cryogenic grinding. J. Food Eng. 51(1), 59–63 (2002)

[11] Cubadda, F., Baldini, M., Carcea, M., Pasqui, L. A., Raggi, A., Stacchini, P.: Influence of laboratory homogenization procedures on trace element content of food samples: an ICP-MS study on soft and durum wheat. Food Addit. Contam. 18, 778–787 (2001)

[12] Demirel, S., Tuzen, M., Saracoglu, S., Soylak, M.: Evaluation of various digestion procedures for trace element contents of some food materials. J. Hazard. Mater. 152, 1020–1026 (2008)

[13] Grembecka, M., Malinowska, E., Szefer, P.: Differentiation of market coffee and its infusions in view of their mineral composition. Sci. Total Environ. 383, 59–69 (2007)

[14] Mansour, S. A., Belal, M. H., Abou-Arab, A. A. K., Gad, M. F.: Monitoring of pesticides and heavy metals in cucumber fruits produced from different farming systems. Chemosphere 75, 601–609 (2009)

[15] Seenivasan, S., Manikandan, N., Muraleedharan, N. N., Selvasundaram, R.: Heavy metal content of black teas from south India. Food Control 19, 746–749 (2008)

[16] Vélez, D., Devesa, V., Súñer, M. A., Montoro, R.: Metal contamination in food. In: Nollet, L. M. (ed.) Residues and Other Food Component Analysis. Handbook of Food Analysis, vol. 2, 2nd edn, pp. 1485–1512. CRC Press, London (2004)

[17] Mader, P., Száková, J., Curdová, E.: Combination of classical dry ashing with stripping voltammetry in trace element analysis of biological materials: review of literature published after 1978. Talanta 43, 521–534 (1996)

[18] Vassileva, E., Dočekalová, H., Baeten, H., Vanhentenrijk, S., Hoenig, M.: Revisitation of mineralization modes for arsenic and selenium determinations in environmental samples. Talanta 54, 187–196 (2001)

[19] Fecher, P., Ruhnke, G.: Cross contamination of lead and cadmium during dry ashing of food samples. Anal. Bioanal. Chem. 373, 787–791 (2002)

[20] Baklanov, A. N., Bokhan Yu, V., Chmilenko, F. A.: Analysis of food products using carbonization and ultrasonic techniques. J. Anal. Chem. 58, 489–493 (2003)

[21] Grembecka, M., Szefer, P.: Quality of confectionery products in view of their mineral composition. Food Anal. Methods (2011). doi:10.1007/s12161-011-9234-0

[22] Gonzálvez, A., Armenta, S., Cervera, M. L., de la Guardia, M.: Elemental composition of seasoning products. Talanta 74, 1085–1095 (2008)

[23] Afonso, C., Lourenc, o, H. M., Dias, A., Nunes, M. L., Castro, M.: Contaminant metals in black scabbard fish (Aphanopus carbo) caught off Madeira and the Azores. Food Chem. 101, 120–125 (2007)

[24] González, S., Flick, G. J., O'Keefe, S. F., Duncan, S. E., McLean, E., Craig, S. R.: Composition of farmed and wild yellow perch (Perca flavescens). J. Food Compos. Anal. 19(6–7), 720–726 (2006)

[25] Kwoczek, M., Szefer, P., Hać, E., Grembecka, M.: Essential and toxic elements in seafood avail-

able in Poland from different geographical regions. J. Agric. Food Chem. 54,3015-3024(2006)

[26] Malinowska,E. ,Szefer,P. ,Falandysz,J. :Metals bioaccumulation by bay bolete,Xerocomus badius,from selected sites in Poland. Food Chem. 84,405-416 (2004)

[27] Dugo,G. ,La Pera,L. ,Turco,V. ,Lo,P. ,Rosina,M. ,Saitta,M. :Effect of boiling and peeling on manganese content of some vegetables determined by derivative anodic stripping chronopotentiometry (dASCP). Food Chem. 93,703-711 (2005)

[28] Nardi,E. P. ,Evangelista,F. S. ,Tormen,L. ,Saint'Pierre,T. D. ,Curtius,A. J. ,Souza,S. S. ,de Barbosa,F. :The use of inductively coupled plasma mass spectrometry (ICP-MS) for the determination of toxic and essential elements in different types of food samples. Food Chem. 112,727-732 (2009)

[29] Lodenius,M. , Tulisalo, E. : Open digestion ofsome plant and fungus materials for mercury analysis using different temperatures and sample sizes. Sci. Total Environ. 176,81-84 (1995)

[30] Gasparics,T. ,Marti'nez,R. M. G. ,Caroli,S. ,Za'ray,G. :Determination of trace elements in Antarctic krill samples by inductively coupled atomic emission and graphite furnace atomic absorption spectrometry. Microchem. J. 67,279-284 (2000)

[31] Achterberg,E. P. ,Braungardt,C. B. ,Sandford,R. C. ,Worsfold,P. J. :UV digestion of seawater samples prior to the determination of copper using flow injection with chemiluminescence detection. Anal. Chim. Acta 440,27-36 (2001)

[32] Lambert, D. F. , Turoczy, N. J. : Comparison of digestion methods for the determination of selenium in fish tissue by cathodic stripping voltammetry. Anal. Chim. Acta 408,97-102(2000)

[33] Recknagel,S. ,Brätter,P. ,Tomiak,A. ,Rösick,U. :Determination of selenium in blood serum by ICP-OES including an on-line wet digestion and Se-hydride formation procedure. Fresenius J. Anal. Chem. 346,833-836 (1993)

[34] Simon,S. ,Barats,A. ,Pannier,F. ,Potin-Gautier,M. :Development of an on-line UV decomposition system for direct coupling of liquid chromatography to atomic-fluorescence spectrometry for selenium speciation analysis. Anal. Bioanal. Chem. 383,562-569 (2005)

[35] Burguera,J. L. , Burguera, M. : Molybdenum in human whole blood of adult residents of the Merida State (Venezuela). J. Trace Elem. Med. Biol. 21,178-183 (2007)

[36] Amarasiriwardena, D. , Krushevska, A. , Argentine, M. , Barnes, R. M. : Vapour-phaseacid digestion of micro samples of biological material in a high-temperature,high-pressure asher for inductively coupled plasma atomic emission spectrometry. Analyst 119,1017-1021(1994)

[37] Tinggi,U. ,Reilly,C. ,Patterson,C. :Determination of manganese and chromium in foods by atomic absorption spectrometry after wet digestion. Food Chem. 60,123-128 (1997)

[38] Muñoz,O. ,Bastias,J. M. ,Araya,M. ,Morales,A. ,Orellana,C. ,Rebolledo,R. ,Velez,D. :Estimation of the dietary intake of cadmium,lead,mercury,and arsenic by the population of Santiago (Chile) using a Total Diet Study. Food Chem. Toxicol. 43,1647-1655 (2005)

[39] Biziuk,M. ,Kuczyńska,J. :Mineral components in food- analytical implications. In: Szefer,P. , Nriagu,J. O. (eds.) Mineral Components in Foods,pp. 1-32. CRC Press/Taylor Francis Group,

London/New York (2007)

[40] Lamble, K. L., Hill, S. J.: Microwave digestion procedures for environmental matrices. Analyst 123,103R-133R (1998)

[41] Sun, D.-S., Waters, J. K., Mawhinney, T. P.: Determination of thirteen common elements in food samples by inductively coupled plasma atomic emission spectrometry: comparison of five digestion methods. J. AOAC Int. 83,1218-1224 (2000)

[42] Gawalko, E. J., Nowicki, T. W., Babb, J., Tkachuk, R., Wu, S.: Comparison of closed-vessel and focused open-vessel microwave dissolution for determination of cadmium, copper, lead, and selenium in wheat, wheat products, corn bran, and rice flour by transverse-heated graphite furnace atomic absorption spectrometry. J. AOAC Int. 80,379-387 (1997)

[43] Dolan, S. P., Capar, S. G.: Multi-element analysis of food by microwave digestion and inductively coupled plasma-atomic emission spectrometry. J. Food Compos. Anal. 15,593-615 (2002)

[44] Krzysik, M., Grajeta, H., Prescha, A.: Chromium content in selected convenience and fast foods in Poland. Food Chem. 107,208-212 (2008)

[45] Official Methods of Analysis of AOAC International, 17th edn. Rev 1, AOAC International, Gaithersburg, MD, USA, Official Method 973.34. Cadmium in Food- Atomic Absorption Spectrophotometric Method (2002)

[46] Official Methods of Analysis of AOAC International, 17th edn. Rev 1, AOAC International, Gaithersburg, MD, USA, Official Method 972.25. Lead in Food- Atomic Absorption Spectrophotometric Method (2002)

[47] Official Methods of Analysis of AOAC International 17th edn. Rev 1, AOAC International, Gaithersburg, MD, USA, Official Method 999.10. Lead, Cadmium, Zinc, Copper, and Iron in Foods- Atomic Absorption Spectrophotometry after Microwave Digestion (2002)

[48] Official Methods of Analysis of AOAC International, 17th edn. Rev 1, AOAC International, Gaithersburg, MD, USA, Official Method 999.11. Determination of Lead, Cadmium, Copper, Iron, and Zinc in Foods- Atomic Absorption Spectrophotometry after Dry Ashing- NMLK-AOAC Method (2002)

[49] Official Methods of Analysis of AOAC International, 17th edn. Rev 1, AOAC International, Gaithersburg, MD, USA, Official Method 985.16. Tin in Canned Foods- Atomic Absorption Spectrophotometric Method (2002)

[50] Official Methods of Analysis of AOAC International, 17th edn. Rev 1, AOAC International, Gaithersburg, MD, USA, Official Method 969.32. Zinc in Food- Atomic Absorption Spectrophotometric Method (2002)

[51] International Organization for Standardization ISO 6636-2, Fruits, Vegetables and Derived Products- Determination of Zinc Content- Part 2: Atomic Absorption Spectrometric Method (1981)

[52] International Organization for Standardization ISO 7952, Fruits, Vegetables and Derived Products- Determination of Copper Content-MethodUsing Flame Atomic Absorption Spectrometry (1994)

[53] International Organization for Standardization ISO 9526, Fruits, Vegetables and Derived Products-

Determination of Iron Content by Flame Atomic Absorption Spectrometry (1990)

[54] International Organization for Standardization ISO 11813, Milk and Milk Products-Determination of Zinc Content- Flame Atomic Absorption Spectrometric Method (1998)

[55] Official Methods of Analysis of AOAC International, 17th edn. Rev 1, AOAC International, Gaithersburg, MD, USA, Official Method 991.25. Calcium, Magnesium, and Phosphorus in Cheese- Atomic Absorption Spectrophotometric and Colorimetric Method (2002)

[56] Official Methods of Analysis of AOAC International, 17th edn. Rev 1, AOAC International, Gaithersburg, MD, USA, Official Method 985.35. Minerals in Infant Formula, Enteral Products, and Pet Foods- Atomic Absorption Spectrophotometric Method (2002)

[57] Da-Col, J. A. , Domene, S. M. A. , Pereira-Filho, E. R. : Fast determination of Cd, Fe, Pb and Zn in food Rusing AAS. Food Anal. Methods 2, 110-115 (2009)

[58] Lombardi-Boccia, G. , Aguzzi, A. , Cappelloni, M. , Di Lullo, G. : Content of some trace elements and minerals in the Italian total-diet. J. Food Compos. Anal. 13, 525-527 (2000)

[59] Yebra, M. C. , Cancela, S. , Cespón, R. M. : Automatic determination of nickel in foods by flame atomic absorption spectrometry. Food Chem. 108, 774-778 (2008)

[60] Kawashima, L. M. , Valente Soares, L. M. : Mineral profile of raw and cooked leafy vegetables consumed in Southern Brazil. J. Food Compos. Anal. 16, 605-611 (2003)

[61] Pandey, J. , Pandey, U. : Accumulation of heavy metals in dietary vegetables and cultivated soil horizon in organic farming system in relation to atmospheric deposition in a seasonally dry tropical region of India. Environ. Monit. Assess. 149, 61-74 (2009)

[62] Tuzen, M. , Soylak, M. : Evaluation of trace element contents in canned foods marketed from Turkey. Food Chem. 102, 1089-1095 (2007)

[63] Saracoglu, S. , Saygi, K. O. , Uluozlu, O. D. , Tuzen, M. , Soylak, M. : Determination of trace element contents of baby foods from Turkey. Food Chem. 105, 280-285 (2007)

[64] Santelli, R. E. , Bezerra de Almeida, M. , de SantAna, O. D. , Cassella, R. J. , Ferreira, S. L. C. : Multivariate technique for optimization of digestion procedure by focused microwave system for determination of Mn, Zn and Fe in food samples using FAAS. Talanta 68, 1083-1088(2006)

[65] Howe, A. , Fung, L. H. , Lalor, G. , Rattray, R. , Vutchkov, M. : Elemental composition of Jamaican foods 1: a survey of five food crop categories. Environ. Geochem. Health 27, 19-30 (2005)

[66] Bellido-Milla, D. , Moreno-Perez, J. M. , Hernandez-Artiga, M. P. : Differentiation and classification of beers with flame atomic spectrometry and molecular absorption spectrometry and sample preparation assisted by microwaves. Spectrochim. Acta B 55, 855-864 (2000)

[67] Abebe, Y. , Bogale, A. , Hambidge, K. M. , Stoecker, B. J. , Bailey, K. , Gibson, R. S. : Phytate, zinc, iron and calcium content of selected raw and prepared foods consumed in rural Sidama, Southern Ethiopia, and implications for bioavailability. J. Food Comp. Anal. 20, 161-168(2007)

[68] Milacic, R. , Kralj, B. : Determination of Zn, Cu, Cd, Pb, Ni and Cr in some Slovenian foodstuffs. Eur. Food Res. Technol. 217, 211-214 (2003)

[69] Khajeh, M. : Optimization of microwave-assisted extraction procedure for zinc and copper determi-

nation in food samples by Box-Behnken design. J. Food Compos. Anal. 22,343-346(2009)

[70] Miller-Ihli, N. J. : Atomic absorption and atomic emission spectrometry for the determination of the trace element content of selected fruits consumed in the United States. J. Food Compos. Anal. 9,301-311 (1996)

[71] dos Santos, J. S., dos Santos, M. L. P., Conti, M. M., dos Santos, S. N., de Oliveira, E. : Evaluation of some metals in Brazilian coffees cultivated during the process of conversion from conventional to organic agriculture. Food Chem. 115,1405-1410 (2009)

[72] Butcher, D. J., Sneddon, J. : A Practical Guide to Graphite Furnace Atomic Absorption Spectrometry. Wiley-Interscience, New York (1998)

[73] International Organization for Standardization ISO 6561, Fruits, Vegetables and Derived Products- Determination of Cadmium Content- Flameless Atomic Absorption Spectrometric Method (1983)

[74] International Organization for Standardization ISO 6633, Fruits, Vegetables and Derived Products- Determination of Lead Content- Flameless Atomic Absorption Spectrometric Method (1984)

[75] International Organization for Standardization ISO 14377, Canned Evaporated Milk-Determination of Tin Content- Method Using Graphite Furnace Atomic Absorption Spectrometry (2002)

[76] International Organization for Standardization ISO 15774, Animal and Vegetable Fats and Oils- Determination of Cadmium Content by Direct Graphite Furnace Atomic Absorption Spectrometry (2000)

[77] International Organization for Standardization ISO 12193, Animal and Vegetable Fats and Oils- Determination of Lead Content- Graphite Furnace Atomic Absorption Method (1994)

[78] International Organization for Standardization ISO 10540-2, Animal and Vegetable Fats and Oils- Determination of Phosphorus Content- Part 2: Method Using Graphite Furnace Atomic Absorption Spectrometry (2003)

[79] International Organization for Standardization ISO 8294, Animal and Vegetable Fats and Oils- Determination of Copper, Iron and Nickel Contents- Graphite Furnace Atomic Absorption Method (1994)

[80] International Organization for Standardization ISO 11212-4, Starch and Derived Products-Heavy Metals Content- Part 4: Determination of Cadmium Content by Atomic Absorption Spectrometry with Electrothermal Atomization (1997)

[81] International Organization for Standardization ISO 11212-3, Starch and Derived Products-Heavy Metals Content- Part 3: Determination of Lead Content by Atomic Absorption Spectrometry with Electrothermal Atomization (1997)

[82] Official Methods of Analysis of AOAC International, 17th edn. Rev 1, AOAC International, Gaithersburg, MD, USA, Official Method 997.15. Lead in Sugars and Syrups- Graphite Furnace Atomic Absorption Method (2002)

[83] Capar, G. S., Cunningham, W. C. : Element and radionuclide concentrations in food: FDA total diet study 1991-1996. J. AOAC Int. 83,157-177 (2000)

[84] Tripathi, R. M., Mahapatra, S., Raghunath, R., Kumar, A. V., Sadasivan, S. : Daily intake of alu-

minium by adult population of Mumbai, India. Sci. Total Environ. 299, 73-77 (2002)

[85] Alberti-Fidanza, A., Burini, G., Perriello, G.: Trace elements in foods and meals consumed by students attending the faculty cafeteria. Sci. Total Environ. 287, 133-140 (2002)

[86] Malmauret, L., Parent-Massin, D., Hardy, J.-L., Verger, P.: Contaminants in organic and conventional foodstuffs in France. Food Addit. Contam. 19, 524-532 (2002)

[87] Larsen, E. H., Andersen, N. L., Møller, A., Petersen, A., Mortensen, G. K., Petersen, J.: Monitoring the content and intake of trace elements from food in Denmark. Food Addit. Contam. 19, 33-46 (2002)

[88] Bratakos, M. S., Lazos, E. S., Bratakos, S. M.: Chromium content of selected Greek foods. Sci. Total Environ. 290, 47-58 (2002)

[89] López, F. F., Cabrera, C., Lorenzo, M. L., López, M. C.: Aluminum content in foods and beverages consumed in the Spanish diet. J. Food Sci. 65, 206-210 (2000)

[90] López, F. F., Cabrera, C., Lorenzo, M. L., López, M. C.: Aluminium content of drinking waters, fruit juices and soft drinks: contribution to dietary intake. Sci. Total Environ. 292, 205-213 (2002)

[91] Cabrera, C., Lloris, F., Giménez, R., Olalla, M., López, M. C.: Mineral content in legumes and nuts: contribution to the Spanish dietary intake. Sci. Total Environ. 308, 1-14 (2003)

[92] Cindric, I. J., Zeiner, M., Steffan, I.: Trace elemental characterization of edible oils by ICP-AES and GFAAS. Microchem. J. 85, 136-139 (2007)

[93] Lendinez, E., Lorenzo, M. L., Cabrera, C., López, M. C.: Chromium in basic foods of the Spanish diet: seafood, cereals, vegetables, olive oils and dairy products. Sci. Total Environ. 278, 183-189 (2001)

[94] Musaiger, A. O., Al-Jedah, J. H., D'souza, R.: Occurrence of contaminants in foods commonly consumed in Bahrain. Food Control 19, 854-861 (2008)

[95] Julshamn, K., Thorlacius, A., Lea, P.: Determination of arsenic in seafood by electrothermal atomic absorption spectrometry after microwave digestion: NMKL collaborative study. J. AOAC Int. 83, 1423-1428 (2000)

[96] Fedorov, P. N., Ryabchuk, G. N., Zverev, A. V.: Comparison of hydride generation and graphite furnace atomic absorption spectrometry for the determination of arsenic in food. Spectrochim. Acta B 52, 1517-1523 (1997)

[97] Veillon, C., Patterson, K. Y.: Analytical issues in nutritional chromium research. J. Trace Elem. Exp. Med. 12, 99-109 (1999)

[98] Dedina, J., Tsalev, D. L.: Hydride Generation Atomic Absorption Spectrometry. Wiley, Chichester (1995)

[99] Official Methods of Analysis of AOAC International, 17th edn. Rev 1, AOAC International, Gaithersburg, MD, USA, Official Method 986. 15. Arsenic, Cadmium, Lead, Selenium, andZinc in Human and Pet Foods- Multielement Method (2002)

[100] Tsalev, D. L.: Vapor generation or electrothermal atomic absorption spectrometry- both! Spectro-

chim. Acta B 55,915-931 (2000)
[101] Yan, X. -P. , Ni, Z. -M. :Vapor generation atomic absorption spectrometry. Anal. Chim. Acta 291,89-105 (1994)
[102] Kabengera, C. , Bodart, P. , Hubert, P. , Thunus, L. , Noirfalise, A. :Optimization and validation of arsenic determination in foods by hydride generation flame atomic absorption spectrometry. J. AOAC Int. 85,122-127 (2002)
[103] Tinggi, U. :Determination of selenium in meat products by hydride generation atomic absorption spectrophotometry. J. AOAC Int. 82,364-367 (1999)
[104] Hussein, L. , Bruggeman, J. :Selenium analysisof selected Egyptian foods and estimated daily intakes among a population group. Food Chem. 65,527-532 (1999)
[105] Scanlon, K. A. , MacIntosh, D. L. , Hammerstrom, K. A. , Ryan, P. B. :A longitudinal investigation of solid-food based dietary exposure to selected elements. J. Expo. Anal. Environ Epidemiol. 9, 485-493 (1999)
[106] Tsuda, T. , Inoue, T. , Kojima, M. , Aoki, S. :Market basket and duplicate portion estimation of dietary intakes of cadmium, mercury, arsenic, copper, manganese, and zinc by Japanese adults. J. AOAC Int. 78,1363-1368 (1995)
[107] Díaz-Alarcón, J. P. , Navarro-Alarcón, M. , López-García de la Serrana, H. , López-Martínez, M. C. :Determination of selenium in meat products by hydride generation atomic absorption spectrometry-selenium levels in meat, organ meats, and sausages in Spain. J. Agric. Food Chem. 44,1494-1497 (1996)
[108] Sawaya, W. N. , Al-Awadhi, F. , Aziz, A. , Al-Rashdan, A. , Mahjoub, B. T. , Al-Amiri, H. :Nutritional profile of Kuwaiti composite dishes:minerals and vitamins. J. Food Compos. Anal. 11,70-88 (1998)
[109] Plessi, M. , Bertelli, D. , Monzani, A. :Mercury and selenium content in selected seafood. J. Food Compos. Anal. 14,461-467 (2001)
[110] Foster, L. H. , Sumar, S. :Selenium concentrations in soya based milks and infant formulae available in the United Kingdom. Food Chem. 56,93-98 (1996)
[111] Yadav, S. K. , Singh, I. , Sharma, A. , Singh, D. :Selenium status in food grains of northern districts of India. J. Environ. Manage. 88,770-774 (2008)
[112] Moretto, L. , Cadore, A. S. :Determination of arsenic in food samples by hydride generation-atomic absorption spectrometry. Microchim. Acta 146,239-244 (2004)
[113] International Organization for Standardization ISO 6637, Fruits, Vegetables and Derived Products- Determination of Mercury Content- Flameless Atomic Absorption Method (1984)
[114] Official Methods of Analysis of AOAC International, 17th edn. Rev 1, AOAC International, Gaithersburg, MD, USA, Official Method 977. 15. Mercury in Fish- Alternative Flameless Atomic Absorption Spectrophotometric Method (2002)
[115] Storelli, M. M. , Marcotrigiano, G. O. :Total mercury levels in muscle tissue of swordfish (Xiphias gladius) and bluefin tuna (Thunnus thynnus) from the Mediterranean Sea (Italy).

J. Food Prot. 64,1058-1061 (2001)

[116] Falandysz,J. ,Jedrusiak,A. ,Lipka,K. ,Kannan,K. ,Kawano,M. ,Gucia,M. ,Brzostowski,A. , Dadej, M. : Mercury in wild mushrooms and underlying soil substrate from Koszalin, Northcentral Poland. Chemosphere 54,461-466 (2004)

[117] Love,J. L. , Rush, G. M. , McGrath, H. : Totalmercury and methylmercury levels in some New Zealand commercial marine fish species. Food Addit. Contam. 20,37-43 (2003)

[118] Storelli,M. M. ,Stuffler,R. G. ,Marcotrigiano,G. O. :Total and methylmercury residues in tunafish from the Mediterranean Sea. Food Addit. Contam. 19,715-720 (2002)

[119] Tahán,J. E. ,Sanchez,J. M. ,Granadillo, V. A. ,Cubillan, H. S. , Romero, R. A. : Concentration of total Al,Cr,Cu,Fe,Hg,Na,Pb,and Zn in commercial canned seafood determined by atomic spectrometric means after mineralization by microwave heating. J. Agric. Food Chem. 43,910- 915 (1995)

[120] Vereda Alonso, E. , Siles Cordero, M. T. , García de Torres, A. , Cañada Rudner, P. , Cano Pavón,J. M. : Mercury speciation in sea food by flow injection cold vapor atomic absorption spectrometry using selective solid phase extraction. Talanta 77,53-59 (2008)

[121] Voegborlo, R. B. , El-Methnani, A. M. , Abedin, M. Z. : Mercury, cadmium and lead content of canned tuna fish. Food Chem. 67,341-345 (1999)

[122] Dabeka,R. W. , Bradley,P. , McKenzie,A. D. : Routine, high-sensitivity, cold vapor atomic absorption spectrometric determination of total mercury in foods after low-temperature digestion. J. AOAC Int. 85,1136-1143 (2002)

[123] Zenebon,O. , Sakuma, A. M. , Dovidauskas, S. , Okada, I. A. , de MaioFranca, D. , Lichtig, J. : Rapid food decomposition by H_2O_2-H_2SO_4 for determination of total mercury by flow injection cold vapor atomic absorption spectrometry. J. AOAC Int. 85,149-152 (2002)

[124] Thomas,R. :Practical Guide to ICP-MS,pp. 1-316. CRC Press,Gaithersburg,MD (2008)

[125] Heitkamper,D. T. ,Caruso,J. A. : Chromatographic sample introduction for plasma mass spectrometry. In:Krull,I. S. (ed.) Trace Metal Analysis and Speciation, pp. 49-70. Elsevier, Amsterdam (1991)

[126] Zbinden,P. ,Andrey,D. :Determination of trace element contaminants in food matrixes using a robust,routine analytical method for ICP-MS. At. Spectrosc. 19,214-219 (1998)

[127] Bhandari, S. A. , Amarasiriwardena, D. : Closed-vessel microwave acid digestion of commercial maple syrupfor the determination of lead and seven other trace elements by inductively coupled plasma-mass spectrometry. Microchem. J. 64,73-84 (2000)

[128] Rädlinger,G. ,Heumann,K. G. :Iodine determination in food samples using inductively coupled plasma isotope dilution mass spectrometry. Anal. Chem. 70,2221-2224 (1998)

[129] Fecher,P. A. ,Goldmann,I. , Nagengast,A. : Determination of iodine in food samples by inductively coupled plasma mass spectrometry after alkaline extraction. J. Anal. At. Spectrom. 13, 977-982 (1998)

[130] Choi, Y. C. , Kim,J. , Lee, H. -S. , Kim, C. , Hwang, I. K. , Park, H. K. , Oh, C. -H. : Selenium

content in representative Korean foods. J. Food Compos. Anal. 22,117-122 (2009)

[131] Whyte, A. L. H. , Raumati, H. G. , Greening, G. E. , Gibbs-Smith, E. , Gardner, J. P. A. : Human dietary exposure to heavy metals via the consumption of greenshell mussels (Perna canaliculus Gmelin 1791) from the Bay of Islands, northern New Zealand. Sci. Total Environ. 407, 4348-4355 (2009)

[132] Bagga, D. K. , Jarrett, K. A. : Risk assessment of growing vegetables in lead contaminated soil: a greenhouse study ICFAI. J. Life Sci. 3, 7-13 (2009)

[133] Chung, S. W. C. , Kwong, K. P. , Yau, J. C. W. , Wong, W. W. K. : Dietary exposure to antimony, lead and mercury of secondary school students in Hong Kong. Food Addit. Contam. 25, 831-840 (2008)

[134] Jorhem, L. , Åstrand, C. , Sundström, B. , Baxter, M. , Stokes, P. , Lewis, J. , Petersson, G. K. : Elements in rice from the Swedish market: 1. Cadmium, lead and arsenic (total and inorganic). Food Addit. Contam. 25, 284-292 (2008)

[135] Leblond, C. , Mephara, J. , Sauvé, S. : Trace Metals (Cd, Co, Cr, Cu, Hg, Ni, Pb, and Zn) in food supplements of marine origin. Hum. Ecol. Risk Assess. 14, 408-420 (2008)

[136] Taylor, V. , Jackson, B. , Chen, C. : Mercury speciation and totaltrace element determination of low-biomass biological samples. Anal. Bioanal. Chem. 392, 1283-1290 (2008)

[137] Llobet, J. M. , Falco, G. , Casas, C. , Teixido, A. , Domingo, J. L. : Concentrations of arsenic, cadmium, mercury, and lead in common foods and estimated daily intake by children, adolescents, adults, and seniors of Catalonia, Spain. J. Agric. Food Chem. 51, 838-842 (2003)

[138] Rose, M. : Bromine and iodine in 1997 UK total diet study samples. J. Environ. Monit. 3, 361-365 (2001)

[139] Noël, L. , Leblanc, J. -C. , Guérin, T. : Determination of several elements in duplicate meals from catering establishments using closed vessel microwave digestion with inductively coupled plasma mass spectrometry detection: estimation of daily dietary intake. Food Addit. Contam. 20, 44-56 (2003)

[140] Gundersen, V. , Bechmann, I. E. , Behrens, A. , Stürup, S. : Comparative investigation of concentrations of major and trace elements in organic and conventional Danish agricultural crops. 1. Onions (Allium cepa Hysam) and Peas (Pisum sativum Ping Pong). J. Agric. Food Chem. 48, 6094-6102 (2000)

[141] Simpkins, W. A. , Louie, H. , Wu, M. , Harrison, M. , Goldberg, D. : Trace elements in Australian orange juice and other products. Food Chem. 71, 423-433 (2000)

[142] Falandysz, J. , Szymczyk, K. , Ichihashi, H. , Bielawski, L. , Gucia, M. , Frankowska, A. , Yamasaki, S. -I. : ICP/MS and ICP/AES elemental analysis (38 elements) of edible mushrooms growing in Poland. Food Addit. Contam. 18, 503-513 (2001)

[143] Kelly, S. , Baxter, M. , Chapman, S. , Rhodes, C. , Dennis, J. , Brereton, P. : The application of isotopic and elemental analysis to determine the geographical origin of premium long grain rice. Eur. Food Res. Technol. 214, 72-78 (2002)

[144] Sahan, Y., Basoglu, F., Gücer, S.: ICP-MS analysis of a series of metals (Namely: Mg, Cr, Co, Ni, Fe, Cu, Zn, Sn, Cd and Pb) in black and green olive samples from Bursa, Turkey. Food Chem. 105, 395-399 (2007)

[145] Toniolo, R., Pizzzariello, A., Tubaro, F., Susmel, S., Dossi, N., Bontempelli, G.: A voltammetric approach to an estimated of metal release from tinplate promoted by ligands present in canned vegetables. J. Appl. Electrochem. 39, 979-988 (2009)

[146] Catarino, S., Curvelo-Garcia, A. S., de Sousa, R. B.: Measurements of contaminant elements of wines by inductively coupled plasma-mass spectrometry: a comparison of two calibration approaches. Talanta 70, 1073-1080 (2006)

[147] Official Methods of Analysis of AOAC International, 17th edn. Rev 1, AOAC International, Gaithersburg, MD, USA, Official Method 984.27. Calcium, Copper, Iron, Magnesium, Manganese, Phosphorus, Potassium, Sodium, and Zinc in Infant Formula- InductivelyCoupled Plasma Emission Spectroscopic Method (2002)

[148] Official Methods of Analysis of AOAC International, 17th edn. Rev 1, AOAC International, Gaithersburg, MD, USA, Official Method 985.01. Metals and Other Elements in Plants and Pet Foods- Inductively Coupled Plasma Emission Spectroscopic Method (2002)

[149] International Organization for Standardization ISO 10540-3, Animal and Vegetable Fats and Oils-Determination of Phosphorus Content- Part 3: Method Using Inductively Coupled Plasma (ICP) Optical Emission Spectroscopy (2002)

[150] Miller-Ihli, N. J.: Trace element determinations in foods and biological samples using inductively coupled plasma atomic emission spectrometry and flame atomic absorption spectrometry. J. Agric. Food Chem. 44, 2675-2679 (1996)

[151] Lomer, M. C. E., Thompson, R. P. H., Commisso, J., Keen, C. L., Powell, J. J.: Determination of titanium dioxide in foods using inductively coupled plasma optical emission spectrometry. Analyst 125, 2339-2343 (2000)

[152] Rao, P. V., Subba, M., Vaibhav, A., Ganesan, K.: Mineral composition of edible seaweed Porphyra vietnamensis. Food Chem. 102, 215-218 (2007)

[153] D'Ilio, S., Alessandrelli, M., Cresti, R., Forte, G., Caroli, S.: Arsenic content of various types of rice as determined by plasma-based techniques. Microchem. J. 73, 195-202 (2002)

[154] Bode, P.: Jadrowe techniki analityczne w badaniach rodowiskowych [w] Namieśnik, J., Chrzanowski, W., Szpinek, P. (red.): Nowe horyzonty i wyzwania w analityce i monitoringuśrodowiskowym, pp. 277-293. CEEAM, Gdańsk (2003)

[155] Fernández, P. L., Pablos, F., Martín, M. J., González, A. G.: Multi-element analysis of tea beverages by inductively coupled plasma atomic emission spectrometry. Food Chem. 76, 483-489 (2002)

[156] dos Santos, E. J., de Oliveira, E.: Determination of mineral nutrients and toxic elements in Brazilian soluble coffee by ICP-AES. J. Food Compos. Anal. 14, 523-531 (2001)

[157] Özcan, M.: Mineral contents ofsome plants used as condiments in Turkey. Food Chem. 84, 437-

440 (2004)

[158] Krejcová, A., Cernohorský, T., Meixner, D.: Elemental analysis of instant soups and seasoning mixtures by ICP-OES. Food Chem. 105, 242-247 (2007)

[159] Jurado, J. M., Alcázar, A., Pablos, F., Martín, M. J., González, A. G.: Classification of aniseed drinks by means of cluster, linear discriminant analysis and soft independent modelling of class analogy based on their Zn, B, Fe, Mg, Ca, Na and Si content. Talanta 66, 1350-1354 (2005)

[160] Ioannidou, M. D., Zachariadis, G. A., Anthemidis, A. N., Stratis, J. A.: Direct determination of toxic trace metals in honey and sugars using inductively coupled plasma atomic emission spectrometry. Talanta 65, 92-97 (2005)

[161] Kucera, J., Randa, Z., Soukal, L.: A comparison of three activation analysis methods for iodine determination in foodstuffs. J. Radioanal. Nucl. Chem. 249, 61-65 (2001)

[162] Ventura, M. G., do Carmo Freitas, M., Pacheco, A., van Meerten, T., Wolterbeek, H.: Selenium content in selected Portuguese foodstuffs. Eur. Food Res. Technol. 224, 395-401 (2007)

[163] Soupioni, M. J., Symeopoulos, B. D., Papaefthymiou, H. V.: Determination of trace elements in bottled water in Greece by instrumental and radiochemical neutron activation analyses. J Radioanal. Nucl. Chem. 268, 441-444 (2006)

[164] Islam, M. T., Islam, S. A., Latif, S. A.: Detection of arsenic in water, herbal and soil samples by neutron activation analysis technique. Bull. Environ. Contam. Toxicol. 79, 327-330 (2007)

[165] Dang, H. S., Jaiswal, D. D., Nair, S.: Daily dietary intake of trace elements of radiological and nutritional importance by the adult Indian population. J. Radioanal. Nucl. Chem. 249, 95-101 (2001)

[166] Lee, J. Y.: Daily dietary intake of elements of nutritional and radiological importance by adult Koreans. J. Radioanal. Nucl. Chem. 249, 39-45 (2001)

[167] Cozzollino, S. M. F.: Determination of mineral constituents in duplicate portion diets of two university student groups by instrumental neutron activation. J. Radioanal. Nucl. Chem. 249, 21-24 (2001)

[168] Waheed, S., Zaidi, J. H., Ahmad, S., Saleem, M.: Instrumental neutron activation analysis of 23 individual food articles from a high altitude region. J. Radioanal. Nucl. Chem. 254, 597-605 (2002)

[169] Arriola, S. H., Monroy, G. F., Cruz, M. M.: Determination of Fe and Al contamination by NAA at preparation of traditional Mexican food. J. Radioanal. Nucl. Chem. 271, 597-598 (2007)

[170] Alamin, M. B., Bejey, A. M., Kucera, J., Mizera, J.: Determination of mercury and selenium in consumed food items in Libya using instrumental and radiochemical NAA. J. Radioanal. Nucl. Chem. 270, 143-146 (2006)

[171] Singh, V., Garg, A. N.: Availability of essential trace elements in Indian cereals, vegetables and spices using INAA and the contribution of spices todaily dietary intake. Food Chem. 94, 81-89 (2006)

[172] González, E. B. , Sanz-Medel, A. : Liquid chromatographic techniques for trace element speciation analysis. In: Caruso, J. A. , Sutton, K. L. , Ackley, K. L. (eds.) Elemental Speciation New Approaches for Trace Element Analysis, pp. 81 – 121. Elsevier Science B. V, Amsterdam (2000)

[173] Buldini, P. L. , Cavalli, S. , Trifirò, A. : State-of-the-art ion chromatographic determination of inorganic ions in food. J. Chromatogr. A 789, 529–548 (1997)

[174] Ambushe, A. A. , McCrindle, R. I. , McCrindle, C. M. E. : Speciation of chromium in cow's milk by solid-phase extraction/dynamic reaction cell inductively coupled plasma mass spectrometry (DRC-ICP-MS). JAAS 24, 502–507 (2009)

[175] Pedrero, Z. , Madrid, Y. : Novel approaches for selenium speciation in foodstuffs and biological specimens: a review. Anal. Chim. Acta 634(2), 135–152 (2009)

[176] Pedrero, Z. , Elvira, D. , Cámara, C. , Madrid, Y. : Selenium transformation studies during Broccoli (Brassica oleracea) growing process by liquid chromatography-inductively coupled plasma mass spectrometry (LC-ICP-MS). Anal. Chim. Acta 596, 251–256 (2007)

[177] Sanz, E. , Muñoz-Olivas, R. , Cámara, C. , Sengupta, M. , Kumar, A. S. : Arsenic speciation in rice, straw, soil, hair and nails samples from the arsenic-affected areas of Middle and Lower Ganga plain. J. Environ. Sci. Health A 42, 1695–1705 (2007)

[178] Sun, G. -X. , Williams, P. N. , Zhu, Y. -G. , Deacon, C. , Carey, A. -M. , Raab, A. , Feldmann, J. , Meharg, A. A. : Survey of arsenic and its speciation in rice products such as breakfast cereals, rice crackers and Japanese rice condiments. Environ. Int. 35, 473–475 (2009)

[179] Vieira, M. A. , Grinberg, P. , Bobeda, C. R. R. , Reyes, M. N. M. , Campos, R. C. : Non-chromatographic atomic spectrometric methods in speciation analysis: a review. Spectrochim. Acta B 64, 459–476 (2009)

[180] Lobinski, R. : Elemental speciation and coupled techniques. Appl. Spectrosc. 51, 260 – 278 (1997)

[181] Nemanič, T. M. , Leskovsek, H. , Horvat, M. , Vriser, B. , Bolje, A. : Organotin compounds in the marine environment of the Bay of Piran, northern Adriatic Sea. J. Environ. Monit. 4, 426–430 (2002)

[182] Quevauviller, P. : Certified reference materials: a tool for quality control of elemental speciation analysis. In: Caruso, J. A. , Sutton, K. L. , Ackley, K. L. (eds.) Elemental Speciation New Approaches for Trace Element Analysis, pp. 531–565. Elsevier Science B. V. , Amsterdam (2000)

[183] Stewart, I. I. : Electrospray mass spectrometry: a tool for elemental speciation. Spectrochim. Acta B 54, 1649–1695 (1999)

[184] Szpunar, J. , Bouyssiere, B. , Lobinski, R. : Sample preparation techniques for elemental speciation studies. In: Caruso, J. A. , Sutton, K. L. , Ackley, K. L. (eds.) Elemental Speciation New Approaches for Trace Element Analysis, pp. 7 – 40. Elsevier Science B. V, Amsterdam (2000)

[185] Van Dael, P. : Trace element speciation in food: a tool to assure food safety and nutritional quali-

ty. In: Ebdon, L. , Pitts, L. , Cornelis, R. , Crews, H. , Donard, O. F. X. , Quevauviller, P. (eds.) Trace Element Speciation for Environment, Food and Health, pp. 233 - 240. RSC, Cambridge (2001)

[186] Adams, F. , Ceulemans, M. , Slaets, S. : GC hyphenations for speciation analysis of organometal compounds. LC GC Europe 14, 548-563 (2001)

[187] Barefoot, R. R. : Distribution and speciation of platinum group elements in environmental matrices. Trends Anal. Chem. 18, 702-707 (1999)

[188] Das, A. K. , de la Guardia, M. , Cervera, M. L. : Literature survey of on-line elemental speciation in aqueous solutions. Talanta 55, 1-28 (2001)

[189] Ferrarello, C. N. , Fernández de la Campa, M. R. , Sanz-Medel, A. : Multielement trace-element speciation in metal-biomolecules by chromatography coupled with ICP-MS. Anal Bioanal. Chem. 373, 412-421 (2002)

[190] Kamnev, A. A. , Antonyuk, L. P. , Smirnova, V. E. , Serebrennikova, O. B. , Kulikov, L. A. , Perfiliev, Y. D. : Trace cobalt speciation in bacteria and at enzymic active sites using emission Mössbauer spectroscopy. Anal. Bioanal. Chem. 372, 431-435 (2002)

[191] Mestek, O. , Komínková, J. , Koplík, R. , Borková, M. , Suchánek, M. : Quantification of copper and zinc species fractions in legume seeds extracts by SEC/ICP-MS: validation and uncertainty estimation. Talanta 57, 1133-1142 (2002)

[192] Muñoz-Olivas, R. , Cámara, C. : Speciation related to human health. In: Ebdon, L. , Pitts, L. , Cornelis, R. , Crews, H. , Donard, O. F. X. , Quevauviller, P. (eds.) Trace Element Speciation for Environment, Food and Health, pp. 331-353. RSC, Cambridge (2001)

[193] Albert, J. , Rubio, R. , Rauret, G. : Arsenic speciation in marine biological materials by LC-UV-HG-ICP/OES. Fresenius J. Anal. Chem. 351, 415-419 (1995)

[194] Albert, J. , Rubio, R. , Rauret, G. : Extraction method for arsenic speciation in marine organisms. Fresenius J. Anal. Chem. 351, 420-425 (1995)

[195] Chen, Z. L. , Lin, J. -M. , Naidu, R. : Separation of arsenic species by capillary electrophoresis with sample-stacking techniques. Anal. Bioanal. Chem. 375, 679-684 (2003)

[196] Gómez-Ariza, J. L. , Sánchez-Rodas, D. , Inmaculada, G. I. , Morales, E. : A comparison between ICP-MS and AFS detection for arsenic speciation in environmentalsamples. Talanta 51, 257-268 (2000)

[197] Gómez-Ariza, J. L. , Caro de la Torre, M. A. , Giráldez, I. , Sánchez-Rodas, D. , Velasco, A. , Morales, E. : Pretreatment procedure for selenium speciation in shellfish using highperformance liquid chromatography-microwave-assisted digestion-hydride generationatomic fluorescence spectrometry. Appl. Organomet. Chem. 16, 265-270 (2002)

[198] Larsen, E. H. , Berg, T. : Trace element speciation and international food legislation- a Codex Alimentarius position paper on arsenic asa contaminant. In: Ebdon, L. , Pitts, L. , Cornelis, R. , Crews, H. , Donard, O. F. X. , Quevauviller, P. (eds.) Trace Element Speciation for Environment, Food and Health, pp. 251-257. RSC, Cambridge (2001)

[199] Larsen, E. H. , Quétel, C. R. , Munoz, R. , Fiala-Medioni, A. , Donard, O. F. X. : Arsenic speciation in shrimp and mussel from the mid-Atlantic hydrothermal vents. Mar. Chem. 57,341–346 (1997)

[200] Richarz, A. -N. , Brätter, P. : Speciation analysis of trace elements in the brain of individuals with Alzheimer's disease with special emphasis on metallothioneins. Anal. Bioanal. Chem. 372, 412–417 (2002)

[201] Ringmann, S. , Boch, K. , Marquardt, W. , Schuster, M. , Schlemmer, G. , Kainrath, P. : Microwave-assisted digestion of organoarsenic compounds for the determination of total arsenic in aqueous, biological, and sediment samples using flow injection hydride generation electrothermal atomic absorption spectrometry. Anal. Chim. Acta 452,207–215 (2002)

[202] Vilanó, M. , Rubio, R. : Determination of arsenic species in oyster tissue by microwaveassisted extraction and liquid chromatography-atomic fluorescence detection. Appl. Organomet. Chem. 15, 658–666 (2001)

[203] Miekeley, N. , Mortari, S. R. , Schubach, A. O. : Monitoring of total antimony and its species by ICP-MS and on-line ionchromatography in biological samples from patients treated for leishmaniasis. Anal. Bioanal. Chem. 372,495–502 (2002)

[204] Petit de Peña, Y. , Vielma, O. , Burguera, J. L. , Burguera, M. , Rondon, C. , Carrero, P. : On line determination of antimony (III) and antimony (V) in liver tissue and whole blood by flow injection- hydride generation- atomic absorption spectrometry. Talanta 55,743–754 (2001)

[205] Smichowski, P. , Madrid, Y. , Cámara, C. : Analytical methods for antimony speciation in waters at trace and ultratrace levels. A review. Fresenius J. Anal. Chem. 360,623–629 (1998)

[206] Chau, Y. K. , Yang, F. , Brown, M. : Determination of methylcyclopentadienyl-manganese tricarbonyl (MMT) in gasoline and environmental samples by gas chromatography with helium microwave plasma atomic emission detection. Appl. Organomet. Chem. 11,31–37(1997)

[207] Cámara, C. , Cornelis, R. , Quevauviller, P. : Assessment of methods currently used for the determination of Cr and Se species in solutions. Trends Anal. Chem. 19,189-194 (2000)

[208] Darrie, G. : The importance of chromium in occupational health. In: Ebdon, L. , Pitts, L. , Cornelis, R. , Crews, H. , Donard, O. F. X. , Quevauviller, P. (eds.) Trace Element Speciation for Environment, Food and Heath, pp. 315–330. RSC, Cambridge (2001)

[209] Kotas, J. , Stasicka, Z. : Chromium occurrence in the environment and methods of its speciation. Environ. Pollut. 107,263–283 (2000)

[210] Marqués, M. J. , Salvador, A. , Morales-Rubio, A. , de la Guardia, M. : Chromium speciation in liquid matrices: a surveyof the literature. Fresenius J. Anal. Chem. 367,601–613 (2000)

[211] Yao, W. , Byrne, R. H. : Determination of trace chromium (VI) and molybdenum (VI) in natural and bottled mineral waters using long pathlength absorbance spectroscopy (LPAS). Talanta 48,277–282 (1999)

[212] Chai, Z. , Mao, X. , Hu, Z. , Zhang, Z. , Chen, C. , Feng, W. , Hu, S. , Ouyang, H. : Overview of the methodology of nuclear analytical techniques for speciation studies of trace elements in the

biological and environmental sciences. Anal. Bioanal. Chem. 372,407-411 (2002)

[213] Olivas, R. M., Donard, O. F. X., Camara, C., Quevauviller, P.: Analytical techniques applied to the speciation of selenium in environmental matrices. Anal. Chim. Acta 286,357-370 (1994)

[214] Quevauviller, P., Morabito, R.: Evaluation of extraction recoveries for organometallic determinations in environmental matrices. Trends Anal. Chem. 19,86-96 (2000)

[215] Crews, H. M.: Speciation of trace elements in foods, with special reference to cadmium and selenium: is it necessary? Spectrochim. Acta B 53,213-219 (1998)

[216] Günther, K., Kastenholz, G. J. B.: Characterization of high molecular weight cadmium species in contaminated vegetable food. Fresenius J. Anal. Chem. 368,281-287 (2000)

[217] Lobinski, R., Edmonds, J. S., Suzuki, K. T., Uden, P. C.: Species-selective determination of selenium compounds in biological materials. Pure Appl. Chem. 72,447-461 (2000)

[218] Crnoja, M., Haberhauer-Troyer, C., Rosenberg, E., Grasserbauer, M.: Determination of Sn and Pb organic compounds by solid-phase microextraction-gas chromatography-atomic emission detection (SPME-GC-AED) after in situ propylation with sodium tetrapropylborate. J. Anal. At. Spectrom. 16,1160-1166 (2001)

[219] Forsyth, D. S., Taylor, J.: Detection of organotin, organomercury, and organolead compounds with a pulsed discharge detector (PDD). Anal. Bioanal. Chem. 374,344-347 (2002)

[220] Jackson, B. P. S., Allen, P. L., Hopkins, W. A., Bertsch, P. M.: Trace element speciation in largemouth bass (Micropterus salmoides) from a fly ash settlingbasin by liquid chromatography-ICP-MS. Anal. Bioanal. Chem. 374,203-211 (2002)

[221] Pyrzyńska, K.: Organolead speciation in environmental samples: a review. Microchim. Acta 122, 279-293 (1996)

[222] Pereiro, I. R., Diaz, A. C.: Speciation of mercury, tin, and lead compounds by gas chromatography with microwave-induced plasma and atomic-emission detection (GC-MIP-AED). Anal. Bioanal. Chem. 372,74-90 (2002)

[223] Salih, B.: Speciation of inorganic and organolead compounds by gas chromatography atomic absorption spectrometry and the determination of lead species after pre-concentration onto diphenylthiocarbazone-anchored polymeric microbeads. Spectrochim. Acta B 55,1117-1127 (2000)

[224] Szpunar-Lobińska, J., Witte, C., Łobiński, R., Adams, F. C.: Separation techniques in speciation analysis for organometallic species. Fresenius J. Anal. Chem. 351,343-477 (1995)

[225] Wasik, A., Namiesnik, J.: Speciation of organometallic compounds of tin, lead, and mercury. Polish J. Environ. Studies 10,405-413 (2001)

[226] Williams, S. P.: Occupational health and speciation using nickel and nickel compounds as an example. In: Ebdon, L., Pitts, L., Cornelis, R., Crews, H., Donard, O. F. X., Quevauviller, P. (eds.) Trace Element Speciation for Environment, Food and Health, pp. 297-307. RSC, Cambridge (2001)

[227] Schedlbauer, O. F., Heumann, K. G.: Development of an isotope dilution mass spectrometric method for dimethylthallium speciation and first evidence of its existence in the

ocean. Anal. Chem. 71,5459-5464 (1999)

[228] Elgethun, K., Neumann, C., Blake, P.: Butyltins in shellfish, finfish, water and sediment from the Coos Bay estuary (Oregon, USA). Chemosphere 41,953-964 (2000)

[229] White, S., Catterick, T., Fairman, B., Webb, K.: Speciation of organo-tin compounds using liquid chromatography- atmospheric pressure ionisation mass spectrometry and liquid chromatography-inductively coupled plasma mass spectrometry as complementary techniques. J. Chromatogr. A 794,211-218 (1998)

[230] Lindemann, T., Range, A., Dannecker, W., Neidhart, B.: Stability studies of arsenic, selenium, antimony and tellurium species in water, urine, fish and soil extracts using HPLC/ICP-MS. Fresenius J. Anal. Chem. 368,214-220 (2000)

[231] De Carvalho, L. M., Schwedt, G.: Sulfur speciation by capillary zone electrophoresis: conditions for sulfite stabilization and determination in the presence of sulfate, thiosulfate and peroxodisulfate. Fresenius J. Anal. Chem. 368,208-213 (2000)

[232] Unger-Heumann, M.: Rapid tests- a convenient tool for sample screening with regard to element speciation. In: Ebdon, L., Pitts, L., Cornelis, R., Crews, H., Donard, O. F. X., Quevauviller, P. (eds.) Trace Element Speciation for Environment, Food and Health, pp. 211-222. RSC, Cambridge (2001)

[233] Bi, S. -P., Yang, X., Zhang, F., Wang, X., Zou, G.: Analytical methodologies for aluminium speciation in environmental and biological samples- a review. Fresenius J. Anal. Chem. 370,984 -996 (2001)

[234] Gómez-Ariza, J. L., Morales, E., Giraldez, I., Sanchez-Rodas, D.: Sample treatment and storage in speciation analysis. In: Ebdon, L., Pitts, L., Cornelis, R., Crews, H., Donard, O. F. X., Quevauviller, P. (eds.) Trace Element Speciation for Environment, Food and Health, pp. 51-80. RSC, Cambridge (2001)

[235] Rosenberg, E., Ariese, F.: Quality control in speciation analysis. In: Ebdon, L., Pitts, L., Cornelis, R., Crews, H., Donard, O. F. X., Quevauviller, P. (eds.) Trace Element Speciation for Environment, Food and Health, pp. 17-50. RSC, Cambridge (2001)

[236] Huo, D., Kingston, H. M., Larget, B.: Application of isotope dilution in elemental speciation: speciated isotope dilution mass spectrometry (SIDMS). In: Caruso, J. A., Sutton, K. L., Ackley, K. L. (eds.) Elemental Speciation New Approaches for Trace Element Analysis, pp. 277-313. Elsevier Science B. V, Amsterdam (2000)

[237] Encinar, J. R.: Isotope dilution analysis for speciation. Anal. Bioanal. Chem. 375,41-43(2003)

[238] Emons, H.: Challenges from speciation analysis for the development of biological reference materials. Fresenius J. Anal. Chem. 370,115-119(2001)

[239] Kabata-Pendias, A., Mukherjee, A. B.: Trace Elements from Soil to Human. Springer, Berlin (2007)

[240] Coelho, N. M. M., Coelho, L. M., de Lima, E. S., Pastor, A., de la Guardia, M.: Determination of arsenic compounds in beverages by high-performanceliquid chromatography-inductively cou-

pled plasma mass spectrometry. Talanta 66,818-822 (2005)

[241] Dietz, C., Sanz, J., Sanz, E., Muñoz-Olivas, R., Cámara, C.: Current perspectives in analyte extraction strategies for tin and arsenic speciation. J. Chromatogr. A 1153,114-129 (2007)

[242] Li, X., Jia, J., Wang, Z.: Speciation of inorganic arsenic by electrochemical hydride generation atomic absorption spectrometry. Anal. Chim. Acta 560,153-158 (2006)

[243] Mandal, B. K., Suzuki, K. T.: Arsenic round the world: a review. Talanta 58,201-235 (2002)

[244] McLaughlin, M. J., Parker, D. R., Clarke, J. M.: Metals and micronutrients- food safety issues. Field Crops Res. 60,143-163 (1999)

[245] Reilly, C.: Pollutants in food- metals and metalloids. In: Szefer, P., Nriagu, J. O. (eds.) Mineral Components in Foods, pp. 363-388. CRC Press/Taylor Francis Group, London/New York (2007)

[246] Sharma, V. K., Sohn, M.: Aquatic arsenic: toxicity, speciation, transformations, and remediation. Environ. Int. 35,743-759 (2009)

[247] Jain, C. K., Ali, I.: Arsenic: occurrence, toxicity and speciation techniques. Water Res. 34,4304-4312 (2000)

[248] Wrobel, K., Wrobel, K., Parker, B., Kannamkumarath, S. S., Caruso, J. A.: Determination of As(III), As(V), monomethylarsonic acid, dimethylarsinic acid and arsenobetaine by HPLC-ICP-MS: analysis of reference materials, fish tissues and urine. Talanta 58,899-907 (2002)

[249] Pearson, G. F., Greenway, G. M., Brima, E. I., Haris, P. I.: Rapid arsenic speciation using ion pair LC-ICPMS with a monolithic silica column reveals increased urinary DMA excretion after ingestion of rice. JAAS 22(4),361-369 (2007)

[250] Chatterjee, A.: Determination of total cationic and total anionic arsenic species in oyster tissue using microwave-assisted extraction followed by HPLC-ICP-MS. Talanta 51,303-314(2000)

[251] McSheehy, S., Pohl, P., Łobinski, R., Szpunar, J.: Investigations of arsenic speciation in oyster test reference material by multidimensional HPLC-ICP-MS and electrospray tandem mass spectrometry (ES-MS-MS). Analyst 126,1055-1062 (2001)

[252] McSheehy, S., Pohl, P., Velez, D., Szpunar, .: Multidimensional liquid chromatography with parallel ICP-MS and electrospray MS-MS detection as a tool for the characterization of arsenic species in algae. Anal. Bioanal. Chem. 372,457-466 (2002)

[253] Coelho, N. M. M., Parilla, C., Cervera, M. L., Pastor, A., de la Guardia, M.: High performance liquid chromatography- atomic fluorescence spectrometric determination of arsenic species in beer samples. Anal. Chim. Acta 482,73-80 (2003)

[254] Pal, A., Chowdhury, U. K., Mondal, D., Das, B., Nayak, B., Ghosh, A., Maity, S., Chakraborti, D.: Arsenic burden from cooked rice in the populations of arsenic affected and nonaffected areas and Kolkata city in West-Bengal, India. Environ. Sci. Technol. 43,3349-3355 (2009)

[255] Roychowdhury, T.: Impact of sedimentary arsenic through irrigated groundwater on soil, plant, crops and human continuum from Bengal delta: special reference to raw and cooked rice. Food

Chem. Toxicol. 46, 2856–2864 (2008)

[256] Williams, P. N., Islam, M. R., Adomako, E. E., Raab, A., Hossain, S. A., Zhu, Y. G., Feldmann, J., Meharg, A. A. : Increase in rice grain arsenic for regions of Bangladesh irrigating paddies with elevated arsenic in groundwaters. Environ. Sci. Technol. 40, 4903–4908 (2006)

[257] Ackerman, A. H., Creed, P. A., Parks, A. N., Fricke, M. W., Schwegel, C. A., Creed, J. T., Heitkemper, D. T., Vela, N. P. : Comparison of a chemical and enzymatic extraction of arsenic from rice and an assessment of the arsenic absorption from contaminatedwater by cooked rice. Environ. Sci. Technol. 39, 5241–5246 (2005)

[258] Kohlmeyer, U., Jakubik, S., Kuballa, J., Jantzen, E. : Determination of arsenic species in fish oil after acid digestion. Microchim. Acta 151, 249–255 (2005)

[259] Vela, N. P., Heitkemper, D. T. : Total arsenic determination and speciation in infant food products by ion chromatography-inductively coupled plasma-mass spectrometry. J. AOAC Int. 87, 244–252 (2004)

[260] Lai, V. W. -M., Cullen, W. R., Ray, S. : Arsenic speciation in scallops. Mar. Chem. 66, 81–89 (1999)

[261] Brisbin, J. A., B'Hymer, C., Caruso, J. A. : A gradient anion exchange chromatographic method for the speciation of arsenic in lobster tissue extracts. Talanta 58, 133–145 (2002)

[262] Villa-Lojo, M. C., Alonso-Rodríguez, E., López-Mahía, P., Muniategui-Lorenzo, S., Prada-Rodríguez, D. : Coupled high performance liquid chromatography-microwave digestion-hydride generation-atomic absorption spectrometry for inorganic and organic speciation in fish tissue. Talanta 57, 741–750 (2002)

[263] Gong, Z., Lu, X., Ma, M., Watt, C., Le, X. C. : Arsenic speciation analysis. Talanta 58, 77–96 (2002)

[264] Eustace, D. J., Walters, M., Riley, S., Anderson, M. : Practical assessment of mercury exposure, contamination and clean-up. Chem. Health Safety 11, 16–23 (2004)

[265] Gochfeld, M. : Cases of mercury exposure, bioavailability, and absorption. Ecotoxicol. Environ Safety 56, 174–179 (2003)

[266] Ozuah, P. O. : Mercury poisoning. Curr. Probl. Pediatr. 30, 91–99 (2000)

[267] Zahir, F., Rizwi, S. J., Haq, S. K., Khan, R. H. : Low dose mercury toxicity and human health. Environ. Toxicol. Pharmacol. 20, 351–360 (2005)

[268] Magalhães, M. C., Costa, V., Menezes, G. M., Pinho, M. R., Santos, R., Monteiro, L. R. : Intra- and inter-specific variability in total and methylmercury bioaccumulation by eight marine fish species from the Azores. Mar. Pollut. Bull. 54, 1654–1662 (2007)

[269] Carro, A. M., Mejuto, M. C. : Application of chromatographic and electrophoretic methodol-ogy to the speciation of organomercury compounds in food analysis. J. Chromatogr. A 882, 283–307 (2000)

[270] Bouyssiere, B., Szpunar, J., Lobinski, R. : Gas chromatography with inductively coupled plasma mass spectrometric detection in speciation analysis. Spectrochim. Acta B 57, 805–828 (2002)

[271] Mizuishi, K., Takeuchi, M., Hobo, T.: Direct GC determination of methylmercury chloride on HBr-methanol-treated capillary columns. Chromatographia 44, 386–392 (1997)

[272] Donard, O. F. X., Krupp, E., Pecheyran, C., Amouroux, D., Ritsema, R.: Trends in speciation analysis for routine and new environmental issues. In: Caruso, J. A., Sutton, K. L., Ackley, K. L. (eds.) Elemental Speciation New Approaches for Trace Element Analysis, pp. 451–493. Elsevier Science B. V, Amsterdam (2000)

[273] Baeyens, W., Leermakers, M., Molina, R., Holsbeek, L., Joiris, C. R.: Investigation of headspace and solvent extraction methods for the determination of dimethyl-and monomethyl-mercury in environmental matrices. Chemosphere 39, 1107–1117 (1999)

[274] Palmieri, H. E. L., Leonel, L. V.: Determination of methylmercury in fish tissue by gas chromatography with microwave-induced plasma atomic emission spectrometry after derivatization with sodium tetraphenylborate. Fresenius J. Anal. Chem. 363, 466–469 (2000)

[275] Rosenkrantz, B., Bettmer, J.: Rapid separation of elemental species by multicapillary GC. Anal. Bioanal. Chem. 373, 461–465 (2002)

[276] Grinberg, P., Campos, R. C., Mester, Z., Sturgeonet, R. E.: Solid phase microextraction capillary gas chromatography combined with furnace atomization plasma emission spectrometry for speciation of mercury in fish tissue. Spectrochim. Acta A 58, 427–441 (2003)

[277] Vanhaecke, F., Resano, M., Moens, L.: Electrothermal vaporisation ICP-mass spectrometry (ETV-ICP-MS) for the determination and speciation of trace elements in solid samples- a review of real-life applications from the author's lab. Anal. Bioanal. Chem. 375, 188–195 (2002)

[278] Liang, L.-N., Jiang, G.-B., Liu, J.-F., Hu, J.-T.: Speciation analysis of mercury in seafood by using high-performance liquid chromatography on-line coupled with cold-vapor atomic fluorescence spectrometry via a post column microwave digestion. Anal. Chim. Acta 477, 131–137 (2003)

[279] Moreno, M. E., Pe'rez-Conde, C., Cámara, C.: The effect of the presence of volatile organo-selenium compounds on the determination of inorganic selenium by hydride generation. Anal. Bioanal. Chem. 375, 666–672 (2003)

[280] Kannamkumarath, S. S., Wrobel, K., Wrobel, K., Vonderheide, A., Caruso, J. A.: HPLC-ICP-MS determination of selenium distribution and speciation in different types of nut. Anal. Bioanal. Chem. 373, 454–460 (2002)

[281] Uden, P. C.: Modern trends in the speciation of selenium by hyphenated techniques. Anal. Bioanal. Chem. 373, 422–431 (2002)

[282] Gallignani, M., Valero, M., Brunetto, M. R., Burguera, J. L., Burguera, M., Petit de Pena, Y.: Sequential determination of Se(IV) and Se(VI) by flow injection-hydride generation-atomic absorption spectrometry with HCl/HBr microwave aided pre-reduction of Se(VI) to Se(IV). Talanta 52, 1015–1024 (2000)

[283] Dernovics, M., Stefánka, Z., Fodor, P.: Improving selenium extraction by sequential enzymatic

processes for Se-speciation of selenium-enriched Agaricus bisporus. Anal. Bioanal. Chem. 372, 473-480 (2002)

[284] Stefánka, Z. , Ipolyi, I. , Dernovics, M. , Fodor, P. : Comparison of sample preparation methods based on proteolytic enzymatic processes for Se-speciation of edible mushroom (Agaricus bisporus) samples. Talanta 55, 437-447 (2001)

[285] Wrobel, K. , Kannamkumarath, S. S. , Wrobel, K. , Caruso, J. A. : Hydrolysis of protein with methanesulfonic acid for improved HPLC-ICP-MS determination of seleno-methionine in yeast and nuts. Anal. Bioanal. Chem. 375, 133-138 (2003)

[286] Tuzen, M. , Saygi, K. O. , Soylak, M. : Separation and speciation of selenium in food and water samples by the combination of magnesium hydroxide coprecipitation-graphite furnace atomic absorption spectrometric determination. Talanta 71, 424-429 (2007)

[287] Gosetti, F. , Frascarolo, P. , Polati, S. , Medana, C. , Gianotti, V. , Palma, P. , Aigotti, R. , Baiocchi, C. , Gennaro, M. C. : Speciation of selenium in diet supplements by HPLC-MS/MS methods. Food Chem. 105, 1738-1747 (2007)

[288] Viñas, P. , López-García, I. , Merino-Meroño, B. , Campillo, N. , Hernández-Córdoba, M. : Determination of selenium species in infant formulas and dietetic supplements using liquid chromatography-hydride generation atomic fluorescence spectrometry. Anal. Chim. Acta 535, 49-56 (2005)

[289] Guerin, T. , Astruc, A. , Astruc, M. : Speciation of arsenic and selenium compounds by HPLC hyphenated to specific detectors: a review of the main separation techniques. Talanta 50, 1-24 (1999)

[290] Pyrzyńska, K. : Analysis of selenium species by capillary electrophoresis. Talanta 55, 657-667 (2001)

[291] Infante, H. G. , Hearn, R. , Catterick, T. : Current mass spectrometry strategies for selenium speciation in dietary sources of high-selenium. Anal. Bioanal. Chem. 382, 957-967 (2005)

[292] Blunden, S. , Wallace, T. : Tin in canned food: a review and understanding of occurrence and effect. Food Chem. Toxicol. 41, 1651-1662 (2003)

[293] Crews, H. M. : The importance of trace element speciation in food issues. In: Ebdon, L. , Pitts, L. , Cornelis, R. , Crews, H. , Donard, O. F. X. , Quevauviller, P. (eds.) Trace Element Speciation for Environment, Food and Health, pp. 223-227. RSC, Cambridge (2001)

[294] Perring, L. , Basic-Dvorzak, M. : Determination of total tin in canned food using inductively coupled plasma atomic emission spectroscopy. Anal. Bioanal. Chem. 374, 235-243 (2002)

[295] Reilly, C. : Metal Contamination of Food: Its Significance for Food Quality and Human Health, pp. 1-266. Blackwell Science, Oxford (2002)

[296] Hoch, M. : Organotin compounds in the environment- an overview. Appl. Geochem. 16, 719-743 (2001)

[297] Bettmer, J. , Buscher, W. , Cammann, K. : Speciation of mercury, platinum and tin- focus of research and future developments. Fresenius. J. Anal. Chem. 354, 521-528 (1996)

[298] Brede, C., Pedersen-Bjergaard, S., Lundanes, E., Greibrokk, T.: Capillary gas chromatography coupled with microplasma mass spectrometry for organotin speciation. J. Chromatogr. A 849, 553–562 (1999)

[299] Caldorin, R., Menegário, A. A.: Speciation analysis of Sn(II) and Sn(IV) using baker's yeast and inductively coupled plasma optical emission spectrometry. Microchim. Acta 157, 201–207 (2007)

[300] Zhu, X., Zhao, L., Wang, B.: Speciation analysis of inorganic tin (Sn(II)/Sn(IV)) by graphite furnace atomic absorption spectrometry following ion-exchange separation. Microchim. Acta 155, 459–463 (2006)

[301] Belfroid, A. C., Purperhart, M., Ariese, F.: Organotin levels in seafood. Mar. Pollut. Bull. 40, 226–232 (2000)

[302] Albalat, A., Potrykus, J., Pempkowiak, J., Porte, C.: Assessment of organotin pollution along the Polish coast (Baltic Sea) by using mussels and fish as sentinel organisms. Chemosphere 47, 165–171 (2002)

[303] Sudaryanto, A., Takahashi, S., Tanabe, S., Muchtar, M., Razak, H.: Occurrence of butyltin compounds in mussels from Indonesian coastal waters and some Asian countries. Water Sci. Technol. 42, 71–78 (2000)

[304] Kannan, K., Falandysz, J.: Butyltin residues in sediment, fish, fish-eating birds, harbor porpoise and human tissues from the Polish coast of the Baltic Sea. Mar. Pollut. Bull. 34, 203–207 (1997)

[305] St-Jean, S. D., Courtenay, S. C., Pelletier, É., St-Louis, R.: Butyltin concentrations in sediments and blue mussels (Mytilus edulis) of the southern Gulf of St. Lawrence, Canada. Environ. Technol. 20, 181–189 (1999)

[306] Marcic, C., Lespes, G., Potin-Gautier, M.: Pressurised solvent extraction for organotin speciation in vegetable matrices. Anal. Bioanal. Chem. 382, 1574–1583 (2005)

[307] Dong, N. V., Thuy, T. X. B., Solomon, T.: The transformation of phenyltin species during sample preparation of biological tissues using multi-isotope spike SSID-GC-ICPMS. Anal. Bioanal. Chem. 392, 737–747 (2008)

[308] Üveges, M., Abrankó, L., Fodor, P.: Optimization of GC-ICPMS system parameters for the determination of butyltin compounds in Hungarian freshwater origin sediment and mussel samples. Talanta 73, 490–497 (2007)

[309] Prieto, A., Zuloaga, O., Usobiaga, A., Etxebarria, N., Fernández, L. A., Marcic, C., de Diego, A.: Simultaneous speciation of methylmercury and butyltin species in environmental samples by headspace-stir bar sorptive extraction-thermal desorption-gas chromatography-mass spectrometry. J. Chromatogr. A 1185, 130–138 (2008)

[310] Szefer, P., Grembecka, M.: Mineral components in foods of animal origin and in honey. In: Szefer, P., Nriagu, J. O. (eds.) Mineral Components in Foods, pp. 163–230. CRC Press/Taylor Francis Group, London/New York (2007)

[311] Barałkiewicz, D., Kozka, M., Piechalak, A., Tomaszewska, B., Sobczak, P.: Determination of cadmium and lead species and phytochelatins in pea (Pisum sativum) by HPLC-ICP-MS and HPLC-ESI-MS^n Talanta 79, 493-498 (2009)

[312] Bermejo, P., Pena, E., Dominguez, R., Bermejo, A., Fraga, J. M., Cocho, J. A.: Speciation of iron in breast milk and infant formulas whey by size exclusion chromatography-high performance liquid chromatography and electrothermal atomic absorption spectrometry. Talanta 50, 1211-1222 (2000)

[313] Grembecka, M., Szefer, P.: Chemometria w badaniu żywności, [w:] Zuba, D., Parczewski, A. (red): Chemometria w analityce: wybrane zagadnienia, pp. 253 - 279. Wydawnictwo Instytutu Ekspertyz Sadowych, Kraków (2008)

[314] Luykx, D. M. A. M., van Ruth, S. M.: An overview of analytical methods for determining the geographical origin of food products. Food Chem. 107, 897-911 (2008)

[315] Szefer, P.: Chemometric techniques in analytical evaluation of food quality. In: Szefer, P., Nriagu, J. O. (eds.) Mineral Components in Foods, pp. 69-122. CRC Press/Taylor Francis Group, London/New York (2007)

[316] Brito, G., Peña-Méndez, E., Novotná, K., Díaz, C., García, F.: Differentiation of heat-treated pork liver pastes according to their metal content using multivariate data analysis. Eur. FoodRes.
Technol. 218, 584-588 (2004)

[317] Heaton, K., Kelly, S. D., Hoogewerff, J., Woolfe, M.: Verifying the geographical origin of beef: the application of multi-element isotope and trace element analysis. Food Chem. 107, 506-515 (2008)

[318] Schmidt, O., Quilter, J. M., Bahar, B., Moloney, A. P., Scrimgeour, C. M., Begley, I. S., Monahan, F. J.: Inferring the origin and dietary history of beef from C, N and S stable isotope ratio analysis. Food Chem. 91, 545-549 (2005)

[319] Astorga-Espana, M. S., Pena-Mendez, E. M., Garcia-Montelongo, F. J.: Application of principal component analysis to the study of major cations and trace metals in fish from Tenerife (Canary Islands). Chemom. Intell. Lab. Syst. 49, 173-178 (1999)

[320] Hellou, J., Zitko, V., Friel, J., Alkanani, T.: Distribution of elements in tissues of yellowtail flounder Pleuronectes ferruginea. Sci. Total Environ. 181, 137-146 (1996)

[321] Szefer, P., Domagała-Wieloszewska, M., Warzocha, J., Garbacik-Wesołowska, A., Ciesielski, T.: Distribution and relationships of mercury, lead, cadmium, copper and zinc in perch (Perca fluviatilis) from the Pomeranian Bay and Szczecin Lagoon, southern Baltic. Food Chem. 81, 73-83 (2003)

[322] Bechmann, I. E., Stürup, S., Kristensen, L. V.: High resolution inductively coupled plasma mass spectrometry (HR-ICPMS) determination and multivariate evaluation of 10 trace elements in mussels from 7 sites in Limfjorden, Denmark. Fresenius J. Anal. Chem. 368, 708 - 714 (2000)

[323] Bermejo-Barrera, P., Moreda-Piñeiro, A., Bermejo-Barrera, A.: Sample pre-treatment fortrace elements determination in seafood products by atomic absorption spectrometry. Talanta 57, 969–984 (2001)

[324] Machado, M. L., Mendez, E. P., Sanchez, M. S., Montelongo, F. G.: Interpretation of heavy metal data from mussel by use of multivariate classification techniques. Chemoshere 38, 1103–1111 (1999)

[325] Szefer, P., Ali, A. A., Ba-Haroon, A. A., Rajeh, A. A., Gełdon, J., Nabrzyski, M.: Distribution and relationships of selected trace metals in molluscs and associated sediments from the Gulf of Aden, Yemen. Environ. Pollut. 106, 299–314 (1999)

[326] Szefer, P., Frelek, K., Szefer, K., Lee, C.-B., Kim, B.-S., Warzocha, J., Zdrojewska, I., Ciesielski, T.: Distribution and relationships of trace metals in soft tissue, byssus and shells of Mytilus edulis trossulus from the southern Baltic. Environ. Pollut. 120, 423–444 (2002)

[327] Szefer, P., Kim, B.-S., Kim, C.-K., Kim, E.-H., Lee, C.-B.: Distribution and coassociations of trace elements in soft tissue and byssus of Mytillus edulis galloprovincialis relative to the surrounding seawater and suspended matter of the southern part of Korean Peninsula. Environ. Pollut. 129, 209–228 (2004)

[328] Szefer, P., Wołowicz, M.: Occurrence of metals in the cockle Cerastoderma glaucum from different geographical regions in view of principal component analysis. SIMO-Mar. Pollut. 64, 253–264 (1993)

[329] Brescia, M. A., Monfreda, M., Buccolieri, A., Carrino, C.: Characterisation of the geographical origin of Buffalo milk and mozzarella cheese by means of analytical and spectroscopic determinations. Food Chem. 89, 139–147 (2005)

[330] Pillonel, L., Badertscher, R., Froidevaux, P., Haberhauer, G., Holzl, S., Horn, P., Jakob, A., Pfammatter, E., Piantini, U., Rossmann, A., Tabacchi, R., Bosset, J. O.: Stable isotope ratios, major, trace and radioactive elements in emmental cheeses of different origins. Lebensm.-Wiss. u-Technol. 36, 615–623 (2003)

[331] Puerto, P., Fresno Baquero, M., Rodriguez Rodriguez, E. M., Darias, M. J., Diaz Romero, C.: Chemometric studies of fresh and semi-hardgoats' cheeses produced in Tenerife (Canary Islands). Food Chem. 88, 361–366 (2004)

[332] Woodcock, T., Fagan, C. C., O'Donnell, C. P., Downey, G.: Application of near and mid-infrared spectroscopy to determine cheese quality and authenticity. Food Biopocess. Technol. 1, 117–129 (2008)

[333] Devillers, J., Doré, J. C., Marenco, M., Poirier-Duchêne, F., Galand, N., Viel, C.: Chemometrical analysis of 18 metallic and nonmetallic elements found in honeys sold in France. J. Agric. Food Chem. 50, 5998–6007 (2002)

[334] Hernandez, O. M., Fraga, J. M. G., Jiménez, A. I., Jiménez, F., Arias, J. J.: Characterization of honey from the Canary Islands: determination of the mineral content by atomic absorption spectrophotometry. Food Chem. 93, 449–458 (2005)

[335] Lachman, J., Kolihová, D., Miholová, D., Kosata, J., Titera, D., Kult, K.: Analysis of minority honey components: possible use for the evaluation of honey quality. Food Chem. 101, 973-979 (2007)

[336] Latorre, M. J., Peña, R., García, S., Herrero, C.: Authentication of Galician (N. W. Spain) honeys by multivariate techniques based on metal content data. Analyst 125, 307-312 (2000)

[337] Latorre, M. J., Peña, R., Pita, C., Botana, A., García, S., Herrero, C.: Chemometric classification of honeys according to their type. II. Metal content data. Food Chem. 66, 263-268 (1999)

[338] Pisani, A., Protano, G., Riccobono, F.: Minor and trace elements in different honey types produced in Siena County (Italy). Food Chem. 107, 1553-1560 (2008)

[339] Terrab, A., González, A. G., Díez, M. J., Heredia, F. J.: Mineral content and electrical conductivity of the honeys produced in Northwest Morocco and their contribution to the characterization of unifloral honeys. J. Sci. Food Agric. 83, 637-643 (2003)

[340] Škrbić, B., Onjia, A.: Multivariate analyses of microelement contents in wheat cultivated in Serbia (2002). Food Control. 18, 338-345 (2007)

[341] Anderson, K. A., Magnuson, B. A., Tschirgi, M. L., Smith, B.: Determining the geographic origin of potatoes with trace metal analysis using statistical and neural network classifiers. J. Agric. Food Chem. 47, 1568-1575 (1999)

[342] Gundersen, V., McCall, D., Bechmann, I. E.: Comparison of major and trace element concentrations in Danish greenhouse tomatoes (Lycopersicon esculentum Cv. Aromata F1) cultivated in different substrates. J. Agric. Food Chem. 49, 3808-3815 (2001)

[343] Mohamed, A. E., Rashed, M. N., Mofty, A.: Assessment of essential and toxic elements in some kinds of vegetables. Ecotoxicol. Environ. Safety 55, 251-260 (2003)

[344] Moros, J., Llorca, I., Cervera, M. L., Pastor, A., Garrigues, S., de la Guardia, M.: Chemometric determination of arsenic and lead in untreated powdered red paprika by diffuse reflectance near-infrared spectroscopy. Anal. Chim. Acta 613, 196-206 (2008)

[345] Padín, P. M., Peña, R. M., García Martín, M. S., Iglesias, R., Barro, S., Herrero, C.: Characterization of Galician (N. W. Spain) quality brand potatoes: a comparison study of several pattern recognition techniques. Analyst 126, 97-103 (2001)

[346] Rivero, R. C., Hernandez, P. S., Rodriguez, E. M. R., Martin, J. D., Romero, C. D.: Mineral concentrations in cultivars of potatoes. Food Chem. 83, 247-253 (2003)

[347] Plessi, M., Bertelli, D., Albasini, A.: Distribution of metals and phenolic compounds as a criterion to evaluate variety of berries and related jams. Food Chem. 100, 419-427 (2007)

[348] Herrador, A. M., Gonzalez, A. G.: Pattern recognition procedures for differentiation of Green, Black and Oolong teas according to their metal content from inductively coupled plasmaatomic emission spectrometry. Talanta 53, 1249-1257 (2001)

[349] Marcos, A., Fisher, A., Rea, G., Hill, S. J.: Preliminary study using trace element concentrations and a chemometrics approach to determine the geographical origin of tea. J. Anal. At. Spectrom. 13,

521-525 (1998)

[350] Moreda-Pineiro, A. , Fisher, A. , Hill, S. J. : The classification of tea according to region of origin using pattern recognition techniques and trace metal data. J. Food Compos. Anal. 16, 195-211 (2003)

[351] Anderson, K. A. , Smith, B. W. : Chemical profiling to differentiate geographic growing origins of coffee. J. Agric. Food Chem. 50, 2068-2075 (2002)

[352] Martín, M. J. , Pablos, F. , González, A. G. : Discrimination between arabica and robusta green coffee varieties according to their chemical composition. Talanta 46, 1259-1264 (1998)

[353] Awadallah, R. M. , Ismail, S. S. , Mohamed, A. E. : Application of multi-element clustering techniques of five Egyptian industrial sugar products. J. Radioanal. Nucl. Chem. 196, 377 - 385 (1995)

[354] Peña-Méndez, E. M. , Hernández-Suárez, M. , Díaz-Romero, C. , Rodríguez-Rodríguez, E. : Characterization of various chestnut cultivars by means of chemometrics approach. Food Chem. 107, 537-544 (2008)

[355] Dugo, G. , la Pera, L. , Pellicanó, T. M. , di Bella, G. , D'Imperio, M. : Determination of some inorganic anions and heavy metals in D. O. C. Golden and Amber Marsala wines: statistical study of the influence of ageing period, colour and sugar content. Food Chem. 91, 355-363 (2005)

[356] Frias, S. , Conde, J. E. , Rodríguez-Bencomo, J. J. , García-Montelongo, F. , Pérez-Trujillo, J. P. : Classification of commercial wines from the Canary Islands (Spain) by chemometric techniques using metallic contents. Talanta 59, 335-344 (2003)

[357] Kment, P. , Mihaljevič, M. , Ettler, V. , Šebek, O. , Strnad, L. , Rohlová, L. : Differentiation of Czech wines using multielement composition- a comparison with vineyard soil. Food Chem. 91, 157-165 (2005)

[358] Marengo, E. , Aceto, M. : Statistical investigation of the differences in the distribution of metals in Nebbiolo-based wines. Food Chem. 81, 621-630 (2003)

[359] Alcázar, A. , Pablos, F. , Martín, M. J. , González, A. G. : Multivariate characterisation of beers according to their mineral content. Talanta 57, 45-52 (2002)

[360] Adam, T. , Duthie, E. , Feldmann, J. : Investigations into the use of copper and other metals as indicators for the authenticity of Scotch whiskies. J. Inst. Brew. 108, 459-464 (2002)

[361] Camean, A. M. , Moreno, I. , López-Artíguez, M. , Repetto, M. , González, A. G. : Differentiation of Spanish brandies according to their metal content. Talanta 54, 53-59 (2001)

第10章
痕量分析在医学诊断和选择药物监控中的应用

10.1 代谢物组学

21世纪初,现代医学面临新的挑战,同样伴随着蓬勃发展的新技术、新颖的诊断和分析手段。为了应对这些挑战并弄清楚疾病过程,像基因组学、转录组学、蛋白质组学和代谢物组学研究得到了极大的发展。这些"组学",创造了一个新的名为"系统组学"的生物学门类。此外,基因组序列的破译促进了基因组学的发展。然而,仍然有很多问题没有明确的答案,尤其是有关基因表达和致病过程的联系仍没有明确的解释。

蛋白质和信使RNA与基因有直接联系,它们是基因的表达产物,相应地,代谢产物可以视为基因编码的中间产物。代谢组学似乎比其他组学更难以分析和进行生物学解释,但是这并不意味着我们无法从代谢组学中得到宝贵的有用信息。代谢组学这个领域提供的一些信息可以补充基因组和病理过程之间关系的知识。

人体中包括3万~5万个基因、15万~30万个转录组、大约100万个蛋白质,但是细胞代谢物(内源性低分子化合物)的数量为3500~10000个。与转录组和蛋白质的数量相比,细胞代谢物的数量相当少,但最终代谢谱中包含的依赖关系的数量是多维的,这种依赖性以及代谢物的数量使代谢组学既有趣又复杂。此外,迄今为止随着分析技术和先进的生物信息学方法的发展,过去那些含量极低的无法被检测的、潜在的重要代谢产物,已经可以成功地进行定量分析和解释了。

获得的代谢组学数据集经过生物信息学评估,补充了有关人体状况的信息,并与蛋白质组学和转录组学一起提供了生物过程及它们之间关系的整体视图。

10.2 系统生物学中的代谢组学

作为系统生物学的一部分,代谢组学的起源可以追溯到20世纪70年代,当时Authur Robinson及其同事在做有关志愿者营养需求的研究,建议考虑和分析他们尿液中代谢物的多条色谱峰[1]。

根据Nicholson等[2]的说法,代谢组学通过分析单个细胞、生物体液或组织的代谢成分,以一种综合的方式详细说明了代谢物组的生化特征和功能。然而,在文献中,对代谢组学的定义很明确,代谢组学是通过对血液、尿液和脑脊液等生物液体的分析,来研究整个生物体的代谢概况。代谢组学尽可能多地同时关注细胞代谢物的定量和定性测定。此外,代谢物组的概念是通过与基因组、转录组和蛋白质组的类比产生的,并将其定义为在一个给定细胞中低分子化合物(代谢物)的总量。此外,Nicholson和Wilson[3]还给代谢物组提供了一个额外的术语,定义其为所有细胞代谢物以及它们在生物体中相互作用的产物的总和。

代谢物组的全面定量和定性分析对于充分了解细胞功能至关重要。研究表明尽管一种酶浓度的微小变化对生化途径的整体变异性影响甚微,但细胞代谢物的水平却会被明显改变。代谢组是蛋白质组的一种衍生物,它直接受蛋白质组变化的影响。因此,作为基因组、转录组和蛋白质组之间相互作用的最终产物,代谢物含量会以更大的规模反映这些变化。可以说代谢组分析可以视为一种比其他组学中高分子量化合物分析更灵敏的工具。此外,对代谢组的研究有助于更容易地了解复杂的生物系统以及疾病期间和应用药物治疗后发生的变化。

每种物组都是创建一个系统生物学的共同基础。然而,代谢组学和基因组学、转录组学和蛋白质组学之间存在着显著的差异。这些差异可以从两个方面考虑,即理论和技术。从理论的角度来看,基因和代谢物之间的关系,就像基因、信使RNA分子和蛋白质之间的关系一样并不简单。单个基因可以决定特定细胞中特定代谢物的存在,但可能不会影响其含量,这取决于特定代谢途径中的酶活性以及作用于每种酶上的效应分子。从技术的角度来看,代谢物的含量取决于不同的外部和内部的因素,如机体的生理和病理状态。

应该强调的是,就技术差异而言,在代谢物组分析中,没有哪种单一的分析技术可以测定所有的低分子化合物,这一点跟基因组和蛋白质组分析不同。在代谢组学中,各种互补的分析技术通常与先进的生物信息学工具结合使用,用于大数据分析。

不同种类的生物体中的代谢物组的数量是不同的。在单细胞生物如面包酵母(酿酒酵母)中,代谢物的数量有600种,仅为基因总量(6000种)的十分之一,而植物的代谢物的数量要多得多,拟南芥(拟南芥属)的代谢物化合物估计为5000种。

总的来说,所有植物物种的代谢物的数量估计为9万~20万种,其中次生代谢物占了很大比例。不同的科学家对人类代谢物的数量有着不同的估计。比如,根据Kell[4]的说法,估计人类代谢物组的数量接近2700种,涉及1100~3300个反应,这仅仅是一个估计值,因为人类的代谢物组还没被完全地计算进来,许多酶的基质也没被发现。此外,在估计代谢物组的适当规模时存在困难,这些困难归因于许多确定的代谢物是外源的(来源于饮食或药物)或来源于内源性细菌菌群。目前,人类在线代谢物组数据库包含了超过4500个在人类血液中检测到的代谢物,以及3995个在人类尿液中检测到的代谢物。这个数据库还在不断地扩展和更新,据估计,它最终将包含超过1万种代谢物。

10.3 代谢组学工具

随着现代分析技术和先进预处理方法的出现,我们能够在相对较短的时间内准确地测定大量的代谢物,并且结果具有良好的可靠性。以前应用的生化检测仅限于测定一种单一的、之前已经确定了的代谢物或已知的代谢物组类群。分析技术的发展允许我们使用像核磁共振法(NMR)和质谱分析法(MS)等先进技术,这反过来又需要使用多元数学方法来计算经分析后获得的大量数据。

由于代谢物的化学结构变化很大,因此实际上几乎不能找到一种能够同时测定生物样品中所有代谢物的分析测试技术。许多代谢物是极性的,不易挥发,并且在它们的结构中没有染色基团,这使得它们很难被发现。因此,现代方法采用多种互相补充的分析技术来对代谢物进行综合分析,从而使大量的代谢物被检测和测定。

一个典型的代谢组学工作流程包括3个主要阶段:样品预处理、分析测定和生物信息学数据计算。

10.3.1 样品预处理

样品预处理是代谢组学研究中极为重要的一步,因为这对最终结果有很大的影响。有时,这一步可被省略或显著简化,这取决于所使用的方法和分析技术。在非靶向性研究如代谢物指纹识别中,尽管有尽可能多的代谢物被测定,但是样品前处理限于除去血液蛋白、离心分离和过滤等步骤。当使用NMR时,样品预处理也会简化,因为样品可以在氘化水(D_2O)中稀释。NMR实际上是无损检测,它允许重复利用相同的样本进行后续分析。这在分析技术中是一个例外,因为通常在分析过程中,测试样品会经历不可逆转的变化,并且不能再用于其他测试方法的测定。

液-液萃取(LLE)、固相萃取(SPE)和固相微萃取(SPME)经常用于样品预处理。萃取的类型与所选择测定的代谢物类型高度相关。在之前的代谢组学研究中,提取主要集中在充分稳定的、可以被连续提取的化合物中,如碳水化合物、酯类、氨基酸或有机酸。

代谢物主要被提取到水或甲醇溶液中,不管它们是亲水的还是亲脂的,都作为单独的部分被分离并进行分析。再次强调,没有哪种提取方法能同时提取所有的代谢物。适用于提取一组代谢物的应用条件可能导致另一组具有不同物理化学性质的代谢产物的降解。例如,生物碱的提取是在基础条件下进行的,然而这种条件却不利于醛类化合物的分离。类似地,在固相萃取条件下,吸附剂层的选择决定了给定代谢物的选择性。阳离子交换吸附剂层可适用于阴离子化合物的提取,而阴离子交换吸附层则可用于提取分离阳离子代谢物。由于在提取的化合物的化学结构中有特有的化学基团,因此也存在其他的吸附剂层对这类化学物质有极强的选择性,例如,核苷通过苯硼酸吸收层进行提取,其特点是在核苷结构中具有1,2-顺势二醇和1,3-顺势二醇,这种物质对苯硼酸具有高亲和力。然而,由于脱氧核苷的化学结构中没有顺势二醇,因此这种方法不适合脱氧核苷的提取。

此外,从分析过的许多代谢物的大量性能来看,提取方法实际上作用很小。通常,它可以被替换,或者像平时对硫醇化合物那样在衍生化之前进行。这些化合物是非常不稳定的,但是可以使用适当的衍生化步骤将它们转化为稳定的衍生品,使它们能够被测定,然而,此时由于进一步的分析只集中在经过化学改性的物质上,因此许多相关的代谢物被忽略了。

10.3.2 分析测定

核磁共振是一种利用磁场来确定有机化合物结构组成的技术。该技术可用于测定大量化学性质不同的细胞代谢物,并识别它们的结构。然而,这种技术对代谢物的痕量分析不够灵敏。经核磁共振分析得到的信息被称为核磁共振波谱,其中每个波段都对应着分析物的精确结构。这种图谱通常被看作分子指纹特征图谱。

一种比 NMR 更灵敏的技术是与各种允许对代谢物预先分离,如高效液相色谱(HPLC)、超高效液相色谱(UHPLC)、毛细管电泳(CE)、气相色谱(GC)相结合的 MS 检测技术。MS 的应用可以根据它们与分离系统的相互作用来分析大量的代谢物,首先是基于它们自身的相对分子量。

气相色谱与质谱联用仪(GC-MS),通常需要一个衍生步骤,以提高测定代谢物的稳定性和数量。分离后的代谢产物,经过衍生化后,被引入质谱仪,在其中它们经过电离,并根据质量-电荷比 m/z 被有选择性地筛选确定。Fiehn 等[5]利用 GC-MS

能够确定紫锥菊植物提取物中的326种代谢物。其中只有一半的代谢物被识别。其余的部分将通过气相色谱与串联质谱联用技术(GC-MS/MS)或NMR来确定。

二维气相色谱(GC/GC)与MS检测联用,有助于分离和识别不能被GC-MS技术分离的代谢产物。二维气相色谱(GC/GC)的优点是它结合了不同极性的色谱柱,并能在不同的温度下进行分析。因此,以前无法测定的代谢物可以用这种方法进行分离和测定。

与气相色谱分析法(GC)相比,用液相色谱-质谱联用分析法(LC-MS)在对样品进行分析时不需要衍生化步骤。利用具有不同理化性质的各种固定相的色谱柱,将代谢物从样品基质中分离出来。液相色谱-质谱联用分析法比气相色谱分析法更常用,因为它更适用于测定不稳定的化合物、难以衍生的化合物和非挥发性化合物[6-7]。因此,可以使用LC-MS来确定更多的具有各种物理化学性质的代谢物。此外,样品预处理过程要简单得多,这对分析变异性的最小化有显著的效果。

与LC-MS或GC-MS相比,毛细管电泳与质谱分析法(CE-MS)联用的相对较少。但是,CE的主要优点是有很高的分离效率,能在相对较短的时间内(10分钟以内)获得10万~100万个理论塔板数;分析样品用量(纳升级)和缓冲液用量(毫升级)均很少,而且从低分子量化合物到生物聚合物的均广泛适用。这对于昂贵和有害的化学物质的低消耗则具有显著的优势,既降低了成本和又符合绿色化学原则。但是,它的灵敏度,当以摩尔浓度计算时(浓度灵敏度)相对较低,而以分子质量计算时,(质量灵敏度)又相对较高质量,这种现象对痕量代谢物的测定尤其明显。此外,与LC方法相比,CE中不稳定的电渗透流造成[8]分析物迁移/保留时间的重复性较差。

傅里叶变换红外光谱(FTIR)基于被分析样品在红外范围内吸收特性进行测定。从获得的光谱中,可以识别特定的官能团和结构。在代谢物组学研究中,FTIR用于测定复杂的混合物,并可与LC和GC技术联合使用[9-10]。

10.3.3 获得数据集的生物信息学分析

分析技术的发展与新颖、先进的计算方法的采用相关,这些计算方法能对多元数据进行预处理和进一步研究。代谢概况的评估,可先使用适当的分离技术进行分离,后采用化学计量方法对其进行分析。然而,原始数据组包含许多由分析变异性引起的畸变(背景噪声、分析迁移/保留时间的变化、基线漂移或随着时间的推移灵敏度的降低)。采用恰当的化学计量预处理方法,能够将不需要的漂移或信号从原始数据集中剔除,而不会有丢失重要信息的风险。应该提到的是,化学计量通常用在数据预处理过程的每个阶段:从去噪和基线校正(Savitzky Golay和Cooley Tukey算法、傅里叶变换)到同步转移信号和被分析物的保留/迁移时间(相关优化翘曲、动态时间扭曲或参数时间扭曲)的各个阶段[11]。通常,用获得的数据集的化学计量预处理方法,通过主成分分析、聚类分析或判别分析来探索和阐明变量之间的关系。

主成分分析能够将大型数据矩阵简化为包含正交相关信息的 2 个或 3 个主要成分。采用这种方式,由许多变量影响的代谢特征的变化可以被测量和比较。随后,使用简化数据矩阵的其他计算程序,可以确定和评估变量(代谢物)的重要性,在这一步分析中,判别或回归计算方法非常重要。

代谢数据分析的另一个重要方面是对结果的形象化。在应用有效的生物信息学工具进行数据形象化之后,可以更容易地比较多变量数据矩阵。

10.4 代谢组学的研究思路

在代谢组学研究中,有 3 种常用的研究思路,分别是目标代谢组学、代谢物图谱和代谢特征指纹分析[12]。

1. 目标代谢组学

目标代谢组学被用于分析已知的、之前确认识别过的代谢物,例如,变化的生化过程或基因突变的标记物。所选择的代谢物,可以作为病理变化过程的基质或产物,使用校准曲线和稳定同位素标记的内部标准来定量测定。因此,测量的水平可以看作体内病理变化过程的指示器。由于代谢物是定量测定,其选择性提取在样品预处理过程中至关重要,因此使用适当的、经过验证的样品预处理步骤可以显著提高分析的灵敏度,这对于测定痕量代谢物(如植物激素)尤为重要。

2. 代谢物图谱

代谢物图谱基于对一组代谢物或某些代谢物序列的定量或筛选方法的分析。这种策略通常用于分析特定生化过程(如,克雷布斯循环、糖酵解)或具有共同物理化学性质的代谢物(如,碳水化合物、氨基酸)。与目标方法类似,样品前处理对所需的代谢物也有选择性,其目的是尽可能减少样品基质的影响。

3. 代谢特征指纹分析

代谢特征指纹分析广泛应用于功能基因组学和临床诊断学中。这一策略的重点是使用各种互补的分析技术来确定分析生物样本中存在的所有代谢物。这种方法是筛选分析的一个例证,在这种分析中,样本可被定性地分析和互相比较。作为分析测定的结果,使用多变量化学计量学方法对获得的光谱进行比较,以找到根据生物来源或身体状态区分分析样品的代谢物。这种方法的局限性是分析方法的低再现性和样本矩阵的影响,有时候随着改变样本分类和分析结果的解释,会引起一些问题。

10.5 代谢组学在低剂量代谢物分析中的应用实例

10.5.1 人体红细胞糖酵解循环代谢物代谢概况的电泳分析

在人体中,红细胞是一种特殊类型的细胞,不包含像细胞核和线粒体这样的细胞器。红细胞正常工作所需的能量几乎完全来自糖酵解过程。糖酵解过程中的代谢产物通常用酶催化方法测定。然而,酶催化法的局限性在于,在一个分析过程中只能确定一种代谢物。采用毛细管间接分光光度检测法,是一种替代酶催化法的方法,也适用于毛细管电泳与质谱耦合法(CE-ESI/MS)[14]。与MS联用检测时,虽然采用间接分光光度法检测的选择性和灵敏度比CE法低,但它具有低复杂性、易于分析、干扰背景成分影响相对较小等优点[15-17]。糖酵解周期的代谢产物,如6-磷酸葡萄糖(G6P)、磷酸果糖(F6P)、1,6-二磷酸果糖(F1,6P)、二羟基丙酮磷酸(DAP)、2,3-磷酸甘油酯(2,3-dpg)、3-磷酸甘油酸(3-PG)、磷酸烯醇丙酮酸(PEP)、丙酮酸(PYR)和乳酸(LA),都是低分子量化合物,且其结构中都没有明显的生色基团。在对方法[13]进行优化后,采用以下电泳分离条件:基础电解质由20mmol/L 2,6-吡啶羧酸酯和4mmol/L 十六烷基三甲基溴化铵(CTAB)组成,pH值为12.3;未熔融二氧化硅毛细管总长为100cm(距离探测器90cm),内部直径为50μm;施加电压为25kV;毛细管温度控制在15℃;样品用20s流动注射加入。

该方法的日内和日间重复性,分别表示被测定的代谢物迁移时间的变化系数(CV%),范围分别为0.3%~1.9%和3.8%~5.1%。高峰区域的日内和日间重复性,分别为2%~10.4%和10.2%~15.8%。该方法的线性范围为12.5~2000.0μM($1M=1mol/dm^3$,其线性方程的特征回归系数为0.997~0.999(除2,3-DPG回归系数为0.991外)。分析代谢物的检出限为2,3-DPG的38.1μg/mL($5×10^{-5}$M)到PEP的1.29μg/mL($6.25×10^{-6}$M)。

经过验证的方法用于22名健康志愿者提取的血液样本中糖酵解循环过程中代谢产物代谢概况的测定。为了分离红细胞。首先样品用离心机分离;然后进行溶解和超细过滤。红细胞中被测定的代谢物平均浓度范围F6P的49.6±23.0μM,到2,3-DPG的(3.1±0.9)mM。

经验证所获得的结果与文献中除使用免疫分析方法[18-20]外所获得的结果进行了比较,对于糖酵解周期中大多数测定的代谢产物而言,其平均浓度与文献中提到的浓度一致。唯一的例外是DAP含量,它的平均浓度是Minakami和Yoshikawa[18]所获得浓度的2倍。然而,在每项研究中,使用免疫分析方法测定得到的 DAP 含量也各不相同,范围为 9μM[20]到 138μM[18]。在电泳图谱中观察到的许多附加的峰中(除了糖酵解循环过程的代谢产物之外),其中一个峰被识别为与无机磷酸盐相对应。此外,测量得到的浓度最高的是 2,3-DPG,它是红细胞糖酵解周期的一个特定代谢物,在血红蛋白与氧的结合和释放过程中扮演着重要的角色。所得到的结果证实了该方法的高特异性和灵敏度。该方法从生物学的角度来看具有重要作用,因为它实现了单次运行中少于 15min 的持续糖酵解循环过程中所涉及的 7 个主要代谢物的测定,这是相比于酶催化方法只允许每次测定一个代谢物的主要优势。此外,该方法似乎适用于对患者和健康志愿者的糖酵解循环代谢物代谢概况进行比较研究。

10.5.2 肝癌患者尿液中新陈代谢概况的测定

肝癌(HCC)是目前世界上第六种最常见的癌症,也是全球第三大癌症死亡原因。肝癌的预诊主要取决于疾病的发展阶段。此外,目前,血清检测使用的 α-甲胎蛋白(AFP),由于较高的假阳性和假阴性结果被认为不够有效。因此,检测生物样本中的 HCC 时,需要一种对 HCC 灵敏度足够高且足够特别的新的生物标记物进行检测。

在样品制备过程中,用 GC-MS-三甲基硅-三氟或-奥乙酰胺(BSTFA)作为衍生剂,采用 GC-MS 方法进行研究。这项研究的目的是比较肝癌患者($n=20$)和健康志愿者($n=20$)的尿液样本的代谢概况,开发一种能将选定的代谢物检测出来的诊断模型,该模型具有潜在的肝癌诊断意义[21]。

样品预处理是获得理想结果的关键。为了有选择性地利用 GC-MS 从生物基质中确定代谢物,有必要进行衍生化操作。为了达到这个目的,尿液样本在 37℃条件下培养了 3min,并离心分离 15s,将 800μL 甲醇和 100μL 的 L-2-氯丙氨酸加入 1mL 的尿液上清液中。将所获得的混合物离心分离 5min 并采用超声处理。之后,使用 0.5M 的 NaOH 将样品的 pH 值调整为 9~10。然后通过内径为 0.45μm 薄膜过滤器过滤,将 100μL 的小份滤液在氮气流下干燥。这些样品使用 100μL 的 BSTFA 和 1% 的 TMCS(三甲基氯硅烷)在 100℃条件下衍生化处理 1h。

在对该方法进行优化后,采用了以下分析条件:具有固定相 HP-5MS 的硅毛细管(内径为 30m×0.25mm,厚度为 25μm);柱状温度从 80℃开始,以 10℃/min 的速度升温到 280℃;注射温度恒定在 250℃;氢气作为一种运载气体,流量恒定为 1mL/min。用单四极的质谱检测器作为分析仪,在 150℃条件下,以 50~800m/z 的扫描速率进行扫描。

根据 GC-MS 的分析结果,确定了 103 种代谢物,其中 66 种被成功识别,18 种用来建立诊断模型。在这 18 种代谢物中,有 5 种(苏贝酸、甘氨酸、L-酪氨酸、L-

苏氨酸和丁二酸)在肝癌患者群体中的水平比在健康志愿者群体中的水平($p<0.05$)要高得多。其他代谢物(草酸、木糖醇、尿素、磷酸盐、丙酸、苏氨酸、庚二酸、丁酸、三羟基戊酸、次黄嘌呤、阿拉伯呋喃糖、羟基脯氨酸二肽和四羟基戊酸)在健康志愿者群体者中显示出较高的水平($p<0.05$)。此外,吴等在尿液样本的代谢组研究中,使用了在血清中使用的 ELISA 测试方法,确定了同一批患者和健康志愿者的 AFP 含量水平。当 AFP 浓度超过 20ng/mL 时,说明结果呈阳性,肝细胞癌存在。在研究群体中,使用 AFP 标记识别物,有 75% 的病人被正确地分到 HCC 组,然而还有健康的志愿者也被诊断为有 HCC,这表明,AFP 不适合作为一种普遍性的标记物,因为有 25% 的肝癌患者并没有显示阳性的结果。这一标记识别物的平均灵敏度和特异性分别为 40%~65% 和 76%~96%。

作者还将代谢组学的结果与从 AFP 的测定中得到的结果相结合。该模型是用线性判别分析创建的。还加入了主成分分析。有了这个模型,才有可能检测出潜在诊断价值的代谢物。此外,对代谢物概况分析的模型大大减少了因使用 AFP 标记物分析,而误将正常人判定为肝癌患者的数量[21]。

10.5.3 利用核磁共振对肌萎缩性脊髓侧索硬化症患者的血浆样本进行代谢分析

肌萎缩性脊髓侧索硬化(amyotrophic lateral sclerosis,ALS)是一种无法治愈的神经退行性疾病,它会导致皮质脊髓、脊髓前角神经元和脑干的退化。由于症状的多样性和病情的逐渐发展,ALS 极难诊断。因此,对病人的诊断可能持续几个月到1年。患者的平均存活时间是 2~4 年。在疾病发展过程中,由于肌肉的消耗,患者逐渐出现系统性的行动能力恶化现象。在后期阶段,患者在说话、吞咽和呼吸方面有困难,最终病人多死于呼吸衰竭。该疾病的病因尚不清楚,但主流的猜想涉及基因突变、蛋白质聚合、谷氨酸兴奋性中毒、氧化应激、线粒体功能失调和微胶质激活。

Kumar 等[22]对 30 名患有 ALS 症的患者,10 名患有平山症的患者(与 ALS 相似的病灶运动神经元疾病,影响上肢)和 25 名健康志愿者进行了血浆检查。测定使用了质子核磁共振(HNMR),分析物含量测定范围为 2.0~2.2ppm。在无法对代谢物特性进行确认的情况下,作者还应用了二维双量子滤波关联能谱法(double quantum filtered correlation spectroscopy,DQF-COSY),以及全相关能谱法(total correlation spectroscopy,TOCSY)。统计分析是使用 U Mann Whitney 测试进行的。利用单向方差分析分析了超过两组的平均显著差异,紧接着进行 Turkey 测试。利用 Spearman 的相关试验,计算了疾病持续时间与代谢物浓度之间的相关性。

作为统计分析的结果,谷氨酸($p<0.001$)、β-羟基丁酸($p<0.001$)、醋酸($p<$

0.01)、丙酮($p<0.05$)和甲酸($p<0.001$)在 ALS 患者中的含量明显高于健康志愿者。相反,在 ALS 患者中,谷氨酰胺($p<0.02$)和组氨酸($p<0.001$)的浓度水平明显低于健康志愿者。此外,结果表明,ALS 组和健康志愿者中丙氨酸、赖氨酸、丙酮酸、柠檬酸盐、葡萄糖、肌酸/肌酐和酪氨酸等代谢物的浓度相当。在患有平山症的患者中,谷氨酸的浓度($p<0.01$)、丙酮酸($p<0.05$)和甲酸盐($p<0.05$)明显高于对照组的浓度[22]。

除此之外,作者研究了代谢物含量与 ALS 疾病的持续时间之间可能存在的相关性。结果,在 ALS 患者和健康受试者中,两种不同水平的代谢物的浓度与 ALS 疾病的持续时间有明显的相关性。谷氨酸在疾病持续时间和代谢浓度之间呈正相关($p<0.001$, $r=0.6487$),而组氨酸则显示出负相关($p<0.001$, $r=0.5641$)[22]。

数据分析还显示,在 ALS 患者和患有平山病患者的血浆中,谷氨酸的浓度明显更高。谷氨酸是最重要的神经递质之一,控制着未成熟神经细胞的生长和迁移过程。这种代谢物被认为可以通过刺激导致兴奋性中毒,这就是在 ALS 疾病中观察到的导致神经元死亡的过程。在 ALS 患者中观察到的谷氨酰胺水平降低可能代表在兴奋中毒期间发生在突触末梢和星形胶质细胞中的谷氨酸-谷氨酰胺转化的周期非平衡性。

在 ALS 患者和平山症患者中,另一种代谢物甲酸盐的水平也升高了。这种代谢物通过抑制细胞色素氧化酶活性,对线粒体电子传输和供能造成干扰。甲酸对细胞色素氧化酶的抑制引发的细胞死亡被认为是 ATP 的部分丧失导致的细胞基本功能所需能量损失的结果。此外,通过甲酸盐对细胞色素氧化酶的抑制,可促进细胞有毒活性氧的生成,这也意味着细胞死亡。

第三种代谢物,组氨酸被看作是一种抗氧化剂,与健康志愿者相比,ALS 患者中的水平较低。因此,在 ALS 患者中观察到的较低水平可能会导致更强的无氧自由基的攻击。此外,据报道,低浓度的组氨酸与能量损耗和炎症有关。

来自患者和健康对照组的血浆样本中代谢物的研究表明了个体代谢物浓度的变化如何反映体内发生的过程。对 ALS 患者和平山症患者的诊断是一个重要的问题,因此,所选择的代谢物可能作为诊断这些疾病的潜在标记物。

10.5.4 利用高效液相色谱与串联质谱法测定潜在癌症标记物尿核苷

核苷是核糖核酸(ribonucleic acid,RNA)的代谢物。由于 RNA 的过度流动,像发生在癌症、炎症和艾滋病等病理状态下,核苷被大量排泄到尿液中。正常的核苷(腺苷、鸟苷、胞苷、尿苷)进一步退化为尿酸、β-丙氨酸或 β-氨基异丁酸或重新利用。另外,修饰过的核苷,尤其是甲基化的核苷,在尿液中没有变化。因此,升高的核苷含量水平似乎可以作为癌症存在的一个指标,并用作癌症愈后的诊断工具[23]。

Struck 等[23]使用 HPLC-ESI-MS/MS 测定健康受试者($n=61$)和泌尿生殖道癌患者($n=68$)尿液中的核苷水平。样品预处理程序基于苯基硼酸吸附剂床的 SPE。作为方法优化的结果，色谱条件设定如下：Zorbax Extend-C18 柱($2.1mm×50mm, 1.8\mu m$)，具有 $2\mu m$ 玻璃低分散在线过滤器；梯度洗脱的两种流动相中流动相 A 由 0.05%甲酸的水溶液组成，流动相 B 由 0.05%甲酸的甲醇溶液组成；流速为 0.3mL/min。针对每种代谢物的碎裂电压、毛细管电压和碰撞能量电压优化了质谱参数；离子源参数也进行了优化。

色谱测定的结果是，作者成功地对 12 个核苷进行了量化，其中 5 个（N_2, N_2-二甲基鸟苷、肌苷、3-甲基尿苷、6-甲基尿苷和 N_2-甲基鸟苷）在研究组中显示出浓度上的显著差异。使用 U Mann 惠特尼测试($p<0.05$)的统计分析证实了这一点。对于包含统计意义的核苷数据，应用了非监督判别多元统计分析，如 k-最近邻算法(K-NN)和偏最小二乘判别分析法(PLS-DA)。另外，作者计算了由 5 个具有统计学意义的代谢物组成的模型的敏感性和特异性。K-NN 模型的灵敏度和特异性分别为 71.43%和 50.00%，而 PLS-DA 模型则分别为 42.86%和 88.89%。

其他出版物也证实了核苷类成分在诊断泌尿生殖道癌潜在的重要作用[24-25]。例如，采用灵敏的测定技术(如 LC-MS/MS)分析健康受试者和癌症患者的生物样本的代谢物，即使代谢物的浓度很小，也可以检测到明显的差异(在核苷存在的条件下)，并且潜在的癌症标记物的选择，也可以用来在疾病发生的早期阶段诊断出疾病。

10.6 小结

代谢组学研究采用既具有灵敏性又具有选择性的分析技术和多元化学统计方法，使其成为测定生物和环境样品中痕量代谢物的潜在的有效工具。合适的代谢组学(如目标或非目标代谢组学)的应用决定了代谢物的分析方法(定量或定性测定)。在定量测量中，甚至可以确定代谢物浓度的微小变化，从而评估研究组之间的确切差异，如健康对照组和患者组的差异。另外，从定性的角度来看，在不同浓度水平下，大量的代谢物需要应用适当的数据预处理，从多维数据集中提取给定代谢物的信息。用这两种方法对所选择代谢物不同含量(甚至是痕量)的评定，可以从统计学上评价它们对某一特定疾病的预诊和诊断价值的灵敏性和特异性。这些信息也可以成功地用来帮助理解发生在体内分子水平上的一些病理过程。

10.7 在生物体液中所选药物及其代谢物的测定

10.7.1 引言

近年来,人们对能够更好地理解导致各种疾病发生和发展的病理现象的研究更加重视。借助新的诊断工具,目前可以识别最复杂异常的早期症状。许多疾病的治疗取得了巨大进步,延长了患者的生命。

在许多疾病的诊断和药物治疗方面同样取得了重大进展。我们正在见证新一代药物的出现,所有的药理作用都影响改善治疗效果的过程。当血液中药物浓度与其反应之间存在相关性,并且当药物的药代动力学复杂且因代谢和其他制剂的差异而使个体患者而异时,在治疗期间应对所使用的药物浓度进行测量。药物治疗越来越全面,并针对不同的病人有所差异,然而,病人服用的药物与其他物质的相互作用,影响其最终使用结果的这一过程,是现代医学研究中最重要的问题之一。

遗传多态性导致的多种代谢,会影响个体对治疗各种疾病的药物的反应、治疗方法及出现有利或不利效果的概率。通过测定药物代谢物的可能性,允许在发生代谢紊乱时,使用个体药物剂量的规格并选择一种替代药物。除了寻找反映在生物系统病理状态中的人体代谢途径中的异常外,临床医生感兴趣的第二个主要领域是寻找疾病标志物以及监测其在生物体液中的浓度。

这些代谢物的浓度水平与疾病的发生阶段相关,可以使人们更容易地就治疗做出决定。由于细胞变化总是先于临床症状,因此在任何疾病症状出现之前须发现代谢物浓度的不规则,通过治疗的早期实施,可以减缓或抑制疾病的进展。

10.7.2 多变量诊断学

多变量诊断(疾病标记物、所用药物的代谢物和内源性化合物、抗氧化剂等饮食成分的代谢物)可以极大地帮助改善临床试验的设计和早期药物监测的效果。

在临床实践、新药开发和诊断方面将这些数据与具体的治疗联系起来非常重要,它不仅仅以标准为基础进行治疗,而且在某种程度上使有效的治疗方法更具针对性。基于这个原因,一个在治疗期间涉及药物监测的项目已经建立,"治疗药物的监测"(TDM)是以几个学科的合作联系为基础的,特别是药物动力学、药效学和分析化学[26]。TDM 包括体液中的药物浓度的测定,这可以帮助开发最有效的药物,同时也为单个患者提供最大限度的安全药物剂量方案。在设计和研究阶段,确定药物的药效学和药代动力学参数,可以在使用之前预测它们在生物体中的分布和代谢。然而,患者的个体特征差异和同时广泛使用多种药物(可能导致相互作用)使药物浓度不完全符合预期,因此在治疗期间需要对药物浓度进行控制。

在重症监护室住院的病人(术后)通常需要进行联合治疗(例如,同时使用不同的药物),这是由于多器官功能障碍患者需要同时控制复杂的临床问题。此外,在需要长期进行基本治疗的慢性疾病治疗过程中,会存在接触病毒或细菌感染、损伤等情况。在这些情况下,需要对这些症状进行额外的针对性的治疗(如抗生素治疗、抗热治疗和镇痛治疗)。在联合治疗的情况下,有必要了解药物之间相互作用的可能性、它们的作用机制以及可能对药物的代谢动力学的影响[27]。多药疗法的现象也会发生在一个病人在没有相应的处方指导而服用了大量药物的情况下,药物之间的相互作用反而会引起毒性增加和严重的治疗并发症。

10.7.3 生物分析在个性化治疗中的作用

对体液中治疗药物的监测是临床化学研究的热点之一。监测治疗是一种最合适的药物治疗方法,特别是对于新开发的药物,其特点、药物动力学和生物利用率尚不完全清楚。通过监测一种药物及其代谢物在生物体液中的浓度,来确定和消除治疗的不良反应。此外,监测还可以跟踪治疗的进展和针对药物代谢在不同个体间的差异来制定药物剂量。药物反应和新陈代谢的个体间差异是环境因素(饮食、疾病、其他药物、接触其他有害物质)、遗传因素(基因类型、性别和种族)和生物利用率相互作用的结果,因此不同个体间的差异非常大[28]。

个性化医疗的概念是基于了解同一疾病患者之间的差异,了解各种疾病的复杂机制。个性化治疗应使治疗方法适应疾病的分子类型,并考虑药物代谢的个体差异(药物基因组学)。换句话说,这种治疗应该是在适当的时间使用有效的药物和服用合适的剂量。根据个体患者的生物学特性做出治疗决定,可以提高治疗的效果,并将毒性和不良影响的风险降到最低。因此,通过诊断-治疗过程的个体化,分子医学为个体患者和基于人群的医疗保健系统创造了新的、更有效的医疗保健机会。1000年前,Hippocrates就阐明,了解什么人得这种病,远比了解这种人得什么病要重要得多[29]。

10.7.4 药物代谢动力学和生物药物可利用率

一个高度发达的研究领域是对药物动力学和生物可利用率的生物分析研究。药物动力学是研究药物对人体的生化和生理作用,主要涉及对吸收、分布、生物转化(新陈代谢)和分泌物的定量测定和数学评估。基于对血液或尿液中药物浓度测量的数学分析的药物代谢动力学,可以确定合理的药物动力学参数,以便能够进行安全的药物治疗。药物动力学的研究通常是通过确定一个生物基质(血液、血清、血浆、尿液或唾液)的总药物浓度或身体在特定时间内产生的代谢产物来进行的。根据测试成分(或组分组)浓度的变化,可以确定给定剂量和这些浓度之间的数学关系,并计算出合适的参数来定性确定试验药物的药物代谢动力学性能。

最重要的药物动力学参数包括生物可利用率、分布量、清除量和生物药物半衰期。

生物利用度是一个决定药物从给药地点吸收到血流中的速度和数量的参数,决定其在作用部位的浓度和治疗效果。生物可利用率被定义为身体内添加药物后随着血管进入体循环的药物剂量的分数(百分比)。药物进入血液的程度和速率用以下参数表征:药物在血液中的血浆浓度曲线下方的面积(AUC),血液中药物的最大浓度(C_{max})和达到最大浓度所需要的时间(T_{max})(如果能测定的话)。为了确定给定剂型的生物利用度,有必要将这3个参数都确定。AUC测量可以计算出生物可利用率的范围(EBA)。影响药物生物可利用率的各种因素可大致分为与年龄相关或与患者相关(个体间的差异,如遗传因素、疾病类型、用药方式和时间、胃填充程度、食物或其他药物的种类)两种。对制药公司和医师来说,EBA的确定具有非常重要的实际意义,能够适当选择原料药的类型、剂型、用药方式和药物剂量。在药物治疗无效的情况下,它有助于做出改变现有治疗方法的决策[30]。

通常,即使是药物在生物体液中的总浓度的测定也是非常困难的,因为其在被测生物体液中的浓度较低,这通常是服用小剂量药物的结果。由于药物中活性物质的生物可利用率降低,再加上首过效应的加强,测定的问题也变得复杂起来,同时体内缓慢吸收导致的低浓度分析物会快速消失。因此,需要将分析样品离析和进一步浓缩,以提高分析方法的灵敏度,从而获得正确的药物或代谢物的药物动力学模型[31]。

10.7.5 药物与其他化合物的相互作用

实际上,了解药物与其他化合物的相互作用非常重要,因为可以利用有利的相互作用发挥药效,同时,尽量减少具有非常危险副作用组合的使用。近年来,人们越来越关注饮食成分对治疗许多疾病所用药物治疗效果的影响。第二种药物的加入由于两者间会发生反应,原来一种药物的药效会发生改变(增强或减弱)。

从临床的角度来看,药物的不良反应不可避免所以显得尤其重要。这些相互作用能减弱或增强药物的作用,导致出现与预期不同的中毒症状或药理活性现象。不良的相互作用需要通过特定的治疗来控制,甚至是剂量的改变,这是各种机制相互作用以及病人个体特征对药物代谢影响的结果。有时完全不熟悉或不寻常的机制是这两种类型反应的基础,有积极的或不利的影响。然而,大多数情况下,这些反应都是重复的,有时是非常熟知的,这些反应的大部分动力学过程(药动学的相互作用)是由抑制或诱导代谢酶引起的。

随着越来越多的可用药物和膳食补充剂,加上密集的广告活动,这些产品消费量显著增加。此外,同时服用多种药物而缺乏对其作用机制的了解,也会导致多药现象的发生,其风险主要是由于它们之间的相互作用导致药物不良反应发生概率(或强度)的增加。同时使用草药和合成药物会通过3种机制增加相互作用的风险:

(1) 对胃肠道吸收药物的影响(药代动力学时期);
(2) 细胞色素P-450酶(药代动力学时期)对药物代谢的影响;
(3) 添加剂和高活性添加剂协同作用(药效学阶段)。

在药剂师和内科医生的日常工作中,仍然低估了膳食成分和膳食补充剂对合成药物生物有效性影响的评价。约75%的药物是口服的,几乎所有的饮食成分都可能影响药物在身体一个阶段或所有阶段的吸收。在同时使用几种药物的情况下,相互作用的数量随着使用的药物数量的增加而增加,如果服用超过5种药物,交互作用就会变得失调[32]。

依赖于使用患者群体参数来预测药物浓度的这种声明是不可靠的。患者对药物使用的个体反应,是由遗传多态性引起的,决定了药物最终效果的差异,就像补品或饮食。在这种情况下,在治疗期间对药物浓度的监测,以及对其新陈代谢的追踪,似乎是最合理的策略,然而,有一种风险是,某些营养物质或补品可以改变药物的作用,这取决于不同的药物动力学过程。这会造成严重的临床后果,使治疗效果减弱或失效,甚至发生危险的相互作用。在治疗期间药物和食物一起服用药物会出现各种临床症状,这意味着处方药的治疗效果不可预测。药物与患者摄入的其他物质相互作用,会影响药物使用的最终结果,这是现代药物治疗中最重要的问题之一。因此,饮食成分和医生同时开具的其他药物对治疗效果的影响越来越受到人们关注。治疗药物的监测是一种确保最佳药物治疗的方法,它考虑了不同的药代动力学和生物利用度。

由于缺乏对药物相互作用、膳食成分和补充剂所产生的风险的了解,因此有必要进行这方面的研究,以确定这种相互作用的潜在后果。同时,对药物和膳食成分的确定可以帮助决策个性化的治疗过程。在有药物代谢干扰的情况下,这样的测定可以排除或引入额外的制剂进行治疗[33]。

植物产品对合成药物代谢的影响是由细胞色素P-450(CYP)酶的抑制或激活引起的。天然植物产品或膳食补充剂的使用,对CYP潜在活性进行评估,对于预测其成分与药物之间的相互作用具有重要意义。因此,人们的注意力集中在可用于食品市场的产品研究,即天然非营养物质对药物吸收的影响。非营养性膳食成分主要是次生植物代谢物,其中包括酚类化合物,如酚酸和类黄酮。非营养物质对健康的影响尚不清楚,目前,还没有关于这些物质被身体吸收和代谢的程度的明确答案,也没有关于这些化合物每日摄入量的信息。这些信息非常重要,某些非营养的天然物质被认为是抗营养因子,主要是因为它们抑制了消化,减少了营养或药物的生物利用率,同时它们也有可能与药物产生不良的相互作用。而非营养性天然物质对健康的积极影响不仅仅是由于它们的抗氧化特性,还在于这些物质与各种代谢过程有关,并增强人体的免疫系统,它又与所有其他生理系统相互作用,如呼吸系统、消化系统、神经系统、泌尿生殖系统和肌肉骨骼[34]。

药物的代谢包含 CYP2C9、CYP2C8、CYP3A4 和 CYP1A2,除此之外它们还可能涉及其他内源性和外源性化合物的生物转化。起重要作用的是 CYP2C9。它的活性可以被许多物质抑制或激发。抑制 CYP2C9 的活性可以减缓药物的新陈代谢,增加其在体内的浓度,而导致不良的副作用。新陈代谢的增加会使药物更快地消化,使其不能达到有效的治疗浓度。在目前使用的治疗物质中,超过 100 种治疗物质被确认为 CYP2C9 的基质。它们占据了所有处方药的 10%~20%[35]。

与上述内容有关,评估自然植物产品和膳食补充剂的用法对 CYP 引起的潜在的抑制,以及预测不同治疗组的膳食成分和药物之间的相互作用非常重要。

面对这些需求,人们越来越重视开发新的分析方法,以便对有机物中低分子量混合物进行快速、高效的分析。一个重要的目标是开发用于测定和发现大量化合物的新的、有效的方法。目前,基于 MS 的高灵敏度和选择性的特点,在这些分析中发挥了重要的作用。同时测定药物和膳食补充剂分析方法的使用,为消除或减少被测定的生物活性化合物之间的相互作用创造了一个真正的机会,也可以为决定治疗过程提供必要的新信息。药物和其代谢物的浓度水平与饮食成分浓度的相关性有助于决定改变饮食和需要控制补充营养。一方面,这类研究为掌控许多疾病的治疗进展提供了工具;另一方面,其为旨在消除药物和饮食中的化合物之间可能发生的相互作用的营养建议提供了科学依据。

10.7.6 分析方法的选择

在目前的临床实践中,仅有少量从组织和生物体液检查中获得的信息可以加以利用。对于代谢途径的失调,会反映在内源性化合物代谢物的定性和定量组成中,通过对这些检查结果进行分析可以验证这些疾病发病原因的假设。

在选择适当的分析方法时,对问题的定义是起决定性作用的,它决定了样本类型和分析方法的选择。基本问题主要涉及测定所需要的精确度、准确性、特异性和检出限。随着这种从患者身上取样及分析测试方法的发展,在定义问题阶段所考虑的限制条件是被测样本的数量和大小,以及样品处理和获得分析结果所需的时间,这在检测婴儿和儿童药物的分布情况时十分重要。此外,在选择测试样本时,还需要分析提供的药物分布信息是否详尽,以及从采集开始到分析为止是否采用了合适的样品储存方法。另一个关键问题是何时取样,这个问题对于门诊患者来说是有真正意义的,体液和组织中物质的量浓度在几个小时后才能达到平衡,一些药物需要在晚上进行服用,这样的病例需要住院治疗,因此会对医院的经济产生影响。总而言之,有必要选择一个大小合适的样本进行分析,从采集到分析期间的储存不能影响分析物的浓度。此外,样品应能提供要求的信息以便于实验室每次的采集[36]。

由于复杂基体的存在,分析样本的类型限制了分析方法的选择。在选择一种分析方法时,同样应考虑特定方法可能的浓度测量范围以及在通常采用的给药方案中实际样本的药物含量。在准备分析的过程中,所分析的物质不应因所服用药物的代谢和体内产生的大量代谢物而发生变化。此外,所选择的方法应能测定药物的剂量形式及其代谢物。所选择的分析方法的应用需要开发匹配的样品制备方法。无论是对于药物还是对于其代谢物,所选择的处理方法都应消除基体的干扰,分析浓度应满足测量范围,而且不引起分析物质结构的任何变化。

只有在选择合适的材料进行处理方法的研究和开发后,才可以对样品进行分析。研究的最后阶段是对结果进行分析,并对病人的病情得出结论。

在相关文献中,描述了用适当灵敏度和选择性的分析方法进行药物及其代谢物的动力学研究。所使用的绝大多数方法都涉及 LC 技术。在许多化合物的测定中,包括属于不同治疗组的药物,越来越多地使用色谱与 MS 的组合,在描述生物液体中药物测定的规程中,使用了分光光度(UV-vis)、荧光和 MS 或 MS/MS 检测器。用于分离的固定相有非极性基团(C18),或者其他亲水相互作用的液相色谱(HILIC)。乙腈、甲醇和水以不同比例混合被用作洗脱液,通常添加酸或氨来改变 pH 值,这对被检测的分析物的质子化或去质子化会有影响。利用手性固定相,以及伴有各种缓冲区的乙腈或甲醇混合物流动相,可以分离出生物液溶液进行色谱分析,以检测药物及其代谢产物。气相色谱法也经常与各种类型的检测器结合使用。传统的色谱分离技术的另一种选择是使用毛细管电泳色谱技术,毛细管区电泳(CZE)的分离特别有效。

由于生物材料具有复杂性,因此有必要将分析物从内源性基体衍生物和其他可能影响分析值的化合物中分离出来。此外,样品制备过程除了从生物材料中分离出分析物外,还应包括分析物的衍生化,以提高该方法的灵敏度。在测定生物液体中低含量的药物时,为了产生荧光或紫外吸收,将分析物化学转化为具有比母体化合物更高能量的适当的衍生物非常重要。因为内源性化合物和其他与分析物具有相似结构的药物也可以进行衍生化,所以有必要通过衍生化对生物基质中药物及其代谢产物进行选择性分离,并从其他可能干扰待测化合物的衍生产品中,分离创建分析物衍生品[37]。

为了获得关于药物的药物动力学的可靠信息,需要用一些新方法和新仪器,来测定药物对富含多酚化合物的饮食的生物利用度水平。这些更加新颖的分析方法会提供更多的信息,有助于制定使用新药的治疗方案,提高许多疾病药物治疗的有效性和安全性。只有了解药代动力学(研究病理条件对药物在体内命运的影响的科学)才能精准地预测药物在血液中的浓度。此外,药物和其他化合物之间的相互作用可以通过观察病人的情况和任何不可预见的副作用来判断,这种相互作用的结果可以让研究者通过绝对测量法获得药代动力学和药效学参数,让药物相互作用的精确数学表征能够对药代动力学/药效学类型进行建模。研究所分析药物的药物代谢动力学特性,有助于消除或尽量减少药物与其他活性物质的相互作用。在患者护理的复杂过程中考虑所选化合物的药代动力学研究结果对于临床目的极为重要,并为个性化治疗奠定基础[38]。

10.7.7 操作规范和分析技巧

由于药物及其代谢物化学结构有很强的相似性,因此测定这两组化合物是一个分析难题。相似的结构也会产生相似的物理化学性质,因此,在色谱系统分离阶段有类似的相互作用。相似的性质得到相似的分析信号参数,例如,用分光光度法检测相似波长的光,用伏安法检测相似的氧化还原的电位。如果代谢物以硫酸盐或葡糖苷酸的形式存在,那么分离就更容易了,因为简单地改变色谱方法中流动相的 pH 值,就能开始或阻止分离过程,这导致了(以中性分子和离子的形式)的化合物与色谱系统中固定相和流动相互作用的差异。同样,质子化和去质子化过程(例如,药物和代谢物中的氨基基团)简化了色谱分离。

现有文献描述了多种测定生物体液中特定药物的方法,但不能同时测定代谢物。而关于药物分析的文献很多,本章仅介绍了同时测定药物及其代谢物的一些方法。表 10.1 描述了用于心血管疾病的消炎药和镇痛剂及其代谢物的测定方法。

这类分析中的一个重要问题是存在基体,基体的成分会造成结果假阳性,或者常规方法无法测定。因此,除了开发合适的分析方法外,还必须去除干扰物,并分离和富集分析物。为了达到这个目的,各种类型的萃取法都被应用,常用的是 LLE 和 SPE,但也有一些其他的萃取方法,如包裹吸附剂的微萃取法(MEPS)和超声辅助乳化微萃取法(USAEME)。

在多药联合治疗的情况下,必须了解药物与药物之间相互作用的可能性、作用机制,以及药物动力学方面可能的影响。由于非处方药的普遍使用,病人除了服用医生开的处方药外,也会服用其他非处方药物,这就造成多重药现象。药物的相互作用会导致毒性的增加和严重的治疗并发症。

上述现象需要运用分析方法,同时监测药物在不同治疗组体液中的存在和浓度。在现有文献中描述了几种方法,如表10-2和表10-3所列。

表10-1 被选药物和代谢物测定的色谱分析方法

分析物/样品	方法	色谱分离条件	样品处理	LOD/LOQ	引用文献
非甾醇类抗炎药					
对乙酰氨基酚,代谢物/血液	HPLC UV	色谱柱:Hypersil C_{18}(75mm×4.6mm,3μm) 流动相:A溶剂:20mM甲酸铵缓冲溶液(pH 为 3.5) B 溶剂:甲醇(梯度洗脱) 检测器:λ=254nm	LLE 萃取法/20mM 甲酸铵缓冲溶液(pH 为 3.5)	LOD: 0.03μg/mL(对乙酰氨基酚); 0.1μg/mL(代谢物)	[39]
对乙酰氨基酚,代谢物/尿液,血浆	HPLC UV	色谱柱:EPS C_{18} (250mm×4.6mm,5μm) 流动相:0.1 M KH_2PO_4、异丙醇、四氢呋喃(100:1.5:0.1 pH 3.7) 经磷酸调节(梯度洗脱) 检测器:λ=254nm	血浆: 蛋白质沉淀(30%高氯酸) 尿液:稀释20倍,离心分离	LOD:0.06μg/mL(对乙酰氨基酚); 0.13μg/mL(代谢物)	[40]
对乙酰氨基酚,代谢物/尿液	HPLC NMR/MS	色谱柱:YMC-Pack J'Sphere H80(250mm×2mm,4μm) 流动相: A 溶剂:0.1%的三氟乙酸水溶液 B 溶剂:丙烯腈(梯度洗脱) 检测器:λ=254nm,MS,ESI 离子化,NMR	在线固相萃取	—	[41]
对乙酰氨基酚,代谢物/尿液	UPLS MS	色谱柱:ChromSpeed(50mm×4.6mm,整体) 流动相:A 溶剂:0.1%的甲酸水溶液 B 溶剂:在丙烯腈(梯度洗脱)中 0.1%的甲酸溶液 检测器:MS,ESI 离子化	稀释5倍	—	[42]

续表

分析物/样品	方法	色谱分离条件	样品处理	LOD/LOQ	引用文献
阿司匹林,水杨酸,龙胆酸/血浆	HPLC UV	色谱柱:YMC Hydrosphere C_{18} (150mm×4.6mm,5μm) 流动相:A 溶剂:0.2%的 TFA 在水和丙烯腈混合液中(1000:10v/v) B 溶剂:0.2%的三氟乙酸(梯度洗脱)在水和丙烯腈混合液中(100:900v/v) 检测器:λ=235nm	在线固相萃取/MCX-SAX 联用	LOQ:60ng/mL	[43]
乙酰水杨酸,酮洛芬,双氯芬,萘普生和布洛芬/尿液	UHPLC UV	色谱柱:Poroshell 120 EC-C18 (100mm×3.0mm;2.7μm) 流动相:A 溶剂:0.05% 的 TFA 水溶液 B 溶剂:丙烯腈(梯度洗脱) 检测器:λ = 221nm,λ = 230nm,λ = 239nm,λ = 255nm,λ = 277nm	MEPS 萃取/C18 吸附剂/淋洗:丙烯腈	LOD: 1.07~16.2ng/mL LOQ: 3.21~48.7ng/mL	[44]
1-羟基-布洛芬,2-羟基-布洛芬,3-羟基-布洛芬,羧基布洛芬/尿液	UHPLC MS/MS	色谱柱:Zorbax Rapid Resolution High Definition(RRHD)SB-C18 (50mm×2.1mm,1.8μm) 流动相:A 溶剂:0.1%的甲酸水溶液 B 溶剂:丙烯腈(梯度洗脱) 检测器:MS/MS,ESI 离子化	USAEME 萃取/正辛醇	LOQ:0.5pg/mL	[45]
氟比洛芬,酮洛芬,乙哚乙酸对映异构体/血浆	HPLC UV	色谱柱:Agilent Zorbax C_{18} (250mm×4.6mm,5μm) 流动相: ACN:KH_2PO_4(pH 为 4.5) (60:40v/v)(无梯度洗脱) 检测器:λ=250nm	LLE 萃取/二氯甲烷衍生化	LOD:0.15μg/mL LOQ:0.5μg/mL	[46]

续表

分析物/样品	方法	色谱分离条件	样品处理	LOD/LOQ	引用文献
酮基布洛芬右旋对映异构体/血浆	HPLC UV	色谱柱：Chiral-HSA（100mm×4mm,5μm） 流动相：0.01 M 磷酸盐缓冲溶液：异丙醇（94∶6v/v）和5mM 的正辛酸,pH 为 5.5（无梯度洗脱） 检测器：λ=260nm	在线 LPE 萃取/C_{18} ADS 多孔玻璃柱	LOQ：16ng/mL	[47]
利尿剂					
安体舒通,坎利酮/血浆	HPLC DAD	色谱柱：Waters Symmetry C18（150mm×4mm,5μm） 流动相：甲醇：水（60∶40v/v）（无梯度洗脱） 检测器：λ=238nm（安体舒通）；λ=280nm（坎利酮）	SPE 萃取/Oasis HLB 吸附剂/甲醇淋洗	LOQ：28ng/mL（安体舒通）；25ng/mL（坎利酮）	[48]
安体舒通,坎利酮,利尿磺胺/血浆	HPLC ESI MS/MS	色谱柱：Waters Sunfire C18（50mm×2.1mm,3.5μm） 流动相：甲醇：水（两者都含有 1mM 的醋酸铵/0.001% 乙酸）（梯度洗脱） 检测器：MS/MS,ESI 电离	离心分离	LOD：62.5ng/mL（利尿磺胺）；25ng/mL（安体舒通,坎利酮）	[49]
安体舒通,坎利酮,利尿磺胺/血浆	HPLC MS/MS	色谱柱：Altech Prevail C18（50mm×2.1mm,3.0μm） 流动相：A 溶剂：0.2% 的甲酸水溶液 B 溶剂：水 C 溶剂：丙烯腈（梯度洗脱） 检测器：MS/MS,APCI 电离	SPE 萃取/ABS ELUT Nexus 柱：甲醇洗脱	LOD：3ng/mL（坎利酮,安体舒通）；12ng/mL（利尿磺胺）	[50]
坎利酮,利尿磺胺/尿液	HPLC MS	色谱柱：Zorbax XDB C8（75mm×4.6mm,3.5μm） 流动相：A 溶剂 0.2mM 醋酸铵水溶液：丙烯腈（95∶5v/v） B 溶剂：0.2mM NH_4Ac 水溶液：CAN（5∶95v/v）（梯度洗脱） 检测器：MS,APCI 电离	SPE 萃取/XAD 吸附柱/洗脱：甲醇	—	[51]

续表

分析物/样品	方法	色谱分离条件	样品处理	LOD/LOQ	引用文献
β受体阻滞剂					
美托洛尔、α-羟基美托洛尔,O-去甲基美托洛尔/血浆	HPLC FL	色谱柱:μBondpak Phenyl(300mm×3.9mm,10μm) 流动相:磷酸盐缓冲溶液(PH为3.5):ACN(85:15v/v)(无梯度洗脱) 检测器:$\lambda_{EX/EM}=277/305nm$	LLE萃取/二乙醚:氯仿(4:1v/v)	LOQ:25ng/mL	[52]
米里诺酮,索塔洛尔,美托洛尔,普萘洛尔和卡维地洛,以及它们的代谢物:5-羟苯基卡维地洛,o-去甲基-卡维地洛,4-羟基-羟甲基醇,α-羟基-美托洛尔,o-去甲基美托洛尔/尿液	UHPLC UV	色谱柱:Hypersil GOLD™(50mm×2.1mm,1.9μm) 流动相: A溶剂:0.05%三氟乙酸水溶液 B溶剂:丙烯腈(梯度洗脱) 检测器:$\lambda=227nm$, $\lambda=240nm$, $\lambda=25nm$, $\lambda=280nm$, $\lambda=324nm$	SPE萃取/Oasis HLB柱/淋洗:甲醇:丙酮:甲酸(4.5:4.5:1v/v/v)	LOD:12.8~37.9ng/mL LOQ:38.4~-113.8ng/mL	[53]
米林酮,索塔洛尔,美托洛尔,普萘洛尔,卡维二醇,5-羟基苯甲醇,邻氨基苯甲醇,4-羟基普萘洛尔,α-羟基美托洛醇,o-去甲基美托洛尔/尿液	UHPLC MS/MS	色谱柱:Hypersil GOL™(50mm×2.1mm,1.9μm) 流动相:A溶剂:0.1%甲酸水溶液 B溶剂:丙烯腈(梯度洗脱) 检测器:MS/MS,ESI电离	蛋白质沉淀	LOQ:0.05~40.00ng/mL	[54]

续表

分析物/样品	方法	色谱分离条件	样品处理	LOD/LOQ	引用文献
米里诺酮,索塔洛尔,美托洛尔,普萘洛尔,卡维蒂洛/尿液	UHPLC UV	色谱柱：Hypersil GOL™（50mm×2.1mm,1.9μm） 流动相： A 溶剂:0.05% TFA 水溶液 B 溶剂：丙烯腈（梯度洗脱） 检测器：λ = 227nm, λ = 254nm, λ = 280nm, λ = 324nm	SPE 萃取/Oasis HLB 吸附柱/淋洗:甲醇:丙酮:甲酸(45:45:10 v/v/v)	LOD:(10.3~23.7)ng/mL LOQ:(30.9~71.0)ng/mL	[55]
普萘洛尔,4-羟基普萘洛尔/血浆	HPLC MS/MS	色谱柱：LiChrospher 60 RP Select B（125mm×4mm,5μm） 流动相:1mM 甲酸铵水溶液（pH 为 3.1）:丙烯腈（20:80v/v）（无梯度洗脱） 检测器：MS/MS,ESI 电离	SPE 萃取/Oasis HLB 柱/淋洗:甲醇:丙酮:甲酸(45:45:10 v/v/v)	LOQ:0.2ng/mL	[56]
美托洛尔,α-羟基美托洛尔对映异构体/血浆	HPLC FL	色谱柱:Chirobiotic T(250mm×4.6mm,5μm) 流动相：丙烯腈:甲醇:二氯甲烷:丙烯酸:TEA(50:30:14:2:2v/v/v/v)（无梯度洗脱） 检测器:λ_{EX/EM}=225/310nm	SPE 萃取/C2 吸附柱/淋洗 0.1 M 盐酸:丙烯腈(50:50 v/v)	LOQ:0.5ng/mL（美托洛尔）;1.0ng/mL（α-羟基美托洛尔）	[57]
美托洛尔对映异构体/血浆	HPLC FL	色谱柱 A：Chiralpak AD(250mm×4.6mm,5μm) 色谱柱 B:Chirarcel OD-H(150mm×4.6mm,5μm) 流动相： 溶剂 A:己烷:异丙醇:DEA(95:5:0.1v/v/v) 溶剂 B:正己烷:乙醇:2-丙醇:二乙醇胺（88:10.2:1.8:0.2v/v/v/v）（无梯度洗脱） 检测器:λ_{EX/EM}=229/298nm	LLE 萃取:二氯甲烷:二异丙醚（1:1/v）SPE 萃取/C18 吸附柱/甲醇洗脱	LOQ:5ng/mL	[58]

续表

分析物/样品	方法	色谱分离条件	样品处理	LOD/LOQ	引用文献
美托洛尔对映异构体/尿液	HPLC FL	色谱柱A：Phenomenex Silica (250mm×4.6mm,5μm) 色谱柱B：Chiracel OD (250mm×4.6mm,5μm) 流动相： 溶剂：正己烷：乙醇：2-丙醇：二乙醇胺(90:5:5:0.5v/v/v/v)(等梯度洗脱) 检测器：$\lambda_{EX/EM}=276/309nm$	LLE萃取/二氯甲烷	LOQ：0.1μg/mL	[59]
美托洛尔对映异构体/尿液	HPLC MS	色谱柱：Chirobiotic T(250mm×4.6mm,5μm) 流动相：甲醇：乙酸：NH_3(100:0.15:0.15v/v/v)(无梯度洗脱) 检测器：MS,ESI电离	LLE萃取/乙酸乙酯	LOQ：0.5ng/mL	[60]
卡维蒂洛，5-羟苯基卡维蒂洛对映异构体/尿液	HPLC FL HPLC MS/MS	色谱柱：CHIRALCEL® OD-RH(150mm×4.6mm;5μm) 流动相：溶剂A：0.05%三氟乙酸和0.05%二乙醇胺水溶液 溶剂B：丙烯腈 检测器：$\lambda_{EX/EM}=254/356nm$ MS/MS,ES电离	SPE萃取/Oasis HLB吸附柱/淋洗：甲醇：丙酮：甲酸(4.5:4.5:1v/v/v)	LOD：4.73~8.07ng/mL LOQ：(14.2~24.2)ng/mL	[61]
普萘洛尔对映异构体/血浆	HPLC MS/MS	色谱柱：Chirobiotic V (250mm×4.6mm,5μm) 流动相：甲醇：乙酸：TEA(梯度洗脱) 检测器：MS/MS：APCI电离	SPE萃取/Oasis MCX吸附柱/2.25% NH_4OH 的甲醇溶液	LOD：0.03ng/mL	[62]
索他洛尔对映异构体/血浆	HPLC FL	色谱柱：Chiral-CBH(150mm×4.0mm,5μm) 流动相：含15%异丙醇和0.05mMEDTA的10mM磷酸盐缓冲溶液(pH为7.0)(无梯度洗脱) 检测器：$\lambda_{EX/EM}=250/312nm$	SPE萃取/LiChroCart C18柱	LOD：18ng/mL LOQ：37ng/mL	[63]

续表

分析物/样品	方法	色谱分离条件	样品处理	LOD/LOQ	引用文献
其他药品					
地塞米松，6-β-羟基-地塞米松/尿液	HPLC DAD	色谱柱：Nova-Pak C_{18}（400mm×3.9mm，4μm） 流动相：溶剂A：0.06%的三氟乙酸和醋酸铵缓冲溶液（0.01 M，pH 为 4.8）：丙烯腈（90：10v/v） 溶剂B：0.06%的三氟乙酸醋酸铵缓冲溶液：乙腈（30：70v/v）（梯度洗脱） 检测器：λ=245nm	SPE 萃取/Oasis HLB 柱/乙酯/乙酸乙酯+二乙醚	LOQ：10ng/mL（地塞米松）25ng/mL（6-β-羟基地塞米松）	[64]
硝苯地平、氧代脱氢硝苯地平/血浆	HPLC MS/MS	色谱柱：Hypersil BDS C18（50mm×2.1mm，3μm） 流动相：甲醇：1%的TFA水溶液（80：20v/v）（无梯度洗脱） 检测器：MS,ESI 电离	LLE 萃取/二乙醚:正己烷	LOQ：0.5ng/mL	[65]
L-精氨酸,代谢物/尿液	HPLC FL	色谱柱：Purospher® STAR RP-18e（250mm×4mm，5μm） 流动相：溶剂A：磷酸氢二钠/磷酸二氢钾，pH 为 6.88 溶剂B：ACN 溶剂C：甲醇（梯度洗脱） 检测器：$λ_{EX/EM}=338/455nm$	SPE 萃取/Oasis MCX 吸附柱/淋洗：25% NH_3：水：甲醇（10：40：50v/v/v）	LOD：（0.05~198.50）pmol/20μL 注射 LOQ：0.17~655.10pmol/20μL 注射	[66]

表 10-2 色谱法测定生物材料中从不同治疗组中所选药物

分析物/样品	方法	色谱分离条件	样品处理	LOD/LOQ	引用文献
苊香豆醇,苯丙香豆醇对映异构体/血浆	HPLC UV	色谱柱：S, S-Whelk-01（250mm×4mm，5μm） 流动相：溶剂A：己烷：无水乙醇与乙酸 溶剂B：己烷：无水乙醇（梯度洗脱） 检测器：λ=310nm	LLE 萃取/甲苯	LOD：5ng/mL	[67]

续表

分析物/样品	方法	色谱分离条件	样品处理	LOD/LOQ	引用文献
苊香豆醇,苯丙香豆醇,华法林对应异构体/血浆	HPLC MS/MS	色谱柱:Chira-Grom-2(250mm×1mm,8μm) 流动相:CAN:甲醇:CH_3COOH(梯度洗脱) 检测器:MS/MS,ESI电离	SPE在线萃取/Poros R2/20吸附柱(2mm×30mm)	LOD:0.5ng/mL LOQ:2ng/mL	[68]
阿力克伦,普拉格雷,利伐沙班/尿液	UHPLC MS/MS	色谱柱:Zorbax Rapid Resolution High Definition SB-C18色谱柱(50mm×2.1mm,1.8μm) 流动相:0.1%的甲酸水溶液:ACN(70:30v/v)(无梯度洗脱) 检测器:MS/MS,ESI电离	MEPS萃取/C8吸附剂/淋洗:甲醇	LOQ:(0.5~5)pg/mL	[69]
阿力克伦、依那普利、羟苯甲酯钠/尿液	UHPLC MS/MS	色谱柱:Poroshell 120 EC-C18(100mm×2.1mm;2.7μm) 流动相: 溶剂A:0.1%甲酸水溶液 溶剂B:CAN(梯度洗脱) 检测器:MS/MS,ESI电离化	MEPS萃取/C8吸附剂/淋洗:甲醇	LOQ:0.01ng/mL	[70]
索塔洛尔,美托洛尔,普萘洛尔,卡维地洛尔,硝苯地平,卡托普利,西拉普里,米里诺,抵克立得,苊香豆醇,呋喃苯胺酸,乙酰水杨酸,水杨酸,布洛芬,萘普生,酮洛芬,双氯芬酸,扑热息痛,安乃近,米屈肼,西地那非,地塞米松,卡马西平,特比萘芬/尿液	UHPLC MS/MS	色谱柱:Zorbax Rapid Resolution High Definition SB-C18(50mm×2.1mm,1.8μm) 流动相: 溶剂A:0.1%甲酸水溶液 溶剂B:甲醇(梯度洗脱) 检测器:MS/MS,ESI电离	蛋白质沉淀	LOQ:(0.05~0.60)ng/mL	[71]

续表

分析物/样品	方法	色谱分离条件	样品处理	LOD/LOQ	引用文献
依那普利,扑热息痛,索他洛尔,安乃近,万古霉素,氟康唑,头孢唑啉,美托洛,阿司匹林,噻氯吡啶,强的松龙,普萘洛尔,迪戈辛,西地那非,呋塞米,地塞米松,卡维地罗,酮洛芬,硝苯地平,特比萘芬,苊香豆醇,安体舒通/尿液	HPLC DAD	色谱柱:LiChroCART® Purospher® STAR,RP-18e（250mm×4mm,5μm）流动相:甲醇:丙烯腈:0.05%三氟乙酸水溶液（梯度洗脱）检测器:DAD λ=200~450nm	调整pH为7.0,蛋白质沉淀	LOD:（0.01~1.44）μg/mL LOQ:（0.04~4.35）μg/mL	[72]
亚胺培南,扑热息痛,安乃近,万古霉素,阿米卡星,氟康唑,头孢唑啉,强的松龙,地塞米松,呋塞米,酮洛芬/尿液	HPLC DAD	色谱柱:LiChroCART Purospher STAR,RP-18e（125mm×3mm,5μm）流动相:甲醇:CAN:0.05%三氟乙酸水溶液（梯度洗脱）检测器:DAD λ=200~450nm	调整pH为8,蛋白质沉淀	LOD:（0.01~1.15）μg/mL LOQ:（0.03~3.75）μg/mL	[73]
阿米卡西娜,头孢唑啉,地塞米松,安乃近,氟康唑,呋塞米,亚胺培南,酮洛芬,扑热息痛,泼尼松龙,西地那非,万古霉素/尿液	HPLC DAD	色谱柱:LiChroCART Purospher STAR,RP-18e（125mm×3mm,5μm）流动相:丙烯腈:缓冲溶液（乙酸-乙酸钠,pH为4.66）:0.05%三氟乙酸水溶液（梯度洗脱）检测器:DAD λ=200~450nm	LLE萃取/醋酸盐:二氯甲烷:三氯甲烷（45:35:20v/v/v）	LOD:（0.01~1.16）μg/mL LOQ:（0.02~3.45）μg/mL	[74]

续表

分析物/样品	方法	色谱分离条件	样品处理	LOD/LOQ	引用文献
阿利克伦,普拉格雷,利伐沙班,泼尼松龙,普萘洛尔,酮洛芬,硝苯地平,萘普生,特比萘芬,布洛芬,双氯芬酸,西地那非,醋硝香豆醇/尿液	UHPLC UV	色谱柱：Poroshell 120 EC-C 18(100mm×3.0mm;2.7μm) 流动相：溶剂 A:0.05% 三氟乙酸水溶液 溶剂 B:丙烯腈(梯度洗脱) 检测器：λ=221nm,λ=228nm,λ=230nm,λ=240nm,λ=250nm,λ=275nm,λ=280nm	SPE 萃取/C₆H₅ 吸附柱/淋洗：甲醇	LOD：(0.003~0.217)μg/mL LOQ：(0.01~0.650)μg/mL	[75]
米利酮,依那普利,卡维地洛,螺内酯,醋硝香豆醇,抵克立得,西拉普利,2-oxo 抵克立得,西拉普利拉,坎利酮,5-羟基卡维地洛,o-去甲基卡维地洛,依那普利拉/尿液	UHPLC UV	色谱柱：Poroshell 120 EC-C 18(100mm×3.0mm;2.7μm) 流动相： 溶剂 A:0.05% 三氟乙酸水溶液 溶剂 B:丙烯腈(梯度洗脱) 检测器：λ=227nm,λ=235nm,λ=255nm,λ=269nm,λ=285nm,λ=324nm	MEPS 萃取/C18 吸附剂/淋洗：甲醇：丙烯腈(50：50v/v)	LOQ：(0.016~0.045)μg/mL	[76]
阿司匹林,咖啡因,扑热息痛/血浆,尿液	HPLC UV	色谱柱：μBondapak C18 (300mm×3.9mm,10μm) 流动相： 溶剂 A:丙烯腈 溶剂 B:水(梯度洗脱) 检测器：λ=280nm	SPE 萃取/C18 色谱柱/淋洗：甲醇	LOD：(0.1~0.2)μg/mL LOQ：(0.15~0.2)μg/mL	[77]
地塞米松,强的松,强的松龙,皮质醇/血清	HPLC MS MS	色谱柱：Symmetry C18 (30mm×2.1mm,3.5μm) 流动相：甲醇：5mM 醋酸盐缓冲液,pH 为 3.25(梯度洗脱) 检测器：MS/MS,ESI 电离	固相萃取/固相萃取柱/淋洗：甲醇	LOD：(0.20~0.58)ng/mL LOQ：5.4~10.7ng/mL	[78]

续表

分析物/样品	方法	色谱分离条件	样品处理	LOD/LOQ	引用文献
索他洛尔,美托洛尔,普萘洛尔,卡维地洛,水杨酸,地塞米松,强的松,酮洛芬/尿液	UHPLC UV	色谱柱:Chromolith Fast Gradient Monolithic RP-18e(50mm×2mm) 流动相: 溶剂A:0.05% 三氟乙酸水溶液 溶剂B:丙烯腈(梯度洗脱) 检测器:λ=227nm,λ=240nm,λ=254nm	SPE 萃取/SDB 吸附柱/淋洗:甲醇	LOD:(11.8~42.2)ng/mL LOQ:(35.6~126.7)ng/mL	[79]
19 利尿剂/尿液	HPLC UV	色谱柱:Hypersil(150mm×3.0mm,5μm) 流动相:十二烷基硫酸钠(SDS):丙二醇:正丁醇:聚芳醚腈:丙烯腈或十二烷基硫酸钠:四氢呋喃:磷酸(梯度洗脱) 检测器:λ=245nm,λ=220nm	过滤(0.45μm)	LOD:(1~39)ng/mL	[80]
头孢吡肟,万古霉素,亚胺培喃/血浆	HPLC UV	色谱柱:Supelcosil LC-18(250mm×4.6mm,5μm) 流动相:0.075 M 醋酸盐缓冲溶液(pH为5.0):丙烯腈(92:8v/v) 检测器:λ=230nm	LLE 萃取/(3)-N-吗啉基丙磺酸	LOD:(0.17~0.38)μg/mL LOQ:(0.4~0.76)μg/mL	[81]
强的松,皮质醇/尿液	HPLC UV	色谱柱:Thermo Hypersil(250mm×4.6mm,5μm) 流动相:二氯甲烷:水:甲醇:四氢呋喃(66.5:30:2.5:1v/v/v/v) 检测器:λ=240nm	SPE 萃取/Oasis HLB 吸附柱/淋洗:甲醇	LOD:(4.8~7.0)ng/mL LOQ:(9.9~11.6)ng/mL	[82]

表 10-3 所选药物的伏安测定法

分析物/样品	方法	伏安法条件	样品处理	LOD/LOQ	引用文献
先锋霉素,头孢唑林/人造血清	DPV	工作电极:玻碳; 参比电极:Ag/AgCl; 支持电解质:0.1 M 磷酸盐缓冲溶液(pH为3.0)	溶于支持电解质中	LOD:1×10^{-6} M LOQ:n/a	[83]

续表

分析物/样品	方法	伏安法条件	样品处理	LOD/LOQ	引用文献
扑热息痛,对乙酰氨基酚葡萄糖醛酸盐,硫酸盐代谢物/模拟溶液	DPV	工作电极:玻碳 参比电极:Ag/AgCl 支持电解质:Britton-Robinson缓冲溶液(pH为3.29)	溶于支持电解质中	LOD:(3.27~5.09)μM LOQ:(1.09~1.70)μM	[84]
普萘洛尔,4-羟基普萘洛尔,4-羟基普萘洛尔硫酸盐/尿液	DPV	工作电极:玻碳,MWCNT-GCE 参比电极:Ag/AgCl 支持电解质:Britton-Robinson缓冲溶液(pH为3)	SPE萃取/Oasis HLB吸附柱/淋洗:甲醇	LOD:(1.10~1.37)μmol/L LOQ:(3.31~4.11)μmol/L	[85]
扑热息痛,速尿灵,安乃近,头孢唑啉和地塞米松/尿液	DPV	工作电极:滴汞电极和石墨 参比电极:Ag/AgCl 支持电解质:Britton-Robinson缓冲溶液(pH为2.4)	SPE萃取/SPE NH₂萃取柱/淋洗:甲醇 LLE萃取/乙酸乙酯	LOD:(0.20~2.57)μM LOQ:(0.53~6.28)μM	[86]
依那普利、赖诺普利/药片	DPP	工作电极:多模式电极 参比电极:Ag/AgCl 支持电解质:硼酸盐缓冲溶液(pH为9和10)	衍生化/2,4-二硝基氟苯	LOD:(0.004~0.600)μg/mL LOQ:(0.14~2.00)μg/mL	[87]
卡维地洛、扑热息痛、西地那非/尿液	DPV	工作电极:玻碳 参比电极:Ag/AgCl 支持电解质:Britton-Robinson缓冲溶液(pH为3.29)	SPE萃取/Oasis MCX吸附柱/淋洗:二氯甲烷:2-丙醇铵(78:20:2v/v/v)	LOD:(2.5~5.0)μg/mL LOQ:—	[88]
抗坏血酸、对乙酰氨基酚/药片	LSV	工作电极:碳糊电极 参比电极:Ag/AgCl 支持电解质:醋酸盐缓冲溶液(pH为4.7)	溶解于水中	LOD:— LOQ:—	[89]

参考文献

[1] Pauling, L., Robinson, A. B., Taranishi, R., Cary, P.: Quantitative analysis of urine vapor and breath by gas-liquid partition chromatography. Proc. Natl. Acad. Sci. 68, 2374-2376 (1971).

[2] Nicholson, J. K., Connelly, J., Lindon, J. C., Holmes, E.: Metabonomics: a platform for studying

drug toxicity and gene function. Nat. Rev. Drug Discov. 405, 153-161 (2002)

[3] Nicholson, J. K. , Wilson, I. D. : Opinion: understanding 'global' systems biology: metabonomics and the continuum of metabolism. Nat. Rev. Drug Discov. 2, 668-676 (2003)

[4] Kell, D. B. : Systems biology, metabolic modelling and metabolomics in drug discovery and development. Drug Discov. Today 11, 1085-1092 (2006)

[5] Fiehn, O. , Kopka, J. , Dormann, P. , Altmann, T. , Trethewey, R. N. , Willmitzer, L. : Metabolite profiling for plant functional genomics. Nat. Biotechnol. 18, 1157-1161 (2000)

[6] Raith, K. , Zellmer, S. , Lasch, J. , Neubert, R. H. H. : Profiling of human stratum corneum ceramides by liquid chromatography-electrospray mass spectrometry. Anal. Chim. Acta 418, 167-173 (2000)

[7] Beaudry, F. , Le Blanc, J. C. Y. , Coutu, M. , Ramier, I. , Moreau, J. P. , Brown, N. K. : Metabolite profiling study of propranolol in rat using LC/MS/MS analysis. Biomed. Chromatogr. 13, 363-369 (1999)

[8] Terabe, S. , Markuszewski, M. J. , Inoue, N. , Otsuka, K. , Nishioka, T. : Capillary electrophoretic techniques toward the metabolome analysis. Pure Appl. Chem. 73, 1563-1572 (2001)

[9] Kuligowski, J. , Quintás, G. , Garrigues, S. , de la Guardia, M. : Application of point-to-point matching algorithms for background correction in on-line liquid chromatography-Fourier transform infrared spectrometry (LC-FTIR). Talanta 80, 1771-1776 (2010)

[10] Berdeaux, O. , Fontagné, S. , Sémon, E. , Velasco, J. , Sébédio, J. L. , Dobarganes, C. : A detailed identification study on high-temperature degradation products of oleic and linoleic acid methyl esters by GC-MS and GC-FTIR. Chem. Phys. Lipids 165, 338-347 (2012)

[11] Hendrinks, M. M. W. B. , Cruz-Juarez, L. , DeBont, D. , Hall, R. D. : Preprocessing and exploratory analysis of chromatographic profiles of plant extracts. Anal. Chim. Acta 545, 53-64 (2005)

[12] Fiehn, O. : Combining genomics, metabolome analysis, and biochemical modelling to under-stand metabolic networks. Comp. Funct. Genomics 2, 155-168 (2001)

[13] Markuszewski, M. J. , Szczykowska, M. , Siluk, D. , Kaliszan, R. : Human red blood cells targeted metabolome analysis of glycolysis cycle metabolites by capillary electrophoresis using an indirect photometric detection method. J. Pharm. Biomed. Anal. 39, 636-642 (2005)

[14] Soga, T. , Ohashi, Y. , Ueno, Y. , Naraoka, H. , Tomita, M. , Nishioka, T. : Quantitative metabolome analysis using capillary electrophoresis mass spectrometry. J. Proteome Res. 2, 488-494 (2003)

[15] Britz-McKibbin, P. , Markuszewski, M. J. , Iyanagi, T. , Matsuda, K. , Nishioka, T. , Terabe, S. : Picomolar analysis of flavins in biological samples by dynamic pH junction-sweeping capillary electrophoresis with laser-induced fluorescence detection. Anal. Biochem. 313, 89-96 (2003)

[16] Markuszewski, M. J. , Britz-McKibbin, P. , Terabe, S. , Matsuda, K. , Nishioka, T. : Determina-tion of pyridine and adenine nucleotide metabolites in Bacillus subtilis cell extract by sweeping borate complexation capillary electrophoresis. J. Chromatogr. A 989, 293-301 (2003)

[17] Markuszewski, M. J. , Otsuka, K. , Terabe, S. , Matsuda, K. , Nishioka, T. : Analysis of carboxylic

acid metabolites from the tricarboxylic acid cycle in Bacillus subtilis cell extract by capillary electrophoresis using an indirect photometric detection method. J. Chromatogr. A 1010, 113-121 (2003)

[18] Minakami, S., Yoshikawa, H.: Thermodynamic considerations on erythrocyte glycolysis. Biochem Biophys. Res. Commun. 18, 345-349 (1965)

[19] Martinov, M. V., Plotnikov, A. G., Vivitsky, V. M., Ataullakhanov, F. I.: Deficiencies of glycolytic enzymes as a possible cause of hemolytic anemia. Biochim. Biophys. Acta 1474, 75-87 (2000)

[20] Harvey, J. W.: The erythrocyte: physiology, metabolism, and biochemical disorders. In: Kaneko, J. J., Harwey, J. W., Bruss, M. L. (eds.) Clinical Biochemistry of Domestic Animals, pp. 157-203. Academic Press, San Diego (1997)

[21] Wu, H., Xue, R., Dong, L., Liu, T., Deng, C., Zeng, H., Shen, X.: Metabolomic profiling of human urine in hepatocellular carcinoma patients using gas chromatography/mass spectrometry. Anal. Chim. Acta 648, 98-104 (2009)

[22] Kumar, A., Lakshimi, B., Kalita, J., Misra, U. K., Singh, R. L., Khetrapal, C. L., Nagesh, B. G.: Metabolomic analysis of serum by (1) H NMR spectroscopy in amyotrophic lateral sclerosis. Clin. Chim. Acta 411, 563-567 (2010)

[23] Struck, W., Siluk, D., Yumba-Mpanga, A., Markuszewski, M., Kaliszan, R., Markuszewski, M. J.: Liquid chromatography tandem mass spectrometry study of urinary nucleosides as potential cancer markers. J. Chromatogr. A 1283, 122-131 (2013)

[24] Szymańska, E., Markuszewski, M. J., Markuszewski, M., Kaliszan, R.: Altered levels of nucleoside metabolite profiles in urogenital tract cancer measured by capillary electrophoresis. J. Pharm. Biomed. Anal. 53, 1305-1312 (2010)

[25] Struck-Lewicka, W., Kaliszan, R., Markuszewski, M. J.: Analysis of urinary nucleosides as potential cancer markers determined using LC-MS technique. J. Pharm. Biomed. Anal. 101, 50-57 (2014)

[26] Burke, M. J., Preskorn, S. H.: Therapeutic drug monitoring of antidepressants: cost implications and relevance to clinical practice. Clin. Pharmacokinet. 37, 147-165 (1999)

[27] Wilkinson, G. R.: The effects of diet, aging and disease-states on presystemic elimination and oral drug bioavailability in humans. Adv. Drug Deliv. Rev. 27, 129-159 (1997)

[28] Woodcock, J.: The prospects for "personalized medicine" in drug development and drug therapy. Clin. Pharmacol. Ther. 81, 164-169 (2007)

[29] Lee, M. S., Flammer, A. J., Lerman, L. O., Lerman, A.: Personalized medicine in cardiovascular diseases. Korean Circ. J. 42, 583-591 (2012)

[30] Gabrielsson, J., Weiner, D.: Pharmacokinetic and Pharmacodynamic Data Analysis: Concepts & Applications. Kristianstads Boktryckeri AB, Sweden (2000)

[31] Bowers, L. D.: Analytical goals in therapeutic drug monitoring. Clin. Chem. 44, 375-380 (1998)

[32] Magiera, S., Uhlschmied, C., Rainer, M., Huck, C., Baranowska, I., Bonn, G.: GC-MS method

for the simultaneous determination of β-blockers, flavonoids, isoflavones and their metabolites in human urine. J. Pharm. Biomed. Anal. 56, 93–102 (2011)

[33] Havsteen, B. H. : The biochemistry and medical significance of the flavonoids. Pharmacol. Ther. 96, 67–202 (2002)

[34] Sica, D. A. : Interaction of grapefruit juice and calcium channel blockers. AJH 19, 768–773 (2006)

[35] Cermak, R. : Effect of dietary flavonoids on pathways involved in drug metabolism. Expert Opin. Drug Toxicol. 4, 17–35 (2008)

[36] Pizzolato, T. M. , de Alda, M. J. L. , Barceló, D. : LC-based analysis of drugs of abuse and their metabolites in urine. Trends Anal. Chem. 26, 609–624 (2007)

[37] Baranowska, I. , Magiera, S. , Baranowski, J. : Clinical applications of fast liquid chromatography: a review on the analysis of cardiovascular drugs and their metabolites. J. Chromatogr. B927, 54–79 (2013)

[38] Mestroni, L. , Taylor, M. R. G. : Pharmacogenomics, personalized medicine, and heart failure. Discov. Med. 11, 551–561 (2011)

[39] Oiviera, E. J. , Watson, D. G. , Morton, N. S. : A simple microanalytical technique for the determination of paracetamol and its main metabolites in blood spots. J. Pharm. Biomed. Anal. 29, 803–809 (2002)

[40] Jensen, L. S. , Valentine, J. , Milne, R. W. , Evans, A. M. : The quantification of paracetamol, paracetamol glucuronide and paracetamol sulphate in plasma and urine using a single highperformance liquid chromatography assay. J. Pharm. Biomed. Anal. 24, 585–593 (2004)

[41] Godejohann, M. , Tseng, L. -H. , Braumann, U. , Fuchser, J. , Spraul, M. : Characterization of a paracetamol metabolite using on-line LC-SPE-NMR-MS and a cryogenic NMR probe. J. Chromatogr. A 1058, 191–196 (2004)

[42] Johnson, K. A. , Plumb, R. : Investigating the human metabolism of acetaminophen using UPLC and exact mass oa-TOF MS. J. Pharm. Biomed. Anal. 39, 805–810 (2005)

[43] Yamamoto, E. , Takakuwa, S. , Kato, T. , Asakawa, N. : Sensitive determination of aspirin andits metabolites in plasma by LC-UV using on-line solid-phase extraction with methylcellulose-immobilized anion-exchange restricted access media. J. Chromatogr. B 846, 132–138 (2007)

[44] Magiera, S. , Gülmez, S. , Michalik, A. , Baranowska, I. : Application of statistical experimental design to the optimisation of microextraction by packed sorbent for the analysis of nonsteroidal anti-inflammatory drugs in human urine by ultra-high pressure liquid chromatography. J. Chromatogr. A 1304, 1–9 (2013)

[45] Magiera, S. , Gülmez, S. : Ultrasound-assisted emulsification microextraction combined with ultra-high performance liquid chromatography-tandemmass spectrometry for the analysis of ibuprofen and its metabolites in human urine. J. Pharm. Biomed. Anal. 92, 193–202 (2014)

[46] Jin, Y. X. , Tang, Y. H. , Zeng, S. : Analysis of flurbiprofen, ketoprofen and etodolac enantiomers by pre-column derivatization RP-HPLC and application to drug-protein binding in human

plasma. J. Pharm. Biomed. Anal. 46,953-958 (2008)

[47] Bayenes, W. R. G., Van der Weken, G., Haustraete, J., Aboul-Enein, H. Y., Corveleyn, S., Remon, J. P., Carcía-Campaňa, A. M., Deprez, P. : Application of the restricted-access precolumn packing material alkyl-diol silica in a column-switching system for the determina-tion of ketoprofen enantiomers in horse plasma. J. Chromatogr. A 871,153-161 (2000)

[48] Sandall, J. M., Millership, J. S., Collier, P. S., Mc Elnay, J. C. : Development and validation of an HPLC method for the determination of spironolactone and its metabolites in paediatric plasma samples. J. Chromatogr. B 839,36-44 (2006)

[49] Deventer, K., Pozo, O. J., Van Eenoo, P., Delbeke, F. T. : Qualitative detection of diuretics and acidic metabolites of other doping agents in human urine by high-performance liquid chromatography-tandem mass spectrometry. Comparison between liquid-liquid extraction and direct injection. J. Chromatogr. A 1216,5829-5827 (2009)

[50] Goebel, C., Trout, G. J., Kazlasuskas, R. : Rapid screening method for diuretics in doping control using automated solid phase extraction and liquid chromatography-electrospray tan-dem mass spectrometry. Anal Chim. Acta 502,65-74 (2004)

[51] Thieme, D., Grosse, J., Lang, R., Mueller, R. K., Wahl, A. : Screening, confirmation and quantitation of diuretics in urine for doping control analysis by high-performance liquid chromatography-atmospheric pressure ionisation tandem mass spectrometry. J. Chromatogr. B 757,49-57 (2001)

[52] Ramenskaya, G. V. : Chromatographic determination of drugs and their metabolites for phenotyping cytochrome P-450 isoenzymes. Pharm. Chem. J. 39,53-56 (2005)

[53] Baranowska, I., Magiera, S., Baranowski, J. : UHPLC method for the simultaneous determination of β-blockers, isoflavones and their metabolites in human urine. J. Chromatogr. B 879,615-629 (2011)

[54] Baranowska, I., Magiera, S., Kusa, J. : Development and validation of UHPLC-ESI-MS/MS method for the determination of selected cardiovascular drugs, polyphenols and their metabolites in human urine. Talanta 89,47-56 (2012)

[55] Baranowska, I., Magiera, S., Baranowski, J. : UHPLC method for the simultaneous determination of β-blockers, isoflavones, and flavonoids in human urine. J. Chromatogr. Sci. 49, 764 - 773 (2011)

[56] Partani, P., Modhave, Y., Gurule, S., Khuroo, A., Monif, T. : Simultaneous determination of propranolol and 4-hydroxypropranolol in human plasma by solid phase extraction and liquid chromatography/electrospray tandem mass spectrometry. J. Pharm. Biomed. Anal. 50,966-976 (2009)

[57] Mistry, B., Leslie, J. L., Eddington, N. D. : Enantiomeric separation of metoprolol and α-hydroxymetoprolol by liquid chromatography and fluorescence detection using a chiral stationary phase. J. Chromatogr. B 758,153-161 (2001)

[58] Lanchote, V. L., Bonato, P. S., Cerqueira, P. M., Pereira, V. A., Cesarino, E. J. : Enantioselective analysis of metoprolol in plasma using highperformance liquid chromatographic direct and

indirect separations:applications inpharmacokinetics. J. Chromatogr. B 738,27-37 (2000)

[59] Kim,K. H. ,Kim,H. J. ,Kang,J. S. ,Mar,W. :Determination of metoprolol enantiomers in human urine by coupled achiral-chiral chromatography. J. Pharm. Biomed. Anal. 22,377-384 (2000)

[60] Jensen,B. P. ,Sharp,C. F. ,Gardiner,S. J. ,Begg,E. J. :Development and validation of a stereoselective liquid chromatography-tandem mass spectrometry assay for quantification of S-and R-metoprolol in human plasma. J. Chromatogr. B 865,48-54 (2008)

[61] Magiera,S. , Adolf, W. , Baranowska, I. : Simultaneous chiral separation and determination of carvedilol and 5'-hydroxyphenyl carvedilol enantiomers from human urine by high performance liquid chromatography coupled with fluorescent detection. Cent. Eur. J. Chem. 11, 2076 - 2087 (2013)

[62] Siluk,D. ,Mager,D. E. ,Gronich,N. ,Abernethy,D. ,Wainer,I. W. :HPLC-atmospheric pressure chemical ionization mass spectrometric method for enantioselective determination of R,S-propranolol and R,S-hyoscyamine in human plasma. J. Chromatogr. B 859,213-221 (2007)

[63] Schlauch,M. ,Fulde,K. ,Frahm,A. W. :Enantioselective determination of (R)-and (S)-sotalol in human plasma by on-line coupling of a restricted-access material precolumn to a cellobiohydrolase I-based chiral stationary phase. J. Chromatogr. B 775,197-207 (2002)

[64] Zurbonsen, K. , Berssolle, F. , Solassol, I. , Aragon, P. J. , Culine, S. , Pinguet, F. : Simultaneous determination of dexamethasone and 6β-hydroxydexamethasone in urine using solid-phase extraction and liquid chromatography:applications to in vivo measurement of cytochrome P450 3A4 activity. J. Chromatogr. B 804,421-429 (2004)

[65] Wang,X. -D. ,Li,J. -L. ,Lu, Y. ,Chen, X. ,Huang, M. ,Chowbay, B. ,Zhou, S. -F. :Rapid and simultaneous determination of nifedipine and dehydronifedipine in human plasma by liquid chromatography-tandem mass spectrometry: Application to a clinical herb-drug interaction study. J. Chromatogr. B 852,534-544 (2007)

[66] Markowski,P. ,Baranowska,I. ,Baranowski,J. :Simultaneous determination of L-arginine and 12 molecules participating in its metabolic cycle by gradient RP-HPLC method. Application to human urine samples. Anal. Chim. Acta 605,205-217 (2007)

[67] Rentsch, K. M. , Gutteck-Amsler, U. , Bührer, R. , Fattinger, K. E. , Vonderschmitt, D. J. : Sensitive stereospecific determination of acenocoumarol and phenprocoumon in plasma by high-performance liquid chromatography. J. Chromatogr. B 742,131-142 (2000)

[68] Vecchione,G. ,Casetta,B. ,Tomaiuolo,M. ,Grandone,E. ,Margaglione,M. :A rapid method for the quantification of the enantiomers of Warfarin,Phenprocoumon and Acenocoumarol by two-dimensional-enantioselective liquid chromatography/electrospray tandem mass spectrometry. J. Chromatogr. B 850,507-514 (2007)

[69] Magiera,S. :Fast,simultaneous quantification of three novel cardiac drugs in human urine by MEPS-UHPLC-MS/MS for therapeutic drug monitoring. J. Chromatogr. B 938,86-95 (2013)

[70] Magiera,S. ,Kusa,J. :Evaluation of a rapid method for the therapeutic drug monitoring of aliskiren, enalapril and its active metabolite in plasma and urine by UHPLC-MS/

MS. J. Chromatogr. B 980,79−87 (2015)

[71] Magiera,S. ,Baranowska,I. :Rapid method for determination of 22 selected drugs in human urine by UHPLC/MS/MS for clinical application. J. AOAC Int. 97,1526−1537 (2014)

[72] Baranowska,I. ,Markowski,P. ,Baranowski,J. :Development and validation of an HPLC method for the simultaneous analysis of 23 selected drugs belonging to different therapeutic groups in human urine samples. Anal. Sci. 25,1307−1313 (2009)

[73] Baranowska,I. ,Markowski,P. ,Baranowski,J. :Simultaneous determination of eleven drugs belonging to four different groups in human urine samples by reversed-phase high-performance liquid chromatography method. Anal. Chim. Acta 570,46−58 (2006)

[74] Baranowska,I. ,Markowski,P. ,Baranowski,J. ,Rycaj,J. :Simultaneous determination of sildenafil, its N-desmethyl metabolite and other drugs in human urine by gradient RP-HPLC method. Chem. Anal. 52,645−671 (2007)

[75] Magiera, S. , Hejniak, J. , Baranowski, J. :Comparison of different sorbent materials for solid-phase extraction of selected drugs in human urine analyzed by UHPLC-UV. J. Chromatogr. B 958,22−28 (2014)

[76] Magiera,S. ,Baranowska,I. :A new and fast strategy based on semiautomatic microextraction by packed sorbent followed by ultra high performance liquid chromatography for the analysis of drugs and their metabolites in human urine. J. Sep. Sci. 37,3314−3320 (2014)

[77] Abu-Qare,A. W. ,Abou-Donia,M. B. :A validated HPLC method for the determination of pyridostigmine bromide, acetaminophen, acetylsalicylic acid and caffeine in rat plasma and urine. J. Pharm. Biomed. Anal. 26,939−947 (2001)

[78] Frerichs,V. A. ,Tornatore,K. M. :Determination of the glucocorticoids prednisone,prednisolone, dexamethasone, and cortisol in human serum using liquid chromatography coupled to tandem mass spectrometry. J. Chromatogr. B 802,329−338 (2004)

[79] Baranowska,I. ,Magiera,S. ,Baranowski,J. :Ultra HPLC method for the simultaneous analysis of drugs and flavonoids in human urine. J. Liq. Chromatogr. Relat. Technol. 34,421−435(2011)

[80] Rosado-Maria, A. , Gasco-Lopez, A. I. , Santos-Montes, A. , Izquierdo-Hornillos, R. :High-performance liquid chromatographic separation of a complex mixture of diuretics using a micellar mobile phase of sodium dodecyl sulphate:application to human urine samples. J. Chromatogr. B 748,415−424 (2000)

[81] López, K. J. V. ,Bertoluci,D. F. ,Vicente,K. M. ,Dell'Aquilla, A. M. ,Santos,S. R. C. J. :Simultaneous determination of cefepime,vancomycin and imipenem in humanplasma of burn patients by high-performance liquid chromatography. J. Chromatogr. B 860,241−245 (2007)

[82] AbuRuz,S. ,Millership,J. ,Heaney,L. ,McElnay,J. :Simple liquid chromatography method for the rapid simultaneous determination of prednisolone and cortisol in plasma and urine using hydrophilic lipophilic balanced solid phase extraction cartridges. J. Chromatogr. B 798,193−201 (2003)

[83] Jamasbi,E. S. ,Rouhollahi,A. ,Shahrokhian,S. ,Haghgoo,S. ,Aghajani,S. :The electrocatalytic

examination of cephalosporins at carbon paste electrode modified with CoSalophen. Talanta 71, 1669-1674 (2007)

[84] Baranowska, I., Koper, M.: The preliminary studies of electrochemical behavior of paracetamol and its metabolites on glassy carbon electrode by voltammetric methods. Electroanalysis 21, 1194-1199 (2009)

[85] Baranowska, I., Koper, M.: Electrochemical behavior of propranolol andits major metabolites, 4'-hydroxypropranolol and 4'-hydroxypropranolol sulfate, on glassy carbon electrode. J. Braz. Chem. Soc. 22, 1601-1609 (2011)

[86] Baranowska, I., Markowski, P., Gerle, A.: Determination of selected drugs in human urine by differential pulse voltammetry technique. Bioelectrochemistry 73, 5-10 (2008)

[87] Razak, O. A., Belal, S. F., Bedair, M. M., Barakat, N. S., Haggag, R. S.: Spectrophotometric and polarographic determination of enalapril and lisinopril using 2, 4-dinitrofluorobenzene. J. Pharm. Biomed. Anal. 31, 701-711 (2003)

[88] Baranowska, I., Koper, M., Markowski, P.: Electrochemical determination of carvedilol, sildenafil and paracetamol using glassy carbon electrode. Chem. Anal. 53, 967-981 (2008)

[89] Săndulescu, R., Mirel, S., Oprean, R.: The development of spectrophotometric and electroanalytical methods for ascorbic acid and acetaminophen and their applications in the analysis of effervescent dosage forms. J. Pharm. Biomed. Anal. 23, 77-87 (2000)

第11章
法医学分析

11.1 刑事学分析

法医用途的各种材料的分析称为痕迹分析。对在犯罪现场或事件现场形成的刑事痕迹进行分析,面临的困难不仅仅是分析物的含量很低,而是可供检测的样品量非常少,其质量通常为毫克级或微克级。现场获得的如皮肤碎片、特殊纤维、灰尘、玻璃和塑料片、土壤颗粒或血滴等各种微量材料都是刑事事件的宝贵信息来源。因此,发现和妥善保管事件现场的材料对于在刑事实验室进行正确检测和解释犯罪情节至关重要。

11.1.1 引言

在刑事学中引入了痕迹和微量的概念,以界定区分在犯罪现场发现的材料特性并对其进行检测。Jan Sehn 定义了刑事痕迹的概念,他说:"刑事观念的痕迹在客观现实中是随事件调查而发生变化的,这些痕迹可以帮助再现实际发生的情况和确定现场事件过程的基础。"[1] 正如这个定义,痕迹是某些行为和现象的结果,因此它与这些行为或现象构成了因果关系。痕迹是事故现场留下的一个客体或一部分客体,如鞋或轮胎的一个压痕、一个工具表面产生的凹痕或刮擦、一个指纹、一个液体污渍(包括血液、油渍)、一根头发、一块玻璃或装饰品碎片,或因为加热、外力引起的物体形状的变化。通过感知或借助放大镜、显微镜、照明器等技术设备来认知痕迹,痕迹因为具有物质性质可以用来检测和研究。

痕迹的重要性主要在于它可能重现事件。根据痕迹,有可能重现某一特定事件的过程,并确定哪些人参与了该事件及其在事件发生时的行为。此外,痕迹甚至可以帮助直接逮捕事件的肇事者;对痕迹的检查也可以界定相关人员是否或如何与司法机构调查的事件有关；痕迹也可以在确定地点、人员和事物方面发挥重要作用。

不同规模的事件现场都能产生犯罪痕迹。随着测量设备的进步、分析化学技术的发展、犯罪痕迹的推理证实研究，以及少量暴露的宏观痕迹研究，使捕获并检测微观痕迹变得特别重要。在物理化学术语中，微观痕迹与宏观痕迹没有区别，微观痕迹可以是毫克量或更少量的物质的颗粒，如土壤颗粒、尘埃、微纤维、皮肤碎片和液体微滴；也可以是一种嗅觉器官无法察觉的气体以及肉眼看不见或几乎无法看见的机械动作的微小痕迹，像抓痕、凹痕或裂纹。Mirosaw Owoc 为此做了如下定义："微观痕迹是犯罪的痕迹，由于它们的尺寸太小或其他的特殊性质，如果没有适当的观测仪器，人们难以察觉或几乎不能察觉到它们，只能运用微量分析方法来检测。"

通常必须使用各种显微技术和微量分析技术才能检测到微观痕迹，杂质和添加剂也是极微量的，必须使用先进的分析方法进行研究。

11.1.2 微观痕迹的特点

微观痕迹不仅有微观特征，而且有普遍性的特征，即不管罪犯如何努力去克服都无法完全避免痕迹的产生，想清除这些痕迹是很困难的。微观痕迹是由于罪犯穿着的衣服、使用的工具/仪器等与周围事物相互作用而产生的。最常见的微观痕迹是从最初较大的物体分离下来的小颗粒，如微纤维、玻璃颗粒、油漆、金属、塑料、土壤或炸药等。在宏观痕迹出现的地方，微观痕迹也同样存在，构成了相互间的一种补充。然而，从犯罪行为的角度来看，由于罪犯或其他人会破坏或清除宏观痕迹，甚至有经验的罪犯不会留下任何宏观痕迹，所以，微观痕迹的价值要高得多，因为微观痕迹总是不可避免地被留在犯罪现场。微观痕迹出现在每个事件的现场，在罪犯使用的每个工具上，或是受害者及其所使用的物品上。灰尘颗粒或纤维容易聚集在基质的凹槽和缝隙中，因此，很难将它们清除。微观痕迹能够抵抗清洗和擦拭等破坏性因素而留存下来，黏附力的作用还有助于微观痕迹保持在基质表面。从化学成分方面来说，微观痕迹通常是许多化合物的混合物。

由于微观痕迹极小，因此它们经常被无意识地破坏，例如，它们可能会意外被毁或转移到另一个基质上，或者在较高的温度下蒸发，或者被火燃烧破坏。从留下痕迹到发现和确保痕迹检查的过程中，微观痕迹属性可能会改变，时间长短以及大气条件都会导致痕迹或发生痕迹的基质发生化学、生物分解，从而使痕迹分析难以进行。

11.1.3 犯罪痕迹检测的难点

对发现的痕迹进行分析可能性研判、分析方法的选择以及对所得结果的解释受许多因素的影响。其中最主要的因素是检测目的和所采用的分析方法。

1. 检测目的

识别定性是痕迹化学分析的首要目的。目前,有许多分析方法和技术能够精准地识别各种各样的材料。无论样本的大小如何,都对其化学成分进行定性测定,确定其种类通常并不困难。就涂料或纤维等材料而言,参考强大的数据库可以确定其生产商。然而,在分析过程中获得的有关材料的信息只会产生团体识别,而不是个体识别。也就是说,这些材料通过分析可以识别判定为相同组分和属性的一类材料,这种类别群体大小是变化的,取决于分析过程中获得的信息量。分析化学方法可以帮助判定被比较的材料是相同的还是不同的,然而这些方法不能确定它们是否属于某一物质或同一物质体系的一部分,以及确定是否是同一种物质。因此,这些方法不能个体识别所获得的材料。当然,个体识别是有可能实现的,但需要的不仅仅是对痕迹的化学成分和物理化学性质的测定。

通过检测化学成分和某些物理化学性质,并与标准物质、数据库或从嫌疑人处收集的比较材料进行对照,可以确定痕迹属性。刑事实验室建立了大量玻璃、油漆、纸张、织物和其他材料的产品和物理特性的数据库,并收集了数以千计的用于对照的样本库。

2. 材料样本

微观痕迹通常是非常小的材料样本,通常在毫克级或微克级别,如油漆碎片、几滴油、单个纤维、一根头发、一块玻璃或塑料碎片。一旦这种材料在事故现场被提取,就无法再次收集检测。因此,它在数量上是有限的,检测时必须满足所有目的的检查,其中一些还应保留下来。此外,提交给实验室的样本经常被污染,而且常常很难与它们发生的基质分离,被研究的材料越小,其结构越精细,基质对分析结果的影响就越大。因此,如果不能将所研究的材料与基质分离,就应该考虑基质对分析结果的影响。例如,检测确认收到款项的签名,是以姓名首字母缩写的,最多由几个字母组成,这些字母用普通墨水或印度绘画墨水书写,墨水量非常少还渗透到纸的结构中。从纸张中提取墨水会损坏该文件,需要司法机构批准检测请求。因此,需要直接在文件上检测墨水,而不将其与纸分离。通过使用红外光谱法(IR)和拉曼光谱法,在获得的谱图中区分墨水和纸张的吸收带,进而正确解读光谱数据并获取有关墨水成分的信息。

用于化学分析的材料样本很少是一个单一纯净系统,通常是含有特殊组分的复合物。在这些情况下,有必要将一个或几个重要的组分从系统中分离出来,并剔除大量对检测没有意义的基体。

通常情况下,被测材料主要成分或部分成分对法院调查员来说并不重要,而与事件有关的附着在较大物体表面裂缝上的物质粒子才是关键。例如,对于从火灾现场收集的材料的研究,它组成了一个各类材料燃烧产物的混合物,分析目的是找到产生火源的引燃物痕迹,它们通常是高度易燃的液体,可能被封闭在烧焦的聚合物材料(地板、地毯)孔隙中。

在法庭案件中,刑事学检测的材料构成了材料证据。对它们的鉴定是其与参考物或提供研究的比对材料进行比较。

3. 检测微观痕迹的作用

尽管微观痕迹具有持久性,但微观痕迹研究的有效性可能受到多种不利因素的影响。微观痕迹在事件现场形成时,由于各种因素的影响可能导致其产生了异质性,不能代表其原始存在的整体客观状态。例如,在弄脏的衣服上发现的泥土的痕迹,只有极小部分土壤或尘埃形成了痕迹,因此留在织物表面的只有这样一个小样本。

影响微观痕迹从形成到被固定期间的因素包括次生变化,如生物分解和腐蚀、飘浮在大气中的尘埃污染、生物材料(如含有枪伤的皮肤碎片)的降解等。这些因素使原始的微观痕迹很难搜寻到,例如,枪伤上的射击残渣。同样,金属钉的腐蚀使侦查员无法在其表面看到工具(如锯片)使用的痕迹。

还应该注意微观痕迹从提取到检测期间的其他影响因素,如环境湿度影响、不正确的固定方法和意外干扰等。比如,在恶劣天气下储存的湿衣服很容易被霉菌覆盖,这使得在衣服表面寻找纤维、油漆颗粒等变得困难甚至不可能。这些不利因素的影响会因时间的流逝而加剧。

11.1.4 痕迹分析方法

为了识别构成犯罪痕迹的材料,有必要研究它们的形态,定性分析它们的化学成分并研究一些物理化学性质。由于形成痕迹的材料样本很小,需要采用微分析方法进行识别,比如使用不破坏样品或使用样品量最少的方法,使用相同或不同的方法进行重复分析。目前,仪器方法在检测痕迹方面发挥着非常重要的作用,它们能够快速获得结果,并且对分析成分具有高灵敏度和低检出限的特征,将检测结果与标准/参考物质获得的数据库进行比对,识别所研究材料的分析结果。所选方法具有高度辨别能力,能够区分不同性质的样本。

在进行检测时数据应用的基本原则是核查,一种方法得到的结果应由其他技术来证实。痕迹检测有一个跨学科的特点,即对提交材料的鉴定需要来自不同学科专家的合作。

目前,即使是在事件现场肉眼可见的痕迹,也因没有足够精确的分析方法,导致许多有效的信息没有获取到。随着微量分析方法得到有效的发展,而且法庭调查员参与到实际案情分析中,痕迹分析逐渐成为一个事件有价值的信息来源,可以提供关于事件的过程和参与其中的人的有用信息。

构成犯罪痕迹的样本量是很小的,因此其应用的分析技术应该是非破坏性的,在完成分析后可以进行样本保存或复原。所有的光学显微镜法和光谱测量技术都是这样的方法。色谱技术广泛应用于对犯罪痕迹的研究,色谱法用的研究样本非常少。

1. 显微镜法

分析犯罪痕迹的基本方法是显微镜法。分析过程总是从显微镜开始,它得到的结果决定了如何选择后续的检测方法。进行显微分析的基本目的是观察样品的形态,确定其结构、厚度和均匀性。在分析过程中,有可能发现可能的夹杂物。有时,显微镜也可以识别如天然来源的纤维、矿物质、枪的残留物等样品或土壤污染物等。显微镜通常用于检测样品的比对,获得两个样本的相似显微图像是推断其相似性或具有共同的来源的基础。显微镜检测是非破坏性的,即使必须对显微样品进行制备,样品也可以回收,以便进行下一步的检查。显微镜有两种类型,光学显微镜和电子显微镜。

光学显微镜能进行几倍到一千倍的放大倍数的样品观测。光学显微镜下样品的照明光源是白光,透射光显微镜是白光穿过样品,反射光学显微镜是利用样品表面的反射光。此外还有利用偏振光的偏振光显微镜,以及短波长光照射使样品产生荧光的荧光显微镜。

电子显微镜利用了一种完全不同于光源的能量,即加速电子。电子显微镜获得的表面图像具有很好的分辨率和景深,这是光学显微镜无法比拟的。在电子显微镜下,电子束不会直接形成图像,通过碰撞激发样本表面而形成图像。激发样本的电子束直径常常很小,大约是微米的几百分之一。由于电子不能穿透样本,因此得到的图像能够反映样品的表面反射或者表层结构。

2. 显微光谱分析法

显微光谱法是光学显微镜和光谱分析的结合,是刑事分析不可或缺的一种技术手段。显微镜创建、记录和解释放大的图像,光谱则利用物质辐射能的发射、吸收和反射来测量其结构、性质和组成。根据辐射能的类型,显微光谱分析法可以分为红外、紫外可见(UV-vis)以及拉曼显微光谱法。显微光谱分析还包括X射线显微光谱分析,该分析方法由电子显微镜代替了光学显微镜。红外和拉曼显微光谱法能够测定和比较研究样品的化学成分;紫外可见分光光度法以独立于观察者的客观方式比较样品的颜色;X射线显微分光光度法可以测定元素组成。

显微光谱分析的优点是可以对极少量的样品进行分析,通常不需要将其与基质分离,分析前没有繁琐的样本制备过程,可以在不破坏样品的情况下进行多次重复测量。显微光谱分析的另一个独特的优势是可以对样品的测量区域进行拍照和归档,这使得它现在被大多数犯罪分析实验室用于事件现场发现的痕迹的检测。然而,从原理上看,显微光谱分析的缺点是只允许微区分析,当样品不均匀和受到污染时会对光谱测量结果产生重大影响。

3. 红外显微光谱法

红外显微镜是一种傅里叶变换红外分光光度计(FTIR)与光学显微镜联用仪器。显微镜放大样品以确定其形态,并选定分析区域;分光光度计通过透射光或反射光的测量来确定化学成分。红外显微光谱测量提供了关于样品的微观结构和光学特性等方面的信息,在对样品观察和光谱测量中均可以采用偏振光。

为了获得高质量的光谱,必须确保有足够的红外辐射能到达检测器,而且要准确地选定样品的分析区域,否则会导致信噪比减小。分析人员用一束白光可以选定样本分析区域,红外衍射可以扩展光谱分析区域,因此,精确划分光谱分析区域对于微小样品获得良好的红外光谱至关重要。

红外显微光谱法在法庭案例中作为材料证据的各种物质痕迹的定性分析、样品的同质性分析、表面夹杂物和污染识别,以及结构缺陷检测都非常有效,它的主要缺点是微小样本的物理性质会影响测量的光谱精度,并造成所获光谱的失真。

红外显微光谱法常用于微观痕迹的识别,如极小量的颜料、塑料、纤维、橡胶和胶水;也用于化学成分的分析,如墨水或调色剂等。

4. 紫外可见显微光谱法

该分析方法不依赖于分析者眼力识别的敏感度和质量,能够以客观方式比对不同材质的微小样品的颜色,如单根纤维、油漆颗粒、在伪造的文件上留下的墨水或圆珠笔墨水痕迹。紫外可见显微光谱仪器由光学显微镜、紫外可见光光谱仪通过模拟数字转换器与计算机连接组成,仪器能获得两个相似颜色的样本之间存在的光谱差异信息,而这种差异信息在光学显微镜下是难以区分的。如果两个样品获得完全一致的光谱,则证明其颜色相同,具有相同的颜料和染料成分。应用适当的微观光谱分析软件可以给色度坐标赋值来精确定义颜色,这些坐标构建了由色度、亮度和饱和度组成的色彩空间,色彩空间中点的重叠证明了样品具有的相同颜色。1931年,国际照明委员会第一次对颜色的测量和描述进行了细致的标准化[2],约40年后,提出了数字色彩的描述和色度坐标的计算。

5. 拉曼显微光谱法

拉曼显微光谱法是红外显微光谱法的一种补充方法。光学显微镜与分光光度计相结合可以测量一个几微米直径大小的样品散射的辐射光。激发样品所用的激光不同可以获得不同成分的信息。该方法的优点包括灵敏度高、光谱分辨率高、测量时间短、识别能力强,以及有对样品中所选成分进行空间成像的可能性。这些优势受到法医专家的青睐。该方法最擅长分析颜料颗粒、单根纤维和文件上的墨水中的痕量色素成分。拉曼显微光谱检测不会侵入和破坏被测样品,是一种用于检测文件真实性的基本技术,能够在被质疑的文件上直接区分墨水和圆珠笔墨水。有时样品会产生荧光,这时需要利用几个激光激发器来获得直读拉曼光谱或使用表面增强共振拉曼散射光谱(SERRS)。

6. X射线显微光谱法

使用扫描电子显微镜和能量色散X射线联用检测系统(SEM-EDX)或使用X射线微荧光光谱仪(μ-XRF)进行X射线显微分析。当电子显微镜中阴极发射加速电子束轰击样品表面时,样品材料发出特征X射线辐射,或者在用电子束激发阳极靶原子的过程中,X荧光光谱仪(XRF)的X射线管中产生X射线辐射。上述仪器分析能够提供有关分析样品元素组成的信息。

电子束的电子能量越大,检测元素的平均原子数越小,进入样本表面就越深。在电子显微镜下,电子渗透的范围为零点几微米到几微米。μ-XRF的检测深度能达到毫米级。

利用电子产生的特征X射线辐射强度与被测元素的浓度之间的线性关系可以定量分析元素。原则上,利用SEM-EDX进行的X射线显微分析是点分析,适于研究在电子束中稳定的非常小的固体样品。而X射线荧光法可检测固体和液体样品,由于到达仪器检测器的信号来自被测样品总量,因此它不是点分析,SEM-EDX方法更灵敏。

7. 裂解气相色谱法

裂解气相色谱法也是一种重要的微分析技术,检测仪器由裂解器、气相色谱质谱仪组成。裂解器将样品分解为简单的挥发性物质,气相色谱质谱仪用于分离和检测挥发性物质。与显微光谱法相比,这种检测技术对样品只有很小程度上的破坏,对于聚合物、塑料、橡胶等高分子材料的化学成分分析是不可或缺的。样品在惰性气体中,高温或电磁辐射使样品化学键断裂而分解,产生稳定的特征碎片并通过色谱柱分离,质谱对化合物种类进行识别并提供初始样本的组成信息。选择合适的热解条件可以控制样品的断裂方式和特征碎片的形成,从而可以区分化学成分相似的样品。通过保持相同的裂解条件和稳定的测量条件,可以从初始样品中以可重复的方式获得相同类型的碎片。在联机系统中利用适当的试剂对样品进行初步衍生化,有利于改善某些化合物的检测。

分析方法所需的样本量较少,根据聚合物的类型和应用仪器的类型,通常为$3\sim5\mu g$,测量精度为$10\%\sim20\%$,可用于涂料、塑料、橡胶、胶水和胶带等聚合物材料的痕迹研究。

11.1.5　测量结果的利用

事实上测量结果都有测量误差,在对整个犯罪痕迹样品的检测过程中,最困难的工作是对所获得的测量结果进行解释。测量结果的可重复性尤其重要,因此需要测量精度必须足够高,且不应该包含系统误差,而是接近真值。选用的方法应该准确、可靠,每个量的测量都需要重复多次,然后观察结果的分散性并确定测量误差。每个测量方法都必须经过验证。

既可以用简单的统计学方法也可以用复杂的化学计量统计方法对测量数据进行处理。显著性检验用于判断两个比对样品的测量结果所存在的细微差异是被测量的实际差异还是由偶然误差造成的。

采用相关系数分析、聚类分析或神经网络分析等化学计量学方法能够依据测量参数将被测材料以适当的产品组进行分类。例如,可根据玻璃的元素组成或折射率对玻璃碎片进行分类。

将检测结果用于刑事解释非常重要。在案件中构成物证材料的痕迹化学检测有助于识别罪犯,因此,一个法医力求使用可靠的检测方法和程序进行痕迹检测并获得被测样品最典型的特征数据,如组成和性能。如果被测材料化学成分和性质经比对后不足以说明它们是同一种物质,那么还需了解材料类型的差异、不符合常规的演变以及日常使用情况,了解犯罪事件的发生过程有利于解释检测结果。一些犯罪学研究实验室建立了各种典型材料的数据库,其中的材料常用于事故现场犯罪痕迹的保留,一般包括油漆、玻璃、塑料和纤维等材料。材料数据库既包含有关产品的技术信息,也包含实验室的检测结果。

将检测结果进行分析处理可以获得以下结论:如果通过分析结果的比对可以认定痕迹证据材料的特性和化学成分与参考物质一致,那么可以认为证据材料与参考物有共同的出处。如油漆、玻璃、塑料和纤维样品,这意味着在案件发生前它们是一个整体;如果是土壤痕迹样品,意味着它们可能来自某一特定区域的同一个地方。绝对的确定是不可能的,比如会有一种可能,被研究的材料来自两个不同的产品,它们虽然属于同一类型,但来自两个生产批次,因此,它们的化学成分只有微小的差别。但是如果被比对材料在性质或成分上有差异,那么我们可以假定两者是显著不同的。

11.1.6　微观痕迹的种类

在刑事学实验室中最为常见的检测对象是接触痕迹,即油漆涂层、玻璃、单纤维、土壤、书写材料的小碎片等。此外,经识别可显示源自火灾现场的碎片,如可燃液体痕迹或所使用的火器痕迹。

1. 油漆

油漆痕迹最常见于汽车事故、抢劫或入室盗窃等事件中,以油漆微碎片的形式出现,通常仅有几平方毫米,甚至更小。有时油漆痕迹会是在事件现场的人的衣服上或其他物体上留下的油漆污点或着色条纹。检测油漆的目的是确定油漆痕迹的来源与嫌疑人的物品,如他的车辆,所使用的工具等之间的相似性程度;也可以确定油漆产品的类型、用途、生产商和生产年份。对油漆的常规检测包括颜色、色度以及油漆碎片的结构和化学成分[3]。

大多数情况下,油漆碎片为多层结构,层厚在 10~50μm,每层都是许多化合物的混合物。油漆污点通常是由一层或两层的油漆材料混合而成,而且油漆会与织物纤维在一起并且嵌入其底部。在光学显微镜下可观察到油漆碎片横截面的层数、颜色和厚度(图 11-1),这是油漆涂层的形态特征。如果样品用偏振光照射,或者被紫外光照射会产生荧光,这些油漆涂层通常会更显著。

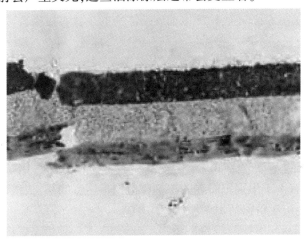

图 11-1 (见彩图)汽车漆掉落碎片的横截面在透射光中的显微图像

颜色是油漆样品最具特色的特征之一,可以通过与油漆生产商标准油漆目录中的颜色进行比对来准确描述。以汽车漆为例,如果油漆样品的颜色和标准油漆目录中的颜色匹配,就可以确定汽车的型号和它的生产商。

然而,色彩的视觉比对是主观的,受光线、检测者视力的清晰度和质量等影响。在可见光范围内,观察者可以利用显微光谱分析技术(MSP-vis)客观地比对微小样本之间的颜色,不需要分析颜料成分。每种颜色都可以用色度、亮度和饱和度三个变量来描述,可认定为这些变量构成的颜色空间中的一个坐标点,颜色空间中重叠的点说明其颜色相同。约 40 年前就建立了一种变量数学统计方法,可根据测量光谱定义的三维颜色参数坐标来评估颜色的相似性,通过建模确定阈值有助于评估颜色的差异性和相似性[4-5]。

每种油漆都含有黏合剂,它由合成树脂、有机颜料和无机颜料、填充剂和装饰效果颜料混合而成。颜料决定了油漆涂层的颜色,填充剂的作用是保证油漆涂层的装饰效果,如覆盖和抛光,以及阻止环境因素对涂层的影响。相同颜色的油漆可以包含相同的聚合基,但是颜料和填充剂的采用取决于油漆的使用者和制作者。聚合物黏合剂的组分通常用红外光谱法来定性检测。采用GC-MS对油漆中的气体挥发物进行分析,能够检测聚合物黏结剂的含量差异。例如在图11-2和图11-3中,3种油漆样品含有相同类型的聚合物黏合剂:苯乙烯-丙烯-聚氨酯黏合剂和相同的无机颜料主体成分:二氧化钛,因此它们的红外光谱图非常相似,但裂解色谱图明显不同,这表明油漆样品的聚合物含量有很大的不同。事实上,它们是3种不同的油漆[6-10]。

无机颜料和填充剂通常采用SEM-EDX或XRF测量油漆元素组成。根据油漆的基本成分和油漆行业中可能使用的颜料数据库可以确定颜料的含量。只有在使用拉曼光谱的情况下,才能检测添加到油漆中的少量有机颜料,通过将样品的拉曼光谱与标准颜料谱图库比对可以识别有机颜料种类。

通过油漆的分析数据可以判断油漆样本是否来自同一涂层,如果不能获得从嫌疑人身上获取的参考材料,则只能确定油漆的种类和涂层的类型。以汽车油漆为例,可以找到事故中涉及的汽车样式。通常分析数据须与数据库进行对比,该数据库包含了欧洲各种型号的油漆涂料及其分层的化学成分信息,该数据库建立在欧洲始于1995年,其后每年都会更新产品信息,来自许多刑事实验室的专家参与了它的创建。通过与该数据库进行比较,可以获得涉事车辆的生产年份、制造、型号等信息,从而有助于识别犯罪者的车辆。但值得注意的是,依据油漆数据库识别车辆的制造和生产年份仅适用于原厂漆车辆。

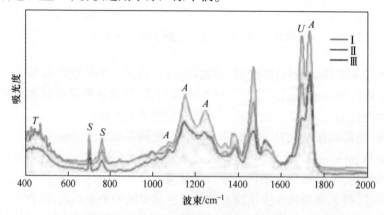

图11-2 (见彩图)3种苯乙烯丙烯酸聚氨酯漆(Ⅰ、Ⅱ、Ⅲ)的红外光谱
S—苯乙烯树脂;A—丙烯酸树脂;U—聚氨酯树脂;T—钛白粉。

在衣服或其他基质上通常只有少量可见的油漆。从它们上面获得的油漆涂层信息量比油漆碎片要小。因此，它的证据价值也比油漆碎片的价值小。

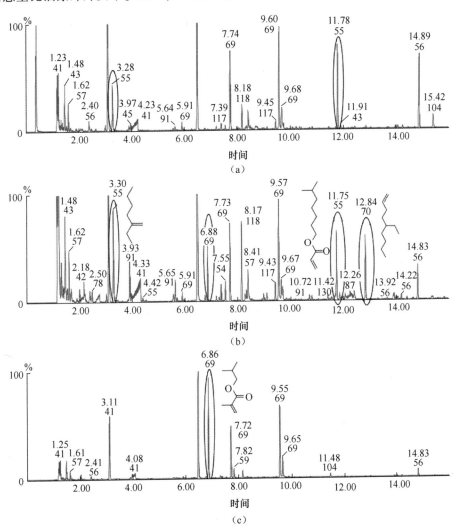

图11-3　（见彩图）3种苯乙烯丙烯酸聚氨酯漆的裂解图（不同处用椭圆标记）

2. 玻璃

众所周知，在房子、商店或汽车里打碎窗户时玻璃碎片会残留在人的衣服上。在交通事故、斗殴、抢劫和枪击事件中，常会发现玻璃碎片。极小的玻璃碎片可以从破碎的玻璃中脱离出几米远，并在衣服纤维之间隐藏很长一段时间。玻璃碎片有各种尺寸，在事件现场发现的是大块碎片，而那些在衣服、头发或人身上发现的通常是尺寸小于1mm的小碎片。玻璃碎片的常规检测包括元素组成和物理性质，如折射率[11-12]。这些数据可以被法医专家用来判断它们是从哪种物体中脱离出来的,从而确定它们的来源,如窗户、瓶子、头灯等。

值得注意的是,不管玻璃的类型和用途如何,玻璃的化学成分和性能都非常相似。玻璃制造技术的进步使不同公司生产的产品在物理和光学性能上的差异更小,同时也降低了同一制造商生产的不同类型产品之间的差异。因此,区分玻璃碎片的能力已经减弱。

用于制造钠钙硅玻璃的主要原料是苏打粉(Na_2CO_3)、生石灰(CaO)和沙子(SiO_2),其他组成成分在不同类型的玻璃中是不同的。玻璃中的主要元素包括Na、Ca、Si 和 Al,都处于几乎相同的含量水平,而差异主要涉及各种添加剂带来的其他元素。添加剂用于改善玻璃的性能,与产品用途有关,或者源自生产过程中使用的原材料中的杂质,它们的含量明显很低,处于痕量水平。

玻璃的化学成分可以由许多方法来测定[13-14],法医专家更喜欢用无损检测方法,用两个或更多的分析方法来检测并验证样品。另一种常用的方法是使用尽可能少的样品材料同时确定多个元素或化学成分。SEM-EDS方法和XRF方法可以满足上述需求。其他的仪器分析技术如电感耦合等离子质谱(ICP-MS)的改进技术——激光烧蚀(LA)与ICP-MS[8-9]联用技术特别有价值,它能够利用激光对玻璃样品进行激发后定量测定30余种痕量元素的含量。

采用SEM-EDS检测主要元素,采用ICP-MS检测微量和痕量元素,并根据玻璃类型进行鉴别和分类,有助于将玻璃归为产品类别,如玻璃薄片、玻璃容器、汽车窗户、汽车头灯或玻璃餐具,采用统计学方法对玻璃元素成分进行定性判定是有必要的。

热浸法是利用浸泡用油的折射率随温度的变化来测量玻璃碎片的折射率,浸入玻璃碎片的油被加热到一定温度时,可以观察到油中玻璃碎片的边缘消失,此时玻璃和油液的折射率相同。显微镜片上浸入油液中的玻璃碎片一般直接放在显微镜的加热台上,然后通过折射率仪测定折射率。

各种玻璃样品的折射率仅有微小差异。因此,对玻璃样品折射率的差异进行评价时须采用各种统计学方法,这样可以分辨观测到的差异是由设备不稳定造成的还是由样品来源不同而造成的。对玻璃微小碎片的比较分析应在折射率和元素组成检测基础上运用统计学方法来评价观测到的差异,最终判定检测的样品是否来自同一个玻璃物体。

3. 纤维

纤维是有价值的犯罪证据。纤维有几毫米长,与衣服、窗帘、地毯或家具覆盖物有些许的联系。两个人之间的每次相互接触都伴随着微纤维从一个人的衣服到另一个人的衣服的转移。纤维主要是在谋杀、抢劫或斗殴等事件的参与人的衣服上发现,也会在强奸受害者的内衣或指甲里发现。在犯罪者强行移动的障碍物

(如窗户、门或栅栏)边缘上,以及在该行为中使用的工具上都会有纤维。在检查汽车事故时,在安全带或座椅套上可以找到纤维并有助于锁定汽车司机。此外,从汽车车身或底盘取回的与受害者衣服相一致的纤维可以确认受害者和汽车之间的联系。

收集纤维证据[15-16]并进行光学检测的目的是对纤维类型进行分类,并确定它们出自哪一个纺织品类型。检测纤维的化学成分和某些物理性质可以采用的方法包括光学显微光谱法、紫外光谱法和拉曼光谱法。光学显微镜在透射和反射模式下能提供形态学信息以及横截面的形状和厚度(图11-4),如果采用偏振光可以观察到结晶度,所获信息能使纤维归属到一个主要类型。一些合成纤维,如聚酰胺(PA)和聚乙烯(PE)在偏振光下具有非常典型的外貌特征,因此,利用偏振光可以分辨出在日光下具有相似颜色和形态的纤维。纤维上色素、添加剂和洗涤剂的荧光的存在也有助于纤维的识别,通常情况下,利用偏振光还可以区分天然纤维和合成纤维,并确定纤维中色素和添加剂的分布。纤维的主要成分是纤维素、酪蛋白、角蛋白等天然成分,或PA、PE、聚丙烯腈、聚烯烃等合成物,它们可以用红外光谱(图11-5)来识别。单一短微纤维通过显微镜或钻石细胞技术获得的光谱可以区分同一化学组中聚合物。例如,不同的聚酰胺具有不同的光谱,范围在(1000 ~ 15000)cm^{-1}。通常纤维的红外光谱中,颜料和染料没有特征谱,须使用拉曼光谱进行鉴别[9],拉曼光谱能够清晰显示出来自颜料、染料和填充剂的峰。

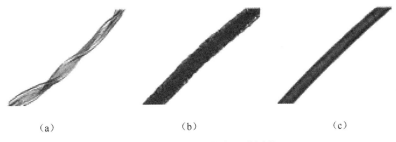

图11-4 普通纤维的显微图像
(a)棉花;(b)羊毛;(c)丙烯酸。

常用UV-vis显微光谱法观察纤维的颜色和深浅,获得一致光谱的纤维证明具有一致的色素含量,但颜料和染料组分是不确定的。黑色和无色的纤维没有显著的纤维颜色特征,所以不能用这种方法进行比较。

分析比对纤维的类型及其成分的信息,目的在于说明已经转移的纤维和嫌疑人服装纤维之间的相似性。信息特征的一致性可说明纤维来源于同一种织物,这意味着纤维从一个人的衣服上被转移到另一个人的衣服上。纤维的证据价值取决于纤维的种类,最流行的纤维如纯棉纤维的证据价值最小,它们可能源自内衣、床

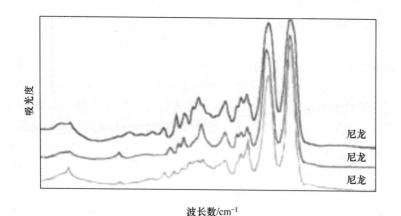

图 11-5 不同聚酰胺纤维的红外光谱

单或灰尘,很难证明它们是从嫌疑人的衣服上转移过来的。同理,流行的牛仔裤纤维的价值也很小,当在环境中很少遇到的纤维出现在现场时其纤维的意义更大。

4. 射击残留物

射击残留物(GSR)的检测对确定涉及使用火器犯罪的某些情况起着重要作用,是对武器和弹药的弹道检测的补充。

离开枪管的粉末气体含有底漆和火药的爆炸反应产物,以及这些材料与弹药筒和武器的其他部件相互作用的产物。GSR 的化学成分和性质取决于弹药生产中所用材料的种类。最具特色的 GSR 是由底漆的成分产生的金属粒子,具有直径为微米级,近似球形的特征形态,也同时具有特殊的化学成分,如铅、锑和钡存在于铅弹中(图 11-6)。在射击孔和射击者的衣服和身体周围发现的 GSR 具有证据价值,通过检测信息可以推断射击距离和使用弹药以及武器的种类。最重要的是,GSR 可以建立嫌犯与枪击事件的联系。

微米级大小的微粒具有与速冷形成液体状态相一致的形态。这种 GSR 粒子包含铅/钡/锑或钡/锑元素的元素组合,是一种独特的弹药周围底漆的爆炸物。

由于 GSR 中的金属粒子体积一般为 $5\sim50\mu m$,人的肉眼无法看到,因此多年来,没有足够有效的方法来识别 GSR 的特征。GSR 可以通过化学反应比色法或诸如原子吸收光谱(AAS)、中子活化分析(NAA)或 X 荧光光谱(XRF)等仪器法间接检测。然而,这些方法并不是特有的,只是记录了所有独立于 GSR 粒子来源的杂质。迄今为止,对 GSR 粒子进行分析的最成功的技术,毫无疑问,是 SEM-EDX[17]。

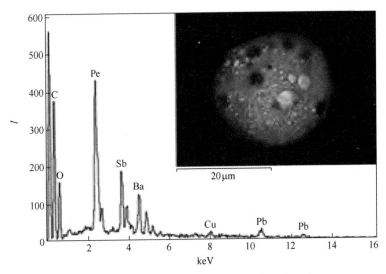

图11-6 射击残留颗粒(嵌入物)及其元素组成

只需要准备一点样品,通常用双面胶带粘接直径为0.5in[1in=2.54(m)]的铝块即可取样。当样品被固定在铝块上时,就可对直径内的球形金属颗粒的化学成分进行检测了。这种方法有很多优点,但操作非常耗时。检测铝块上具有特定特征的粒子可使用自动搜索的合适软件来显著缩短检测时间[17]。

严谨的评估分析可以按以下方式进行:首先确定最典型的纯金属粒子如Pb、Sb、Ba;其次检测伴随纯金属离子大量出现的单、双组分的颗粒,如Sb、Pb、Pb/Sb。从嫌疑人的材料中发现GSR并确认其参与枪击事件,就可推断此人可能正在射击,或出现在枪击的附近,或接触到被GSR高度污染的物体。

由于弹药生产技术已经发展为无铅方案,因此,GSR的证据价值应该根据具体情况建立针对性的分析方法。其中,了解GSR在环境中的耐久性和分散特征至关重要[17-18]。

可以采用红外光谱鉴定在发射枪周边环境中发现的除金属物以外的固体颗粒,如火药颗粒和转化产物,一般可以识别无烟推进剂的主要成分,即硝化纤维素或硝化甘油。通过拉曼光谱法可以识别小剂量的成分,如二苯胺或聚乙胺。这些信息可用于确定射击距离、弹孔的进口和出口,以及区分不同类型弹药残留物[18]。

推进剂燃烧的挥发物和气态产物会在枪管和弹壳中存在一段时间,无论枪击发生在测定之前3天、2~3周还是3周以上[19-20]。采用气相和液相色谱分析方法,包括热解GC-MS、GC-热能分析、LC-MS/MS等,可以对其进行定性和定量检测。

5. 火灾残留物

鉴定可能引起火灾的各种易燃液体是刑事实验室的任务之一。这些液体产品通常是石油产品,包括汽油、燃料油、润滑油和稀释剂,由于其具有挥发性和易燃性等物理化学性质因此经常被犯罪者使用。

流出的易燃液体渗入土壤、纺织品、木材等物质,燃烧只发生在材料表面,在更深处的液体几乎不蒸发并且燃烧得很慢,有些液体在火灾中封闭在多孔材料中。这就是尽管它们燃烧殆尽[22-23],但仍能在火灾残骸中找到易燃液体的原因。

火灾过后的现场会被仔细检查,如果怀疑是人为纵火则从可能的火灾来源处收集实验室分析样本。通常,这种检查仅限于寻找易燃液体的痕迹,这些液体通常用作助燃剂。虽然对火灾残骸的分析不是一个新课题,但对易燃液体的检测和鉴定仍然很困难。可用于点火的商品种类繁多,而且大部分是多组分混合物,是含有与各种溶剂和易燃液体(如乙醚、乙醇、松节油)混合的石油产品,并通过添加树脂或塑料等物质而浓缩。通常含有苯、煤油、机油、醇类和溶剂等的液体混合物在燃烧时会不同程度地挥发,其中一些会热分解[22],这导致样品中存在大量的干扰物,使得分析更加困难。因此,存在于火灾碎片样本中的助燃剂只有很少的量是可用的,而且它们的化学成分与未燃烧液体的成分有很大的不同。另一个影响助燃剂检测有效性的因素是各种灭火剂的使用,这些灭火剂把额外的物质引入分析材料中。

在分析火灾残骸时采用的主要方法是GC法,检测分为三个阶段:首先是将助燃剂从基质中分离并收集起来,然后分离出特定的成分并进行色谱分析,最后确定可能的助燃剂。第一阶段的效果决定了识别被分离和吸附有机化合物的可能性。第一阶段操作失败可能会导致无法识别可疑物质。

石油产品的分离和收集可以通过几种方式进行。最有效的方法是被动吸附,样品和一个装满 Tenax TA 吸附剂的管子被放置在一个严格封闭的 60℃~70℃ 恒温容器中,如广口瓶,并放置 10 h 以上。在此条件下,样品顶端的化合物和吸附在聚合物吸附剂上的样品之间建立了平衡。吸附的化合物经过热解、解吸连同载体气体一起进入一个 GC 柱中进行分离和识别。这一方法便于检测和识别痕量的石油产品。利用 Tenax TA 被动吸附进行的顶空分析通常用于分析物的分离和收集。气相色谱法与自动热解质谱(ATD-GC-MS)联用技术是分离多组分混合物和识别样品组分的最佳技术。GC-MS 能够对可燃液体的痕量物质进行定性和定量分析。

解析色谱图必须考虑到易燃液体在火灾中发生的变化:第一种变化是从被检测材料中蒸发的易挥发性化合物,导致易挥发性成分的含量下降,而挥发性较低的成分含量增加;第二种变化是液体和燃烧材料的热分解产生了可燃混合物中原本并不存在的挥发性物质,使得识别用于点火的液体变得更加困难。这些挥发性物质也有可能一开始就存在于燃烧的材料中,并且是燃烧材料的成分。

检测人员对各种灭火剂对火灾现场采集样品中痕量助燃剂检测的干扰进行了

研究。泡沫灭火剂会将一些干扰物质引入样品中,尽管这些物质可能在分析过程中被检测到,但是只要能够正确解释所获得的结果,它们的存在就不会阻碍对助燃剂的识别。

只要从火灾现场正确地收集样品,并保存在密封的容器中,然后直接送到刑事实验室,目前常用的分析方法都可以很容易地回收痕量的易燃液体。图11-7所示为一块从办公室收集的火灾碎片和在嫌疑犯房间里发现的一个空塑料容器里的类似材料的分析结果,分别获得两幅色谱图,谱图的特征峰明显与蒸发50%的汽油特征峰相一致。研究得出的结论是,用于点火的易燃液体是汽油。

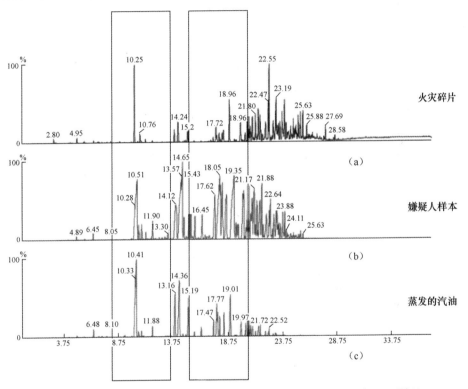

图11-7 火焰碎片、嫌疑人样本和蒸发50%的汽油的色谱图(特征峰用矩形标记)

6. 书写材料

墨水、墨粉等书写材料的化学分析是对有质疑的文献进行刑事检测的重要组成部分,其目的通常是验证一份文件的真伪或确定其年龄或来源。应用非破坏性的分析方法,如显微技术和光学技术,对墨水进行常规墨水参数检查,如颜色、辐射物的发光和吸收,都可以使用这些方法来测定[24]。使用(220~900)nm的波长范围内的透射、反射和荧光光谱,对在纸表面上的墨水、圆珠笔墨水和印刷涂料样品进行鉴别,比用比色计来比较它们的颜色更加客观。

然而,这种光学和光谱测量方法不提供墨水的所有成分的信息,不能依据其化学成分的信息识别墨水。它们只考虑对特定范围内的电磁辐射有强烈反应的成分,因此它们仅仅显示比对样本之间的差异。圆珠笔墨水含有一种或多种染料的黏性液体,是含有天然或合成聚合物混合物的油或油脂,还含有酸性化合物,以减少写作过程中的摩擦系数,同时还有抑制干燥、确保其适宜黏度的糊精物质,其详细的配方是受专利保护的。在文献中没有关于中性墨水成分的信息,只知道除了其他成分外,它们还含有不溶性色素。

一般来说,在利用化学方法对书写材料进行检测时需要先制备一份研究样本。纸层析法、薄层色谱法和毛细管电泳法是常用的实验方法。这些方法主要是分离在检墨水中所含的染料,并对墨水样品进行鉴别。这些技术简单易用,仅需要少量的样本进行检测且具有选择性,并提供可重复的结果。它们的主要缺点是,必须将已准备好的被检测文件上的墨水与基质(如纸)分离。溶剂的提取经常会造成文件的局部损坏。

光谱测量方法,如红外光谱法,提供了被检测样品的主要成分,包括染料、树脂和油性液体等的信息。在墨水的红外光谱中很容易检测到主要的颜料。由于其非破坏性的性质,在刑事调查中拉曼光谱法用于直接识别文件上的墨水,以及确定手写线条的顺序,可以利用这种方法对墨水的老化过程进行检测。此外,HPLC、GC-MS以及元素定量分析方法用于识别某些墨水成分。每种用于墨水检测的方法都有其优点和局限性。在实际操作过程中,需要几种分析方法对墨水进行定性和识别[24]。

基于红外光谱特征峰的位置和相对强度来区分油墨,基于拉曼光谱中的背景曲线的行程和形状判断被测材料的荧光强度也是很重要的参考信息。

元素成分对无机颜料和有机染料的定性很有用。然而,需要一个数据库来识别它们。在所有的圆珠笔墨水中都存在硫、铜、硅和磷等元素成分,一些样品还含有锌、氯、溴和钙等元素。在黑色墨水中,也发现了元素铬和铅。样品在元素成分方面,定量分析和定性分析的结果差异很大。在中性墨水中观察到元素含量的更大变化。

墨粉是用于打印机和复印机的另一种书写材料。它们是黑色或彩色的,内含具有可塑性特征和静电特性的合成聚合物颗粒。在印刷过程中,墨粉的颗粒与纸张相连,但在不破坏纸张表面的情况下,很容易将其去除。它们的大小、形状和成分取决于墨粉的种类。利用红外光谱可以识别如苯乙烯、聚乙烯、丙烯酸酯和甲酰基等聚合物。使用SEM-EDX可以检测铁磁性墨粉中的元素组成,确定铁是主要元素(图11-8和图11-9)。在非铁磁性墨粉中,主要元素为锌或铬,铁含量较少。聚合物和元素的测定可以把墨粉划分为几个组,有时还能识别出打印机的类型[25-27]。

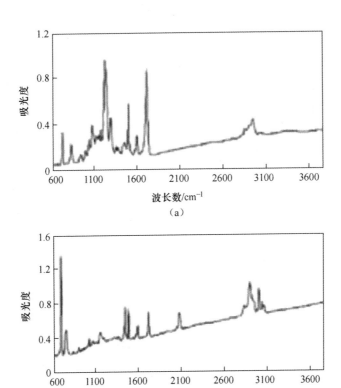

图 11-8 两种不同墨粉的红外光谱

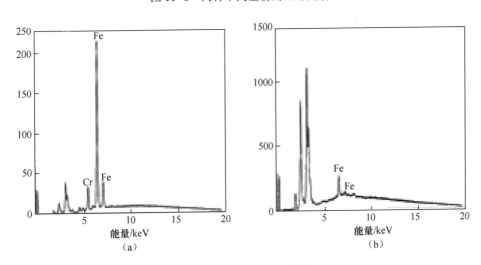

图 11-9 墨粉的 XRF 光谱

(a)铁磁性墨粉；(b)非铁磁性墨粉。

11.1.7 小结

检测不同刑事材料的各种分析方法应能够对毫克级或微克级的少量样品进行分析,并且是非破坏性的,以便可以使用相同或不同的方法进行重复分析。这些方法快速和简单,而且在分析之前不需要进行耗时的样品处理,就能够提供被分析样品的化学成分以及某些物理化学特性的信息,以便对被检样品进行分组和识别。刑事化学家通过被检样品的微量成分或明显特征,能够区分来自同一化学组的其他样品。因此,在刑事检测中,检测少量样品并确定痕量成分是必要的,因为样品中主成分通常不具有犯罪痕迹特性。

刑事化学应用新的、灵敏的分析方法来识别痕迹,收集各种材料的技术信息,以创建数据库。对于所获得信息的评估,统计方法的应用能够评估检查样本之间观察到的差异的显著性以及统计误差。

11.2 毒理学痕量分析

11.2.1 引言

毒理学是一门关于毒药的学科。19世纪初,它与刑事医学分离,被确立为一门独立的学术学科。当时,它只研究在解剖尸体材料中检测毒物,换言之,它协助法医就中毒事件的死因发表意见。从那时起,毒理学逐步在多领域发展起来,目前它是一门跨学科的科学。它以基础学科为基础,结合各种应用学科,一些特定的研究领域被分离出来,如药物、杀虫剂和食物的毒理学领域,环境、工业、临床和刑事的毒理学领域,以及各种子学科如毒物学分析领域等。现代毒理学用定性和定量分析研究化学物质对活体生物的有害影响。

人体内越来越多的低浓度的有毒物质被检测到,现代毒理学分析开始研究的毒理学元素是在司法机构要求下进行毒理学分析的最相关元素,研究的重点在于毒物的类型和来源,化合物进入生物体内的途径对中毒过程的影响,证明存在的哪一种化合物类型必须被确认或排除,以及毒理分析的范围、方向以及研究方法。

目前,至少有1200名科学家,同时也是国际法医毒理学家协会(TIAFT)的成员,为满足司法机构的需要而从事毒理分析领域的工作。有两本著作在法医毒理学家的工作中占有重要的地位:一本是通用性著作《克拉克的药物和毒药分析》(*Clarke's Analysis of Drugs and Poisons*),以一位英国的著名毒理学家命名,目前已发行到第4版;另一本著作是《人类有毒药物和化学品的处置》(*Disposition of toxic drugs and chemicals in man*),每隔几年更新一次,目前已发行到第10版。

11.2.2 中毒的类型

影响人类和其他形式的生化物质会对人体造成伤害,这种伤害在人类活动的任何环境和场所都有可能发生,包括工作场所。中毒可能是意外发生的,也可能是人为故意造成的,如自杀、谋杀。根据有毒化合物在有机体内造成伤害的发展速度定义中毒等级,急性中毒是指服用一定剂量的有毒化合物后中毒所致的伤害快速发展;亚急性中毒变化慢一些;慢性中毒是指服用小剂量的有毒化合物后需要较长一段时间才表现出的中毒反应。从这些定义来看,我们显然正面临着处理大量的有毒物质的问题。大剂量的有毒化合物通常会在机体中产生高浓度,反之小剂量的有毒化合物会导致有毒化合物和/或其代谢物产物的浓度较低。用于确认或排除存在于人体器官内的某一化学物质的分析方法必须具有高精确度、灵敏度和特异性。

11.2.3 原始毒药

自古以来,毒药的有害化学物质就伴随着人类而存在。原始毒药源自天然的物质,即由植物产生的毒素、由动物产生的毒液和天然矿物质,包括 As、Sb、Sn 和 Cu 等。从 19 世纪末和 20 世纪初开始,随着化学物质合成技术的不断发展以及人类文明的快速进步,大量合成化学物质进入人类环境。因此,目前我们正面对着越来越多的源自天然的和合成的有毒物质。

古时人们就已经意识到化学物质的双重特性。许多物质在较低剂量使用时可以治疗疾病,而在高剂量时就被当作毒药。并不是所有进入生物体中或者在体内产生的化学物质造成的伤害都是相同的。毒性极强物质仅需几滴或数十毫克的剂量就会导致有机体的功能失调或死亡;相对无害或几乎无毒性的物质将致命剂量定义在大于或等于 1kg 的水平上。16 世纪,毒物学之父 Theophrastus Bombastus von Hohenheim Paracelsus 在第一次定义毒药中,是这样描述剂量和毒性之间的关系的:"所有的东西都是毒药,没有东西没有毒性,只要达到中毒的剂量就会中毒。"这一定义今天仍然适用,其不仅考虑到用药或吸收剂量,而且还考虑到用药的途径,包括口服、吸入、静脉或皮下注射,以及用药频率是一次还是多次。

11.2.4 现代毒药

利用毒理学分析结果来解释分析中毒过程的影响因素,单一物质或高剂量的低毒性化学物质很少引起中毒。常见的是多种物质的混合物进入机体内,这些物质的相互作用导致了中毒。由于物质之间的相互作用,许多物质的作用与预期结果产生了质或量的变化,这些变化产生于混合物中每个成分带来的影响的叠加。因此,小剂量的各种成分各自产生较低浓度的毒性,累加在一起会导致严重的中毒。检测生物体液和/或内脏器官样本中引起中毒的混合物的所有成分,对于评估中毒的严重程度是很重要的,需要应用复杂的、非常灵敏的、特定的和选择性的分析方法。

对消费者健康甚至生命构成威胁的物质或制剂都可能是毒理学分析的主题。随着当前对非生物体的天然成分化学物质的检测能力越来越强大,检出限越来越低,所获得的分析结果对解释毒理过程更加具有特别的意义。毒理解释必须考虑生物体内化学物质的浓度和受害者的个体灵敏度。个体灵敏度因体内化学物质的特性差异而异,也和生物体内在的个体因素,如种族和性别差异而不同。此外,个体灵敏度也会随着年龄变化,与遗传缺陷、先天和后天普遍存在的疾病状况,包括过敏反应、既往疾病等有关。目前,普遍存在的毒品上瘾、药物成瘾、多毒癖和滥用危害健康的药物等经常导致依赖性、耐受性和脱瘾综合征。使用几种药物同时治疗的多疗法已被广泛应用,并建立了许多不同类型的药物体系。

药物治疗是指在一种有害制剂或药物和/或有毒物质的混合物引起中毒的情况下,不同化学特性的化合物和多种代谢物出现在生物体材料中。同时服用几种药物后,中毒过程包括对毒性物质的吸收、在生物中的分布、在生物体内转化、产生作用和排泄的所有阶段,在此过程中可能发生各种类型的相互作用,包括协同作用、加成、过度加成增强和抑制作用。中毒的复杂性和严重性很难解释,以致无法采取合理有效的治疗方法。在不咨询医生的情况下,独自服用药物越来越多,进入身体中的药物的种类和剂量就难以知晓。了解各种化合物的类型和毒性,有助于选择适当的方法和合适的材料进行分析。

目前,比氰化物、As、Tl 等剧毒物质毒性强很多的物质被广泛应用于医学治疗中并且证明是有效的。最典型的例子就是肉毒杆菌毒素,它是一种毒性异常强大的毒药,但具有双重特性,被用于临床治疗痉挛状态的不自觉的肌肉收缩和过度出汗,甚至用在化妆品领域以暂时性地减少皱纹并达到美容的目的。法医毒理学家会逐渐较多采用有毒药剂用于治疗危重病人,每剂毒药剂在没有医疗监督的情况下就引入人体中都是致命的。在临床使用这些药物时,需要关注是否已经超出了药物的治疗剂量。

环境毒物是指由环境和工业毒物构成的另一类有毒的化合物。二噁英就属于此类,它是人类在家中燃烧各类废弃物产生的。由于对工作安全的要求越来越严格,工业中毒的发生经常是由意外事故或疏忽造成的。尽管金属、过渡金属和阴离子的无机化合物急性中毒变得越来越罕见,但它在当代毒理学上的重要性不容忽视。由于环境污染,生活在高度工业化地区的人们,更易受到低剂量的工业和环境毒物的影响。来自这些地区的食物、大气和水不断地暴露在含有小剂量有毒金属环境中,这些金属可以在人体内不断富集。排除在环境或工业污染环境中可能发生的慢性中毒,需要了解金属含量的正常水平,即它们在人器官中无毒性的浓度。毒性金属可分为以下几类:剧毒金属,如 As、Cd、Cr、Pb、Hg 和 Ni;对人类机体的正常功能至关重要,但具有高度潜在毒性作用的金属,如 Co、Cu、Mn、Se 和 Zn;低毒性金属,如 Sb、Ba、In、Mg、Ag、Te、Tl、Sn、Ti、U 和 V。这样的分类对评价人类暴露在有毒金属环境中的危害是很有用的。另一类毒性金属来自它们在治疗中的应用,如 Al、Bi、Ga、Au、Li 和 Pt。毒性阴离子的检测,如亚硝酸盐(Ⅲ)和硝酸盐(Ⅴ)、氟化物、草酸、氯化物、磷化物和硫化物也是毒理学家关注的领域。其中许多用作人工肥料,甚至用于日常生活中的食品工业。

  作为当代毒理学分析对象的物质的无限来源是麻醉剂等毒品市场。这些物质长期被国际管控,毒品包括大量的合成物质如阿非他命和许多衍生物与植物来源的物质,如大麻毒品、大麻制剂、致幻蘑菇。本书为便于讨论,将这些药剂统归为"经典的精神类药物"。近年来,波兰毒品市场从滥用药物替代品到目前毒品场所的有效毒品的变化,已广为人们关注[30-31]。然而,有一大类新化合物即"设计药"是受控毒品的结构类似物,其中,最突出的是苯乙胺衍生物、卡西酮衍生物、哌嗪衍生物、色胺衍生物和合成大麻素。这些物质中有许多是走私得来的,但也可以在大型的地下秘密实验室中生产。还有一类毒品是由自制物质组成的,例如,没有处方的药物[32-33]。许多新出现在药品市场上的物质有越来越强烈的毒性。不可预测的已知物质的混合物已成为许多不知情的使用者中毒甚至致命的原因[34]。2006 年,药物市场上出现了转基因大麻,其中含有少量(低于 0.20%)的 9-四氢大麻酚(9THC)和高含量的 9-四氢大麻酚-2-羧酸(9THCA-A),在吸食过程中 9THCA-A 转化为 9THC。为了确定大麻是否符合警方的《预防药物成瘾法》中纤维或麻醉品种的界定,分别采用高效液相色谱(HPLC)和气相色谱(GC)或两者一起测定其中所含的每种化合物[35]。

  网购药品脱离了任何形式的管控,从合成代谢类固醇、兴奋剂到各种减肥产品,尤其是源自中国的天然产品,任何东西都可以放进购物车。常被宣传为有助于减肥的植物来源的安全制剂,往往含有大量具有苯丙胺衍生物结构和活性的合成化合物。此外,一些中药制剂有相当高含量的铅,如果长期服用,会产生铅中毒的症状。这些产品中的农药残留同样如此。

  "新精神类物质"(NPS)一词是用于描述各种类型的制剂或含有这些制剂的物质,属于合法毒品、助推剂,它们被当作收藏品、浴盐、河道石头清理剂、植物护理剂、熏香或者看着像可售卖的产品,却不是直接用于人类消费,实际上是以吸毒为目的的。这些产品经常被宣传为"合法的药物替代品"。这些物质有吸引人的名字,如能量片、兴奋丸、迷幻药、丹参(占卜者的圣人)、魔法花园、毒蝇伞(飞伞菌)和印度战士,它们可以在实体店和网络商店中买到,甚至通过电话订单送货上门,这些都鼓励了很多年轻人去尝试,最终吸毒上瘾。自 2010 年 11 月以来,许多此类产品的精神活性成分已被列入《预防药物成瘾法》的管控物质清单。从粉末、药片到草药混合物的胶囊,各种各样的制剂过去和现在都在销售。实验室研究表明,在大多数制剂中,植物材料仅构成存储合成大麻素的载体(JWH-型号,跟随一个数字符号,如 081、007、018 或类似的)[36]。2008 年至今,在世界各地的合法毒品中已经确定含有 30 多种合成大麻素。过去 10 年,在欧洲发现了大约 100 种具有精神类活性的新物质。根据"欧洲药物和药物成瘾监测中心"(EMCDDA)的报告,最大的一类精神活性的物质是苯乙胺,多达 32 种化合物,其后是色胺(22 种)、卡辛酮(15 种)和哌嗪(12 种)[37-38]。一项联合研究[39]中阐明了这些新化合物的质谱以及商业制剂中各种添加剂的光谱。

  俗称的"约会强奸"药物包括大约 70 种药物或精神类活性物质,其会导致强奸或抢劫犯罪行为。一般是往受害者的饮料中偷偷加入这样一种物质,受害者饮用后很快就意识不到正在发生的事情,记忆模糊。在恢复意识后失忆症状伴随着缺乏空间定位、头晕、嗜睡、移动困难、恶心和有时出现幻觉等生理障碍。有时,即使受害者看到施暴者也无法保护自己,因为服用的物质使其移动困难或暂时性麻痹。通常,受害者对事件的回忆非常不清楚,甚至不能作为法院和其他司法机构认定的可靠证人。只有经体液化学毒理学分析才能证实有药物的使用。通常用于测试的样品都是在服用药物几天后才能收集,因此采用的方法必须灵敏并且可检测许多化合物及其代谢物[40-41]。

  与乙醇作用相似的药剂作用于中枢神经系统上能够改变人的行为,影响驾驶能力。这就是《道路交通安全法》为什么规定禁止驾驶者饮酒或服用酒精类似物质后在精神不正常的状态下驾驶汽车。在被警察拦下的地方,可以使用不需要实验室设施的方法来检测司机体内是否存在这些物质,呼气酒精测量仪用来证实司机体内酒精是否存在,这些设备可以在呼出气体中检测酒精的存在,检测结果能够作为证据。控制驾驶员并观察其具有酒精类似物所产生的症状,以及现场提取唾液进行仪器检测以确定这些物质的存在。这样的测试识别检测唾液中的 5 类非法化合物,即鸦片类、四氢大麻酚、可卡因及其代谢物、苯丙胺和苯二氮䓬类物质。现场唾液检测仪器检测出阳性结果后,还必须使用具有各种类型检测器的测试仪器对驾驶员血液进行检测确认,特别是带有质谱检测器的 GC 或 HPLC 等实验室方法。只有这样的毒理学分析结果才能作为起诉一名驾驶员的证据。所使用的实验室方法必须符合 2014 年 7 月 16 日"卫生部令"规定的定量限(LOQ)。所有的LOQ 值都确定在每毫升血液中纳克级含量水平上,1ng/mL 的 9-四氢大麻酚、10ng/mL 的吗啡和可卡因、25ng/mL 的安非他命及其衍生物、100ng/mL 的苯甲酚。

苯二氮类药物还没有建立 LOQ,因为其剂量在(2~25)mg 范围存在许多的衍生物,导致它在血液中的浓度范围非常宽[42]。

毒素和毒液逐渐引起人类的兴趣,尽管这种兴趣不一定有益于人类历史发展,法医毒理学家在日常工作中所能遇到的毒液和毒素的种类越来越多。不同种类的植物中含有几百种毒素,在多种多样的海洋生物中含有数百种毒液,我们所熟知的有毒生物有毒鱼、毒蛇、扁虱、蝎子和蜘蛛等。此外,有些鸟的羽毛只要被触摸就可能引起毒性反应。有时,由于样本的来源不明造成了毒理学测试工作的复杂性,例如,在国外旅行时得到的或在互联网上购买的有毒动物,之后因不具备饲养条件,有时被遗弃,然后咬伤受害者,此时从受害者身上提取体液进行毒理学测试分析就会非常困难。

11.2.5 中毒的途径

有毒物质可以通过各种途径进入人体,中毒严重程度主要取决于药物的毒性、剂量等。中毒程度与用药的途径密切相关,静脉注射用药是将整个药物剂量快速引入生物体中,其所用药剂量最小。口服给药的外源性剂量并不总是被机体完全吸收,因此给药剂量并不是作用剂量。机体对中毒的自然防御机制使部分用药的剂量减轻,如呕吐反应。许多制剂的给药剂量都是先从胃部通过,经静脉血液通过门静脉进入肝脏,然后在肝脏中经过新陈代谢,以代谢物的形式进入循环系统,这一过程称为"先通效应"。如果代谢物的毒性低于给药本身毒性则毒性作用就会降低;相反地,如果代谢物比给药毒性更大,中毒的情况就会更加严重。挥发性制剂,如有毒气体和溶剂,通过呼吸系统进入人体,并伴随空气进入肺泡,肺泡里充满了血液,并且有很大的表面积。大剂量的制剂可以瞬间发生变化,但这并等于会转化为血液中的高浓度制剂。许多毒素是通过皮肤吸收的,如有机磷农药,不仅对健康有负面影响,而且对生命也有威胁。其他的用药途径,如局部、直肠和肌肉注射,在医学上具有重要意义,但也会在自杀、故意中毒的案例中遇到,例如,腐蚀性或燃烧物质中毒。将制剂引入有机体的途径不同,分析人员可能会面对低浓度或更高浓度的制剂和/或其代谢物的检测任务。

11.2.6 中毒症状

有毒化合物对生物体作用的影响是伴随中毒症状。实际上,对某一特定化合物不会出现特定的症状,同一类特定化合物会有比较类似的中毒临床特征。最常见的中毒综合征有:抗胆碱能综合征,涉及抗组胺药、莨菪碱、阿托品、曼陀罗、三环类抗抑郁药等;胆碱能综合征,涉及有机磷化合物、氨基甲酸衍生物、伞菌;幻觉综合征,涉及安非他命衍生物、迷幻药、迷幻蘑菇、占卜者、可卡因、大麻酚;鸦片综合症,涉及鸦片、类鸦片、芬太尼;镇静催眠综合征,涉及巴比妥酸盐、苯二氮卓类、乙醇等;兴奋剂综合征,涉及安非他命及其链衍生物、可卡因。能闻到气味和可察觉的变化对指导毒理学分析也很有用,例如:有杏仁、大蒜、腐烂的鱼或臭鸡蛋化学气味可能表明有氰化物、胂、磷化氢或硫化氢等有毒气体;皮肤出现发红、充血等变色现象则表明可能有硝酸盐(Ⅲ)和硝酸盐(Ⅴ);血液和嘴唇变成巧克力色则表示有高铁血红蛋白原化合物;尿液变色呈现红橙色、蓝绿色、粉红色、深黄色、深棕色甚至黑色则可推断有各种药物或铁化合物。此外,皮肤、口腔黏膜、眼睛出现灼伤则是酸和碱的象征;腹泻则是大多由真菌或金属导致。瞳孔的大小和它们对光线的反应也是重要的提示:瞳孔放大则表明有安非他命、阿托品和大麻醇,但也可能是鸦片戒断综合征;瞳孔缩小表示有鸦片剂、巴比妥酸盐、有机磷、氨基甲酸盐和吩噻嗪。出现褥疮暗示使用了巴比妥酸盐。酸碱平衡分析结果出现酸中毒现象预示有甲醇和乙二醇的影响。

11.2.7 毒理学分析结果的解释

毒理分析者对制剂浓度检测后,最终的任务是对结果进行解释。对结果进行解释必须与事件实际发生情况一致,即中毒行为所引起的症状,或者是在致命中毒情况下的尸检结果。最常见的情况是,分析人员分析并解释被测化合物、治疗用参照化合物的浓度或当药物中未添加毒药时的低浓度检测、常规金属浓度、中毒后被测物浓度,以及致命中毒时的浓度。

以司法判决为目的的毒理学分析结果的解释包含了非常广泛的议题。除了确认是否使用或接触有毒化合物及其用量外,毒物学家还需要解决以下问题:确定中毒者是主动还是被动接触精神类物质,如9HTC,安非他命、可卡因等;确定有毒化合物的来源及导致中毒的原因,如医疗、饮食或蓄意下毒;确定最后一次的服用时间,特别是在经常服用精神活性物质的情况下进行追溯性解释,如确认在采样前几小时血液内酒精的浓度和推断死因为毒品类药物,如安非他命衍生品、迷幻药等;确定尸检的必要性和可行性。

毒理学分析人员的解释范围还包括评估非法秘密实验室生产受控物质的生产能力,以及确定用于生产滥用药物的种植作物的产量,如大麻。

毒理学家的分析任务不仅是证实或排除中毒,以及识别和定量分析警方获得的非生物材料的主要成分,还要根据材料的物理特性和化学成分进行分类和比较,描述毒品样本之间的关联,这种对在样品中微量成分的详细识别称为剖析。剖析可给予许多对司法主体有意义的事实解释,如两个或多个样本之间的关系,经销商和吸毒者之间的联系,样品在当地的分销网络,国内或国际水平之间在微量级别的相似性,样本和合成方法的起源。此外,对物质类型的剖析有助于确定毒品交易的地理位置,跟踪国际控制下的新物质的来源,评估地下实验室的规模和活动,区分来自非法和合法渠道的物质[43]。

死亡化学过程是指发生在尸体中的腐烂分解过程,在解释尸检材料的毒理学分析结果时非常重要。随着时间的推移,死亡后发生渐进变化的方向和强度取决于许多因素,其中起重要作用的是死者的身体状况、所患疾病、受伤状况、经受的疼痛和环境因素。此外,在同一具尸体的不同器官中,降解转化会以不同的方式发生。发生在器官表面的有氧降解转化过程比内部的厌氧降解速度快。当尸体腐败分解时,产生一些毒理学认为重要的物质,如乙基、甲基及更高基的醇类、氰化物。这些物质是内源性的,因此在死亡前不会进入生物体。目前,与发生中毒事件的量相比,还没有发现死后血液样本中会产生高浓度的氰化物离子。另外,生物降解也会使生前摄入的许多外源有毒化合物分解,如高醇的酯类、大量挥发性有机化合物。许多药物,特别是四氨基结构的化合物也会发生分解。氨基酸是生物体的天然成分,在生物体腐烂分解阶段,氨基酸会随之降解导致环境碱性化,因此很难发现酸中毒,尤其是盐酸中毒[44]。

对法医毒理学家来说,对生物体固有的有毒化合物的毒理学分析结果进行解释,可能是一个巨大的挑战。

乙醇的生理水平是0.01‰,而对于糖尿病患者或饥饿状态的人,其含量可能会更高,但永远不会超过一个国家规定的法定饮酒量值,即0.2‰。然而,由于尸体中内源性乙醇的存在,因此对乙醇来源的推断有时是非常困难的。为了确定酒精是死后产生的还是生前喝了酒精饮料进入体内的,尸检应该对两类样本进行分析:血液样本,因为内源性酒精在死亡后很快在血液中产生;体液,如尿液,酒精在其中分解较慢。

甲醇是一种通过代谢过程发挥作用的毒素,它是"波兰烈酒专卖公司"的子公司生产的所有酒精饮料的一种成分,并且天然存在于活体生物中。内源性甲醇的来源还没有明确解释。它的浓度为0.1~3.4mg/kg,比导致中毒事件发生的甲醇浓度低几个数量级。

血红蛋白即羧血红蛋白(HbCO)的一氧化碳衍生物,其生理水平为2%~7%,它是血红蛋白分解的产物。HbCO也可以因为吸入环境中的一氧化碳而产生。在吸烟的情况下,HbCO的生理水平可达到11%~13%的浓度水平,或者通过死后分解过程达到5%~7%的浓度水平。但严重的一氧化碳中毒后,HbCO的浓度可能非常低,不超过2%。

γ-羟基丁酸(GHB)也是活体生物的天然成分。在过去用于医药领域,目前这种化合物作为一种约会强奸药物越来越多地被滥用于娱乐。它在生物体内的浓度会极速降低,无论生物样品取自活体还是尸检,样品在储存过程中GHB浓度都会发生变化。此外,即使服用高剂量的情况下,它也会在体内迅速消失,在血液中 8 h 内是可检测到的,在尿液中 12h 以内是可以检测的。因此,分辨 GHB 来自体外摄入还是体内是非常困难的[45]。

法医毒理学分析人员在其能力范围内进行力所能及的推断和解析。专门从事临床毒理学研究的医生可以解释药物对病人行为的影响。解决与刑事责任有关的问题,包括处罚类型、犯罪、控告、起诉、禁止等,显然属于法律裁决机构的职权范围,也就是在前述两个专家意见相左时,由法院发布对结论有效性的实质性评价。

11.3 现代毒理学分析

毒理学分析是临床和法医毒理学的基石。法医在没有尸检材料毒理学分析结果的情况下,一般不会对死亡原因发表意见。如果不分析体液,就不可能对是否中毒发表绝对的意见。在临床毒理学中,分析检测常用于诊断中毒和监测治疗效果,如果对各种犯罪活动的受害者和行凶者进行治疗,则分析结果也服务于司法行政的需要。毒理学分析结果是否能够作为合法的、科学的和满足预期目的而成为具有证据价值的结果,这取决于测定所用的方法和用于研究所收集的材料类型。

现代毒理学分析生物和非生物材料时分为两个阶段进行,第一个阶段筛选方法,第二个阶段确认方法。分析人员使用筛选测量方法获得初步分析结果,初步分析结果不是绝对的要么否定或要么肯定的结果。其后,通过确定的方法来验证并给出肯定的结果。筛选方法的目的是尽可能宽范围地分析各种化合物。确定的方法更明确,具有比筛选方法更低的检出限(LOD)和定量限(LOQ)的特征。

法医毒理学分析过程包括从收集样本的那一刻起到获得检测结果,直至结果被证实,都应该是受控的,流程标准化是确保过程受控的重要方式。因此,每种用于筛选、确认、定性和定量检测的新方法都必须根据国际要求确定验证参数。根据"法医毒理学科学工作组"(SWGTOX)的规定,生物样本分析中使用的所有方法,必须测定精密度、稀释完整性、干扰性研究、LOD、背景和稳定性等参数。对于定量方法,需要测定偏差、校准模型和 LOQ[46]等参数。回收率、再现性和方法灵敏度、方法选择性和干扰影响等参数被认为是应该被测定的附加参数。使用 LC-MS 方法时要研究基体效应,换言之就是离子化抑制或增强效应,尤其是电喷雾离子化(ESI)室中[47]。用分析物的氘化衍生物作为内标,通过对参考物质的分析和参与实验室间比对来检验方法的正确性,促进了结果不确定度的持续优化。

虽然方法验证不是主要的研究工作,却是现代分析实验室遵守实验室管理规范并对测试结果质量控制过程中不可分割的一部分,也是实验室测定结果的国际评审和认证的必要条件。方法验证无疑延长了一种方法的开发周期,但对于确定实现预期分析目标的有效性是至关重要的。这个过程并不与爱因斯坦的说法相冲突,他说:"所有的事情都应该尽可能简单,但不应更简单。"因此,方法验证是不能被简化的部分。

测试样本的选择应主要考虑从服用有毒物质到样本采集之间的时间,以及进行测试的地点,或在临床或司法实验室,或在事故现场。例如,在路边对司机酒驾的检查。化合物存在于各种样本的母本、活性的和非活性的代谢物中。活性代谢物会影响生命过程,生物体中非活性代谢物的存在可以证明很久以前的一种物质代谢。现行的有效的分析技术用于检测、识别和确定传统生物样本,如血液、尿液和内脏器官,以及选择性样本,如头发、唾液和汗液,中的化学物质。近年来,大量研究表明唾液和血液中各种化合物的浓度之间存在相关性[48]。假设一种有毒物质是通过汗液排出的,那么枕头套也可以是一种有用的、间接的毒理学检测材料。有些分析人员从事沾附在各种材料上的或新或旧血迹的分析。值得一提的是,现代分析方法可以通过分析人体遗骸上的苍蝇幼虫生长过程,得出人体中毒死亡的原因[49-50],也用于估计死亡时间,而不受炎热的[51]或寒冷的气候影响[52]。

有时,由于各种原因,用于检查的材料数量非常有限。在研究方法时,应充分考虑这种情况。通常只收集 1mL 的血液样本用于筛查分析,其中只有 0.1mL 的血液样本用于目标或确定性分析。

分析的方向和程序取决于问题的类型。一个事件的未知情况或一个未知的毒性因素需要系统的毒理学分析(STA),以便分析结果包含尽可能多的有毒物质。在服用的有毒化合物已知的情况下,首先对该化合物进行分析,一个阳性结果必须通过另一种独立的方法来确认。当研究一种只有未知毒性因素引起的症状时,使用补充技术以及药物、药理学和药物动力学方面的知识的能力是特别重要的[53]。

11.3.1 筛选方法

免疫化学方法是利用免疫化学反应(ICh)、酶免疫分析(EIA)、放射免疫测定(RIA)、荧光偏振免疫测定法(FPIA)和溶液中微粒的动力学相互作用进行的商业化测试,经常用作筛选方法。它们被设计用来检测明确的但数量较少的药物类型,如鸦片剂、大麻酚、苯丙胺衍生物、三环类抗抑郁药和苯二酚类药物,或更罕见的单一化合物,如地高辛。这些检测方法能够快速分析尿、血清或唾液等体液样品。检测结果与整组化合物类型相关,被定义为整组阳性或整组阴性。使用可见的或电子的读数分析仪检测方法可以得到定量、半定量或定性的结果。

ICh 的优点主要有灵敏、快速,不需要对被测生物材料进行预处理,所需样品量很少,一般为 0.01~0.10mL。它的主要缺点是特征性低。

免疫化学测试中所有的 ICh 测试都使用抗体或其他结合蛋白、抗原和标记物。

这些测试的基本原理是来自生物样本的药物在无标记抗原和有标记抗原之间结合部位的竞争。根据标记抗体或抗原的标记物类型,包括放射性标记物、化学标记物和荧光标记物来决定所采用的检测技术。在 RIA 测试中,标记物是被引入抗原、抗体或酶中的核素(如 ^3H、^{14}C、^{125}I、^{131}I)。放射性测量方法是非常灵敏的。当使用荧光标记物或发光基团时分别测量偏振光(FPIA)或化学发光的荧光变化。各种类型的酶也可以是标记物,被测物中主要成分是样品的成分、用特定酶标记的抗原、一种针对抗原的特定抗体和基质,该基质当参与酶催化反应时,会引起可测量的光学信号的变化。在涉及吸光度测量的试验中,与样品共存的具有高摩尔吸光度系数特征的化合物影响测量结果,会导致假阳性结果的产生。颜色在溶液或指示剂区产生变化的化学化合物或染料也可以作为标记物使用,最简单的测试方法是使用浸泡在尿液中的试纸(如美国前线测试)测试样本,在试纸上,样品的成分遵循色谱法分离原理,抵达放置了抗体的区域。尿液中的药物与抗体结合,剩余的游离抗体抵达有固定抗原的第二区域,抗原捕获所有游离的抗体。只有样品中与药物结合的抗体颗粒才会经历进一步的色谱过程。这些结合体到达标记区,通常含有胶体金作为标记物。这个区域的颜色发生变化,红色的带状表示结果呈阳性。在其他条形带的测试中(如 Hydrex),尽管有相同的标记物,但彩色带的出现表明结果呈阴性。

配备各类检测器的色谱检测仪,如薄层色谱(TLC)、GC 和 HPLC,当与 MS 结合后更加具有普遍性和特效性。这些方法,包括 MS,被定义为开放性的方法,即它们允许连续的新化合物进入先前开发的分析程序中。使用 GC 和 HPLC 串联含有选择性离子监测器(SIM)、选择反应监测器(SRM)或多重反应监测器(MRM)的质谱仪(MS/MS)等筛选方法完成对一些化合物的分离和检测。由于设备的局限性,在一个分析过程中可以同时记录到适当强灵敏度分析信号的数量是有限的。这就是为什么近年来开发的具有严格定义的化合物类型筛选方法取代所谓的通用筛选方法。例如,一种检测影响司机精神运动性能的物质的方法,以及一种检测用于引发强奸或抢劫的物质的方法。尽管所有的筛选方法都不如确定的方法或针对特定化合物方法灵敏,但这种途径确保了筛选方法具有令人满意的灵敏度。

值得注意的是,由于被分析物种类的多样化,痕量浓度的分析需求越来越多,各种各样的分析方法上的缺陷也开始显现[53]。

光谱筛选方法不仅仅局限于有机化合物,为研究生物材料中的金属含量,除了建立了火焰(F-AAS)和冷汞原子蒸气等无焰原子吸收光谱法(CV-AAS)外,还有其他技术,如电感耦合等离子体发射光谱法(ICP-OES)或电感耦合等离子体质谱法(ICP-MS)也被越来越频繁地使用。后两种技术可以同时分析大约 70 种元素,具体元素取决于分析人员可获得的标准物质的数量。测定 Hg、As 和 Se 元素的方法可以选择带有氢化物发生器的原子吸收法(HG-AAS),而含有电热原子化器

(ET-AAS)的AAS不仅能测定正常范围内的微量重金属,还可以测定在慢性中毒和某些急性中毒情况的微量重金属,如Tl、Pb或Se等。为了排除或确认中毒,特别是有毒金属、半金属和非金属的慢性中毒,了解某种生物材料中所含微量元素的正常浓度水平是至关重要的。

系统性的毒理学分析是指在一个事件的情形完全未知的情况下,例如在森林中发现的一具尸体或在公园里发现的一个无意识的人,需要系统考虑可能导致死亡或伤害的有毒化合物种类,需要检测的化合物数量会不断增加。采用某个分析过程不可能包含所有在毒理学领域中具有重要意义的化学成分。

有毒物质可以按各种方式分类,如可以按字母顺序分类;也可以按药理活性作用分类,如三环类抗抑郁药、抗惊厥药物、抗高血压药物;或根据化学结构分类,如苯二氮卓衍生物、巴比妥酸、吩噻嗪。就系统性的毒理学分析(STA)目的而言,最佳方法应该是依据从各种生物材料中的化合物提取技术类别将它们分组,可以分为6个基本组[53]。下面按提取时间顺序进行分组,在进行毒理学分析时应该按照同样顺序分组。

(1) 气体和挥发性化合物,可以通过顶空扩散进行分离。

(2) 有毒阴离子,如硝酸盐(Ⅲ)、硝酸盐(Ⅴ)、磷化物和草酸盐,通过渗析分离。

(3) 难挥发有机化合物,最合适的方法是控制pH条件下用有机溶剂筛选技术进行分离,如液-液萃取法、LLE或固相萃取法、SPE等。

(4) 农药类,大多数都需要一个特定的分离程序,因为一般的程序是无效的。

(5) 金属和非金属,需要应用各种矿化技术,包括湿法、灰化、微波辅助等。

(6) 毒素和大量化合物,如季铵碱类物质和二噁英,需要特殊的分离技术,使用离子对或离子交换树脂、形成衍生品、连续提取、沉淀和浓缩。

检测或排除每组生物材料中(图11-10)广泛存在的化合物,需要应用越来越灵敏的仪器分析技术。

11.3.2 确认方法

现代仪器分析技术,特别是GC-MS、LC-MS或LC-MS/MS与各类电离发生装置耦合,并连同阴极和阳极离子的各种离子检测器共同形成分析系统,其中电离发生方式包括电子电离(EI)、化学电离(CI)、电喷雾电离(ESI)光化电离(APPI);离子监测方式包括总离子流(TIC)、选择性离子监测(SIM)、母离子扫描(PIS)、子离子扫描(DIS)、选择性反应监测(SRM)。分析系统不仅能够分离混合化合物和识别单个成分,而且可以在几十微升的血液、血清、尿液中或几十毫克的毛发等生物样品中测定皮克(pg)级分析物浓度。使用这些仪器分析技术需要生物样品的制

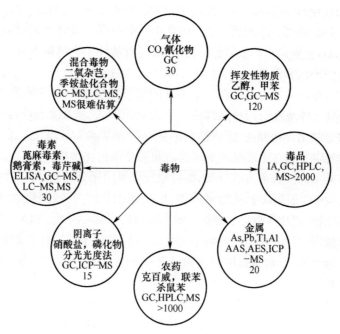

图 11-10 用于分析目的的毒药分类及其最常用测定方法和预估的
毒理学相关化合物数量。

备、多重预处理,因此使用合适的标准物质对整个分析过程进行确认和质量控制研究。这些分析工作非常耗时和耗力。近年来,作为检测器的质量分析器不断改进,商品化的四极杆(Q)、离子阱(IT)、三重四极杆(Q3)、飞行时间(TOF)、傅里叶变换离子回旋共振(FT-ICR)等新型的技术能使 LOD 降到很低的值。

11.3.3 识别系统

每个实验室都会建立自己的分析程序,这些程序可以是基于商业的或已发布的检测系统,但是在不同实验室条件下必须确定方法的测量重复性、灵敏度和稳定性。目前,已经开发出许多检测系统对有机化合物进行筛选分析,大量的识别系统用于最古老的色谱方法(TLC),每种测量方法的测量范围都可以涵盖上千种不同的化合物,包括药物和它们的代谢物[54]。对于具有传统检测器的 GC 方法,有两种分析方法:第一种方法用于筛选分析,以发现少量的挥发性有机化合物,通常使用两种类型的检测器,火焰离子化检测器(FID)和氮磷探测器(NPD),同时含有 4500 种化合物的保留指数的数据库为检测提供了基础[55];第二种方法是识别可溶性的挥发性有机化合物[56]。二极管阵列高效液相色谱仪(HPLC-DAD)与计算机毒品识别系统联用,在相对保留时间和分光光谱的基础上,可以识别酸性和碱性化合物,以及介于其间的中性化合物。与来自不同公司的高性能色谱系统相兼容的 HPLC-DAD 系统,除了可以检测上面提到的识别元素之外,还可以检测利用菜单中的 2682 种化合物以及通过比较 1600 个发色团或它的组合[57-58]选出的分子结构。

色谱技术与质谱法可以有多种组合,如 GC-MS、GC-MS/MS、LC-MS 和 LC-MS/MS。质谱图参考文库是每种类型仪器的一个组成部分,通过与已识别的化合物相似的谱图进行比对。对于 GC-MS-EI 技术来说,有两个独立的库,即广泛使用的 Wiley 和 NIST 2008(W8/N08)。W8/N08 文库包含 562000 种 EI 光谱、5308 种服从 MS/MS 技术的母离子碎片光谱、超过 200 万个化合物名称和它们的同义词、35000 个结构式和 43000 个 GC 保留指数。自动质谱分析反卷积和识别系统(AMDIS)也集成到这个库中。Pfleger/Maurer/Weber 文库对毒理学分析是最有用的,它包含 7800 种药物、农药、农药代谢物、衍生化产品(特别是甲基和甲硅烷基衍生品)和人工制品(可能在分析过程中产生化合物,例如,在高温影响下色谱仪的注射室内产生的物质)的光谱。这个库还包含化合物的其他数据,比如 Kovats 保留指数、结构或经验公式、摩尔质量、化学文摘服务注册号码、分类化合物的药理学组名称、生物样品的类型和样品制备方法的描述[47]。尽管如此,通过一个分析程序,也不能把所有母体化合物及其代谢物包含在这个文库中。

在开发 LC-MS 分析程序时,应注意在 ESI 室中某一化合物离子化后或在 APCI 条件下单一碎片化后获得的质量谱,与 GC 中的 EI 类型光谱相比,只有很少的片段。在 APCI 模式中获得的假分子离子具有较低的识别价值,在应用了多重分裂和子离子光谱聚焦后,光谱的识别价值增加。对 ESI 来说,在样品中共存的化合物可以减少分析物的离子化,这称为离子抑制,抑制效应会导致样本中浓度很低但毒性很强的化合物被忽视。考虑到上述情况,利用碰撞诱导离解的方法,通过内源分裂的 ESI 类型子质谱(MS/MS)库是可以有效解决问题的[59],库中包含了在低、中、高碰撞能量下的超过 800 种药物制剂的光谱图。此外,针对 LC-MS-Q 方法创建了一个质量谱库,而 Schreiber[60]建立了一个用于识别杀虫剂和爆炸化合物的 ESI 和 APCI 类型光谱库,这些文库已经商业化。对于 LC-MS 或 LC-MS/MS 方法,到目前为止,大多数文库都是在单独的实验室中以自制的方式创建的。这些文库适用于特定的仪器或相同类型的仪器。

11.3.4 分析方案

通常在分析检测过程中,需要预处理程序处理样本,再采用各种检测技术来出具最终结果,如 GC、HPLC,以及目前越来越少用的 TLC。GC-MS 或 LC-MS 联用技术具有灵敏度高、线性范围宽的优点,甚至可达 3 个数量级,针对这些检测优势,检测人员会用一种样品处理方法,如提取和衍生化,处理所有分析物,尽管有时处理样品程序对特定的分析物来说并不是最优的,如从一个碱性介质提取酸性化合物的离析过程效率很低。迄今为止,广泛使用的由 Maurer 等开发的方法是利用 GC-MS-EI 技术[61],在 pH 值为 8~9 的尿液提取物中同时检测和识别超过 2000 种化合物,它们来自 20 个药理学组的药物及其代谢物。现代液相色谱技术可以通过预设测量参数和条件,如在两个碰撞电压下,可以记录质谱信息。根据不同类型检测器的特点,如 MS/MS、电子俘获检测器(ECD)和离子化器(FID、EI、APCI 和 ESI),采用选择或联合互补的测试方法进行检测。许多分析人员常用酸提取的方法,在仪器分析或者分别分析之前将其提取物组合起来。在许多实验室中,为研究一种确定的生物材料,如血液、尿液、唾液或毛发等,以及研究特定的药理学组如苯二酮类药物[63-64]、抗抑郁药物[65]、β-受体阻滞药[66]等,开发出了含有最大后验概率(MAP)[62]的设计方法。越来越多的开发方法涉及特定的问题和生物材料。例如,LC-MS-APCI 方法用于检测和确定在血液[41]和尿液[67]中导致强奸、抢劫、盗窃等犯罪行为发生的物质。LC-MS-ESI 方法检测司机血液中与酒精之类的物质,如检测血浆[69-70]中的源自植物的苯烷基胺,苯烷基胺被认为是属于 2C 组的设计药物,更准确地说,是在苯环的 2 号和 5 号位置上连接两个二甲氧基;还有在尿液[71]中的死亡上限毒素,α-鹅膏蕈碱和 β-鹅膏蕈碱。联用检测技术能够使传统的用于筛选分析、识别和定量分析的方法变成整体分析过程的一部分。例如,在 LC-MS 检测过程中,从生物样品中提取萃取物的过程是连续筛选分析阶段,在该阶段可疑化合物被标记,然后使用总离子流模式(TIC)扫描识别这些化合物,定量分析被识别的化合物。迄今为止,使用 LC-MS 技术开发的最常用的筛选方法是 LC-MS/MS-QTrap 法,可筛选血液和尿液中的 301 种化合物,以及 LC-MS/MS-ESI 法[72],可筛选血液中的 238 种药物。Alder 的团队[73]研究并比较了 LC-MS/MS 法和 GC-MS-EI 法,用这两种方法识别了 500 种常用的杀虫剂。Pang 等[74]应用凝胶层析法进行初步分离,对 660 种农药进行了确认,然后利用 GC-MS 和 LC-MS/MS 对植物保护剂的 437 个活性成分进行了定量分析,并将这些活性成分分为 4 类。

11.3.5 小结

由于越来越多的样本要接受毒理学分析,因此分析过程的自动化是本领域的发展方向,通过开发软件实现自动调谐、收集、编辑和数据归档、创建报告,文库检索等定量分析的全过程。关于这方面的研究可以检索到大量的文献和互联网出版物。通过对分析结果的内部和外部系统的质量控制来验证并持续改进分析程序。

这些程序能使法医毒理学分析者在缺乏一个特定事件信息的情况下,通过在设备齐全的实验室的日常工作,筛选大约3000种化合物。具有重大毒理学意义的残余化合物的数量约为10万种,需要对特定化合物采用特定的分析程序。

分析人员了解分析标准和方法,并通过不断改进和克服方法缺陷,应用可普遍接受的分析方法来寻找和识别毒物。此外,他们还借助生物材料中具有显著性、有效性和敏感性的生物标记物检测毒物,能在出现症状初期通过检测血液、唾液和尿液等,或在中毒症状减弱后的中毒后期通过检测尿液、头发和汗液等确认是否中毒及中毒类别。分析者所具有的经验、敏锐度、辨别事实之间逻辑关系的能力以及机缘决定着其在事件调查研究中所发挥的作用。

目前,对生物材料的毒理学分析越来越多地体现在微量有毒化合物的检测、识别和确定等方面。化合物检测标准也在随着毒理学领域中知识和要求的更新而发生变化。显然,最严格的标准要求通常和测量结果会导致的法律制裁相关。毒理学分析家要结合特殊领域的丰富经验和知识,满足所有严谨的测试条件。定性识别的结果应该描述为"声明:没有特殊情况,就是这种化合物",定量结果应描述为"确定浓度的化合物可能导致健康紊乱或威胁到生命,但它不是唯一的诱因条件"。分析者在其严格限制的职权范围内与法医合作为司法机构提供分析结论和建议。没有宣布中毒死亡原因就做出裁决目前是很难想象的,至少要对尸检材料进行毒理学分析,事实调查结果、尸体解剖的图片和毒理学分析结果都具有一致性,授权法医才会对死因发表意见。毒理学分析者可以评估滥用物质或精神活性物质的有效单剂的数量,估计用于生产滥用药物的作物产量,如大麻,或非法实验室生产受控物质的生产能力,如安非他命。但是,对这些关系到刑事责任的调查结果的实质性评估显然属于裁决机构的职权范围。

参考文献

[1] Wójcikiewicz, J. (ed.): Forensic Ekspertise. Wolters Kluwer Polska (2007)

[2] Hunt, R.W.G., Pointer, M.R.: Measuring Colour. Wiley, Chichester (2011)

[3] Caddy, B. (ed.): Forensic Examination of Glass and Paint. Taylor & Francis, London (2001)

[4] Zięba-Palus, J., Trzcińska, B., Kościelniak, P.: Comparative analysis of car paint traces in terms of color by Vis microspectrometry for forensic needs. Anal. Lett. 43(3), 436–445(2010)

[5] Trzcińska, B., Zięba-Palus, J., Kościelniak, P.: Application of microspectrometry in visible range to differentiation of car solid paints for forensic purposes. J. Mol. Struct. 924–926,393–399 (2009)

[6] Milczarek, J., Dziadosz, M., Zięba-Palus, J.: Way to distinguish car paint traces based on epoxy layers analysis by pyrolysis-gas chromatography - mass spectrometry (Py-GC/MS). Chemia

Analityczna 54, 173–185 (2009)

[7] Milczarek, J., Zięba-Palus, J.: Examination of spray paints on plasters by the use of pyrolysis-gas chromatography-mass spectrometry for forensic purposes. J. Anal. Appl. Pyrol. 86, 252–259 (2009)

[8] Zięba-Palus, J., Zadora, G., Milczarek, J.: Differentiation and evaluation of evidence value of styrene acrylic urethane topcoat car paints analysed by pyrolysis gas chromatography. J. Chromatogr. A 1179, 47–58 (2008)

[9] Zięba-Palus, J., Borusiewicz, R., Kunicki, M.: PRAXIS-combined μ-Raman and μ-XRF spectrometers in the examination of forensic samples. Forensic Sci. Int. 175, 1–7 (2008)

[10] Zięba-Palus, J., Michalska, A., Wesełucha-Birczyńska, A.: Characterisation of paint samples by infrared and Raman spectroscopy for criminalistic purposes. J. Mol. Struct. 993, 134–141 (2011)

[11] Pawluk-Kołc, M., Zięba-Palus, J., Zadora, G.: Differentiation of glass fragments on the basis of refractive index values determined by the thermoimmersion method. Application in criminalistics. Probl. Forensic Sci. 56, 48–64 (2003)

[12] Pawluk-Kołc, M., Zięba-Palus, J., Parczewski, A.: The effect of annealing on the distribution of refractive index in windscreen and windowpane. Forensic Sci. Int. 174, 222–228 (2008)

[13] Zadora, G.: The role of statistical methods in assessing the evidential value of physico-chemical data. Probl. Forensic Sci. 65, 91–103 (2006)

[14] Zadora, G.: Classification of glass fragments based on elemental composition and refractive index. J. Forensic Sci. 54, 49–59 (2009)

[15] Wąs-Gubała, J.: Selected aspects of forensic examination of textile traces. Fibres Text.East.Eur. 17(4), 26–29 (2009)

[16] Wąs-Gubała, J., Machnowski, W.: Application of Raman spectroscopy for differentiation among cotton and viscose fibers dyed with several dye classes. Spectrosc. Lett. 47(7), 527–535 (2014)

[17] Brożek-Mucha, Z.: Comparison of cartridge case and airborne GSR-a study of their elemental contents and morphology by means of SEM EDX. X-Ray Spectrom. 36, 398–407(2007)

[18] Brożek-Mucha, Z.: On the prevalence of gunshot residue in selected populations-an empir-ical study performedwith SEM-EDX analysis. Forensic Sci. Int. 237, 46–52 (2014)

[19] Lopez-Lopez, M., Delgado, J.J., Garcia-Ruiz, C.: Ammunition identification by means of the organic analysis of gunshot residues using Raman spectroscopy. Anal. Chem. 84, 3581–3585 (2012)

[20] Bueno, J., Sikirzhytski, V., Lednev, I.K.: Attenuated total reflectance-FT-IR spectroscopy for gunshot residue analysis: potential for ammunition determination. Anal. Chem. 85, 7287–7294 (2013)

[21] Brożek-Mucha, Z.: Chemical and morphological study of gunshot residue persisting on the shooter by means of scanning electron microscopy and energy dispersive X-ray spectrometry.Mi-

crosc. Microanal. 17, 972-982 (2011)

[22] Borusiewicz, R., Zadora, G., Zięba-Palus, J.: Application of head-space analysis with passive adsorption for forensic purposes in the automated thermal desorption-gas chromatographymass spectrometry system. Chromatographia 60, 133-142 (2004)

[23] Borusiewicz, R., Zięba-Palus, J.: A comparison of effectiveness of Tenax TA and Carbotrap 300 in concentration of flammable liquids compounds. J. Forensic Sci. 52(1), 70-74(2007)

[24] Fabiańska, E., Trzcińska, B.: Differentiation of ballpoint and liquid inks a comparison method in use. Probl. Forensic Sci. 66, 383-400 (2001)

[25] Zięba-Palus, J., Kunicki, M.: Application of microinfrared and Raman spectrometry in examination of inks. Forensic Sci. Int. 158, 164-172 (2006)

[26] Trzcińska, B.M.: Analysis of writing inks in changed documents. A preliminary study with TLC. Chem. Anal. 46, 507-513 (2001)

[27] Trzcińska, B.M.: Analytical differentiation of black powder toners of similar polymer composition for criminalistic purposes. Chem. Anal. 51, 147-157 (2006)

[28] Moffat, A.C., Osselton, M.D., Widdop, B., Watts, J. (eds.): Clarke's Analysis of Drugs and Poisons, 4th edn. Pharmaceutical Press, London (2011)

[29] Baselt, R.C.: Disposition of Toxic Drugs and Chemicals in Man, 10th edn. Biomedical Publications, Seal Beach (2014)

[30] Kała, M.: Substancje powodujące uzależnienie w praktyce Instytutu Ekspertyz Sądowych w Krakowie. Przegl. Lek. 54, 430-437 (1997)

[31] Kała, M.: Scena narkotykowa w Polsce z punktu widzenia toksykologa sadowego. Przegl. Lek. 67, 594-597 (2010)

[32] Janowska, E., Chudzikiewicz, E., Lechowicz, W.: Ephedrone-new street drug obtained from Proasthmin. Probl. Forensic Sci. 39, 44-53 (1999)

[33] Zuba, D.: Medicines containing ephedrine and pseudoephedrine as a source of methcathinone. Probl. Forensic Sci. 71, 323-333 (2007)

[34] www.emcdda.eu.int, Intoxications with cocaine adulterated with atropine in four EU Member States. Information from the EMCDDA and REITOX Early Warning System (Nov./Dec. 2004- Feb. 2005)

[35] Stanaszek, R., Zuba, D.: A comparison of developed and validated chromatographic methods (HPLC, GC MS) for determination of delta-9-tetrahydrocannabinol (δ^9-THC) and delta-9-tetrahydrocannabinolic acid (δ^9-THCA-A) in hemp. Probl. Forensic Sci. 71, 313-322 (2007)

[36] Zuba, D., Byrska, B., Maciów, M.: Comparison of 'herbal highs' composition. Anal. Bioanal. Chem. 400, 119-126 (2011)

[37] Byrska, B., Zuba, D., Stanaszek, R.: Determination of piperazine derivatives in 'legal highs'. Probl. Forensic Sci. 81, 101-113 (2010)

[38] Stanaszek, R., Zuba, D.: 1-(3-chlorophenyl)piperazine (mCPP)-a new designer drug that is still a legal substance. Probl.Forensic Sci. 66, 220-228 (2006)

[39] Zuba, D., Byrska, B., Pytka, P., Sekuła, K., Stanaszek, R.: Widma masowe składnikówaktywnych preparatów typu dopalacze. Wydawnictwo Instytutu Ekspertyz Sądowych, Kraków(2011)

[40] Adamowicz, P., Kała, M.: Date-rape drugs scene in Poland. Przegl. Lek. 62, 572–575 (2005)

[41] Adamowicz, P., Kała, M.: Screening for drug-facilitated sexual assault by means of liquid chromatography coupled to atmospheric pressure chemical ionisation-mass spectrometry (LC APCI MS). Probl. Forensic Sci. 76, 403–411 (2008)

[42] Kała, M.: Środki działające podobnie do alkoholu w organizmie kierowcy. Paragraf na drodze 11, 41–68 (2004)

[43] Byrska, B., Zuba, D.: Profiling of 3,4-methylenedioxymethamphetamine by means of high-performance liquid chromatography. Anal. Bioanal. Chem. 390, 715–722 (2008)

[44] Kała, M., Chudzikiewicz, E.: The influence of post-mortem changes in biological material on interpretation of toxicological analysis results. Probl. Forensic Sci. 54, 32–59 (2003)

[45] Kasprzak, K., Adamowicz, P., Kała, M.: Determination of gamma-hydroxybutyrate (GHB) in urine by gas chromatography-mass spectrometry with positive chemical ionisation (PCI GC MS). Probl. Forensic Sci. 67, 289–300 (2006)

[46] Scientific Working Group for Forensic Toxicology (SWGTOX): Standard practices for method validation in forensic toxicology. J. Anal. Toxicol. 37, 452–474 (2013)

[47] Maurer, H.H.: Hyphenated mass spectrometric techniques-indispensable tools in clinical and forensic toxicology and in doping control. J. Mass Spectrom. 41, 1399–1413 (2006)

[48] Chudzikiewicz, E., Adamowicz, P., Kała, M., Lechowicz, W., Pufal, E., Sykutera, M.,Sliwka,K.: Possibilities of using saliva for testing drivers for estazolam, doxepin and promazine. Probl. Forensic Sci. 57, 166–177 (2005)

[49] Gunn, J., Shelly, C., Lewis, S.W., Toop, T., Archer, M.: The determination of morphine in the larvae ofCalliphora stygia using flow injection analysis and HPLC with chemiluminescence detection. J. Anal. Toxicol. 30, 519–523 (2006)

[50] Kintz, P., Godelar, B., Tracqui, A., Mangin, P., Lugnier, A.A., Chaumont, A.J.: Fly larvae: a new toxicological method of investigation in forensic medicine. J. Forensic Sci. 35, 204–207(1990)

[51] van Wyk, J. M.C., van der Linde, T.C., Hundt, H.K.L.: Determination ofethanol in dipterous maggots on decomposing carcasses. In: Kovatsis, A.V., Tsoukali-Papadopoulou, H. (eds.) Aspect on Forensic Toxicology, The 33rd International Congress on Forensic (TIAFT) and 1st on Environmental Toxicology (Gretox'95),August 27–31, 1995, pp. 439–443. Tecnika Studio, Thessaloniki (1995)

[52] Matoba, K., Terazawa, K.: Estimation of the time of death of decomposed or skeletonised bodies found outdoors in cold season in Sapporo city, located in the northern district of Japan. Leg. Med. 10, 78–82 (2008)

[53] Adamowicz, P., Kała, M.: Complex intoxications. Analytical and interpretational problems. Probl. Forensic Sci. 72, 433–449 (2007)

[54] de Zeeuw, R.A., Franke, J.P., Degel, F., Machbert, G., Schütz, H., Wijsbeek, J.: (eds.): Thin-layer chromatographic R f values of toxicologically relevant substances on standardized systems. Report XV II of the DFG Commission for Clinical-Toxicological Analysis, Special Issue of the TIAFT Bulletin, 2nd edn. VCH Verlagsgesellschaft, Weinheim (1992)

[55] de Zeeuw, R.A., Franke, J.P., Maurer, H.H., Pfleger, K. (eds.): Gas chromatographic retention indices of toxicologically relevant substances on packed or capillary columns with dimethyl-silicone stationary phases. Report X V III of the DFG Commission for Clinical-Toxicological Analysis, Special Issue of the TIAFT Bulletin, 3rd edn. VCH Verlagsgesellschaft, Weinheim (1992)

[56] de Zeeuw, R.A., Franke, J.P., Machata, G., Möller, M., Müller, M.R., Graefe, A., Tiess, D., Pfleger, K., Geldmacher-von Mallinckrodt, M. (eds.): Gas chromatographic retention indices of solvents and other volatile substances for use in toxicological analysis. Report X I X of the DFG Commission for Clinical-Toxicological Analysis, Special Issue of the TIAFT Bulletin. VCH Verlagsgesellschaft, Weinheim (1992)

[57] Pragst, F., Herzler, M., Herre, S., Erxleben, B.T., Rothe, M.: UV-Spectra of Toxic Compounds. Database of Photodiode array UV Spectra of Illegal and Therapeutic Drugs, Pesti-cides, Ecotoxic Substances and Other Poisons. Dieter Helm, Heppenheim (2001). Book and CD

[58] Herzler, M., Herre, S., Pragst, F.: Selectivity of substance identification by HPLC DAD in toxicological analysis using a UV spectra library of 2682 compounds. J. Anal. Toxicol. 27, 233–242 (2003)

[59] Müller, C.A., Weinmann, W., Dresen, S., Schreiber, A., Gergov, M.: Development of a multi-target screening analysis for 301 drugs using a QTrap liquid chromatography/tandem mass spectrometry system and automated library searching. Rapid Commun. Mass Spectrom. 19, 1332–1338 (2005)

[60] http://www.chemicalsoft.de/index-ms.htm

[61] Pfleger, K., Maurer, H., Weber, A.: Mass Spectral and GC Data of Drugs, Poisons, Pesticides, Pollutants and Their Metabolites, 3rd edn. VCH, Weinheim (2000)

[62] Peters, F.T., Schaefer, S., Staack, R.F., Kraemer, T., Maurer, H.H.: Screening for and validated quantification of amphetamines and of amphetamine- and piperazine-derived designer drugs in human blood plasma by gas chromatography/mass spektrometry. J. Mass Spectrom. 38, 659–676 (2003)

[63] Smink, B.E., Mathijssen, M.P., Lusthof, K.J., de Gier, J.J., Egberts, A.C., Uges, D.R.: Comparison of urine and oral fluid as matrices for screening of thirty-three benzodiazepines and benzodiazepine-like substances using immunoassay and LC MS MS. J. Anal. Toxicol. 30, 478–485 (2006)

[64] Kratzsch, C., Tenberken, O., Peters, F.T., Weber, A., Kraemer, T., Maurer, H.H.:

Screening, library-assisted identification and validated quantification of 23 benzodiazepines, flumazenil, zaleplone, zolpidem and zopiclone in plasma by liquid chromatography/mass spectrometry with atmospheric pressure chemical ionization. J. Mass Spectrom. 39, 856–872 (2004)

[65] Kirchherr, H., Kühn-Velten, W.N.: Quantitative determination of forty-eight antidepressants and antipsychotics in human serum by HPLC tandem mass spectrometry: a multi-level, single-sample approach. J. Chromatogr. B Analyt. Technol. Biomed. Life Sci. 843, 100–113 (2006)

[66] Maurer, H.H., Tenberken, O., Kratzsch, C., Weber, A., Peters, F.T.: Screening for library-assisted identification and fully validated quantification of 22 beta-blockers in blood plasma by liquid chromatography-mass spectrometry with atmospheric pressure chemical ionization. J. Chromatogr. A 1058, 169–181 (2004)

[67] Adamowicz, P., Kała, M.: Simultaneous screening for anddetermination of 128 date-rape drugs in urine by gas chromatography-electron ionization-mass spectrometry. Forensic Sci. Int. 198, 39–45 (2010)

[68] Lechowicz, W., Kała, M., Walker, J., Screening and quantification of the twenty-four drugs in oral fluid relevant for road traffic safety by means of LC MS MS ESI. Proceedings of ICADTS and TIAFT Meeting, Seattle, 26 – 30. 08. 2007. http://www.icadts2007.org/print/169screen24drugs.pdf

[69] Beyer, J., Peters, F.T., Kraemer, T., Maurer, H.H.: Detection and validated quantification of nine herbal phenalkylamines and methcathinone in human blood plasma by LC MS MS with electrospray ionization. J. Mass Spectrom. 42, 150–160 (2007)

[70] Habrdová, V., Peters, F.T., Theobald, D.S., Maurer, H.H.: Screening for and validated for quantification of phenethylamine-type designer drugs and mescaline in human blood plasma by gas chromatography/mass spectrometry. J. Mass Spectrom. 40, 785–795 (2005)

[71] Maurer, H.H., Schmidt, C.J., Weber, A.A., Kraemer, T.: Validated electrospray LC MSassay for determination of the mushroom toxins alpha- and beta-amanitin in urine after immunoaffinity extraction. J. Chromatogr. B Biomed. Sci. Appl. 748, 125–135 (2000)

[72] Gregov, M., Ojanperä, I., Vuori, E.: Simultaneous screening for 238 drugs in blood by liquid chromatography-ion spray tandem mass spectrometry with multiple-reaction monitoring. J. Chromatogr. B Analyt. Technol. Biomed. Life Sci. 795, 41–53 (2003)

[73] Alder, L., Greulich, K., Kempe, G., Vieth, B.: Residue analysis of 500 high priority pesticides: better by GC MS or LC MS MS? Mass Spectrom. Rev. 25, 838–865 (2006)

[74] Pang, G.F., Cao, Y.Z., Zhang, J.J., Fan, C.L., Liu, Y.M., Li, X.M., Shi, Y.Q., Wu, Y.P., Guo, T.T.: Validation study 660 pesticide residues in animal tissue by gel permeation chromatography cleanup/gas chromatography-mass spectrometry and liquid chromatography-tandem mass spectrometry. J. Chromatogr. A 1125, 1–30 (2006)

第三部分
痕量分析的特殊应用

第三章

隋唐代科技的回顧

第12章
无机形态和生物无机形态分析的问题和前景

12.1 形态和形态分析

形态是指一种元素存在于现实材料中的不同物理化学形式,元素的存在形式是其同位素组成、电子结构,或氧化数,或分子,或络合物结构的不同表现[1]。形态分析是指确定某个特定元素的存在形式及其含量的过程,这一定义也简明地说明了形态分析的目的。然而,化合物形态分析水平通常取决于现有设备状况、对待测元素生物化学知识认知以及分析人员的经验。对大多数人来说,元素形态分析是一项很有难度的工作,首先,定性检测技术的灵敏度较低,对痕量水平的元素化合物检测很困难;其次,被测化合物有时尚未被证实或没有文献记载,此时,分析过程必须分离出目标物质,并使其纯度达到标准物质的水平。一般包括以下几个步骤。

(1) 选取代表性样品,用于实验分析和样品稳定性研究;
(2) 从样品中提取或浸取所要研究的元素的化合物;
(3) 用半制备或制备液相色谱法分离萃取化合物,然后用电感耦合等离子体质谱法(ICP-MS)或电感耦合等离子体发射光谱法(ICP-OES)检测分离组分中的元素;
(4) 浓缩分离含待测元素的化学组分,采用色谱法进行纯化;
(5) 对获得的组分进行定性鉴别并确定化学结构,检测技术包括核磁共振、红外、基质辅助激光电离、电喷雾电离、常压化学电离质谱法等;
(6) 用 HPLC-ICP-MS 测定化学组分纯度。

如果有适当的标准方法,则上述过程可简化并按照以下步骤进行。
(1) 选取有代表性样品进行实验分析和样品稳定性研究;
(2) 从样品中提取或浸取含待测元素的化合物;
(3) 选择一种分辨率高的方法对提取到的化合物进行色谱分离,常用 HPLC-

ICP-MS；

(4) 通过加内标的方法定性检测分离组分并初步半定量；

(5) 用色谱法准确定性及定量；

(6) 考虑到基体效应,用两种色谱法对分离的化合物进行定量。

分析过程较复杂就容易出现操作失误。在定性分析前首先需要取样和制备样品,然后用高洗脱能力的溶液进行多级分馏,在此过程中可能会引入杂质导致样品性质变化,此后所检测的化合物可能是分析过程中发生的生化反应或物理转化的产物。形态分析根据检测目标可定义为:将检测结果概念性推断为其可能的初始状态,通常与生物学、物理学、毒理学密切相关。但是,检测目标在具体情况下会不同,如在土壤分析中,为了确定某一特定元素的植物吸收能力,重点应该考虑植物,而不是元素化合物。针对各种各样的材料,如环境、食品、体液等,及其化合物的形态分析过程都极其复杂又不尽相同,所以检测的关键是实验室实际操作的正确性,而非科学知识和定性检测技术的局限性。

目前,形态分析的研究对象大约有 20 种元素(主要为 Al、As、Cd、Co、Cu、Pb、Hg、Pt、Se) 和 4 类化合物,分别是卤素、挥发性有机物以及含肽、蛋白质的化合物或糖[2]。研究中最常见的过渡金属元素之一是砷,因为砷具有毒性,而且广泛存在于水、植物和海鲜中。砷在环境中以各种物理化学形式出现,有无机盐、有毒酸(三价或五价砷)、无毒的砷甜菜碱(AsB) 或者与多肽形成的复合物等,30 年来已经发现了 50 多种由植物合成的含砷化合物。因此,对砷的形态分析旨在鉴别和确定它的存在形式,以及不同存在形态会引发的一系列问题[3-8]。

环境中的砷有两个基本来源:①天然的,例如波兰 Kvodzko 山谷中有大量的富砷岩石,这导致了该地区饮用水中含有大量的砷[9-10];②人为的,人类活动使砷进入土壤或水域,其存在形式为除草剂或农药[10-11]、杀菌剂、电子产品和电子废弃物(如半导体材料)、制革工业产生的废弃物[11-12]等。不同的砷化合物所具有的毒性不同,并且在生物体内的富集能力也不一样,由于生物体是人类饮食的基础,所以这是开展砷元素形态分析的主要原因。此外,建立了统一的标准化程序,用于检测地质[11-14]、环境[13-16]、生物[15-18]、临床[17-21]和食品[9,16,20-24]样品中的砷的存在形态。

形态分析须考虑样品的多样性,不仅由于样品基体具有复杂特性,而且由于存在潜在的、能自发转化成有毒化合物的生化反应,这通常是由样品来源决定的。例如,在测试地质样品时,有三价砷或五价砷的无机化合物[25-26],地下水中可能含有生物甲基化过程中产生的高毒性五价砷酸甲基衍生物[27]和低毒性三价砷酸衍生物[28]。由于植物体合成的砷糖、砷脂、砷甜菜碱和砷甲硫氨酸,砷元素的存在形式更多。砷和磷具有相似的性质,导致砷原子和磷原子可以相互置换。在海洋生物中,有 4 种常见的通过连续烷基化和酰基化过程形成的砷糖,分别是砷糖 A,3

-[5′-脱氧-5′-(二甲基二甲酰基)-b-核糖氧基]-2-羟基丙磺酸、砷糖 B,3-[5′-脱氧-5′-(二甲基二甲酰基)-b-核糖氧基]-2-羟基丙二醇、砷糖 C,3-[5′-脱氧-5′-(二甲基二甲酰基)-b-核糖氧基]-2-羟丙基硫酸氢盐、砷糖 D,3-[5′-脱氧-5′-(二甲基二甲酰基)-b-核糖氧基]-2-羟丙基-2,3-羟丙基磷酸盐。表 12-1 列出了常见的砷化合物。

砷主要以离子形式存在,通常用配有原子化或离子化检测器(如 ICP-OES 或 ICP-MS)的高效离子交换色谱对其进行分离,但是,在这一步骤之前还有许多样品处理过程,可能会产生严重的错误结果。

表 12-1 常见的砷化合物

名称	缩写	化学式	毒性	参考文献
砷	As		致癌	[25-26]
无机化合物				
亚砷酸离子	As(Ⅲ)	$AsO(OH)_2^-$		
砷酸离子	As(Ⅴ)	$AsO_2(OH)_2^-$		
有机化合物				
甲基胂	MMA(Ⅴ)	$CH_3AsO(OH)_2$	致癌	[27]
二甲基胂	DMA(Ⅴ)	$(CH_3)_2AsO(OH)$	致癌	[30]
砷甜菜碱	AsB	$(CH_3)_3As^+CH_2COOH$	无毒	[31]
砷胆碱	AsC	$(CH_3)_3As^+(CH_2)_2COOH$	无毒	[32]
一氧化三甲胂	TMAsO	$(CH_3)_3AsO$	暂无数据	
四甲基砷离子	Me_4As^+	$(CH_3)_4As^+$	暂无数据	
三甲基胂		$(CH_3)_3As$	暂无数据	
砷糖		不同结构	暂无数据	
砷脂			暂无数据	

12.2 取样和储存

分析过程的第一步首先是选取有代表性的样品,这对最终结果的质量至关重要,必须确保被测元素的损失最小,并最大限度地防止污染,防止引入有可能打破各种存在形式之间平衡的微量杂质。

在形态分析中选择合适材质和颜色的实验器具非常重要,确保取样、储存过程中的样品洁净,在这个阶段,主要考虑化合物与空气接触时对紫外线辐射和氧化的敏感度因素造成的被测元素污染或损失;其他因素包括元素蒸发、形态转化,以及在容器壁上的吸收或吸附,因此,所有实验容器都应该用去离子水或蒸馏水彻底清洗,对样品没有影响的情况下可使用 10% 硝酸清洗。

采集气体样品通常采用分离技术(如管子、安瓿瓶、塑料袋或内衬为非活性聚合物的罐子)或者吸附技术(如固体吸附剂、溶液或低温冷冻技术),这些操作有可能引起光照、水解、氧化、容器壁吸附而导致元素损失,通过控制适当的温度、光照、湿度或含氧量等参数,可以将损失最小化[33]。进行形态分析的气体样品最安全的储存方法是液氮低温储存,可以保存几天[34]。

液体样品用容器采集和储存时,容器的材质不应溶解到样品中。对金属元素的测定,建议采用聚碳酸酯容器,但当分析含汞化合物时,应选择玻璃容器。

天然水分析时,基体误差是主要的潜在误差来源。由于其与土壤、岩石和大气的不断接触,天然水中几乎含有全部的化学元素。即便是经过处理的饮用水也是一个不均匀的体系,含有水合离子、微粒、胶体、甚至是极小的固体悬浮物,还有其他无机络合物,如碳酸盐、氯化物、硫酸盐和磷酸盐,以及有机化合物,如氨基酸、糖、尿素、腐殖酸和富里酸等。对深层地下水来说,可能还含有大量的一氧化碳和少量的氧气。样品储存过程中物理化学条件的改变会导致被测元素的存在形式发生变化,因此有必要将水样保持在自然低温和较高气压下。

液体生物材料的储存存在风险,因为这种样品通常含有系列生化活性成分,如酶、细菌等,能导致所研究化合物的物理化学形态转化。急速冷冻或冻干可以从液态材料中去除水和某些挥发性有机物,还有紫外线照射等方法可以防止样品转化的发生,样品在避光、4℃条件下可以短期储存。

形态分析最有难度的环节之一是在保持样品原有特性的前提下的固体样品取样[36-37]。从液氮中采集冷冻的生物样品时,应储存在-20℃的中性气体中,用聚乙烯或聚碳酸酯容器收集,应尽可能缩短此过程所用时间。例如,温度、光照、pH等其他因素也会引起被测元素形式转化或者分析物与基体组分、包装物之间的反应。

总之,样品必须以不影响被测物特性的方式储存,不论是新采集的样品还是经浓缩或纯化过程的提取物。要注意的是,取样和样品保存没有通用方法,因此,针对所研究的化合物,需要研究优化方法,下面以含砷化合物为例进行讲解。

砷主要存在于各种类型的水(包括饮用水、地表水、地下水、雨水和废水)、土壤、尘埃、植物和生物流体(血液、奶、尿液)中[38],含砷样品应储存在棕色玻璃瓶、聚乙烯或聚丙烯瓶中。与其他元素相比,砷化合物相对稳定,含砷化合物溶液储存的主要问题是,As(Ⅲ)化合物很容易氧化,在室温下24h可完全氧化[39-40]。As(Ⅲ)化合物存在于每个生物体中,因为它是砷(Ⅴ)酸衍生物的生物甲基化的中间产物,实验证明氧化反应同时生成谷胱甘肽,并伴随着二羟基砷(Ⅲ)酸盐阴离子的甲基化反应[41]。紫外照射是使液体样品稳定的标准方法,同时也加速了氧化过程。另一种常用的方法是硝酸酸化,与ICP测试中的处理方式类似,然而,硝酸也是一种氧化剂,会使砷的形态平衡向砷(Ⅴ)转化。另外,盐酸不具有氧化性,但会在检测阶段出现问题,氯化物是四级杆质谱中化学干扰的主要来源。

铁、锰离子可以很容易地参与到氧化还原反应中[42],在含有铁、锰离子的溶液中可观察到砷化合物的氧化过程。可以通过加入二乙胺四乙酸(EDTA)或其他能够与这些金属离子络合的物质[43],因为配位化合物能够阻止无机砷和铁离子形成不溶物[44]。在铁、锰离子络合的酸性环境中使用抗坏血酸作为还原剂[45],或者在容器密封之前用惰性气体冲洗溶液[46]都可以提高样品的稳定性。

砷形态分析中的一个重要问题是,上述提到的氧化态砷与铁、锰共沉淀形成难溶化合物,或者是形成砷酸铁和砷酸钙[47]。这种现象在酸性(pH<2)环境中并不明显,当溶液中加入 10mM 磷酸和 EDTA 时,溶液最稳定。当锰、铁离子的浓度为 100mg/L 时,溶液在室温下可以保存 4 天;如果温度降至 4℃(但不结冰)和避光条件,可以使样品储存期延长至 28 天;3 个月后砷化合物含量变化不超过 10%[48]。储存所有含铁、锰或钙离子的溶液样品,特别是天然水和土壤提取物都必须执行上述储存程序[49]。对于生物样品(如酵母)水溶液的储存,通常是将温度降低到 4℃,并确保避免光照。砷糖、AsB 和 AsC 具有良好的稳定性,即使在室温条件下,高于 0.5mg/mL 含量的砷化合物[51]也要在储存 9 个月后才能观察到降解产物(图 12-1)[50]。

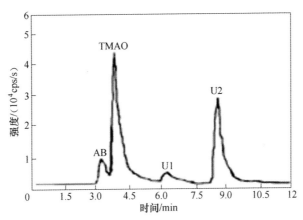

图 12-1　标准物质提取物在 4℃保存 9 个月时的 HPLC-ICP-MS 色谱图
AB—砷甜菜碱;TMAsO——氧化三甲胂;U1、U2—不确定的砷形态。

储存过程中的大多数问题是由砷和植物螯合素形成的复合物引起的,冷冻溶液会使肽发生氧化过程,分解复合物。此时,酸化和往样品中加络合物是不行的,最安全的方法是用惰性气体冲洗溶液或加入具有还原性的物质(如 β-巯基乙醇、二硫苏糖醇或 3-(2-羧乙基)膦),防止硫化物形成[52]。

了解被测化合物的稳定性,对设计正确的贮存、分离、纯化和测定条件起着关键作用。砷的例子表明,尽管多年大量地研究是为了确定其衍生物的特性,但如果实验人员的经验不足,在采集样品的储存过程中就会产生许多问题。

12.3 元素总量的测定

很多情况下,形态分析过程必须先测定样品中某个特定元素的总含量,这样才能评估提取过程中的回收率,或者评价色谱分离过程中被测元素化合物与固定相之间相互作用的程度。总量测量首先采用湿法消解、萃取、离心分离、过滤、色谱柱分离和固相萃取(SPE)等方法获得样品溶液;然后再应用光谱技术如原子吸收光谱(AAS)和发射光谱(OES)测定元素总量,如果使用前置氢化物发生器,可将检出限从 50ng/mL 降至 1ng/mL,这是因为许多元素受到砷元素信号的干扰,生成挥发性物质有利于排除干扰,如元素 Al(在 AAS 和 OES 中)、Cr、Co、V、Cd(在 OES 中)。电化学方法如阳极溶出伏安法(ASV)、吸附转移溶出伏安法(AdTSV)和电位溶出伏安法(PSA)依然很重要[53-54],不过,在痕量分析中最典型的方法是 ICP-MS 法,尽管会受到设备成本的限制。ICP-MS 法具有良好的同位素灵敏度,能够确保检测的选择性,但其主要缺点是存在干扰,尽管比在 AAS 和 OES 中干扰发生频率少,但仍然是产生误差的主要来源。

ICP-MS 有两种类型的干扰:物理干扰和同量异位素干扰[55]。物理干扰出现在样品导入检测器的阶段,主要是雾化效率,受到溶液的密度、粘度及表面张力的影响;相反,同量异位素干扰是具有相同的质荷比(m/z)的单原子或多原子离子产生的信号重叠,既可由等离子气体引起,也可由样品组分引起[56]。消除四极杆 ICP-MS 系统干扰的方法有很多,一种方法是选择被测元素的其他无干扰同位素进行测定[57],但没有同位素的元素如 As、Co 不适用。当测定轻质元素时,可以降低等离子体温度[58]。然而,这样做会降低检测灵敏度,如 As 元素具有很高的电离电位。

通过优化雾化气流量[59]或者与接口组件有关的燃烧器位置,可以显著减少氧化物、氢氧化物和多电荷离子,而灵敏度仅略有降低[60]。不过这不适用于砷元素(^{75}As),因为由等离子体氩气和样品基体中的氯产生的离子(^{40}Ar^{35}Cl$^+$)会造成干扰,而氩和氯是主体组分,砷在样品中处于痕量级,所以测量结果可能与实际不符。当待测溶液具有低离子强度时,可以通过添加 1%~2% 的少量甲醇来减少氯对分析结果的影响,但这种优化可能导致取样锥上积碳而造成系统污染[61]。

砷测定可以使用校正曲线方程来消除上述同位素离子的干扰。氩氯对 ^{75}As 砷离子信号值的贡献可以计算出来,因为氯有两种同位素:^{35}Cl 和 ^{37}Cl,后者在等离子体中形成 ^{40}Ar^{37}Cl$^+$ 离子在 m/z 77 处有信号响应。可以采用适当的数学方程分析和修正 m/z 75 处的离子信号强度[24,62]。但只有当样品中 As 和 Cl 含量的比不超过 100 时才适用[24,62],而且,校正曲线方程的适用范围无法外延。

消除同位素干扰的另一种方法是在质量分析器之前加一个碰撞反应

池[63-65]，池内充满气体，气体有两种：①惰性碰撞气体，如氦气[66-67]。由于碰撞，可以降低干扰离子的能量，使之无法到达四极杆分析器，或引起它们的衰变。②反应气体，如甲烷[68]、氢气[59,69]、氨气，通过与ArCl⁺离子反应，形成新的具有不同质量数的多原子离子[57]，即

$$^{40}Ar^{35}Cl + H_2 \rightarrow {}^{40}Ar^1H_2^+ + HCl \qquad (12.1)$$

甲烷在砷的测定中作为反应气，可以生成 $^{75}As^{12}C^1H^1H^+$（$m/z=89$），具有比 ^{75}As 同位素更高的灵敏度[70]。

如果使用有碰撞单元的仪器，那么引入的反应气会提高多原子离子形成的风险，也会干扰被测元素的测定。在分析环境样品时，其他元素，如 $Ca^{[71-72]}$、Fe、Cu 是主要误差的潜在来源，仪器的部件，如镍锥，也会造成干扰，最常见的误差还包括使用干扰元素，例如锗，作为内标[63]。这些元素及其离子的对砷测定干扰影响如下[73]：

$^{58}Ni^{16}O^1H^+ > {}^{60}Ni^{14}N^1H^+ > {}^{59}Co^{16}O^+ > {}^{74}Ge^1H^+ > {}^{58}Fe^{16}O^1H^+ > {}^{40}Ca^{35}Cl^+ > {}^{63}Cu^{12}C^+$

消除氯离子干扰的有效途径是在用 ICP-MS 分析砷元素之前进行色谱分离[9][61]，所用的洗脱液为含有磷酸根离子的溶液，或不太常见的碳酸根离子[74]，该方法也可用于控制干扰的发生。图 12-2 为同时在 m/z 35 和 75 处水样的 ICP-MS 色谱图。图中 A 表明样品中含有大量的氯离子；B 表明样品中并不存在干扰离子 $^{40}Ar^{35}Cl^+$，这说明在优化的等离子体条件下不会形成 $^{40}Ar^{35}Cl^+$。如果无法消除多原子离子，那么 HPLC 法可以分离砷化合物中的氯离子。

尽管有上述不足，但 ICP-MS 仍然具有很好的灵敏度，并广泛应用于测定样品中的元素总量，然而，在实验室条件下，只有具有丰富经验的工作人员才能发挥其作用。

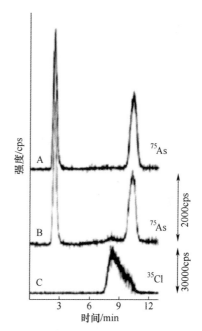

图 12-2 同在 m/z 35 与 75 处水样的 ICP-MS 色谱图

A，不含氯离子的 10ng/mL As(Ⅲ) 和 As(Ⅴ)；
B，含 400μg/mL 氯离子的 10ng/mL As(Ⅲ) 和 As(Ⅴ)；
C，400μg/mL 氯离子。

12.4 固体样品的萃取和浓缩

萃取是指将物质从固相转移到液相,或者从一种液相转移到与之不互溶另一液相的过程。从实际操作过程来看,萃取可以通过浸提来实现,包括化合物从固相到液相的固-液萃取(SLE)、直接萃取的液-液萃取(LLE)、间接萃取的固相萃取(SPE)、固相微萃取(SPME)[75]。化合物对两个相的亲和力不同,所以浸提效率用提取百分数表示,实际上,常用的概念是萃取率,可理解为物质从一个相到另一个相的转化程度,以百分数表示。确定萃取率的方法有很多,可以分为两种。

(1) 基于所提取元素的量相对于样品消解得到的总量,或者相对于全部总量。

(2) 采用标准加入法[76]测出不同形态被测元素的萃取率,或分析具有相似或相同基体的标准物质(CRM)。

在第二种情况下,如果有标准物质建议使用标准物质,因为内标的加入可能会导致化合物与基体之间由于结合程度和结合方式的差异出现重现性差的问题,表现为萃取率偏高。然而,在第一种情况下,分析过程简单明了,可以由浸出和提取来确定萃取率。

根据所用萃取剂的类型,浸提过程可以按以下几种方式进行。

(1) 使用酸溶液[77]、碱溶液[78]、缓冲溶液或者具有强氧化性或强还原性的水溶性化合物。提取过程往往是一个长期而且不完全的过程,所以要使用超声波或微波加热的方法来提高萃取率[79]。

(2) 最常见的将极性有机物或金属有机物提取出来的有机溶剂萃取方法是索氏提取[80],冷却、浓缩的溶剂蒸汽将样品中挥发性或热不稳定的化合物提取出来。该过程也可以采用加速溶剂萃取(ASE)技术来实现,也称加压液相萃取(PLE)技术,即在升高的温度和压力下使用溶剂[81],高压条件为(30~200)atm(1atm=1.01325×10^{5}Pa),温度为50~200℃[82]。

(3) 活性酶蛋白包括蛋白酶、脂肪酶、胰蛋白酶、胃蛋白酶、细胞分裂初期或纤维素酶及其混合物的溶液,固体生物样品基体性质匹配[79,83-84]。将蛋白质、多糖和长链脂肪分解成氨基酸、糖或短链脂肪,通过使用超声波可以大幅提高基体酶的分解,例如,可以提高酵母细胞壁的分解效率,从而使硒的萃取率提高20%[85]。

(4) 超临界流体,最常见的是碳(Ⅳ)氧化物,也有添加少量极性溶剂,如甲醇、乙腈或水进行改性的。水是超临界流体萃取(SFE)技术最常用的试剂,因为在超临界压力下将温度从50℃增加到400℃,就有可能实现萃取剂从亚临界态到超临界态的转变,从而使极性和中极性化合物浸出[86]。

最佳浸提介质的选择取决于分析化合物的性质、样品基体以及样品组分之间物化结合程度。基体的疏水性水平显著影响萃取率,即便被浸提物质具有强极性,通过溶解、物理或化学作用以及酶分解等破坏基体的"晶格",可以使化合物有效地转移到溶液中。对于植物样品最简单的方法是将其放置在液氮中(低温-78℃)破坏细胞壁结构,或者加入表面活性剂,如十二烷基硫酸钠(SDS),部分改变细胞膜中的磷脂结构,释放出疏水性蛋白质。具有较低热稳定性的植物中产生的黄酮类化合物可溶解在有机溶剂中,通常用索氏提取器进行萃取[87]。

通常是从土壤、植物或海洋动物体中浸提砷元素。生物样品的提取一般用甲醇或甲醇和水的混合物,经离心、过滤后的溶液用水稀释后装入离子交换柱。该方法适用于砷酸盐和砷糖,它们在水中的溶解性比在甲醇中好。而且,这种提取的有毒砷(Ⅲ)和砷(Ⅴ)盐比甲基化衍生物、AsB 或 AsC 少得多,毒性明显降低[88]。

提取液中甲醇含量越高,效果越好,既不影响砷(Ⅲ)、砷(Ⅴ)盐和砷糖的释放率,又能提高有机砷衍生物的提取率[89]。可通过超声波处理或加压提取来缩短提取时间[90],但多次提取无法提高工艺效率,因为固体材料中砷的含量仅取决于溶剂保留体积,即通常在第一个循环中,而不是重复浸提次数[91]。

有时,砷的提取采用软化水,但只适用于不同 pH 的溶液中可释放稳定化合物的情况,例如,比对智利阿塔卡马沙漠中发现的 1000 岁木乃伊的头发和印度市区当代居民头发的状况时进行的砷污染检测中发现,两者含有等量的二甲基胂酸[DMA(Ⅴ)]和甲基胂酸[MMA(Ⅴ)][92]。

对于海产品、动物组织或体液中的砷形态分析,不建议使用热水萃取,因为有毒的砷(Ⅲ)酸甲基化衍生物是中间产物,稳定性低,特别是在体外条件下[93-97]。

由于砷的化学性质与磷类似,所以它能够在磷脂中代替磷而形成砷脂。对于藻类的研究发现,多达 50% 的砷元素以非极性溶剂可提取脂的形式存在,然而,关于这些化合物的定性结论报告还是很少的[98]。对于海鲜样品,需要用丙酮、乙醚或氯仿进行初步脱脂以便可以得到更均匀的样品,但会造成 12% ~ 30% 的砷损失[99-100],如果用胰蛋白酶破坏海鲜中的蛋白质,则具有更高的回收率[101-103]。

研究土壤中的砷含量一般是用软化水提取,采用振动或微波处理[104-105],后一种处理方法也适用于生物材料的分析[106-107],通过添加磷酸和抗坏血酸[105],提高提取物中 As(Ⅴ)化合物的稳定性。

有时,元素在不同溶剂或不同酸性水溶液中具有不同的稳定性,此时可以采用连续浸提。随着离子强度的提高、pH 降低,或引入络合剂,提取液的化学活性逐渐增加。对于砷的形态分析不建议使用氧化性或还原性的物质。当选择恰当的溶剂时,连续浸提或许能重现自然环境中的发生过程。例如,如果分析目的是确定元素在土壤[108]或植物(表12-2)中的迁移水平,则不必将待测元素完全转移到溶液中,尤其是对痕量分析。

表12-2 连续浸提使用的溶液

土壤中的金属浸提	植物中的金属浸提
土壤中通过离子作用结合的金属离子 1M $MgCl_2$ 或 CH_3COONH_4,pH 为 7	细胞质与液泡中的肽 10mM Tris-HCl,pH 为 7.4
难溶土壤基体表面以碳酸盐形式存在或吸附的金属离子 1M CH_3COONa 或 0.05M EDTA,pH 为 5	细胞壁中的肽和多糖 2%崩溃酶或纤维素酶与 10mM Tris-HCl 的混合液,pH 为 7.4
难溶金属氧化物 0.1M NH_2OH,pH 为 2 或 0.2M $NH_4C_2O_4$,pH 为 3	疏水性蛋白质 2% SDS 与 10mM Tris-HCl 的混合液,pH 为 7.4
与有机化合物强结合的金属离子 30% H_2O_2	柠檬酸、苹果酸等有机酸 10mM CH_3COONH_4,pH 为 4.5

浸提得到的溶液或者水溶液在进行分析前需要浓缩。最常用的方法是冷冻干燥,利用升华作用使溶剂在降低压力和温度的条件下蒸发,或者在惰性气体中加速蒸发。这些方法通常要求单独保存,即使是热不稳定的化合物。但是,当蒸馏温度高于120℃时,As(Ⅲ)甲基化衍生物会转化成 As(Ⅴ)衍生物以及其他降解形式[109]。

使用 SPE 浓缩应具有温和的条件,也可用 SPME 简化程序。

SPE 最早由 Zhang Z. 等[110]提出,最初用于分离水中的有机化合物,这项技术使分析过程简化为两个阶段:固体吸附和少量洗脱液解吸[111]。SPE 柱包含极性、疏水性和离子交换等各种典型的液相色谱柱材料,以这种方式制得的浓缩溶液可直接进入 HPLC 或气相色谱(GC)[75]。对于生物样品来说,使用 SPE 能够获得良好的浸提效果,因为 SPE 柱中充满了均匀的固体材料,通常是植物、真菌或酵母与标准基体的混合物,这种"基质 SPE"是最常用的分离有机化合物的方法[112]。预先在超声波清洗机中填充好适当的洗脱剂,采用这样的 SPE 柱,可以提高提取率。该方法将提取和预纯化阶段合二为一[113-114],通过联合使用"基质 SPE"和 PLE 系统自动获取待测元素[115]。

SPME 装置类似于一支"微量进样器",在一根熔融石英纤维上涂有活性吸附材料的固定相,常用的材料是聚二甲基硅氧烷、聚丙烯酸酯、聚酰亚胺,或气相、液相色谱柱材料。涂有活性物质的纤维头可以浸入样品溶液中或顶空气体中,该方法主要与 GC 联用,纤维上的化合物在进样口受热脱附,然后直接导入 GC 柱中,从而大大减少样品损失[116]。非挥发性物质,包括金属有机化合物,需要衍生化以得到挥发性衍生物[117]。

涂有吸附剂的纤维也可以改用含有疏水性溶剂(单滴微萃取,SDME)的微量注射器,将注射器针头置于极性样品溶液中,挤出并悬挂一滴疏水性溶剂,待萃取完成后,将液滴收回。在此萃取模式下,待测物在体积为(1~3)μL的溶剂中吸收,可以直接导入 GC 或 HPLC 进样器[117-118]。该方法与 SPME 互为补充,也可以用来提取无机化合物(见 SDME[119-120])。

上述每种方法都能够使样品快速浓缩,可用于痕量或超痕量化学分析。

至少在 20 年前,分析砷化合物主要是用液相萃取法(LLE),通过还原无机砷盐来分离有机砷和无机砷,形成的卤素金属酸或杂多酸用甲苯或三氯甲烷萃取,总砷与上述方法提取的无机砷之差则被认为是有机砷[121]。As(Ⅲ)的测定省略了还原过程[122],因为这个过程很繁琐,是非选择性的,而且极不准确。

SPE 的应用解决了上述一部分问题,特别是对于浓缩水溶液样品中的砷[123],SPE 具有比 LLE 更好的选择性,因为它可以从活性炭床上连续洗脱化合物,从而分离 As(Ⅲ)和 As(Ⅴ)的无机物以及砷酸的苯基(PAS)和二甲基(DMA)衍生物[124]。由于提取过程很短,所以 SPE 组件能够与 ICP-MS 检测器联用[125]。但是,无论采用何种改性二氧化硅,砷的回收率都较低,甚至低于 50%,这是因为在分离物质与硅醇基团之间形成了氢键[114,126-127]。

使用 SPME 技术提取、富集无机砷化合物,并用于选择性测定 As(Ⅲ)和 As(Ⅴ)是不可取的,因为砷在萃取过程中会转化成挥发性氢化物[128-129]。SDME 也有同样的问题,在顶空模式下,砷转化成了挥发性氢化物[130]。可以通过控制溶液的酸度、选择还原剂的类型和浓度,或使用络合剂来实现部分选择性[131]。通过优化实验条件,使分析能够达到 pg/mL 水平。

以上方法可以使待测化合物至少浓缩 10 倍,但是这个过程通常也将基体一起浓缩,基体是检测噪声的潜在来源。即使去除基体能够提高灵敏度,但是信噪比(S/N)仍然会干扰样品中不同形态之间的平衡。因此,提取物应该在最短的时间内测试,可以通过具有高灵敏度检测器的色谱技术来实现。

12.5 应用联用技术测定与鉴别元素形态

联用技术是无机形态和生物无机形态分析的基本工具,专用的高灵敏度检测器与高性能分离技术联用具有充分的检测优势,主要包括 GC、HPLC、CE 以及超临界流体色谱(SFC)。

12.5.1 气相色谱联用技术

气相色谱法主要用于分离挥发性和中等挥发性的化合物,其原理是依据物质的沸点不同以及与固定相之间相互作用的程度差异。GC应用的基本前提条件是待测物质在300℃下具有挥发性,金属和类金属的烷基衍生物以及一些氨基酸能够满足这一条件。环境分析中重要的金属有机化合物,如烷烃或苯基衍生物必须进行衍生化生成挥发性衍生物。用于形成挥发性氢化物的试剂除了$NaBH_4$之外,主要还用到了四烷基硼酸盐[132]:$NaBEt_4$用来制备汞化合物的乙基衍生物;$NaBPr_4$用于在水介质中制备锡、汞、铅的丙基衍生物;$NaBBu_4$用于在格氏试剂存在下的非水介质中形成丁基衍生物[133]。烷基化反应通常不是依据甲基化来定量的,但是,甲基化过程可以通过使用含有稳定同位素的标准物质富集样品来控制[134]。

使用硼酸四丙酯可以同时测定 Sn、Hg、Pb 的化合物[135],在胺基酰基化反应后[136],硒蛋氨酸的羧基与异丙醇发生酯化反应,可以使分子结构略有改变,从而采用 GC-MS 根据分子特异性进行定性。

极易挥发的沸点低于100℃的金属有机化合物或氢化物,在填充柱等度洗脱条件下可以实现分离。但这种方法分辨率较差,无法分离某些金属形态。同时,在使用气相色谱-电感耦合等离子体质谱(GC-ICP-MS)检测系统中,过量的有机化合物会降低等离子体能量,甚至会导致等离子体熄灭。

毛细管气相色谱仪具有更高的分辨率,可以同时分析多种金属的烷基化衍生物,但是毛细管柱很容易"超载",所以待测物含量必须比柱子容量小。由于氧含量相对较高,以及毛细管柱到等离子体之间容易形成气溶胶使分析灵敏度变差。可以采用多根短柱来提高毛细管气相色谱的灵敏度和分辨率,并大大缩短分析时间。一束毛细管柱能够分离大量的待测物质,而且只需最小稀释量的载气。较短的分析时间和高载气流量可以实现不同沸点化合物的等温分离,这反过来又降低了设备成本和最终的测试成本[137]。

实际上,金属有机化合物的形态分析主要使用三个联用系统:GC 分别与微波诱导等离子体发射光谱(MIP-OES)、ICP-MS,以及电子轰击离子源质谱(EI-MS)联用系统,这三种方法的灵敏度都很高。汞的甲基化衍生物也可以用 AAS 和 AFS 法检测,锡的烷基化衍生物可以用火焰光度检测器(FDP)测定。多元素分析最好用 ICP-MS,特别是配有碰撞反应池时能够提高仪器的选择性;也可以采用 MIP-OES 分析轻质元素,如 S、F、Cl 元素序号之前的元素。ICP-MS 的检测灵敏度能达到1fg水平,而且其同位素的检测特性能够减少基体效应。由于元素具有天然同位素比,因此使用同位素稀释剂与洗脱液混合后进行分析,可以有很好的测量精度,并且不需要使用标准物质,还可以通过缩短测量时间来提高检测精度。因此,气相色谱不仅能与四极杆质谱检测器联用,还能与飞行时间(TOF)质谱检测器或多通道扇形磁场质谱检测器联用。通过减少或消除电离和生成离子对不同同位素的信号干扰,增加色谱峰出峰时间的测量次数等方式来提高测量精度。

可以通过同位素信号强度校准 ICP-MS,一般有以下几种方法。

(1) 同位素比值法,即样品中被测元素的某一特定同位素与加标准后同位素信号强度的比值,通常是在溶液中加入外标。

(2) 同位素稀释法,在样品中加入待测元素的浓缩稳定同位素。将不同功率等离子体条件下提取、衍生和离子化后的待测物质化学转化并引入分析系统中,从而控制标准物质的状态。

同位素比值法主要用于测定铅、锡、硒等挥发性化合物,而同位素稀释法的适用性依赖同位素标准物质,目前,同位素稀释法已被用于测定汞的甲基衍生物和锡的丁基衍生物。同位素比值法的有效性在很大程度上取决于气相色谱仪与ICP-MS之间的连接组件[138]。

在ICP-MS中,样品在室温下以气溶胶的形式引入等离子体区电离,色谱仪和ICP-MS之间的连接器必须防止所分离的化合物蒸气冷凝。通常,用以下几种方式实现电阻加热不锈钢毛细管与检测器的连接。

(1) 直接连接:系统中没有死体积,从而确保了分辨率和灵敏度[139]。

(2) 液体进样器连接:液相色谱洗脱液与进样器固定液混合,形成气溶胶,该方法用于同位素比的流量校准,因为稳定同位素的标准物质常见于水溶液中。洗脱液的稀释降低了该方法的灵敏度,但不会导致由连接器死体积引起的分辨率降低[140]。

(3) 流动室连接:通过扩散室的膜扩散的元素标准物质富集到洗脱液中[141]。

当分析以有机溶液形式引入GC柱中的样品时,需要向洗脱液中加氧,使有机基体燃烧,形成碳氧化物。该方法是为了防止系统积碳,如取样锥污染,提高灵敏度,消除溶剂干扰[142]。

气相色谱在形态分析中的主要缺点是需要进行衍生化,使元素形成挥发性金属衍生物,例如汞。但由于汞生成的产物是有毒的,因此必须确保采用适当的自动制样器和柱载。可以采用可烷基化或还原反应的流动反应器,使反应生成挥发性Pb(Ⅱ)、Hg(Ⅱ)氢化物和Se(Ⅳ)、As(Ⅲ)化合物。另一种新兴的方法是通过水洗将产物转移到毛细管中,在-100℃下进行冷聚焦浓缩,然后释放至毛细管束中分离,该方法能够将分析时间从几小时缩短至几分钟[143]。

砷是较少用气相色谱进行形态分析的元素之一,因为砷是单同位素元素,这就排除了用同位素比值法进行测定,而且砷的衍生化能力较差。GC的目的是从无机砷化合物中形成挥发性氢化物[144],却掩盖了其最初的形态信息。在用MS、AAS、AFS或MIP-OES检测之前,一些有机砷与L-半胱氨酸[145]的衍生化需要通过冷聚焦技术来浓缩毛细管中的分离物质。由于砷的极性,不适合采用GC对其进行形态研究,然而,这一特性使砷在水或水/甲醇混合溶液中具有很好的溶解性,因此,液相色谱是最合适的分离方法。

12.5.2 液相色谱联用技术

液相色谱法是一种无须衍生化就能分离不挥发性和高分子量化合物的方法,由于待测物对固定相和流动相的亲和力不同,因此能够选择性地进行分离,并且可以通过选择流动相的组成或改变固定相来调节该方法的适用性。

形态分析中最常用的方法有以下几种。

尺寸排阻色谱法,也称凝胶渗透色谱法(SEC),是分离不同分子量和不同分子大小化合物的一种方法。由于待测物与固定相之间的空间作用相对较弱,所以不稳定态的金属可以从更稳定的络合物以及稳定的离子化合物中分离出来。然而,待测物与固定相之间的吸附和离子作用会使金属元素的回收率降低 50%,而且这些相互作用也会导致保留时间不稳定[146]。该分离过程可以在水环境中进行,也可以在有机溶剂中进行,因为此方法不具有选择性,所以主要用作多维色谱法的第一阶段[147]。

离子色谱法(IC)主要是以离子的形式在水溶液中分离物质。该方法使用的流动相中含有大量的盐,用来保持 pH 稳定并决定待测物组分的保留顺序。离子色谱法仅能够分离阳离子或阴离子,未被流动相分离的阳离子或阴离子将被洗脱,在下一步二维色谱中载入适当的离子交换柱中[148]。

反相液相色谱法(RPLC)能够分离具有不同疏水性和极性的物质,具有良好的选择性。该方法使用的流动相中含有有机溶剂和少量无机盐[149],待测物质的疏水性特性是有效使用该方法的基础,但是,分离之前带电物质必须先转化成中性物质,如在流动相中加入相应的反离子。

离子对液相色谱法(IPLC)是 RPLC 的一种,流动相中的低浓度无机盐与反离子形成竞争或将固定相改性[150],因而保持了良好的选择性。但是,这种方法不适用于常规分析,因为重现性较差。

正相液相色谱法(NPLC)和亲水作用液相色谱法(HILIC)是基于待测物质与固定相之间极性不同来进行分离的,待测物质通过离子作用、偶极作用或氢键结合。使用极性有机溶剂、水及其混合物与挥发性盐一起控制 pH[151],HILIC 可以分离 RPLC 所不能分离的极性化合物。

在痕量形态分析中,液相色谱常与灵敏度高的 MS 技术联用,通过使用短软管将 HPLC、SEC、IC、IPLC 等各类色谱柱的出口与分配器连接,通常采用 HPLC-ICP-MS 联用仪进行金属元素形态的常规测定。如果流动相中含有有机溶剂,如在 RPLC 中,则必须将气溶胶冷却,通氧使碳燃烧成碳氧化物,接口使用铂锥[152]。通常,还需要用较小的死体积来代替雾室,以提高分辨率。RPLC-ICP-MS 的另一个缺点是梯度洗脱的重复性较差,这是由于洗脱液中有机溶剂的含量会随着时间的增加而增加。常规分析一般用 IC-ICP-MS,筛选分析可以采用 SEC-ICP-MS 进行[153]。利用色谱质谱联用仪进行基础研究,取决于所研究化合物的类型和性质。目前,ICP-MS 越来越多的用作研究金属元素同位素,采用同位素稀释法获得更高的测量精度。

在定量分析之前,必须通过比较样品的保留时间与标准保留时间来验证化合物的结构,可以通过加内标来实现。将保留时间与所记录的标准保留时间进行比较,或者使用选择性检测模式,这一过程通常在 HPLC-ESI-MS 中进行,即将液相色谱与 ESI-MS 联用。由于挥发性有机溶剂能够提高检测效率,因此 ESI-MS 也常与 RPLC 联用[154],目前已有广泛应用,例如,鉴定动植物组织中的砷代谢产物的结构。这种方法可以用植物螯合剂(PC)和谷胱甘肽[$AsPC_3$, $GSAsPC_2$, $As(PC_2)_2$]混合物测定砷配合物的结构[155],这是因为在 ESI 电离的色谱条件下,含有有机溶剂和三氟乙酸的流动相能够分离肽及其稳定配合物。然而,鉴定 IC-ICP-MS 所分离的化合物的结构并不是那么简单,因为电喷雾离子化过程不能耐受洗脱液中大量的盐,在这种情况下,需要从离子色谱柱中收集洗脱液,先进行脱盐,然后将其导入检测器[156-157]。

色谱峰纯度经常被忽视,但仍是评估系统性能的重要参数。通过测量谱图上两个特征峰的信号强度的比($\lambda_{1max}/\lambda_{2max}$, m_1/z_1 和 m_2/z_2 的比)来评估,如果是纯色谱峰,这些信号强度比值应该是恒定的。使用色谱软件可以准确、快速地评估色谱峰纯度。因此,在每个连续样品的分析过程中都应该检查色谱峰的纯度。

在砷的形态分析中,液相色谱法因为不需要衍生化,看似比气相色谱法简单但也存在一些潜在的误差来源。色谱法的选择需要对所分离化合物的溶解性有初步认识,大多数已知的含砷化合物都是极性的,这不仅有利于离子化过程,也可用于离子交换色谱。这些化合物一般都是弱酸,很容易去质子化,当生成 As(V)有机衍生物时,去质子化趋势会降低,例如砷(V)酸的 pK 为 2.2,而 MMA 的 pK 增大到 3.6;砷(Ⅲ)酸的酸性较弱,pK_a 为 9.22,但其甲基衍生物则有所增强,DMA(Ⅲ)的 pK 为 6.2。砷化合物的解离常数为 2.2~12.2,如此宽的 pK 值是因为它们是多聚酸,并且这一显著特性是因为砷化合物的去质子化程度会影响其与碱性固定相相互作用的强度,即方法的保留性和选择性[158-159]。流动相的 pH 必须使分离的化合物以至少 90%的效率电离(pK±1),由于该物质对固定相的亲和力较大,所以必须使用高离子强度的流动相。砷化合物的阴离子交换色谱通常使用弱离子交换剂,如叔铵盐和 pH 为 6.0 或 8.5 的流动相,主要通过四乙基氢氧化铵[160]、磷酸盐缓冲溶液或碳酸盐缓冲溶液[161]来调节 pH。由于环境中的碱性对砷(Ⅲ)酸的电离来说是不够的,因此需要调节 pH,但是,调节过程中会通过沉淀/共沉淀作用减少其他形态的损失。在这种情况下,色谱柱死体积中的砷(Ⅲ)酸与不带电化合物,如 AsB、AsC、砷糖 B,将一起被洗脱,因为它们是碱性的。

阴离子交换色谱是一种分析地下水的有效方法,其局限在于地表水样品中化合物的分离。生物样品的分析也应通过阳离子交换色谱进行,因为 AsB、AsC、砷糖等砷化合物结构中大多存在 As=O 基团,容易质子化。在 pH 约为 3 的溶液中,砷(Ⅴ)酸和砷糖 A、砷糖 C 均不发生质子化反应,从色谱柱死体积中被洗脱。而且,由于流动相的 pH 接近许多化合物的 pK 值,所以它们的色谱峰相对较宽,出现不对称拖尾,有时甚至是双峰。阳离子交换色谱的分辨率虽不如阴离子交换色谱的分辨率,但该技术可用于检测牡蛎和其他海洋生物中大量存在的无毒 AsB,测定其含量可以制备一个砷化合物的物质平衡,并确定其形态中的少部分无机金属。

用丙二酸和四乙基氢氧化铵的混合物作为砷化合物质子化和去质子化的反离子,可以将 HPLC 的适用范围扩大到 IPLC。该方法分辨率最佳,但测量不确定度相当大,因为固定相会随着时间的变化而改变[162]。另一个缺点是与中性化合物,如 MMA(Ⅴ)与砷(Ⅴ)酸,生成离子对而产生共洗脱的风险,由于平衡时间过长,此方法不能采用梯度洗脱[163]。事实上,上述方法都不是通用的,也没有一种方法能分离十几种具有不同酸性和碱性的含砷化合物,因此,特别是当待测金属具有复杂的生物基质时,需要同时使用几种方法,例如测定致癌性的砷(Ⅲ)酸,可与下列物质进行共洗脱。

(1) 在阴离子交换色谱中与 DMA、AsB、砷糖 B 和四甲基胂离子共洗脱,可用于分析含多种砷糖的水和海洋植物。

(2) 在阳离子交换色谱中与砷糖 A、砷糖 C 和砷糖 D 共洗脱,可用于分析尿液和海产品。

(3) 在 IPLC 中与 AsC 共洗脱,可作为参考方法。

(4) 在 SEC 中与 AsC 和 DMA(Ⅲ) 共洗脱,可用于分离高分子量化合物。

针对有特殊要求的形态分析,选择适合的方法时,需要仔细验证方法的选择性,很少用直接分析标准物质混合物的方法。显然,尽管迄今为止已鉴定出 20 多种砷糖,但只有少数砷糖是商业上可用的。所以,考虑到色谱峰纯度,建议采用间接方法,主要有两种情形:在相对简单的情形下,延长洗脱梯度的步长,并检查峰是否分裂;在耗时更长的情形下,从柱中收集对应于不同峰的馏分,浓缩,然后用另一种色谱法分离。在这种情况下,该过程可以是自动化的,并且特别适合分析从生物组织中获得的提取物,由于消除了基体效应的影响,因此具有更好的测量精度和灵敏度。

对于洗脱时间很短,接近死时间的化合物分离,保持方法的选择性是比较困难的。在这种情况下,需要对设置的色谱条件进行初步检查,包括所分析样品的组成。例如,在阴离子交换色谱中使用典型的洗脱剂(pH=8.5),目的是促进分离化合物的解离。如果忽略了在中性溶液中导入柱中的物质达到酸/碱平衡所需的时间,就会导致它们在死体积中被洗脱。例如,对于MMA(V),其连续解离常数pK_1为3.6和pK_2为8.22[164]。

分析高离子强度的溶液时,也会出现保留时间减少的现象。例如用浓度为1mM的磷酸盐缓冲溶液进行梯度洗脱,当氯化钠含量为(0.1~1)M时,待分离的化合物与固定相没有充分的相互作用。该问题最常见于尿液中含有超过100mM浓度的氯化钠的分析。待测溶液的高盐度和溶剂化在RPLC和存在与固定相的竞争作用的HILIC中也存在同样问题,因此通常需要脱盐,可由SPE以及其他方法进行[165]。

影响分辨率的另一个因素是常见的阴离子表面活性剂SDS。SDS经常存在于生物材料的提取物中,用于将疏水性蛋白质转移到水溶液中,或用于凝胶电泳法释放蛋白质。然而,在某些情况下,有意将SDS添加到胶束液相色谱的流动相中,可以从无机砷中分离出砷结合肽。在该情形下,添加SDS可使砷(V)酸和氯化钠分离,在ICP-MS中就检测不到同位素干扰[166]。

如果待测物质不能充分地与固定相作用,则它们会接近死时间而被洗脱;相反,如果相互作用过强则很难从柱中完全洗脱。通过评估待测化合物的回收率可以确定这一现象。回收率被理解为洗脱物质的量与装入柱中的物质的量的比例。回收率可以评估是否选择了适用、正确的色谱分离条件,并可以确定是否还有其他未知物质吸附在固定相表面。如果发现某化学成分回收率较低,则有理由假定该成分以某些形式吸附在柱上,并且只有改变洗脱剂(pH)才能洗脱。例如,在标准物质中检测到未知的砷化合物[167],或者在羊尿中发现的另一种新化合物2-二甲基胂硫代乙酸,[168]都是该类研究结果。

低回收率也可能是由于固定相的表面积过大而引起待测物的量变,这是痕量分析中的一个普遍倾向。例如,青蛙体内的含砷量为mg/kg水平,从其体内提取的砷(V)酸及其甲基化衍生物MMA、DMA、三甲胂氧化物(TMAO)和砷(Ⅲ)酸的回收率为101%~104%,而含量以ng/kg计的鱼体内提取物的回收率只有20%[169]。

当分析样品中组分含量明显不同时,有必要正确绘制相对窄的浓度范围曲线,这种"多级校准"是一种提高测量准确度的方法,因为曲线斜率随着元素检出量的减少而增加。

对于与固定相具有较强作用的化合物,可以通过改变洗脱剂的强度和梯度洗脱来提高回收率,从而确保良好的分辨率。在离子色谱中,有效的解决方案包括改变洗脱液的pH,弱化分析物与固定相的相互作用,以及官能团改性。然而,一个重要的工艺优化标准是防止在柱上出现强烈的吸附形式,如由MMA(V)形成的聚合阴离子[170]或许存在微溶形式。不适宜的色谱条件不仅延长停留时间,而且降低回收率,增加标准曲线的斜率。另一个常见问题是宽峰,特别是在等度洗脱中,宽峰会阻碍分析物的结合。进行砷形态分析需要使用标准物质,通过信号的峰面积来确定某种形态的含量也是主要的误差来源。

待测化合物的洗脱效率有时会随着分析柱的数量而增加,这时,同等量的待测物会产生更大的峰面积,这是由于被吸附的分析物逐渐使固定相改性;造成的另一个影响是在盲样分析过程中可能出现待测化合物的峰,即使延长柱子的冲洗时间,也不能完全消除改性的影响,使结果出现正误差[171]。类似的误差还可以由基体效应产生,但是,通过分析不同稀释度的样品,可以比较容易地识别这类误差。

前文所述的液相色谱分离元素形态分析方法,会让人觉得形态分析很容易具有准确和精确的检测结果。然而,这仅仅是一部分事实,因为前文主要讨论样本的分离,而未涉及在形态分析过程中的其他阶段。虽然其他阶段的全面概述超出了本书的范围,但是我们可以简要地说明它们对测试结果的重要性。例如,以均匀的牡蛎粉末标准物质中大约 30mg/kg AsB 作为分析物进行实验室间研究[172],发现存在 AsB 的快速光氧化和未知砷与 AsB 的共洗脱的现象。其他形态砷化合物的测定也存在较大误差,如亚砷酸(Ⅲ)被氧化,砷酸(Ⅴ)低于大多数方法的检出限等。AsC 只在单一实验室中进行了鉴别和检测,两种砷糖在多个实验室中进行了检测时,即使矿物质中砷的总量是确定的,但在提取后的沉积物中砷的含量检测结果却出现显著差异。

这些例子都清楚地说明了形态分析的困难程度,需要针对每个案例进行分析,其中,典型误差表(表 12-3)是很有用的。但是应该知道,每个分析过程都是极其复杂的,化合物的稳定性会导致许多问题,分离和检测水平在很大程度上取决于实验室设备、HPLC 柱材料修正及其填充物的改进。不仅如此,熟知所研究的元素形态,将会为出现的任何问题提供重要的参考信息。

表 12-3 形态分析中的典型误差

样品制备与储存
采集的样品量不足,例如低于平均精度或均匀性水平;
用于样品储存或转移到溶液时的清洗方法和调节容器的方法不当;
在样品中加入过量的标准物质,改变了样品基体的物理化学性质,引起物质失衡;
加入的内标与待测化合物相互作用,形成络合、离子对、氧化还原反应、加成或水解反应;
在不可控 pH 条件下,或压力和温度升高时萃取,引起水解或烷基化反应;
未从生物材料中去除活性酶蛋白,导致化合物的分解失控;
储存样品的光照、温度、气体等条件不恰当

续表

ICP-MS 检测
在加入形态不匹配的无机标准物质时,评价待测元素的回收率; 使用不恰当的校正方程; 加入的内标和待测元素存在互相干扰; 未考虑基体对测量的影响,例如,未对样品存在的大分子进行初步半定量分析; 在样品中加入干扰物质,如盐酸或高氯酸
HPLC-ICP-MS 检测
未根据样品含量变化调整方法的分辨率导致选择性不够,未经标准物质校准或验证色谱方法; 未检查不同稀释度下样品的基体效应; 离子色谱中流动相的温度失控; 流动相 pH 控制不够; 未能控制干扰化合物的保留时间; 使用普通试剂而不是标准物质来绘制校准曲线; 未检查色谱柱中元素的回收率; 使用较宽范围而不是多级校准的标准曲线; 未能控制色谱峰纯度; 基于过量加入的标准物质来验证化合物结构

12.5.3 超临界流体色谱技术

超临界流体色谱(SFC)兼有液相色谱和气相色谱大部分的优点,方法选择性取决于所使用的流体类型,常用的流体有中性超临界二氧化碳(Ⅳ)[173],该方法的分离效率接近气相色谱法,化合物样品不需要挥发性的,但是应该有足够的极性,以便溶解在水或甲醇中,可以省略 GC 中必要的衍生化过程。SFC-ICP-MS 联用中连接模块的主要任务是确保适当的温度,防止输送到等离子体之前流动相发生相变[174]。必须进行洗脱液分离,因为随着温度升高,超临界流体的体积会增加[175],可以在燃烧器之前或燃烧器中进行分流。如果在前者情况下,更容易控制流动相的温度,但是洗脱液热损失较大;后者的情况通常具有更好的灵敏度,但测量精度较差。SFC-ICP-MS 很少用于常规分析,在低分子化合物的基础研究中,该方法可以取代 GC 和 HPLC,在锡[176]、铅[177]、锑、汞、砷[178]等有机金属烷基衍生物的形态分析中具有重要地位。

12.5.4 毛细管电泳技术

毛细管电泳(CE)是分离水中离子化合物的技术。化合物在电场中沿熔融石英制成的毛细管以不同速度迁移,它们的速度取决于电泳迁移率,在很大程度上与离子的质荷比(m/z)有关。最常用的是毛细管区带电泳(CZE),离子随电渗流向负电极或正电极移动(反向 CZE)[179]。分离的效果取决于电解质类型,电解质应具有适当的离子流强度,能够给待测离子提供"泵吸"功率。该方法使用的电解质一般是缓冲液,这有助于维持溶液的 pH,保证分析物以离子形式存在。CE 可以通过连接组件与 ICP-MS 联用,连接组件关闭电路并补充毛细管中的洗脱液,使其适用于 ICP 的喷雾条件[180],这种类型的联用常用于砷化合物的分析,因为它可以同时分离带负电和带正电的化合物。CE 的另一个优点是样品量小,仅约为 30nL,但降低了方法灵敏度。砷化合物,最典型的是亚砷(Ⅲ)酸和砷(Ⅴ)酸,以及 MMA 和 DMA,在阳离子表面活性剂存在条件下,通常在弱碱性环境的水中分离,旨在改变毛细管表面的电荷,从而改变电渗流方向[181-182]。可分离的形态数量主要取决于连接 CE 和 ICP-MS 的洗脱液补充模块。使用一个可以轻微抽吸辅助液体的连接器,保持直流式毛细管,这样可以分离检测 AsB、AsC 和苯基胂酸(Ⅲ)酸[183]。以上设备已成功应用于饮用水[184]、尿液和土壤提取物[185]中砷的形态分析。

CE 与 ICP-MS 联用技术比较复杂,易造成微管中毛细管阻塞,降低测量准确度,另一个问题是需要将毛细管引到系统外部,这会增加毛细管内的温度,这些操作上的问题限制了该方法在基础研究以及常规分析中的应用,如无机砷化合物与组氨酸[186]的相互作用。

12.6 小结

当代形态分析主要有两个方向:一是新元素形态的研究,二是将现代化学计量理论引入实际应用中。前者研究的目的是释放和鉴别以前从未描述过的、迄今为止尚未阐明的生命过程中所涉及的物质。后者需要具备良好的实验室经验,包括实际生活中样品的常规分析,即元素已有形态分析。对定量结果的正确分析尽管看似很重要,但并不总是如此。

定量形态分析主要利用 HPLC/GC-ICP-MS 联用技术,它们具有良好的分离效果、高灵敏度及检测器的同位素特异性。定性分析可采用 ESI-MS 与 RPLC 联用,并且越来越多地与 HILIC 联用,有时这会成为常量化合物定量分析的替代方法。

待测元素的多样性使形态分析的过程变得复杂,这些元素在各种环境材料中表现出不同的物理化学性质,其特征在于物质、来源和含量不同。

这种复杂的分析过程需要认真设计,考虑到所有可能的误差来源,包括从样品选取的代表性到正确处理结果及评估不确定度。只有充分利用现有的专业知识和当前的认知水平,才能使形态分析在无机生物化学的发展中占据重要地位,而且形态分析是一个基本的研究工具。

缩略语

AAS	原子吸收光谱法	IPLC	离子对液相色谱
AdTSV	吸附转移溶出伏安法	IR	红外分析法
AFS	原子荧光光谱法	LLE	液-液萃取
AsB	砷甜菜碱	MEKC	胶束电动力学色谱
ASE	加速溶剂萃取法	MIP-OES	微波诱导等离子体发射光谱法
ASV	吸附溶出伏安法		
CE	毛细管电泳	MMA	一甲基胂酸
CEC	胶束电动色谱法	MS	质谱分析法
CRM	有证标准物质	NPLC	正相液相色谱
CZE	毛细管区电泳	OES	发射光谱法
DMA	二甲基胂酸	PLE	加压液相萃取
EI	电子电离	PSA	电位溶出伏安法
ESI	电喷雾电离	RPLC	反相液相色谱
FPD	火焰光度检测器	SDME	单滴微萃取
GC	气相色谱法	SDS	十二烷基硫酸钠
HILIC	亲水作用液相色谱	SEC	尺寸排除色谱
HPLC	高效液相色谱	SFC	超临界流体色谱
HPLC-ESI	高性能液相色谱-电喷雾电离	SFE	超临界流体萃取
		SLE	固-液萃取
MS	质谱分析法	SPE	固相萃取
IC	离子色谱	SPME	固相微萃取
ICP-MS	电感耦合等离子质谱法	Tris	三(羟甲基)氨基甲烷

参考文献

[1] Szpunar, J., Łobiński, R.: Hyphenated Techniques in Speciation Analysis. RSC, Cambridge (2003)

[2] Cornelis, R.: Introduction. In: Cornelis, R. (ed.) Handbook of Elemental Speciation Ⅱ. Species in the Environment, Food, Medicine and Occupational Health. Wiley, West Sussex (2005)

[3] Vela, N.P., Olson, L.K., Caruso, J.A.: Elemental speciation with plasma mass spectrometry. Anal. Chem. 65, 585-597 (1993)

[4] Zoorob, G.K., McKiernan, J.W., Caruso, J.A.: ICP-MS for elemental speciation studies. Mikrochim. Acta 128, 145–168 (1998)

[5] B'Hymer, C., Brisbin, J.A., Sutton, K.L., Caruso, J.A.: New approaches for elemental speciation using plasma mass spectrometry. Am. Lab. 32, 17–39 (2000)

[6] Caruso, J.A., Klaue, B., Michalke, B., Rocke, D.M.: Group assessment: elemental speciation. Ecotoxicol. Environ. Saf. 56, 32–44 (2003)

[7] Montes-Bayon, M., DeNicola, K., Caruso, J.A.: Liquid chromatography-inductively coupled plasma mass spectrometry. J. Chromatogr. A 1000, 457–476 (2003)

[8] Caruso, J.A., Montes-Bayon, M.: Elemental speciation studies-new directions for trace metal analysis. Ecotoxicol. Environ. Saf. 56, 148–163 (2003)

[9] Smedley, P.L., Kinniburgh, D.G.: A rewiev of the sources, behaviour and distribution of arsenic in natural waters. Appl. Geochem. 17, 517–568 (2002)

[10] Połeć-Pawlak, K., Abramski, K., Hołdak, M., Latour, T., Drobnik, M., Lulek, J., Jarosz, M.: Badanie specjacji arsenu w naturalnych wodach mineralnych metoda HPLC-ICP-MS. LAB 9, 28–32 (2004)

[11] Morita, M., Edmonds, J.S.: Determination of arsenic species in environmental and biological samples. Pure Appl. Chem. 64, 575–590 (1992)

[12] Murer, A.J.L., Abildtrup, A., Poulsen, O.M., Christensen, J.M.: Effect of seafood consumption on the urinary level of total hydride-generating arsenic compounds. Instability of arsenobetaine and arsenocholine. Analyst 117, 677–680 (1992)

[13] Tossell, J.A.: Theoretical studies on arsenic oxide and hydroxide species in minerals and in aqueous solution. Geochim. Cosmochim. Acta 61, 1613–1623 (1997)

[14] Dousova, B., Martaus, A., Filippi, M., Kolousek, D.: Stability of arsenic species in soils contaminated naturally and in an anthropogenic manner. Water Air Soil Pollut. 187, 233–241 (2008)

[15] Mattusch, J., Wennrich, R., · Schmidt, A.C., Reisser, W.: Determination of arsenic species in water, soils and plants. Fresenius J. Anal. Chem. 366, 200–203 (2000)

[16] Agusa, T., Takagi, K., Iwata, H., Tanabe, S.: Arsenic species and their accumulation features in green turtles (Chelonia mydas). Marine Poll. Bull. 57, 782–789 (2008)

[17] Schmidt, A.C., Mattusch, J., Wennrich, R.: Distribution of arsenic species in different leaf fractions-an evaluation of the biochemical deposition of arsenic in plant cells. Microchim. Acta 151, 167–174 (2005)

[18] Lin, C.J., Wu, M.H., Hsueh, Y.M., Siu-Man, S.S., Cheng, A.L.: Tissue distribution of arsenic species in rabbits after single and multiple parenteral administration of arsenic trioxide: tissue accumulation and the reversibility after washout are tissue-selective. Cancer Chemother. Pharmacol. 55, 170–178 (2005)

[19] Kurttio, P., Komulainen, H., Hakala, E., Kahelin, H., Pekkanen, J.: Urinary excretion of arsenic species after exposure to arsenic present in drinking water. Arch. Environ. Contam. Toxi-

col. 34, 297–305 (1998)

[20] Sun, Y.C., Chen, Y.J., Tsai, Y.N.: Determination of urinary arsenic species using an on-line nano-TiO$_2$ photooxidation device coupled with microbore LC and hydride generation-ICP-MS system. Microchem. J. 86, 140–145 (2007)

[21] Sur, R., Dunemann, L.: Method for the determination of five toxicologically relevant arsenic species in human urine by liquid chromatography-hydride generation atomic absorption spectrometry. J. Chromat. B 807, 169–176 (2004)

[22] Vetter, J.: Arsenic content of some edible mushroom species. Eur. Food Res. Technol. 219, 71–74 (2004)

[23] Pizarro, I., Gómez, M., Palacios, M.A., Cámara, C.: Evaluationof stability of arsenic species in rice. Anal. Bioanal. Chem. 376, 102–109 (2003)

[24] B'Hymer, C., Caruso, J.A.: Arsenic and its speciation analysis using high-performance liquid chromatography and inductively coupled plasma mass spectrometry. J. Chromat.A 1045, 1–13 (2004)

[25] Venupopal, B., Luckey, T.D.: Metal Toxicity inMammals 2. Plenum Press, New York (1978)

[26] Ritsema, R., Dukan, L., Navarro, T.R.I., van Leeuwen, W., Oliveira, N., Wolfs, P., Lebret, E.: Speciation of arsenic compounds in urine by LC-ICP MS. Appl. Organomet. Chem. 12, 591–599 (1998)

[27] Penrose, W.R.: Arsenic in the marine and aquatic environments: analysis, occurrence, and significance. Crit. Rev. Environ. Control. 5, 465–482 (2000)

[28] Mandal, B.K., Suzuki, K.T.: Arsenic round the world: a review. Talanta 58, 201–235 (2002)

[29] Dembitsky, V.M., Rezanka, T.: Natural occurrence of arseno compounds in plants, lichens, fungi, algae species, and microorganisms. Plant Sci. 165, 1177–1192 (2003)

[30] Brown, J., Kitchin, K., George, M.: Dimethylarsinic acid treatment alters six different rat biochemical parameters: relevance to arsenic carcinogenesis. Teratog. Carcinog. Mutagen. 17, 71–84 (1997)

[31] Marafante, E., Vahtner, M., Dencker, L.: Metabolism of arsenocholine in mice, rats and rabbits. Sci. Total Environ. 34, 223–240 (1984)

[32] Foa, V., Colombi, A., Maroni, M., Buratti, M.: Arsenic. In: Alessio, L., Berlin, A., Bori, M., Roi, R. (eds.) Biological Indicators for the Assessment of Human Exposure to Industrial Chemicals. CEC ISPRA, Luxembourg(1987)

[33] Haas, K., Feldmann, J.: Sampling of trace volatile metal(loid) compounds in ambient air using polymer bags: a convenient method. Anal. Chem. 72, 4205–4211 (2000)

[34] Pecheyran, C., Quetel, C.R., Lecuyer, F.M.M., Donard, O.X.F.: Simultaneous determination of volatile metal (Pb, Hg, Sn, In, Ga) and nonmetal species (Se, P, As) in different atmospheres by cryofocusing and detection by ICP MS. Anal. Chem. 70, 2639–2645 (1998)

[35] Ahmed, R., Stoeppler, M.: Decomposition and stability studies of methylmercury in water using cold vapour atomic absorption spectrometry. Analyst 111, 1371–1374 (1986)

[36] Guy, P.M.: Sampling of Heterogenous and Dynamic Material Systems. Elsevier, Amsterdam (1992)

[37] Kateman, G.: Chemometrics-Sampling Strategies. Springer, Berlin (1987)

[38] Hoenig, M.: Preparation steps in environmental trace element analysis-facts and traps. Talanta 54, 1021–1038 (2001)

[39] Caruso, J.A., Heitkemper, D.T., B'Hymer, C.: An evaluation of extraction techniques for arsenic species from freeze-dried apple samples. Analyst 126, 136–140 (2001)

[40] Roig-Navarro, A.F., Martinez-Bravo, Y., Lopez, F.J., Hernandez, F.: Simultaneous determination of arsenic species and chromium(VI) by high-performance liquid chromatography-inductively coupled plasma-mass spectrometry. J. Chromatogr. A 912, 319–327 (2001)

[41] Challenger, F.: Biosynthesis of organometallic and organometalloidal compounds. In: Organometals and organometalloids: occurrence and fate in the environment. In: Brinkman, F.E., Bellama, J.M. (eds.) ACS Symposium Series 82, pp. 1-22. American Chemical Society, Washington, DC (1978)

[42] Yang, Y., Bowadt, S., Hawthorne, S.B., Miller, D.J.: Subcritical water extraction of polychlorinated biphenyls from soil and sediment. Anal. Chem. 67, 4571–4576 (1995)

[43] Tyson, J.F.: High-performance, flow-based, sample pre-treatment and introduction procedures for analytical atomic spectrometry. J. Anal. At. Spectrom. 14, 169–178 (1999)

[44] Kim, M.J.: Separation of inorganic arsenic species in groundwater using ionexchange method. Bull. Environ. Contam. Toxicol. 67, 46–51 (2001)

[45] Day, J.A., Montes-Bayon, M., Vonderheide, A.P., Caruso, J.A.: A study of method robustness for arsenic speciation in drinking water samples by anion exchange HPLC-ICP-MS. Anal. Bioanal. Chem. 373, 664–668 (2002)

[46] Edwards, M., Patel, S., McNeill, L., Chim, H., Frey, M., Eaton, A.D., Antweiler, R.C., Taylor, H.E.: Considerations in As analysis and speciation: a modified field technique can quantify particulate As, soluble As(III), and soluble As(V) in drinking water. J. Am. Water Works Assoc. 90, 103–114 (1998)

[47] Larsen, E.H., Pritzl, G., Hansen, S.H.: Speciation of eight arsenic compounds in human urine by high-performance liquid chromatography with inductively coupled plasma mass spectrometric detection using antimonate for internal chromatographic standardization. J. Anal. At. Spectrom. 8, 557–563 (1993)

[48] Daus, B., Weiss, H., Mattusch, J., Wennrich, R.: Preservation of arsenic species in water samples using phosphoric acid-limitations and long-term stability. Talanta 69, 430–434 (2006)

[49] Garcia-Manyes, S., Jimenez, G., Padro, A., Rubio, R., Rauret, G.: Arsenic speciation in contaminated soils. Talanta 58, 97–109 (2002)

[50] Le, S.X.C., Cullen, W.R., Reimer, K.J.: Speciation of arsenic compounds in some marine organisms. Environ. Sci. Technol. 28, 1598–1604 (1994)

[51] Jokai, Z., Hegoczki, J., Fodor, P.: Stability and optimization of extraction of four arsenic spe-

cies. Microchem. J. 59, 117–124 (1998)

[52] Cornelis, R., Caruso, J., Crews, H., Heumann, K.: Handbook of Elemental Speciation: Techniques and Methodology. Wiley, West Sussex (2003)

[53] Estela, J.M., Tomás, C., Cladera, A., Cerda, V.: Potentiometric stripping analysis: a review. Crit. Rev. Anal. Chem. 25, 91–141 (1995)

[54] Florence, T.M.: Electrochemical approaches to trace element speciation in waters; a review. Analyst 111, 489–505 (1986)

[55] Evans, E.H., Giglio, J.J.: Interferences in inductively coupled plasma mass spectrometry. A review. J. Anal. At. Spectrom. 8, 1–18 (1993)

[56] Bulska, E., Krata, A.: Instrumentalne metody spektralne stosowane w analizie próbek środowiskowych. In: Richling, A., Lechnio, J. (eds.) Z problematyki funkcjonowania krajobrazów nizinnych. Wydawnictwo Uniwersytetu Warszawskiego, Warszawa (2005)

[57] Krata, A., Bulska, E.: Wykorzystanie reakcji w fazie gazowej do eliminacji interferencji izobarycznych w ICP MS. Analityka 2, 25–29 (2002)

[58] Tanner, S.D.: Characterization of ionization and matrix suppression in inductively coupled 'cold' plasma mass spectrometry. J. Anal. At. Spectrom. 10, 905–921 (1995)

[59] Hinojosa, R.L., Marchante-Gayón, J.M., García Alonso, J.I., Sanz-Medel, A.: Quantitative speciation of selenium in human serum by affinity chromatography coupled to post-column isotope dilution analysis ICP-MS. J. Anal. At. Spectrom. 18, 1210–1216 (2003)

[60] Wang, R.Y., Hsu, Y.L., Chang, L.F., Jiang, S.J.: Speciation analysis of arsenic and selenium compounds in environmental and biological samples by ion chromatography-inductively coupled plasma dynamic reaction cell mass spectrometer. Anal. Chim. Acta 590, 239–244 (2007)

[61] Thaylor, H.E.: Inductively Coupled Plasma-Mass Spectrometry: Practices and Techniques. Academic, San Diego (2001)

[62] Polya, D.A., Lythgoe, P.R., Abou-Shakra, F., Gault, A.G., Brydie, J.R., Webster, J.G., Brown, K.L., Michailidis, K.M.: IC-ICP-MS and IC-ICP-HEX-MS determination of arsenic speciation in surface and groundwaters: preservation and analytical issues. Mineral. Mag. 67, 247–261 (2003)

[63] Du, Z., Houk, R.S.: Attenuation of metal oxide ions in inductively coupled plasma mass spectrometry with hydrogen in a hexapole collision cell. J. Anal. At. Spectrom. 15, 383–388 (2000)

[64] Darrouzes, J., Bueno, M., Lespes, G., Holeman, M., Potin-Gautier, M.: Optimisation of ICPMS collision/reaction cell conditions for the simultaneous removal of argon based interferences of arsenic and selenium in water samples. Talanta 71, 2080–2084 (2007)

[65] Niemela, M., Peramaki, P., Kola, H., Piispanen, J.: Determination of arsenic, iron and selenium in moss samples using hexapole collision cell, inductively coupled plasma-mass spectrometry. Anal. Chim. Acta 493, 3–12 (2003)

[66] Xie, Q., Kerrich, R., Irving, E., Liber, K., Abou-Shakra, F.: Determination of five arsenic

species in aqueous samples by HPLC coupled with a hexapole collision cell ICP-MS. J. Anal. At. Spectrom. 17, 1037-1041 (2002)

[67] Nakazato, T., Tao, H., Taniguchi, T., Isshiki, K.: Determination of arsenite, arsenate, and monomethylarsonic acid in seawater by ion-exclusion chromatography combined with inductively coupled plasma mass spectrometry using reaction cell and hydride generation techniques. Talanta 58, 121-132 (2002)

[68] Sloth, J.J., Larsen, E.H.: Application of inductively coupled plasma dynamic reaction cell mass spectrometryfor measurement of selenium isotopes, isotope ratios and chromatographic detection of selenoamino acids. J. Anal. Atom. Spectrom. 15, 669-672 (2000)

[69] Leonhard, P., Pepelnik, R., Prange, A., Yamada, N., Yamada, T.: Analysis of diluted seawater at the ng L level using an ICP-MS with an octopole reaction cell. J. Anal. At. Spectrom. 17, 189-196 (2002)

[70] Feldmann, I., Jakubowski, N., Thomas, C., Stuewer, D.: Application of a hexapole collision and reaction cell in ICP-MS Part II: analytical figures of merit and first applications. Fresenius J. Anal. Chem. 365, 422-428 (1999)

[71] Brown, R.J.C., Yardley, R.E., Brown, A.S., Milton, M.J.T.: Sample matrix and critical interference effects on the recovery and accuracy of concentration measurements of arsenic in ambient particulate samples using ICP-MS. J. Anal. At. Spectrom. 19, 703-705 (2004)

[72] Dufailly, V., Noel, L., Guerin, T.: Optimisation and critical evaluation of a collision cell technology ICP-MS system for the determination of arsenic in foodstuffs of animal origin. Anal. Chim. Acta 611, 134-142 (2008)

[73] Ritsema, R., Dukan, L., Roig, I., Navarro, T., Van Leeuwen, W., Oliveira, N., Wolf, P., Lebret, E.: Speciation of arsenic compounds in urine by LC-ICP MS. Appl. Organomet. Chem. 12, 591-599 (1998)

[74] Wu, J., Mester, Z., Pawliszyn, J.: Speciation of organoarsenic compounds by polypyrrolecoated capillary in-tube solid phase microextraction coupled with liquid chromatography/electrospray ionization mass spectrometry. Anal. Chim. Acta 424, 211-222 (2000)

[75] Bouyssiere, B., Szpunar, J., Potin-Gautier, M., Łobin'ski, R.: Sample preparation techniques for elemental speciation studies. In: Cornelis, R. (ed.) Handbook of Elemental Speciation. Techniques and Methodology. Wiley, West Sussex (2003)

[76] Hardy, S., Jones, P.: Development of a capillary electrophoretic method for the separation and determination of trace inorganic and organomercury species utilizing the formation of highly absorbing water soluble dithizone sulphonate complexes. J. Chromatogr. A 791, 333-352 (1997)

[77] Rapsomanikis, S., Craig, P.J.: Speciation of mercury and methylmercury compounds in aqueous samples by chromatography-atomic absorption spectrometry after ethylation with sodium tetraethylborate. Anal. Chim. Acta 248, 563-567 (1991)

[78] Potin-Gautier, M., Gilon, N., Astruc, M., De Gregori, I., Pinochet, H.: Comparison of selenium extraction procedures for its speciation in biological materials. Int. J. Environ. Anal. Chem.

67, 15-25 (1997)

[79] Gómez-Ariza, J.L., Sánchez-Rodas, D., Giráldez, I., Morales, E.: Comparison of biota sample pretreatments for arsenic speciation with coupled HPLC-HG-ICP-MS. Analyst 125, 401-407 (2000)

[80] Schmidt, A.C., Reisser, W., Mattusch, J., Popp, P., Wennrich, R.: Evaluation of extraction procedures for the ion chromatographic determination of arsenic species in plant materials. J. Chromatogr. A 889, 83-91 (2000)

[81] Richter, B.E., Jones, B.A., Ezzell, J.L., Porter, N.L., Avdalovic, N., Pohl, C.: Accelerated solvent extraction: a technique for sample preparation. Anal. Chem. 68, 1033-1039 (1996)

[82] Gallagher, P.A., Wei, X., Shoemaker, J.A., Brocknoff, C.A., Creed, J.T.: Detection of arsenosugars from kelp extracts via IC-electrospray ionization-MS-MS and IC membrane hydride generation ICP-MS. J. Anal. At. Spectrom. 14, 1829-1834 (1999)

[83] Ceulemans, M., Witte, C., Łobin'ski, R., Adams, F.C.: Simplified sample preparation for GC speciation analysis of organotin in marine biomaterials. Appl. Organomet. Chem. 8, 451-461 (1995)

[84] Pannier, F., Astruc, A., Astruc, M.: Determination of butyltin compounds in marine biological samples by enzymatic hydrolysis and HG-GC-QFAAS detection. Anal. Chim. Acta 327, 287-293 (1996)

[85] Guzmn Mar, J.L., Hinojosa Reyes, L., Mizanur Rahman, G.M., Skip Kingston, H.M.: Simultaneous extraction of arsenic and selenium species from rice products by microwave-assisted enzymatic extraction and analysis by ion chromatography-inductively coupled plasma-mass spectrometry. J. Agric. Food Chem. 57, 3005-3013 (2009)

[86] Gopalakrishnan, N., Narayanan, C.S.: Carbon dioxide extraction of Indian jasmine concrete. Flav. Fragrance J. 6, 135-138 (2006)

[87] Segovia Garcia, E., Garcia Alonso, J.I., Sanz Medel, A.: Determination of butyltin compounds in sediments by means of hydride generation/cold trappinggas chromatography coupled to inductively coupled plasma mass spectrometric detection. J. Mass Spectrom. 32, 542-549 (1997)

[88] Edmonds, J.S., Shibata, Y., Prince, R.I.T., Francesconi, K.A., Morita, M.: Arsenic compounds in tissues of the leatherback turtle, Dermochelys coriacea. J. Mar. Biol. Assoc. U.K. 74, 463-466 (1994)

[89] Byrne, A.R., Slejkovec, Z., Stijve, T., Fay, L., Gœcossler, W., Gailer, J., Irgolic, K.J.: Arsenobetaine and other arsenic species in mushrooms. Appl. Organomet. Chem. 9, 305-313 (1995)

[90] Gallagher, P.A., Shoemaker, J.A., Wei, X., Brockhoff-Schwegel, C.A., Creed, J.T.: Extraction and detection of arsenicals in seaweed via accelerated solvent extraction with ion chromatographic separation and ICP-MS detection. Fresenius J. Anal. Chem. 369, 71-80 (2001)

[91] Francesconi, K.A.: Toxic metal species and food regulations-making a healthy choice. Analyst. 132, 17-20 (2007)

[92] Raab, A., Channock, J., Bahrami, F., Feldmann, J.: Arsenic Species in Hair and Skin of Pre-columbian Mummies UsingXANES/EXAFS and HPLC-ICP-MS. European Winter Conference on Plasma Spectrochemistry, Hungary, XXX (2005)

[93] Biswas, D., Banerjee, M., Sen, G., Das, J.K., Banerjee, A., Sau, T.J., Pandit, S., Giri, A.K., Biswas, T.: Mechanism of erythrocyte death in human population exposed to arsenic through drinking water. Tox. App. Pharm. 230, 57–66 (2008)

[94] Le, X.C., Ma, M.S., Lu, X.F., Cullen, W.R., Aposhia, H.V., Zheng, B.S.: Determination of monomethylarsonous acid, a key arsenic methylation intermediate, in human urine. Environ. Health Perspect. 108, 1015–1018 (2000)

[95] Kitchin, K.T.: Recent advances in arsenic carcinogenesis: modes of action, animal model systems and methylated arsenic metabolites. Toxicol. Appl. Pharm. 172, 249–261 (2001)

[96] Lunde, G.: The synthesis of fat and water soluble arseno organic compounds in marine and limnetic algae. Acta Chem. Scand. 27, 1586–1594 (1973)

[97] Morita, M., Shibata, Y.: Isolation and identification of an arseno-lipid from a brown alga Undaria pinnatifida (wakami). Chemosphere 17, 1147–1152 (1988)

[98] Morita, M., Shibata, Y.: Chemical form of arsenic in marine macroalgae. Appl. Organometal. Chem. 4, 181–190 (1990)

[99] Larsen, E.H., Pritzl, G., Hansen, S.H.: Arsenic speciation in seafood samples with emphasis on minor constituents: an investigation using high-performance liquid chromatography with detection by inductively coupled plasma mass spectrometry. J. Anal. At. Spectrom. 8, 1075–1084 (1993)

[100] Ybanez, N., Velez, D., Tejedor, W., Montoro, R.: Optimization of the extraction, clean-up and determination of arsenobetaine in manufactured seafood products by coupling liquid chromatography with inductively coupled plasma atomic emission spectrometry. J. Anal. At. Spectrom. 10, 459–465 (1995)

[101] Branch, S., Ebdon, L., O'Neill, P.: Determination of arsenic species in fish by directly coupled high-performance liquid chromatography-inductively coupled plasma mass spectrometry. J. Anal. At. Spectrom. 9, 33–37 (1994)

[102] Lamble, K.J., Hill, S.J.: Arsenic speciation in biological samples by on-line high performance liquid chromatography-microwave digestion-hydride generation-atomic absorption spectrometry. Anal. Chim. Acta 334, 261–270 (1996)

[103] McKiernan, J.W., Creed, J.T., Brockhoff, C.A., Caruso, J.A., Lorenzana, R.M.: A comparison of automated and traditional methods for the extraction of arsenicals from fish. J. Anal. At. Spectrom. 14, 607–613 (1999)

[104] Hansen, S.H., Larsen, E.H., Pritzl, G., Cornett, C.: Separation of seven arsenic compounds by high-performance liquid chromatography with on-line detection by hydrogen-argon flame atomic absorption spectrometry and inductively coupled plasma mass spectrometry. J. Anal. At. Spectrom. 7, 629–634 (1992)

[105] Ruiz-Chancho, M.J., Sabe, R., Lopez-Sanchez, J.F., Rubio, R., Thomas, P.: New approaches to the extraction of arsenic species from soils. Microchim. Acta 151, 241-248 (2005)

[106] Ackley, K.L., B'Hymer, C., Sutton, K.L., Caruso, J.A.: Speciation of arsenic in fish tissue using microwave-assisted extraction followed by HPLC-ICP-MS.J. Anal. At. Spectrom. 14, 845-850 (1999)

[107] Quaghebeur, M., Rengel, Z., Smirk, M.: Arsenic speciation in terrestrial plant material using microwave-assisted extraction, ion chromatography and inductively coupled plasma mass spectrometry. J. Anal. At. Spectrom. 18, 128-134 (2003)

[108] Gleyzes, C., Tellier, S., Astruc, M.: Fractionation studies of trace elements in contaminated soils and sediments: a review of sequential extraction procedures. Trends Anal. Chem. 21, 1-17 (2002)

[109] Dagnac, T., Padro, A., Rubio, R., Rauret, G.: Speciation of arsenic in mussels by the coupled system liquid chromatography -UV irradiation-hydride generation-inductively coupled plasma mass spectrometry. Talanta 48, 763-772 (1999)

[110] Zhang, Z., Pawliszyn, J.: Headspace solid-phase microextraction. Anal. Chem. 65, 1843-1852 (1993)

[111] Mester, Z., Sturgeon, R., Pawliszyn, J.: Solid phase microextraction as a tool for trace element speciation. Spectrochim. Acta Part B 56, 233-260 (2001)

[112] Ramos, L., Kristenson, E.M., Brinkman, U.A.T.: Current use of pressurised liquid extraction and subcritical water extraction in environmental analysis. J. Chromatogr. A 975, 3-29 (2002)

[113] Kristenson, E.M., Ramos, L., Brinkman, U.T.A.: Recent advances in matrix solid-phase dispersion. Trends Anal. Chem. 25, 96-111 (2006)

[114] Ramos, L., Ramos, J.J., Brinkman, U.T.A.: Miniaturization in sample treatment for environmental analysis. Anal. Bioanal. Chem. 381, 119-140 (2005)

[115] Barker, S.A.: Matrix solid phase dispersion (MSPD). J. Biochem. Biophys. Methods 70, 151-162 (2007)

[116] Zhang, Z., Yang, M.J., Pawliszyn, J.: Solid-phase microextraction. Anal. Chem. 66, 844A-853A (1994)

[117] Drozd, J., Novak, J.P.: Chemical Derivatization in Gas Chromatography. Elsevier Scientific, Amsterdam (1981)

[118] Romero, J., Lopez, P., Rubio, C., Batlle, R., Nerín, C.: Strategies for single-drop microextraction optimisation and validation: application to the detection of potential antimicrobial agents. J. Chromatogr. A 1166, 24-29 (2007)

[119] Psillakis, E., Kalogerakis, N.: Developments in single-drop microextraction. Trends Anal. Chem. 21, 53-63 (2002)

[120] Fan, Z.: Determination of antimony (III) and total antimony by single-drop microextraction combined with electrothermal atomic absorption spectrometry. Anal. Chim. Acta 585, 300-304 (2007)

[121] Xia, L.B., Hu, B., Jiang, Z.C., Wu, Y.L., Li, L., Chen, R.: 8-Hydroxyquinoline-chloroform single drop microextraction and electrothermal vaporization ICP-MS for the fractionation of Al in natural waters and drinks. J. Anal. At. Spectrom. 20, 441–446 (2005)

[122] Fitchett, A.W., Daughtrey, E.H., Mushak, J.P.: Quantitative measurements of inorganic and organic arsenic by flameless atomic absorption spectrometry. Anal. Chim. Acta 79, 93–99 (1975)

[123] Münz, H., Lorenzen, W.: Selective determination of inorganic and organic arsenic in foods by atomic-absorption spectroscopy. Fresenius' Z. Anal. Chem. 319, 395–398 (1984)

[124] Sperling, M., Yan, X., Welz, B.: Electrothermal atomic absorption spectrometric determination of lead in high-purity reagents with flow-injection on-line microcolumn preconcentration and separation using a macrocycle immobilized silica gel sorbent. Spectrochim. Acta Part B 51, 1875–1889 (1996)

[125] Latva, S., Hurtta, M., Peraniemi, S., Ahlgren, M.: Separation of arsenic species in aqueous solutions and optimization of determination by graphite furnace atomic absorption spectrometry. Anal. Chim. Acta 418, 11–17 (2000)

[126] Packer, A.P., Ciminelli, V.S.T.: A simplified flow system for inorganic arsenic speciation by preconcentration of As(V) and separation of As(III) in natural waters by ICP-MS. At. Spectrosc. 26, 131–136 (2005)

[127] Barker, S.A.: Matrix solid-phase dispersion. J. Chromatogr. A 885, 115–127 (2000)

[128] Barker, S.A.: Applications of matrix solid-phase dispersion in food analysis. LC–GC Int. 11, 719–724 (1998)

[129] Mester, Z., Sturgeon, R.E., Lam, J.W.: Sampling and determination of metal hydrides by solid phase microextraction thermal desorption inductively coupled plasma mass spectrometry. J. Anal. At. Spectrom. 15, 1461–1465 (2000)

[130] Mester, Z.: Gas phase sampling of volatile (organo)metallic compounds above solid samples. J. Anal. At. Spectrom. 17, 868–871 (2002)

[131] Chamsaz, M., Arbab-Zavar, M.H., Nazari, S.: Determination of arsenic by electrothermal atomic absorption spectrometry using headspace liquid phase microextraction after in situ hydride generation. J. Anal. At. Spectrom. 18, 1279–1282 (2003)

[132] Bermejo-Barrera, P., Moreda-Pineiro, J., Moreda-Pineiro, A., Bermejo-Barrera, A.: Selective medium reactions for the arsenic(III), arsenic(V), dimethylarsonic acid and monomethylarsonic acid determination in waters by hydride generation online electrothermal atomic absorption spectrometry with in situ preconcentration on Zr-coated graphite tubes. Anal. Chim. Acta 374, 231–240 (1998)

[133] Liu, W.P., Lee, K.: Chemical modification of analytes in speciation analysis by capillary electrophoresis, liquid chromatography and gas chromatography. J. Chromatog. A 834, 45–63 (1999)

[134] Leal-Granadillo, I.A., Alonso, J.I.G., Sanz-Medel, A.: Determination of the speciation of or-

ganolead compounds in airborne particulate matter by gas chromatography-inductively coupled plasma mass spectrometry. Anal. Chim. Acta 423, 21-29 (2000)

[135] Fernandez, R.G., Bayon, M.M., Alonso, J.I.G., Sanz-Medel, A.: Comparison of different derivatization approaches for mercury speciation in biological tissues by gaschromatography/ inductively coupled plasma mass spectrometry. J. Mass Spectrom. 35, 639-646 (2000)

[136] Da Smaele, T., Moens, L., Dams, R., Sandra, P., Van ser Eycken, J., Vandyck, J.: Sodium tetra(n-propyl)borate: a novel aqueous in situ derivatization reagent for the simultaneous determination of organomercury, -lead and-tin compounds with capillary gas chromatography-inductively coupled plasma mass spectrometry. J. Chromatogr. A 793, 99-106 (1998)

[137] Palaez, M.V., Bayon, M.M., Alonso, J.I.G., Sanz-Medel, A.: A comparison of different derivatisation approaches for the determination of selenomethionine by GC-ICP-MS. J. Anal. Atomic Spectrom. 15, 1217-1222 (2000)

[138] Bouyssiere, B., Szpunar, J., Lespes, G., Łobin'ski, R.: Gas chromatography with inductively coupled plasma mass spectrometric detection in speciation analysis. Spectrochim. Acta Part B 57(5), 805-828 (2002)

[139] De Smaele, T., Verrept, P., Moens, L., Dams, R.: A flexible interface for the coupling of capillary gas chromatography with inductively coupled plasma mass spectrometry. Spectrochim. Acta Part B 50, 1409-1416 (1995)

[140] Rodriguez, I., Mounicou, S., Łobin'ski, R., Sidelnikov, V., Patrushev, Y., Yamanaka, M.: Species selective analysis by microcolumn multicapillary gas chromatography with inductively coupled plasma mass spectrometric detection. Anal. Chem. 71, 4534-4543 (1999)

[141] Montes-Bayón, M., Gutiérrez Camblor, M., García Alonso, J.I., Sanz-Medel, A.: An alternative GC-ICP-MS interface design for trace element speciation. J. Anal. At. Spectrom. 14, 1317-1322 (1999)

[142] Wasik, A., Pereiro, I.R., Dietz, C., Spunar, J., Łobiński, R.: Speciation of mercury by ICP-MS after on-line capillary cryofocussing and ambient temperature multicapillary gas chromatography. Anal. Commun. 35, 331-335 (1998)

[143] Namieśnik, J.: Trends in environmental analytics and monitoring. J. Crit. Rev. Anal. Chem. 30, 221-269 (2000)

[144] Fukui, S., Hirayama, T., Nohara, M., Sakagami, Y.: Gas chromatographic determination of dimethylarsinic acid in aqueous samples. Talanta 28, 402-404 (1981)

[145] Howard, A.G., Salou, C.: Cysteine enhancement of the cryogenic trap hydride AAS determination of dissolved arsenic species. Anal. Chim. Acta 333, 89-96 (1996)

[146] Szpunar, J., McSheehy, S., Połeć, K., Vacchina, V., Mounicou, S., Rodriguez, I., Łobiński, R.: Gas and liquid chromatography with inductively coupled plasma mass spectrometry detection for environmental speciation analysis-advances and limitations. Spectrochim. Acta Part B 55, 779-793 (2000)

[147] Lecchi, P., Gupte, A.R., Perez, R.E., Stockert, L.V., Abramson, F.P.: Size-exclusion

chromatography in multidimensional separation schemes for proteome analysis. J. Biochem. Biophys. Methods 56, 141–152 (2003)

[148] Suzuki, K.T., Mandal, B.K., Ogra, Y.: Speciation of arsenic in body fluids. Talanta 58, 111–119 (2002)

[149] Buszewski, B., Bocian, S., Felinger, A.: Excess isotherms as a new way for characterization of the columns for reversed-phase liquid chromatography. J. Chromatogr. A 1191, 72–77 (2008)

[150] Dickman, M.J.: Effects of sequence and structure in the separation of nucleic acids using ion pair reverse phase liquid chromatography. J. Chromatogr. A 1076, 83–89 (2005)

[151] Hemstrom, P., Irgum, K.: Hydrophilic interaction chromatography. J. Sep. Sci. 29, 1784–1821 (2006)

[152] Szpunar, J.: Trace element speciation analysis of biomaterials by high-performance liquid chromatography with inductively coupled plasma mass spectrometric detection. Trends Anal. Chem. 19, 127–137 (2000)

[153] Ruzik, R., Lipiec, E., Połeć-Pawlak, K.: Investigation of Cd(II), Pb(II) and Cu(I) complexation by glutathione and its component amino acids byESI-MS and size exclusion chromatography coupled to ICP-MS and ESI-MS. Talanta 72, 1564–1572 (2007)

[154] Caruso, J.A., Montes-Bayon, M.: Elemental speciation studies-new directions for trace metal analysis. Ecotox. Env. Safety 56, 148–163 (2003)

[155] Raab, A., Feldmann, J., Meharg, A.A.: The nature of arsenic-phytochelatin complexes in Holcus lanatus and Pteris cretica. Plant Physiol. 134, 1113–1122 (2004)

[156] McSheehy, S., Marcinek, M., Chassaigne, H., Szpunar, J.: Identification of dimethylarsinoylriboside derivatives in seaweed by pneumatically assisted electrospray tandem mass spectrometry. Anal. Chim. Acta 410, 71–84 (2000)

[157] McSheehy, S., Pohl, P., Łobiński, R., Szpunar, J.: Complementarity of multidimensional HPLC-ICP-MS and electrospray MS-MS for speciation analysis of arsenic in algae. Anal. Chim. Acta 440, 3–16 (2001)

[158] Larsen, E.H.: Method optimization and quality assurance in speciation analysis using high performance liqudi chromatography with detection by inductively coupled plasma massspectrometry. Spectrochim. Acta. Part B 53, 253–265 (1998)

[159] Leeemakers, M., Baeyens, W., De Gieter, M., Smedts, B., Meert, C., De Bisschop, H.C., Morabito, R., Quevauviller, P.: Toxic arsenic compounds in environmental samples: speciation and validation. Trends Anal. Chem. 25, 1–10 (2006)

[160] Shiobara, Y., Ogra, Y., Suzuki, K.T.: Animal species difference in the uptake of dimethylarsinous acid (DMA(III)) by red blood cells. Chem. Res. Toxicol. 14, 1446–1452 (2001)

[161] Sheppard, B.S., Caruso, J.A., Heitkemper, D.T., Wolnik, K.A.: Arsenic speciation by ion chromatography with inductively coupled plasma mass spectrometric detection. Analyst 117, 971–975 (1992)

[162] Kohlmeyer, U., Jantzen, E., Kuballa, J., Jakubik, S.: Benefits of high resolution IC-ICP-MS for the routine analysis of inorganic and organic arsenic species in food products of marine and terrestrial origin. Anal. Bioanal. Chem. 377, 6–13 (2003)

[163] Wróbel, K., Wróbel, K., Parker, B., Kannamkumarath, S.S., Caruso, J.A.: Determination of As(III), As(V), monomethylarsonic acid, dimethylarsinic acid and arsenobetaine by HPLC-ICP-MS: analysis of reference materials, fish tissues and urine. Talanta58, 899–907 (2002)

[164] Shuvaeva, O.V., Koscheeva, O.S., Beisel, N.F.: Arsenic speciation in waters using HPLC with graphite furnace atomic absorption spectrometry as detector. Anal. Sci. 17, 179–181 (2001)

[165] Ding, H., Wang, J., Dorsey, J.G., Caruso, J.A.: Arsenic speciation by micellar liquid chromatography with inductively coupled plasma mass spectrometric detection. J. Chromatogr. A 694, 425–431 (1995)

[166] Wahlen, R., McSheehy, S., Scriver, C., Mester, Z.: Arsenic speciation in marine certified reference materials Part 2. the quantification of water-soluble arsenic species by high-performance liquid chromatography-inductively coupled plasma mass spectrometry. J. Anal. At. Spectrom. 19, 876–882 (2004)

[167] Raab, A., Feldmann, J.: Arsenic speciation in hair extracts. Anal. Bioanal. Chem. 381, 332–338 (2005)

[168] Hansen, H.R., Pickford, R., Thomas-Oates, J., Jaspars, M., Feldmann, J.: 2-dimethylarsinothioyl acetic acid identified in a biological sample: the first occurrence of a mammalian arsinothioyl metabolite. Angew. Chem. Int. Ed. 43, 337–340 (2004)

[169] Schaeffer, R., Francesconi, K.A., Kienzl, N., Soeroes, C., Fodor, P., Varadi, L., Raml, R., Goessler, W., Kuehnelt, D.: Arsenic speciation in freshwater organisms from the river Danube in Hungary. Talanta 69, 856–865 (2006)

[170] Gailer, J., Irgolic, K.J.: The ion-chromatographic behavior of arsenite, arsenate, methylarsonic acid and dimethylarsinic acid on the hamilton PRP-X100 anion-exchange column. Appl. Organomet. Chem. 8, 129–140 (1994)

[171] Oehme, M., Berger, U., Brombacher, S., Kuhn, F., Koliker, S.: Trace analysis by HPLC-MS: contamination problems and systematic errors. Trends Anal. Chem. 21, 322–331 (2002)

[172] Morabito, R., Massanisso, P., Camara, C., Larsson, T., Frech, W., Kramer, K.J.M., Bianchi, M., Muntau, H., Donard, O.F.X., Łobinski, R., McSheehy, S., Pannier, F., Potin-Gautier, M., Gawlik, B.M., Bowadt, S., Quevauviller, P.: Towards a new certified reference material for butyltins, methylmercury and arsenobetaine in oyster tissue. Trends Anal. Chem. 9, 664–676 (2004)

[173] Witkiewicz, Z.: Podstawy Chromatografii. WNT, Warszawa (2000)

[174] Carey, J.M., Caruso, J.A.: Plasma-spectrometric detection for supercritical-fluid chromatography. Trends Anal. Chem. 11, 287–293 (1992)

[175] Lee, M., Markides, K.: Analytical Supercritical-Fluid Chromatography and Extraction, vol. 91, pp. 172-178. Chromatography Conferences, Provo (1990)

[176] Vela, N.P., Caruso, J.A.: Comparison of flame ionization and inductively coupled plasma mass spectrometry for the detection of organometallics separated by capillary supercritical-fluid chromatography. J. Chromatogr. 641, 337-345 (1993)

[177] Vela, N.P., Caruso, J.A.: Determination of tri-and tetra-organotin compounds by supercritical-fluid chromatography with inductively coupled plasma mass spectrometric detection. J. Anal. Atom. Spectrom. 7, 971-977 (1992)

[178] Kumar, U.T., Vela, N.P., Caruso, J.A.: Multi-element detection of organometals by supercritical-fluid chromatography with inductively coupled plasma mass spectrometric detection. J. Chromatogr. Sci. 33, 606-610 (1995)

[179] Schaumloffel, D.: Capillary liquid separation techniques with ICP MS detection. Anal. Bioanal. Chem. 379, 351-354 (2004)

[180] Liu, Y., Lopez-Avila, V., Zhu, J.J., Wiederin, D.R., Beckert, W.F.: Capillary electrophoresis coupled on-line with inductively coupled plasma mass spectrometry for elemental speciation. Anal. Chem. 67, 2020-2025 (1995)

[181] Kirlew, P.W., Castillano, M.T.M., Caruso, J.A.: Evaluation of ultrasonic nebulizers as interfaces for capillary electrophoresis of inorganic anions and cations with inductively coupled plasma mass spectrometric detection. Spectrochim. Acta Part B53, 221-237 (1998)

[182] Prange, A., Schaumloffel, D.: Determination of element species at trace levels using capillary electrophoresis-inductively coupled plasma sector field mass spectrometry. J. Anal. Atom. Spectrom.14, 1329-1332 (1999)

[183] Magnuson, M.L., Creed, J.T., Brockhoff, C.A.: Speciation of selenium and arsenic compounds by capillary electrophoresis with hydrodynamically modified electroosmotic flow and on-line reduction of selenium (VI) to selenium (IV) with hydride generation inductively coupled plasma mass spectrometric detection. Analyst 122, 1057-1062 (1997)

[184] Hsieh, M.-W., Liu, C.-L., Chen, J.-H., Jiang, S.-J.: Speciation analysis of arsenic and selenium compounds by CE-dynamic reaction cell-ICP-MS. Electrophoresis13, 2272-2278 (2010)

[185] Michalke, B., Schramel, P.: Selenium speciation by interfacing capillary electrophoresis with inductively coupled plasma-mass spectrometry. Electrophoresis19, 270-275 (1998)

[186] Chen, T.S., Liu, C.Y.: Histidine-functionalized silica and its copper complex as stationary phases for capillary electrochromatography. Electrophoresis22, 2606-2015 (2001)

第13章
生物和环境样品中贵金属的定量分析

13.1 概述

贵金属由元素周期表第Ⅷ族的6个元素:钌(Ru)、铑(Rh)、钯(Pd)、锇(Os)、铱(Ir)、铂(Pt)[以上又称为铂族金属(PGM)]和第Ⅰ$_B$族的金(Au)元素组成,这些金属元素因具有特定物理化学性质,如稳定性、硬度、延展性、电阻性、化学惰性以及优良的催化活性,而被广泛应用,例如,各种化学过程中的催化剂,催化转化器中的自动催化剂,电气、电子和玻璃工业,珠宝投资行业等[1]。PGM还广泛用作汽车催化剂元件(占2012全年用量的50.4%)[2]。在车辆运行过程中,催化剂的热磨损和机械磨损导致这些金属超痕量释放,这是它们在环境中的首要来源[2-10]。贵金属化合物的医学应用是它们在环境中的第二人为来源(仅次于催化剂)[11-13]。1978年,首次将顺铂、顺二氯铂作为一种有效的抗癌药物用于化疗,其后铂和其他贵金属化合物的多种络合物,被广泛应用于抗癌活性的检查[14-26]。第二代铂化合物(草酸铂和奥沙利铂)已在全世界得到应用,可用于常规治疗,洛巴铂、奈达铂和七肽分别在中国、日本和韩国获得当地批准。目前,顺铂仍是一种先进的药物,占抗癌化疗药物的(50~70)%[17,19],还有许多其他金属化合物应用在临床试验中,如某些钌基化合物,与顺铂相比具有更好的抗肿瘤活性和更少的副作用,因而更有前景[15-18,21-23,25,26]。

评估PGM对环境和人类的影响,需要检测大量不同基体和复杂痕量金属的多种样品,并且研究发生在不同环境区域和生物有机体中的生物转化和生物富集过程。在各种氧化剂和络合剂的作用下发生的转化过程,可导致金属元素转变为更可溶的和生物可用的形态。生物富集过程会导致某些材料中这些金属元素的含量升高,并增加潜在的健康风险。在慢性暴露下(如催化剂生产过程中)[27]观察到的铂金属盐对人的致敏和致突变作用的数据表明,Pt和Pd对细胞系统的毒性作用与 Cd(Ⅱ)和 Cr(Ⅵ)相当[28]。环境中 PGM 生物积累的趋势使得这些来源的金属元素可能产生一定的健康风险[2-8,10-13,29-32]。形态分析是环境与人类危害评价的重要内容,其目的在于识别和定量在不同条件下形成的 PGM 的各种化学形态,并评价其生物活性。检测体液和组织中的金属含量以及不同络合物与生物分子的相互作用,对于评估环境来源的金属和通过化疗引入人体的金属对人体的危害是必不可少的。监测药物在人体中的分布和代谢是了解金属配合物作为药物的治疗效果和副作用的基础。

本章描述了用于化疗的 Pt 和 Ru 以及作为自动催化剂组分的 Pt、Pd 和 Rh 的生物(体液和组织)、环境(植物、空气中颗粒物和尘埃、土壤和沉积物)样品的最具潜力的分析方法。

13.2 样品制备

生物和环境材料分析结果的质量可靠性一般取决于所采用的样品制备技术,特别是测定复杂基体中低含量金属元素和低稳定性形态,当浓度低于现有仪器的检出限(DL)或基体组分有干扰时,需要进行分离和预浓缩。有些仪器分析技术可用于固体材料直接检测,这更有利于降低样品制备对测量结果的影响。在分析贵金属样品,特别是源自复杂环境的贵金属样品时,基体组分的干扰和被测样品的不均匀性,使这项技术的使用受到很大限制。

在环境材料检测中,由于贵金属呈不均匀分布的趋势,因此选取代表性样品至关重要,增大用于分析的样品量可以使被测材料的不均匀性影响最小化。这些金属元素对化学反应的惰性高、基体十分复杂,在所使用的条件下,将金属转变成稳定的、严格定义的络合物十分困难。需要特别注意样品的定量消解,使分析物转化成适合后续步骤(分离、预浓缩和检测)的形态,也要注意采样和储存条件确保被测形态的稳定性。可以采用特殊的预防措施来避免化学试剂和容器的污染,尽量减少引入被测样品中的化学试剂数量以及缩减样品制备步骤。贵金属容易在容器壁上吸附,并且其络合物在弱酸性和中性条件下易水解,因此,有可能造成贵金属损失。容器材料对吸附动力学的影响取决于待测物的浓度,石英是最适合样品消化和储存的容器材料,因为与其他材料(如聚四氟乙烯)相比,石英的吸附较低,在所有分析过程中需仔细清洗容器,以减少空白值。可以查阅用于分析各种材料中贵金属含量样品制备方法(取样、消解、分离和预浓缩)的详细综述[33-37]。

13.2.1 临床样品

选择取样和存储条件以确保临床样品中待测化合物的稳定性,这对于生物活性物质的定性、定量和鉴别非常重要。最基本的是选择接近的生理条件和分析条件。在取样和存储下产生的生物应激性会影响被测样品的自然组成,并显著降低其代表性。任何样品的处理过程都需要对被测化合物在应用条件下的行为有深入的了解,并验证影响其性质和浓度的参数[38],这对于校准过程很重要,特别是对于稳定性差的化合物的检测。储存和培养条件应当保持被测化合物的药理活性形式,使用的校准溶液的组成与实际生物基体相匹配,特别是当生物活性形式是生理条件下发生的生物转化过程的产物时。例如,顺铂的水合物形态 $PtCl(NH_3)_2(H_2O)^{+[39]}$,Ru(Ⅲ)的配合物 HIn 反式 $[RuCl_4(Ind)_2]$(KP1019)[40],需要采取特殊的预防措施来避免仪器和化学试剂对样品的污染。

血液样品建议在取样和离心后直接检测[41-42],超滤物应在取样后直接检测。样品可在短时间内低温储存,如小于或等于0℃(血液)、小于-10℃(超滤物)和-20℃(-80℃)(尿液)[43-45]。临床样品在检测金属元素之前,通常需要稀释或消解。体液稀释可以降低干扰检测的固体含量,可以使用水、稀盐酸、稀硝酸、氯化钠和盐酸的稀溶液以及曲拉通 X-100(聚乙二醇辛基苯基醚)与 0.2%硝酸、10%盐酸、EDTA 或水的混合物作为稀释介质。介质的种类取决于基体和检测技术。但是,原样品的大量稀释会导致灵敏度降低,蛋白质在样品稀释下产生沉淀也会引起问题。若检测金属总量,则体液消解通常是比较好的方法。在测定金属元素之前将有机基体消解,采用浓硝酸、王水、HNO_3+HClO_4、$HNO_3+H_2O_2$ 和 $HNO_3+H_2O_2$+王水的混合液作为消解液。紫外光分解可以提高消解过程的有效性,采用电感耦合等离子体质谱(ICP-MS)技术对未暴露人群的血液[46]和尿液[47]进行生理性 Pd、Pt(Ir、Au)分析,证实了紫外光分解的优越性。与经典的酸消解相比,使用较低量的化学试剂,可降低空白值。只有在 H_2O_2 存在的条件下尿液样品才可以在紫外光下完全分解[47]。

13.2.2 环境样品

由于环境样品性质复杂、对痕量贵金属测定的需要以及基质成分在大多数仪器分析技术中存在干扰,因此需要能够定量回收金属的消解和分离程序。样品制备工艺与检测方法的合理耦合是获得可靠结果的关键,湿法酸处理、火试金法(FA)、氯化法和氧化熔融法等被广泛应用于环境样品的分解,被测样品的初步干燥、粉碎、研磨(粉化)也是必要的。

在检测各种环境样品,如植物、空气中颗粒物、灰尘、土壤和沉淀物的金属总量时,通常采用以酸(王水;HNO_3、HCl 和 HF 的混合酸;HCl 和 $HClO_4$ 的混合酸)和微波高压辅助处理的方法。在硅酸盐基体中,可能需要使用 HF 来分解硅酸盐颗粒中包覆的贵金属[48-51],尽管对于复杂样品(如灰尘)在有/无 HF 的王水处理下的分析结果具有良好的一致性[52-53]。王水处理的两个步骤(王水、HF/HClO$_4$/王水)对于消解极其复杂的空气颗粒物是有效的[54],但是使用湿法化学处理方法从粉尘样品中定量回收 Rh 可能还存在问题[55]。用 HF 消解会使基体元素转化成可溶性物质的量增加,并增加它们对 PGM 检测的干扰。用稀盐酸对被测样品进行预处理,可使样品完全消解后得到的溶液中某些非贵金属的含量降至最低。所用酸的纯度对于降低高的空白值十分重要。

火试金法是测定复杂的、非均质材料(例如,土壤和沉积物)中极低含量 PGM 的一项很有效的技术[56-62]。该技术可用于检测大量样品,达(5~100)g,从而最大限度地减少金属不均匀分布的影响。火试金法的优点是在一个很小的试金扣中富集贵金属。硫化镍试金法是从各种样品中回收所有贵金属的有效方法。分析过程是将样品与 Ni、S、Na$_2$B$_4$O$_7$、Na$_2$CO$_3$ 和 SiO$_2$ 粉末在大约 1000℃下熔融,用 HCl 处理试金扣,去除常见元素,然后用无机混合酸(HNO$_3$+HCl)溶解贵金属。但是,引入样品中的大量化学试剂可能影响溶液中金属元素检测,并产生较高的空白值。

氯气腐蚀氯化法是测定复杂样品中贵金属的一种有效方法,所有贵金属都会受到氯的侵蚀,溶解在稀酸中的二元氯化物或盐是氯化的产物。用氯气通过样品,或单独或与少量氯化钠混合,在 500~600℃ 的开口管中进行干氯化,是最有前景的方法。因为引入样品中的化学物质量少,背景值低,并且可以与检测技术直接结合,文献[63]报道了在岩石分析中该方法与火试金法相比的有效性。

碱性氧化熔融法是处理金属粉末,特别是无法直接湿法酸处理(Ru、Os 和 Ir)金属的有效方法,但是由于回收率低(例如,硅酸盐样品中铂、钯和金的回收率仅为 34%~84%,很少用于复杂贵金属样品的分解[64]。由于在熔体溶解(水、盐酸)下形成的配合物的稳定性低,并且很难将待测物定量转化成具有严格限定组成(适于随后的分离)的化合物,因此碱熔法受到了限制。而且,在溶液中容易形成的羟基络合物可能导致金属元素的定量分离和预富集出现问题,特别是在使用离子交换色谱时。

在复杂环境样品的分析中,通常需要在检测之前对 PGM 进行分离和富集,因为检测浓度很低,大量的基体元素会干扰检测[34-36,65-66]。最常用的技术是用适当的捕收剂,联合离子交换树脂/螯合树脂液相色谱(LC)、吸附-固相萃取(SPE)以及共沉淀法。离子交换色谱法使用阳离子和阴离子交换剂,能够很容易地使溶液中的贵金属(以稳定的阴离子络合物形式)与常见元素的阳离子形式分离。贵金属的阴离子氯化物保留在阴离子树脂上,而轻金属以阳离子形式通过树脂。如果使用阳离子柱,贵金属络合物就会定量地进入洗脱液,而带正电荷的非贵金属元素被保留。将 PGM 定量转化成适当的阴离子络合物,并将非贵金属元素转化成阳离子,这是 PGM 有效分离的基础。以阴离子形式存在于被测溶液中的常见金属在色谱中具有与贵金属的阴离子络合物相似的过程,如铌、铪在消解样品所用的氢氟酸存在的条件下容易形成阴离子氟化物[67],对 ICP-MS 测铂有干扰作用[68-70]。与纯氯化物介质相比,铪与桑色素的初步螯合能使其在常用的 Dowex 50W-X8 阳离子树脂上与铂分离[71]。钇和锆的阴离子形态会对离子交换法分离钯造成困难[68]。贵金属的阴离子络合物和阴离子树脂的强亲和性使它们的定量回收很困难,使用相对苛刻的条件(如 83℃下 12mol/L HNO_3)有助于 Pt 的洗脱(回收率>95%)[72]。

含有官能团的螯合吸附剂能够与被测溶液中的贵金属形成稳定的络合物,它们的分离和预浓缩效果比离子交换树脂更好。贵金属容易与含硫(如双硫腙、二硫代氨基甲酸酯、硫代卡巴肼、2-巯基苯并噻唑和硫脲衍生物)和含氮(如胺、酰胺和杂环含氮化合物)的配体形成稳定的配合物。特别是难以从吸附剂中定量洗脱金属时,在干灰化和高温高压酸处理后直接用 X 射线荧光法测定固体吸附剂中的金属元素都是可行的。在双硫腙吸附剂上吸附,用硫脲和浓硝酸洗脱后,测定土壤样品中铂和钯(Pt 和 Pd 的回收率分别为 98%和 95%)[73]。文献报道土壤中的铂,用硫代卡巴肼衍生物固定,在 Dowex 1X8 树脂和硅胶上吸附的富集因子分别为 14.0[74]和 41.7[75]。铂、钯、铑与罗丹宁衍生物(如 4-羧苯基-硫代罗丹宁)的复合物也可用于柱分离和富集[76]。螯合吸附剂可用于分析含有大量干扰检测的基体元素的复杂样品,以及不能由一个步骤完全分离的样品,如离子色谱。使用二维色谱(阳离子 AG 50W-X8 和 C18 树脂,用 N,N-二乙基-N'-苄基硫脲固定)能够大大提高 Pd 与其他干扰 ICP-MS 检测的大量元素(如 Sr、Rb 和 Y)的分离效率[77]。仅使用阳离子树脂时,样品中大量的轻金属元素是不能定量保留的。

碳基吸附剂是一种新材料,可用于分析不同来源的贵金属样品[78-84]。碳基吸附剂用于从环境样品中提取贵金属的实例有:用氧化碳纳米管从水、粉煤灰和道路灰尘样品中分离、富集钯(预富集系数为 165)[83];用富勒烯 C_{60} 负载二硫代氨基甲酸酯,从道路灰尘样品中分离、富集钯(吸附效率为 99.2%)[78];用聚丙烯腈改性多壁碳纳米管分离、富集铑(预富集系数为 120)[80]。已有文献研究,在测定食品和环境样品中的 Au(Ⅲ)和 Pd(Ⅱ)之前,预先在多壁碳纳米管与聚丙烯胺树枝状大分子接枝的复合材料上进行吸附[84]。近年来,离子印迹聚合物利用各种螯合配合物对环境样品中的贵金属(如 Pt[85]和 Pd[86])进行固相萃取的研究得到了广泛的应用。疏水性贵金属配合物在浊点萃取系统下进行萃取分离,如用 N,N-二己基-N'-苄基硫脲-Triton X-114 从海水和尘埃样品中提取 Pt、Pd、Au[87]。

在检测复杂的环境样品时,由于很难完全分离大量的常见金属,它们可能会部分进入沉淀,可能造成 PGM 损失,因此用捕收剂(Te、Se、As、Hg、Cu)还原共沉淀贵金属时会产生不确定性。例如,用 Te 共沉淀分离后,土壤样品中 Pt 的回收率为 55%~87%[88]。分析时间较长也是该方法的缺点之一。

在分析环境样品时,可通过电解被测溶液和在石墨管热解涂层内表面上电沉积贵金属(如空气颗粒物中的 Pd 和 Pt[89],道路灰尘中的 Pt、Pd、Rh 和 Ru[90])来获得较高的预富集因子。

13.3 铂族金属元素仪器检测技术

若测定生物和环境样品中的铂族金属元素(PGM)含量,特别需要高灵敏度、高选择性、易于与样品制备过程联用的检测技术。电感耦合等离子体质谱法(ICP-MS)、电热原子吸收光谱法(ET-AAS)、吸附伏安法(AV)和中子活化分析法(NAA)在直接测定被测样品中的总金属含量和与仪器分离技术联用方面都有广泛应用。质谱与电喷雾电离(ESI)和毛细管电泳(CE)联用技术(如 ESI-MS^n、LC-ESI-MS^n、LC-ICP-MS、CE-MS^n 和 CE-ICP-MS),为金属的形态分析提供了强有力的支持。质谱已广泛用于检测各种材料中的金属元素分布(元素分析)以及解析金属配合物的分子结构。分子数据对于研究各种金属配合物与生物分子之间的相互作用具有重要意义,是评价其生物活性的基础。

ICP-MS 的检出限可达 ppb 级和亚 ppb 级,具有高选择性、多元素适用性、测量同位素比的独特能力,以及宽动态范围和高操作速度等特点,因此,它是用于各种材料元素分析的优选检测技术。对特定元素(如杂原子)的定性和定量分析,可以用来研究经等离子体破坏的有机化合物,这有助于检测金属与生物分子的相互作用。直接利用该方法定量检测复杂样品,特别是环境样品中的 PGM,会受到基体组分的强烈干扰限制(表 13-1)[48,91-93]。例如,在等离子体条件下产生的铪氧离子(HfO$^+$)严重影响了铂所有同位素的 ICP-MS 信号;铜(ArCu$^+$)、钇(YO$^+$)、锶(SrO$^+$)、锆(ZrO$^+$)的多原子形态是钯检测的主要光谱干扰来源;对铑检测的主要干扰来自铜(ArCu$^+$)和铷(RbO$^+$)的多原子形态以及 Pb^{2+} 离子。在检测未知样品中的 PGM 之前,只有定性、定量和消除干扰,才能获得可靠的结果。被测溶液中干扰物质的含量、所使用的质谱仪的类型和测量条件都会影响干扰的强度。双聚焦扇形磁场质谱仪(0.001Da)(1Da = 1.60054×10^{-27}kg)具有比广泛使用的四极杆(1Da)更好的分辨率,并且测定复杂样品时,如空气中颗粒物、灰尘、土壤和沉积物中的 Pt、Pd、Rh,具有更好的选择性[53,94-95]。改进样品引入技术以获得"干"等离子体条件,其中含有 O$^-$ 和 OH$^-$ 的分子离子的强度与"湿"等离子体相比显著降低,可有助于克服在测定复杂基体中的金属时产生的干扰。为此,可采用电热蒸发、热喷涂雾化、激光烧蚀和膜去溶等技术,并利用数学校正来消除这些干扰。动态反应池能够将干扰离子转化为非活性形态,可以大大减少干扰信号(如在检测 Pt 和 Pd 时 Hf 和 Zr 的干扰)并改善分析物的检出限[96-97]。采用预化学分离步骤,可以有效地克服贵金属检测中的光谱干扰。化学分离在分析含有过量干扰成分的土壤和沉积物等环境材料时是很有用的在富集 PGM 的同时,确定其极低的浓度。样品制备步骤会显著地影响 ICP-MS 结果的选择性和灵敏度。被测溶液中的高盐浓度可导致分析物信号的抑制或增强。在定量检测痕量金属时,应特别注意试剂的纯度、所用的容器和仪器分析路径,以免发生污染和可能的记忆效应,这可能导致所得结果的严重误差。

表 13-1　ICP-MS 法测量 Rh、Pd、Pt 的光谱(同量异位素和多原子离子)干扰

单位:%

同位素	丰度	干　扰
^{103}Rh	100.0	^{206}Pb、^{86}Sr^{16}O^{1}H、^{87}Sr^{16}O、^{63}Cu^{40}Ar、^{87}Rb^{16}O
^{105}Pd	22.2	^{88}Sr^{16}O^{1}H、^{89}Y^{16}O、^{65}Cu^{40}Ar
^{106}Pd	27.3	^{106}Cd、^{90}Zr^{16}O、^{89}Y^{1}H
^{108}Pd	26.7	^{108}Cd、^{92}Zr^{16}O、^{92}Mo^{16}O
^{110}Pd	11.8	^{110}Cd、^{94}Zr^{16}O、^{94}Mo^{16}O
^{194}Pt	32.9	^{178}Hf^{16}O
^{195}Pt	33.8	^{179}Hf^{16}O
^{196}Pt	25.3	^{196}Hg、^{180}Hf^{16}O、^{180}W^{16}O、^{180}Ta^{16}O

原子吸收光谱法(AAS)可用于临床和环境材料中 PGM 的测定。石墨炉原子吸收光谱比火焰原子吸收光谱有更好的检出限,并且可以直接检测相对简单基体的材料,如临床样品,特别是化疗后含有大量金属元素的样品。样品的稀释和离心(体液)能够减少基体干扰(细胞组分、蛋白质、盐和脂肪)。在检测极其复杂的环境材料中的贵金属之前,通常要进行消解。从基体中分离和预富集金属元素,优选 SPE 和共沉淀法。汞在 ET-AAS 法检测中易挥发、干扰小,是一种很好的 PGM 捕收剂。

使用吸附伏安法(AV)可以达到与 ICP-MS 相当甚至更好的检出限[98]。该方法可用于血液样品中 Pt(≤0.8~6.9ng/L)和尿液样品中 Pt(0.5~15)ng/L 的生理水平测定[99]。用 ICP-MS 法检测铑的检出限(2500ng/kg)是吸附伏安法检测限(33ng/kg)的 100 倍[100]。大多数伏安法测定 Pt 和 Pd 是利用 Pt(Ⅱ)与甲醛、Pd 与二甲基乙二肟的复合物沉积在悬汞滴电极表面。但封闭还原电位和严重的基体效应限制了 AV 法直接应用于 PGM 的同时检测。优化电解质组分可以同时检测 Pt、Pd、Rh[101]和 Pt、Rh[102]。AV 法对有机质非常敏感,在检测之前建议通过破坏有机质来限制碳含量(小于或等于 0.1%)。前文已经介绍了在 AV 法检测之前分离、富集各种基体中的 PGM[103]。

中子活化分析法(NAA)基于特定元素放射性核素的产生并检测,由于检测的高灵敏度和无空白,因此其是一项很有吸引力的技术[104]。铂、钯、铑是通过同位

素中子活化进行测定的,$^{198}Pt(n,\gamma)^{199}Pt\rightarrow^{199}Au(t_{1/2}=3.15d)$、$^{108}Pd(n,\gamma)^{109}Pd$ ($t_{1/2}=13.5h$)、$^{104}Rh(t_{1/2}=42s)$[105-107]。该方法需要具备用于产生特定放射性核素的辐照源,由于基体元素的干扰,直接检测极低 PGM 含量的复杂环境样品可能存在问题,特别是需要长时间辐照和用高中子通量密度实现足够的灵敏度时。在生物和环境样品的分析中,应采用适当的放射化学法从被测样品中分离金属元素。将 FA 法与 NAA 法联用,可用于复杂样品(如土壤和沉积物)的检测。目前,核分析技术在生物和环境样品的形态分析中也有应用[108]。

13.4 临床样品中的铂和钌

1978 年,顺铂(图 13-1)作为一种对抗肺癌、卵巢癌、膀胱癌、头颈癌有效的药物应用于化疗。但是,由于其在某些肿瘤的临床治疗过程中有严重的副作用和耐药性,因此顺铂的应用受到限制,这就促使全世界研究者对具有更好抗肿瘤特性和安全特性的第二代化合物进行深入研究。在大量化合物中,卡铂和奥沙利铂(分别在 1992 年和 2003 年)已经在全世界范围内用于临床治疗,其他一些药物(如洛

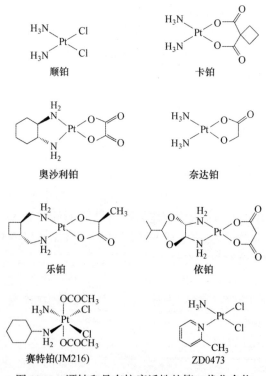

图 13-1 顺铂和具有抗癌活性的第二代化合物

巴铂、奈达铂和七肽)已经在当地(分别是中国、日本和韩国)获得批准,许多化合物(2008年大约有40种[21])也已经通过临床试验,例如,赛特铂(JM216)和ZD0473(图13-1)。关于不同组成、不同配体、不同结构的经典和新的Pt(Ⅱ)、Pt(Ⅳ)基化合物,其治疗活性研究的详细描述参考文献[14,17,19-21,109-110]。

钌化合物对某些肿瘤,特别是对顺铂产生耐药性的肿瘤的治疗已引起人们的关注,它们是下一代抗癌药物最有可能的候选药。在众多被研究对象中,钌的各种配体(如胺、二甲亚砜、多胺聚羧酸酯、杂环)化合物,以及钌与咪唑(Im)和吲唑(In)的复合物是最有希望和选择用于临床开发的。例如,[HIm][trans-[RuCl$_4$(Im)$_2$](KP418)、[HIn][trans-RuCl$_4$(In)$_2$](KP1019)、[HIm][trans-RuCl$_4$(DM-SO)Im](NAMI-A)以及Na[trans-RuCl$_4$(In)$_2$](KP1339)](图13-2)。与顺铂相比,钌化合物的副作用更低,并且它们抑制转移扩散的能力促使人们对这类化合物展开深入的研究[14-16,18,22,25,111-113]。

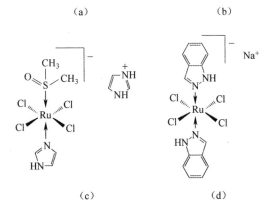

图13-2 用于抗癌临床研究的钌化合物

目前,临床和药物研究中最具挑战的是解释药物的细胞毒性活性机制、药物向癌细胞的输送、药物在人体内的分布和代谢、药物与蛋白质和DNA(作为靶药物分子)的相互作用以及副作用[114-118]。体液和人体组织中元素含量的分析结果为药物在人体内的分布提供数据,形态分析对于解释药物与靶生物分子的相互作用具有重要作用。在生理条件下对化合物与生物配体进行体外培育实验,对于研究它们的作用机制具有重要作用,这种模拟实验要求体外基体的组成尽可能与所研究的生物材料的组成接近。

13.4.1 临床样品中铂的测定

无论是在临床治疗期间还是在暴露于环境之后,测定体液和人体组织中的总铂含量都需要有足够检出限和选择性的仪器分析技术。ICP-MS 为生物样品中铂的测定提供了很好的检出限,例如,顺铂、卡铂和奥沙利铂化疗后人血浆超滤液中的 Pt 为 7.50ng/L[119];奥沙利铂和 5-氟脲嘧啶化疗后血液、血清和血浆超滤液中的 Pt 为 0.1μg/mL[120];不同暴露时间和不同浓度顺铂化疗后卵巢癌细胞 DNA 中 Pt 为 26pg/g[121];顺铂化疗后外围血液单细胞和组织 DNA 中 Pt 为 0.75pg[122]、血清中 Pt 为 1.0μg/L、血浆超滤液中 Pt 为 0.1μg/L、尿液中 Pt 为 2μg/L[123]。ICP-MS 还能够检测未暴露的人体内 Pt 的生理水平,如血液中 Pt 为 0.3~1.3ng/L(检出限为 0.3ng/L)[46]、尿液中 Pt 为 0.48~7.7ng/L(检出限为 0.24ng/L)[47]、肺中 Pt 为 0.778ng/g、肝脏中 Pt 为(0.031~1.42)ng/g、肾组织中 Pt 为 0.051~0.422ng/g(检出限分别为 20pg/g、20pg/g 和 34pg/g 干重)[124]。体液和人体组织中铂的总含量也可以通过 ET-AAS[44,125-135]、AV[99,136-138] 和 NAA[139-141] 测定。AAS 法可用于测定高蛋白质样品超滤液中不同肽组分(<5000Da,<50000Da 和<100000Da)的 Pt 含量[142]。毛细管电泳法分离制备的 DNA 组分中的 Pt[143],以及测定长期储存 5~78 个月条件下卡铂的消解产物(HPLC 法)[133]。AV 法的检出限极低,可用作长期(10 年以上)使用 Pt 药物治疗的患者的检查[136-137]。Pt 在人体和动物体内的检测可用 NAA 法[139],先在铌电极上电解分离金[141],然后萃取二乙基二硫代氨基甲酸金配合物[140]、将金沉淀后提取 Mo(Mo 可在癌细胞中富集)。顺铂在肿瘤细胞内的生物富集系数为 1.8~3.8。

分离技术,特别是离子色谱和毛细管电泳,配上元素特异性分子检测器,能够研究生理条件下产生的各种铂形态,这是解释这些铂类药物代谢和治疗特性的基础[42,116,144-152]。顺铂的一水合化合物是癌细胞在低氯浓度下的水解产物,与 DNA 最具反应性,将繁殖的活细胞(间皮瘤细胞模型 P31)与临床前相关浓度的顺铂培养 24h 后,用两种互补的 LC-ICP-MS 技术,定性分析出一小部分低分子量的铂形态[153]。应用 HPLC-ICP-MS 和 HPLC-MS/MS 技术的实例有测定人血浆中完整的顺铂及其一水合化合物[154],以及尿中药物 ZD0472 的代谢产物[155]和动物血浆超滤液[156]。以二甲基甲酰胺为流动相调节剂,优化 LC-ICP-MS 测定顺铂及其水合物的检出限[157],与 HPLC-ESI(TOF)MS 相比,采用 HPLC-ICP(Q)MS 测定患者尿样中卡铂的灵敏度和不确定度有显著提高(100 倍)[158]。

HPLC-ICP-MS、ESI-MS、MALDI-MS 和串联 MS 技术在这类研究中的实例有铂类药物与血蛋白、白蛋白和铁传递蛋白相互作用的研究[159],大肠癌患者血样中草酸铂和血红蛋白相互作用的研究[160],ZD0743 与尿蛋白加合物的相互作用的研究[161],以及顺铂、反铂和草酸铂与低分子量蛋白质的相互作用的研究[162]。药物与含硫配体(蛋氨酸和半胱氨酸)的相互作用会产生比母体化合物毒性更强的加合物,可通过 ICP-MS 对 Pt:S 比值的评价来检测这类加合物[163]。顺铂、顺式-$[PtCl(NH_3)_2(H_2O)]^+$、顺式-$[Pt(NH_3)_2(H_2O)_2]^{2+}$ 和 4 种蛋氨酸的加合物的检出限分别为 0.31μg/L、0.25μg/L、3.83μg/L、1.07μg/L、0.56μg/L、0.82μg/L 和 2.38μg/L。液相色谱与元素和分子 MS 检测器联用,对于研究药物与活细胞中 DNA 的相互作用具有重要意义[164-166]。毛细管电泳能够在相关生理条件下(含水介质、pH 和温度)研究其相互作用,对此类研究也是很有价值的[146,149,167-172],凝胶电泳与 ICP-MS 和 MALDI-TOF-MS 联用,可用于测定顺铂与寡核苷酸的相互作用[173]。

13.4.2 临床样品中钌的测定

钌的氧化态及其在生理条件下的配体类型和生物转化过程对抗癌复合药物的药理活性有很大影响。钌在生理条件下以 Ru(Ⅱ)、Ru(Ⅲ)和 Ru(Ⅳ)的形式存在,其中 Ru(Ⅱ)、Ru(Ⅲ)化合物倍受关注,因为 Ru(Ⅲ)比 Ru(Ⅱ)和 Ru(Ⅳ)更具生物惰性,而 Ru(Ⅱ)的生物活性最高。Ru(Ⅱ)是 Ru(Ⅲ)在癌细胞中的还原产物,与健康条件相比,癌细胞中较低的 pH(5~6)和较低的氧分子浓度,以及还原剂(抗坏血酸、谷胱甘肽)的存在,使 Ru(Ⅲ)还原成 Ru(Ⅱ)。Ru(Ⅲ)细胞抑制化合物在癌细胞中被还原为 Ru(Ⅱ),有利于降低药物的毒性。Ru(Ⅲ)化合物、KP1019 及其 Na^+ 盐 KP1339、NAMI-A 和 Ru(Ⅱ)与 RAPTA 配体的有机金属配合物 $[Ru(\eta^6-C_6H_5Me)(PTA)Cl_2]$[174-177] 是最突出的抗代谢化合物,具有比铂化合物更好的抗癌活性和选择性。

对各种化合物的药理活性的分析主要集中在测定钌在人体内(体液和人体组织)的分布、进入癌细胞的转运机制、与血浆蛋白的相互作用、水解和氧化还原反应,以及与 DNA 的相互作用上。目前,已知复合物与转铁蛋白的相互作用,转铁蛋白在癌细胞转运中起主导作用,癌细胞的转移受体数量比健康细胞高 2~12 倍[174],碳酸氢根在转铁蛋白与某些钌化合物的结合作用中的协同作用值得关注[178-179]。研究表明,KP1019 复合物与血浆蛋白、白蛋白和转铁蛋白的结合比核苷酸更强,白蛋白是量最大的血浆蛋白,大多数 Ru 都会与白蛋白结合。根据二维体积排阻/阴离子交换色谱-ICP-MS 的测量结果,在等摩尔的白蛋白和转铁蛋白混合物中,KP1019 配合物与人血浆中转铁蛋白的结合不到 20%。然而在生理条件下,白蛋白(5×10^{-5} mol/L)与转铁蛋白(5×10^{-6} mol/L)含量相差十倍时,这种结合低于 2%[180]。另一种 Ru 化合物 NAMI-A 对白蛋白(电化学研究)的强亲和性也被认可[181]。人们发现许多技术可用于研究药理活性的钌化合物与血清蛋白的结合机制,如荧光光谱法、超滤-紫外-可见分光光度法和毛细管电泳法(用于 KP1039 和 KP1339 与白蛋白的结合)[182]、圆二色谱法和 ESI-MS 法(用于 KP1019 与铁传递蛋白的结合)[179]、CE-UV 法[183]和 CE-ICP-MS 法[184]以及电磁共振法[185]。采用串联体积排阻色谱(SEC/SEC)与 ICP-MS 联用技术,鉴定了细胞液中主要的初始结合配体(KP1019 和 KP1339)通过钌分布到可溶性蛋白部分(40kDa 以下)的大型复合蛋白/聚合物(约 700kDa)[186]。CE 和 ESI-MS/MS 技术对于研究钌基化合物的生物作用机制特别有用[111,187-188]。由于分析步骤和基体的检出限无法严格地相互对应,因此体外培育与生理条件下得到的结果的相关性可能还存在问题。用 ICP-MS 或 ET-AAS 直接(稀释或消化后)检测临床样品是评估钌总含量的最常用方法。

13.5 环境样品中的铂、钯和铑

许多综述文章和书籍讨论了存在于各种环境区域下的 PGM 和人类暴露于 PGMs 下产生的问题[2-8,10-13,189]。铂、钯和铑是最常用于此类目的的材料,本节描述了植物,大气颗粒物和尘埃,土壤和沉积物中 Pt、Pd、Rh 含量的分析方法。

13.5.1 植物

植物有富集金属的能力,食用植物也会导致 PGM 进入人体内。因此,植物在旨在评估人体暴露的分析研究中占主要部分。在靠近公路的土壤里(8.6μg/kg Pt、1.9μg/kg Pd 和 1μg/kg Rh)种植的菠菜、水芹、菌丝和刺荨麻等植物中的金属含量明显高于未受污染的土壤(含量接近金属元素的检出限)[190],这激发了研究者对植物中 PGM 含量测定的兴趣。引入催化剂后的结果表明,公路附近种植的植物中有重金属的生物富集,与 Pt 和 Rh 相比,Pd 在植物中的生物富集潜力最强(Pd>Pt≥Rh),与 Cu 和 Zn 以及某些情况下的 Cd 相比,具有相似的生物富集性。植物对金属元素的吸收呈递减顺序:根<茎<叶。藓类常用作评价 PGM 对环境影响的生物样品。苔藓中各种金属含量已有研究,如一项研究表明苔藓中含有 6.4~27.4ng/g 的 Pt 和 1.2~4.6ng/g 的 Rh[191];另一项研究表明苔藓中含有 30ng/g 的

Pt、2.4ng/g 的 Pd 和 5.4ng/g 的 Rh[192]。

ICP-MS 法和 AV 法能够在样品消解后直接测定溶液中金属元素的含量,通过将其他检测技术与样品预富集过程联用,可以优化该方法的检出限。例如,将 Pt 电沉积到石墨管[193]或将样品蒸发[194]与 AAS 法联用。AV 法具有检出限低、仪器和操作成本低等优点,常用作植物样品的检测,如月桂叶中 Pt[(0.085±0.004) μg/g]和 Pd(0.096±0.005μg/g)[195]、甘菊中 Pt(0.157~0.240μg/kg)[196]、云杉枝条中 Pt(543pg/g)和 Rh(22pg/g)[197]、草样中 Pt(0.03~10μg/kg)和 Rh(0.03~21)μg/kg)[100]。用玻碳固体电极伏安法测定了草中的 Pt,测定值范围为叶中的(19.1±1.6)ng/g 至根中的(136±2)ng/g[198]。

植物中金属生物分子种类的识别和定量是阐明其代谢过程和人类暴露的一个具有挑战性的问题[199]。由于很难从样品中分离原有形态以及现有仪器技术的检出限不足,因此这些数据仍然有限。已经公布的植物中金属的无机形态和有机形态的数据,如接触到[Pt(NH$_3$)$_4$](NO$_3$)$_2$溶液的草样中有 90% 是无机 Pt 和 10% 是有机 Pt[200];在大于 10kDa 的肽组分中有 23% 的无机 Pd(凝胶渗透色谱和 X 射线荧光法)[201];在大约 1kDa 的馏分中鉴定出超过 90% 的无机 Pt,其余的存在于 19kDa~>1000kDa 的馏分中(SEC 法和 UV 法)[202]。

13.5.2 大气颗粒物和粉尘

评估大气颗粒物和粉尘中的 PGM 的含量是很重要的,因为它们可能被吸入和富集在人的肺中。汽车催化器释放的纳米颗粒会被输送到环境的各个部分(水、植物、土壤和沉积物),并转化成更多的生物可利用形态。有数据表明,隧道粉尘中铂的溶解度高于转化器中无机形态的铂[30]。金属的分布和富集取决于交通密度、与道路的距离和气象条件(风、雨)。汽车催化器的年限和速度条件直接影响从催化转化器中释放的纳米颗粒的数量。这会导致不同地区样品中金属含量的大范围变化,并且使所测地区的金属含量的相关性降低(例如,与郊区相比,城市的高密度交通区)[203-204]。不同粒径的空气颗粒物中金属含量和分布有显著差异[205-206]。

大气颗粒物和尘埃是很复杂的,难以完全消解,并且大量的基体元素干扰仪器分析技术对 PGM 的检测,这些都是分析过程所面临的挑战。Hf、Cu、Pb 的浓度超过 PGM 浓度几个数量级,导致 ICP-MS 很难检测出 PGM[48-49],可以用 HCl 进行初步处理,使最终待测溶液中轻金属含量降低,但不会造成 PGM 的损失[93]。使用数学校正法[48,205]、改进样品引入等离子体的条件(干等离子体)[53,205,207-208]、采用动态反应池[97]、扇形磁场[53]和化学分离[50,209]等手段来消除 ICP-MS 检测大气颗粒物中金属元素的干扰。有文献表明,在用 ICP-MS 检测之前,可以采用浊点萃取法从尘埃样品(CRM BCR-723)中定量分离和富集Pt[210]。

在使用AAS检测时,通常需要预处理来分离和富集PGM,例如,在石墨管上电沉积Pt(82.8μg/g)、Pd(61.3μg/g)、Rh(19.6μg/g)和Ru(<DL)[90];在负载N,N-二乙基-N'-苯甲酰硫脲的C18微柱上吸附Pt(20~34)pg/m^3[211]和Pd(4~16)pg/m^3[212];在负载DDTC(179.2ng/g)的富勒烯C60上富集Pd[78];以及在硅胶上与二甲基乙二肟0.22~0.23μg/g络合富集[213]。通过将合适的萃取条件与雾化步骤联用,可以显著提高测定道路粉尘中Pt、Pd、Rh(BCR-723)的方法特征性[214]。在NAA检测之前需要进行分离和富集[215-217],使用阴离子柱Dowex 1X8离子交换色谱法测定道路粉尘14~16ng/g[216]及CW7[(50.4±2.0)μg/kg]、CW8[(76.8±2.7)μg/kg]标准物质候选物中的Pt[217]。使用AV法测定空气颗粒物和道路粉尘中的Pt具有极低的检出限(分别为0.5pg/m^3[218]和0.5ng/g[219]),根据取样地点不同,其Pt含量也不同(如3.0~33.0pg/m^3[218]和0.05~5.1pg/m^3[220])。

13.5.3 土壤和沉积物

土壤和沉积物可以积累大量的PGM。Pd对这种基体的亲和力最强,特别是那些高黏土含量和具有可交换性的金属组分[221]。Pt在这些样品中占主导地位,是由于汽车催化器中有高含量的Pt。德国法兰克福-曼海姆路边土壤中Pt的浓度比原始的地理背景值高70倍[60],德国地区Pt、Pd和Rh的背景值分别为Pt<1μg/kg,Pd≤0.5μg/kg和Rh<0.1μg/kg[222]。土壤和沉积物中金属的积累在道路附近地区增加,并且随着与道路距离的增加和采样深度的增加而减少。土壤和沉积物的物理化学性质(pH、氧化还原电位、盐度和络合剂)促使金属向可溶性更强、可流动性更强和生物可利用形态转化。通过微生物作用使金属发生化学氧化、络合和生物化学转化,进而提高其迁移率。

在测定痕量PGM的分析过程中,复杂的土壤和沉积物基体和过量的基体元素需要特别注意,因为基体组分的大量干扰,基本上限制了直接测定样品消解溶液中的金属元素。扇区磁场(SF)-ICP-MS技术对于复杂样品中PGM的检测具有较高的潜力,适用于FA法消解城市湖泊沉积物后直接测定溶液[62]。但是,即使使用SF-ICP-MS法,土壤和沉积物中Pd的测定也存在困难[223]。有数据表明,直接用ICP-MS法(700ng/g)测定沉积物中Pd的含量比用二乙基二硫代氨基甲酸酯萃取分离后测定的量(21ng/g)[192],或用激光烧蚀ICP-MS(80ng/g)与样品消解后SF-ICP-MS(68ng/g)测得的量都大[224]。消除ICP-MS检测Rh中SrO$^+$和RbO$^+$的干扰是很困难的,会导致测量结果不准确。色谱技术或与Te的共沉淀技术通常用于待测物的分离,如果使用阴离子树脂,可用同位素稀释法进行金属的定量分析,但阴离子与 PGM 配合物的完全洗脱存在一定的困难。动态反应池可以优化金属的测定,用氧气作为反应气体将 Zr 转化为高价氧化物,可以消除 Pd 检测过程中 Zr 的干扰[96]。

13.6 质量控制

贵金属样品(特别是环境来源)的复杂性质、待测金属的极低浓度和大量干扰成为影响测量结果可靠性的根本问题。是否有足够可用的、具有严格一致的基体和类似待测物浓度的有证标准物质(certified reference material, CRM)仍是一个亟待解决的问题。地质 CRM 通常含有贵金属,浓度为 mg/g 级[225-227],这限制了它们用于评价环境样品(ng/g 级)和生物样品(pg/g 级)测量结果的可靠性。2001 年发布的道路粉尘标准物质(BCR-723),含有 Pt(81.3±2.5)μg/kg、Pd(6.0±1.9)μg/kg 和 Rh(12.8±1.3)μg/kg[49,228],广泛用于环境样品(如空气中颗粒物、灰尘、土壤和沉积物)分析结果的质量控制。当缺乏合适的 CRM 时(如临床样本的检测),建议进行不同分析方法和不同实验室间的结果比对,在这些分析中建议使用含实际基体(例如,唾液、血浆、超滤液和肺液)的标准来代替合成溶液。被测样品中不同金属形态的定性和定量是很困难的,这使得检测结果的可靠性存在很大问题。因此,应采用多种仪器技术结合对特定样品进行分析。例如,应用色谱、质谱和电化学法[199]、HPLC-ICP-MS 和 HPLC-MS/MS[156]、ESI-MS 和 MALDI[162]、胶束电动色谱、NMR 和 MS[167]、AAS、ESI-MS 和 CD 光谱[179]、SEC-IC-ICP-MS 和 LC-ESI-MS[180],以及 NMR 和 HPLC[229] 等。

参考文献

[1] Platinum 2013 Interim Review, Johnson Matthey PLC, England
[2] Sobrova, P., Zehnalek, J., Vojtech, A., Beklova, M., Kizek, R.: The effects on soil/water/plant/animal systems by platinum group elements. Cent. Eur. J. Chem. 10, 1369-1382 (2012)
[3] Rosner, G., König, H.P., Coenen-Stass, D., WHO: Platinum. Environmental Health Criteria Series, No. 125. International Programme on Chemical Safety. WHO, Geneva (1991)
[4] Zereini, F., Alt, F. (eds.): Anthropogenic platinum-group element emissions. Their impact on men and environment. Springer, Berlin (2000)
[5] Melber, Ch., Keller, D., Mangelsdorf, I., WHO: Palladium. Environmental Health Criteria Series, No. 226. International Program on Chemical Safety. WHO, Geneva (2002)
[6] Ravindra, K., Bencs, L., Van Grieken, R.: Platinum group elements in the environment and

their health risk.Sci. Total Environ. 318, 1–43 (2004)

[7] Zereini, F., Alt, F. (eds.): Palladium emissions in the environment. Analytical methods, environmental assessment and health effects. Springer, Berlin (2006)

[8] Dubiella-Jackowska, A., Kudlak, B., Polkowska, Z., Namieśnik, J.: Environmental fate of traffic-derived platinum group metals. Crit. Rev. Anal. Chem. 39, 251–271 (2009)

[9] Cooper, J., Beecham, J.: A study of platinum group metals in three-way autocatalysts. Platin. Met. Rev. 57, 281–288 (2013)

[10] Reith, F., Campbell, S.G., Ball, A.S., Pring, A., Southam, G.: Platinum in earth surface environments. Earth Sci. Rev. 131, 1–21 (2014)

[11] Kümmerer, K., Helmers, E., Hubner, P., Mascart, G., Milandri, M., Reinthaler, F., Zwakenberg, M.: European hospitals as a sourcefor platinum in the environment in comparison with other sources. Sci. Total Environ. 225, 155–165 (1999)

[12] Lenz, K., Hann, S., Koellensperger, G., Stefanka, Z., Stingeder, G., Weissenbacher, N., Mahnik, S.N., Fuerhacker, M.: Presence of cancerostatic platinum compounds in hospital wastewater and possible elimination by adsorption to activated sludge. Sci. Total Environ. 345, 141–152 (2005)

[13] Goulle, J.P., Saussereau, E., Mahieu, L., Cellier, D., Spiroux, J., Guerbet, M.: Importance of anthropogenic metals in hospital and urban wastewater: its significance for the environment. Bull. Environ. Contamin. Toxicol. 89, 1220–1224 (2012)

[14] Keppler, B.K. (ed.): Metal complexes in cancer chemotherapy. VCH, Weinheim (1993)

[15] Allesio, E., Mestroni, G., Bergamo, A., Sava, G.: Ruthenium antimetastatic agents. Curr. Top. Med. Chem. 4, 1525–1535 (2004)

[16] Ang, W.H., Dyson, P.J.: Classical and non-classical ruthenium based anticancer drugs: towards targeted chemotherapy. Eur. J. Inorg. Chem. 4003–4018 (2006)

[17] Dyson, P.J., Sava G.: Metal-based antitumor drugs in the post genomic era. Dalton Trans. 1929–1933 (2006)

[18] Kostova, I.: Ruthenium complexes as anticancer agents. Curr. Med. Chem. 13, 1085–1107 (2006)

[19] Lippert, B. (ed.): Cisplatin: chemistry and biochemistry of a leading anticancer drug. Wiley-VCH, Weinheim (2006)

[20] Galanski, M.: Special Issue: Anticancer platinum complexes. State of the art and future prospects. Anti-Cancer Agents Med. Chem. 7, 1–138 (2007)

[21] Jakupec, M.A., Galanski, M., Arion, V.B., Hartinger, Ch.G., Keppler, B.K.: Antitumour metal compounds: more than theme and variations. Dalton Trans. 183–194 (2008)

[22] Levina, A., Mitra, A., Lay, P.A.: Recent developments in ruthenium anticancer drugs. Metallomics 1, 458–470 (2009)

[23] van Rijt,S.H., Sadler, P.J.: Current applications and future potential for bioinorganic chemistry in the development of anticancer drugs. Drug Discov. Today 14, 1089–1097 (2009)

[24] Wheate, N.J., Walker, S., Craig, G.E., Oun, R.: The status of platinum anticancer drugs in the clinic and in clinical trials. Dalton Trans. 39, 8113–8127 (2010)

[25] Bergamo, A., Gaiddon, C., Schellens, J.H.M., Beijnen, J.H., Sava, G.: Approaching tumour therapy beyond platinum drugs: status of the art and perspectives of ruthenium drug candidates. J. Inorg. Biochem. 106, 90–99 (2012)

[26] Muhammad, N., Guo, Z.: Metal-based anticancer chemotherapeutic agents. Curr. Opin. Chem. Biol. 19, 144–153 (2014)

[27] Parrot, J.L., Hébert, R., Saindelle, A., Ruff, F.: Platinum and platinosis. Allergy and histamine release due to some platinum salts. Arch. Environ. Health 19, 685–691 (1969)

[28] Schmid, M., Zimmerman, S., Krug, H.F., Sures, B.: Influence of platinum, palladium and rhodium as compared with cadmium, nickel and chromium on cell viability and oxidative stress in human bronchial epithelial cells. Environ. Int. 33, 385–390 (2007)

[29] Merget, R., Rosner, G.: Evaluation of the health risk of platinum group metals emitted from automotive catalytic converters. Sci. Total Environ. 270, 165–173 (2001)

[30] Ek, K.H., Morrison, G.M., Rauch, S.: Environmental routes for platinum group elements to biological materials- a review. Sci. Total Environ. 334–335, 21–38 (2004)

[31] Gagnon, Z.E., Newkirk, C.E., Hicks, S.: Impact of platinum group metals on the environment: a toxicological, genotoxic and analytical chemistry study. J. Environ. Sci. Health 41, 397–414 (2006)

[32] Wiseman, C.L., Zereini, F.: Airborne particulate matter, platinum group elements and human health: a review of recent evidence. Sci. Total Environ. 407, 2493–2500 (2009)

[33] Balcerzak, M.: Sample digestion methods for the determination of traces of precious metals by spectrometric techniques. Anal. Sci. 18, 737–750 (2002)

[34] Prasada, R.T., Daniel, S.: Preconcentration of trace and ultratrace amounts of platinum and palladium from real samples. Rev. Anal. Chem. 22, 167–189 (2003)

[35] Godlewska-Żyłkiewicz, B.: Preconcentration and separation procedures for the spectrochemical determination of platinum and palladium. Microchim. Acta 147, 189–210 (2004)

[36] Mokhodoeva, O.B., Myasoedova, G.V., Kubrakova, I.V.: Sorption preconcentration in combined methods for the determination of noble metals. J. Anal. Chem. 62, 607–622 (2007)

[37] Resano, M., Garcl'a-Ruiz, E., Belarra, M.A., Vanhaecke, F., McIntosh, K.S.: Solid sampling in the determination of precious metals at ultratrace levels. Trends Anal. Chem. 26, 385–395 (2007)

[38] Lim, M.D., Dickherber, A., Compton, C.C.: Before you analyze a human specimen think quality, variability, and bias. Anal. Chem. 83, 8–13 (2011)

[39] Barefoot, R.R.: Speciation of platinum compounds: a review of recent applications in studies of platinum anticancer drugs. J. Chromatogr. B 751, 205–211 (2001)

[40] Küng, A., Pieper, T., Wissiack, R., Rosenberg, B., Keppler, B.K.: Hydrolysisof the tumor-inhibiting ruthenium(III) complexes HIm trans-[$RuCl_4(im)_2$] and HInd trans-[$RuCl_4(ind)_2$]

investigated by means of HPCE and HPLC-MS. J. Biol. Inorg. Chem. 6, 292-299 (2001)

[41] Barefoot, R.R., Van Loon, J.C.: Determination of platinum and gold in anticancer and antiarthritic drugs and metabolites. Anal. Chim. Acta 334, 5-14 (1996)

[42] Yang, Z., Hou, X.D., Jones, B.T.: Determination of platinum in clinical samples. Appl. Spectrosc. Rev. 37, 57-88 (2002)

[43] Kinoshita, M., Yoshimura, N., Ogata, H., Tsujino, D., Takahashi, T., Takahashi, S., Wada, Y., Someya, K., Ohno, T., Masuharra, K., Tanaka, Y.: High-performance liquid-chromato-graphic analysis of unchanged cis-diamminedichloroplatinum (cisplatin) in plasma and urine with post-column derivatization. J. Chromatogr. B 529, 462-467 (1990)

[44] Van Warmerdam, L.J.C., van Tellingen, O., Maes, R.A.A., Beijnen, J.H.: Validated method for the determination of carboplatin in biological fluids by Zeeman AAS. Fresenius J. Anal. Chem. 351, 777-781 (1995)

[45] Koellensperger, G., Hann, S.: Ultra-fast HPLC-ICP-MS analysis of oxaliplatin in patient urine. Anal. Bioanal. Chem. 397, 401-406 (2010)

[46] Begerow, J., Turfeld, M., Dunemann, L.: Determination of physiological palladium, platinum, iridium and gold levels inhuman blood using double focusing magnetic sector field inductively coupled plasma mass spectrometry. J. Anal. At. Spectrom. 12, 1095-1098 (1997)

[47] Begerow, J., Turfeld, M., Dunemann, L.: Determination of physiological palladium and platinum levels in urine using double focusing magnetic sector field ICP-MS. Fresenius J. Anal. Chem. 359, 427-429 (1997)

[48] Gómez, M.B., Gómez, M.M., Palacios, M.A.: Control of interferences in the determination of platinum, palladium and rhodium in airborne particulate matter by inductively coupled plasma mass spectrometry. Anal. Chim. Acta 404, 285-294 (2000)

[49] Zischka, M., Schramel, P., Muntau, H., Rehnert, A., Gomez, G.M., Stojanik, B., Wannemaker, G., Dams, R., Quevauviller, P., Maier, E.A.: A new certified reference material for the quality control of palladium, platinum and rhodium in road dust, BCR-723. Trends Anal. Chem. 21, 851-868 (2002)

[50] Gómez, M.B., Gómez, M.M., Palacios, M.A.: ICP-MS determination of platinum, palladium and rhodium in airborne and road dust after tellurium coprecipitation. J. Anal. At. Spectrom. 18, 80-83 (2003)

[51] Kanitsar, K., Köllensperger, G., Hann, S., Limbeck, A., Puxbaum, H., Stingeder, G.: Determination of Pt, Pd and Rh by inductively plasma sector field mass spectrometry (ICP-SFMS) in size-classified urban aerosol samples. J. Anal. At. Spectrom. 18, 239-246 (2003)

[52] Müller, M., Heumann, K.G.: Isotope dilution inductively coupled plasma quadrupole mass spectrometry in connection with a chromatographic separation for ultra-trace determinations of platinum group elements (Pt, Pd, Ru, Ir) in environmental samples. Fresenius J. Anal. Chem. 368, 109-115 (2000)

[53] Fragnière, C., Haldimann, M., Eastgate, A., Krähenbühl, U.: A direct ultratrace

determination of platinum in environmental, food and biological samples by ICP-SFMS using a desolvation system. J. Anal. At. Spectrom. 20, 626–630 (2005)

[54] Mukai, H., Ambe, Y., Morita, M.: Flow injection inductively coupled plasma mass spectrometry for the determination of platinum in airborne particulate matter. J. Anal. At. Spectrom. 5, 75–80 (1990)

[55] Niemelä, M., Kola, H., Perämäki, P., Piispanen, J., Poikolainen, J.: Comparison of microwave-assisted digestion methods and selection of internal standards for the determination of Rh, Pd and Pt in dust samples by ICP-MS. Microchim. Acta 150, 211–217 (2005)

[56] Heinrich, E., Schmidt, G., Kratz, K.L.: Determination of platinum-group elements (PGE) from catalytic convertors in soil by means of docimasy and INAA. Fresenius J. Anal. Chem. 354, 883–885 (1996)

[57] Farago, M.E., Kavanagh, P., Blanks, R., Kelly, J., Kazantzis, G., Thornton, I., Simpson, P.R., Cook, J.M., Delves, H.T., Hall, G.E.M.: Platinum concentrations in urban road dust and soil, and in blood and urine in the United Kingdom. Analyst 123, 451–454 (1998)

[58] Schäfer, S., Eckhardt, J.D., Berner, Z.A., Stüben, D.: Time-dependent increase of trafficemitted platinum-group elements (PGE) in different environmental compartments. Environ. Sci. Technol. 33, 3166–3170 (1999)

[59] Hutchinson, E.J., Farago, M.E., Simpson, P.R.: Changes in platinum concentrations in soils and dusts from UK cities. In: Zereini, F., Alt, F. (eds.) Anthropogenic platinum-group element emissions. Their impact on men and environment, pp. 57–64. Springer, Berlin (2000)

[60] Zereini, F., Skerstupp, B., Rankenburg, K., Dirksen, F., Beyer, J.M., Claus, T., Urban, H.: Anthropogenic emission of platinum-group elements (Pt, Pd and Rh) into the environment: concentration, distribution and geochemical behaviour in soils. In: Zereini, F., Alt, F. (eds.) Anthropogenic platinum-group element emissions. Their impact on men and environment, pp. 73–84. Springer, Berlin (2000)

[61] de Vos, E., Edwards, S.J., MsDonald, I., Wray, D.S., Carey, P.J.: A baseline survey of the distribution and origin of platinumgroup elements in contemporary fluvial sediments of the Kentish Stour, England. Appl. Geochem. 17, 1115–1121 (2002)

[62] Rauch, S., Hemond, H.F., Peucker-Ehrenbrink, B.: Recent changes in platinum group element concentrations and osmium isotopic compositionin sediments from an urban lake. Environ. Sci. Technol. 38, 396–402 (2004)

[63] Perry, B.J., Van Loon, J.C., Speller, D.V.: Dry-chlorination inductively coupled plasma massspectrometric method for the determination of platinum group elements in rocks. J. Anal. At. Spectrom. 7, 883–888 (1992)

[64] Enzweiler, J., Potts, P.J.: The separation of platinum, palladium and gold from silicate rocks by the anion exchange separation of chloro complexes after a sodium peroxide fusion: an investigation of low recoveries. Talanta 42, 1411–1418 (1995)

[65] Akatsuka, K., McLaren, J.W.: Preconcentration/separation methods for the determination of

trace platinum in environmental samples by ICP-MS. In: Zereini, F., Alt, F. (eds.) Anthropogenic platinum-group element emissions. Their impact on men and environment, pp. 123–131. Springer, Berlin (2000)

[66] Balcerzak, M.: Analytical chemistry of noble metals. In: Meyers, R.A. (ed.) Encyclopedia of analytical chemistry, pp. 8958–8984. Wiley, Chichester (2000)

[67] Schumann, D., Fischer, St., Taut, St., Novgorodov, A.F., Misiak, R., Lebedev, N.A., Bruchertseifer, H.: Sorption of microamounts of Hf- and Ta-nuclides on Dowex 50x8 and Dowex 1x8 from HCl/HF containing aqueous solutions. J. Radioanal. Nucl. Chem. Lett. 187, 9–17 (1994)

[68] Ely, J.C., Neal, C.R., O'Neill Jr., J.A., Jain, J.C.: Quantifying the platinum group elements (PGEs) and gold in geological samples using cation exchange pretreatment and ultrasonic nebulization inductively coupled plasma-mass spectrometry (USN-ICP-MS). Chem. Geol. 157, 219–234 (1999)

[69] Balcerzak, M.: Platinum and hafnium. Analytical couple in ICP-MS technique. Analityka 1, 12–20 (2009) (in Polish)

[70] Balcerzak, M.: Methods for the elimination of hafnium interference with the determination of platinum in environmental samples by ICP-MS technique. Chem. Anal. (Warsaw) 54, 135–149 (2009)

[71] Mikołajczuk, A., Balcerzak, M.: Platinum and hafnium. The separation on Dowex 50W-X8 resin. Analityka 2, 4–5 (2009) (in Polish)

[72] Hodge, V., Stallard, M., Koide, M., Goldberg, E.D.: Determination of platinum and iridium in marine waters, sediments, and organisms. Anal. Chem. 58, 616–620 (1986)

[73] Chwastowska, J., Skwara, W., Sterlinska, E., Pszonicki, L.: Determination of platinum and palladium in environmental samples by graphite furnace atomic absorption spectrometry after separation on dithizone sorbent. Talanta 64, 224–229 (2004)

[74] Gonzalez, G.M., Sanchez Rojas, F., Bosch Ojeda, C., Garcia de Torres, A., Cano Pavon, J.M.: On-line ion-exchange preconcentration and determination of traces of platinum by electrothermal atomic absorption spectrometry. Anal. Bioanal. Chem. 375, 1229–1233 (2003)

[75] Bosch, O.C., Sanchez, R.F., Cano, P.J., Garcia de Torres, A.: Automated on-line separation-preconcentration system for platinum determination by electrothermal atomic absorption spectrometry. Anal. Chim. Acta 494, 97–103 (2003)

[76] Lin, H., Huang, Z.J., Hu, Q., Yang, G., Zhang, G.: Determination of palladium, platinum, and rhodium by HPLC with online column enrichment using 4-carboxyphenyl-thiorhodanine as a precolumn derivatization reagent. J. Anal. Chem. 62, 58–62 (2007)

[77] Rudolph, E., Limbeck, A., Hann, S.: Novel matrix separation—on-line pre-concentration procedure for accurate quantification of palladium in environmental samples by isotope dilution inductively coupled plasma sector field mass spectrometry. J. Anal. At. Spectrom. 21, 1287–1293 (2006)

[78] Leśniewska, B.A., Godlewska, I., Godlewska-Żyłkiewicz, B.: The study of applicability of di-

thiocarbamate-coated fullerene C60for preconcentration of palladium for graphite furnace atomic absorption spectrometric determination in environmental samples. Spectrochim. Acta B 60, 377–384 (2005)

[79] Liang, P., Zhao, E., Ding, Q., Du, D.: Multiwalled carbon nanotubes microcolumn preconcentration and determination of gold in geological and water samples by flame atomic absorption spectrometry. Spectrochim. Acta B 63, 714–717 (2008)

[80] Ghaseminezhad, S., Afzali, D., Taher, M.A.: Flame atomic absorption spectrometry for the determination of trace amount of rhodium after separation and preconcentration onto modified multiwalled carbon nanotubes as a new solid sorbent. Talanta 80, 168–172 (2009)

[81] Afzali, D., Ghaseminezhad, S., Taher, M.A.: Separation and preconcentration of trace amounts of gold(III) ions using modified multiwalled carbon nanotube sorbent prior to flame atomic absorption spectrometry determination. J. AOAC Int. 93, 1287–1292 (2010)

[82] Cheng, X.L., Chen, S.H., Wang, X., Liu, C.: Single-walled carbon nanotubes as solid-phase extraction adsorbent for the preconcentration and determination of precious metals by ICP-MS. At. Spectrosc. 31, 75–80 (2010)

[83] Yuan Ch, G., Zhang, Y., Wang, S., Chang, A.: Separation and preconcentration of palladium using modified multi-walled carbon nanotubes without chelating agent. Microchim. Acta 173, 361–367 (2011)

[84] Behbahani, M., Gorji, T., Mahyari, M., Salarian, M., Bagheri, A., Shaabani, A.: Application of polypropylene amine dendrimers (POPAM)-Grafted MWCNTs hybrid materials as a new sorbent for solid-phase extraction and trace determination of gold(III) and palladium(II) in food and environmental samples. Food Anal. Methods 7, 957–966 (2014)

[85] Leśniewska, B., Kosińska, M., Godlewska-Żyłkiewicz, B., Zambrzycka, E., Wilczewska, A.Z.: Selective solid phase extraction of platinum on an ion imprinted polymers for its electrothermal atomic absorption spectrometric determination in environmental samples. Microchim. Acta 175, 273–282 (2011)

[86] Daniel, S., Praveen, R.S., Rao, T.P.: Ternary ion-association complex based ion imprinted polymers (IIPs) for trace determination of palladium(II) in environmental samples. Anal. Chim. Acta 570, 79–87 (2006)

[87] Meeravali, N.N., Jiang, S.J.: Interference free ultra trace determination of Pt, Pd and Au in geological and environmental samples by inductively coupled plasma quadrupole mass spectrometry after a cloud point extraction. J. Anal. At. Spectrom. 23, 854–860 (2008)

[88] Tresl, I., Mestek, O., Suchanek, M.: The isotope-dilution determination of platinum in soil by inductively coupled plasma mass spectrometry. Collect. Czech. Chem. Commun. 65, 1875–1887 (2000)

[89] Komárek, J., Krásensky, P., Balcar, J., Řehulka, P.: Determination of palladium and platinum by electrothermal atomic absorption spectrometry afterdeposition on a graphite tube. Spectrochim. Acta B 54, 739–743 (1999)

[90] Matusiewicz, H., Lesiński, M.: Electrodeposition sample introduction for ultra trace determinations of platinum group elements (Pt, Pd, Rh and Ru) in road dust by electrothermal atomic absorption spectrometry. Int. J. Environ. Anal. Chem. 82, 207–223 (2002)

[91] Perry, B.J., Barefoot, R.R., Van Loon, J.C.: Inductively coupled plasma mass spectrometry for the determination of platinum group elements and gold. Trends Anal. Chem. 14, 388–397 (1995)

[92] Zischka, M., Wegscheider, W.: Reliability of and measurement uncertainty for the determination of Au, Pd, Pt and Rh by ICP-MS in environmentally relevant samples. In: Zereini, F., Alt, F. (eds.) Anthropogenic platinum-group element emissions. Their impact on men and environment, pp. 201–214. Springer, Berlin (2000)

[93] Simitchiev, K., Stefanova, V., Kmetov, V., Andreev, G., Sanchez, A., Canals, A.: Investigation of ICP-MS spectral interferences in the determination of Rh, Pd and Pt in road dust: assessment of correction algorithms via uncertainty budget analysis and interference alleviation by preliminary acid leaching. Talanta 77, 889–896 (2008)

[94] Köllensperger, G., Hann, S., Stingeder, G.: Determination of rhodium, palladium and platinum in environmental silica containing matrices: capabilities and limitations of ICP-SFMS. J. Anal. At. Spectrom. 15, 1553–1557 (2000)

[95] Barefoot, R.R.: Determination of platinum group elements and gold in geological materials: a review of recent magnetic sector and laser ablation applications. Anal. Chim. Acta 509, 119–125 (2004)

[96] Simpson, L.A., Thomsen, M., Alloway, B.J., Parker, A.: A dynamic reaction cell (DRC) solution to oxide-based interferences in inductively coupled plasma mass spectrometry (ICP-MS) analysis of the noble metals. J. Anal. At. Spectrom. 16, 1375–1380 (2001)

[97] Kan, S.F., Tanner, P.A.: Determination of platinum in roadside dust samples by dynamic reaction cell-inductively coupled plasma-mass spectrometry. J. Anal. At. Spectrom. 19, 639–643 (2004)

[98] Locatelli, C.: Voltammetric analysis of trace levels of platinum group metals-principles and applications. Electroanalysis 19, 2167–2175 (2007)

[99] Messerschmidt, J., Alt, F., Tölg, G., Angerer, J., Schaller, K.H.: Adsorptive voltammetric procedure for the determination of platinum baseline levels in human body fluids. Fresenius J. Anal. Chem. 343, 391–394 (1992)

[100] Helmers, E., Mergel, N.: Platinum and rhodium in a polluted environment: studying the emissions of automobile catalysts with emphasis on the application of cathodic-stripping voltammetry (CSV) rhodium analysis. Fresenius J. Anal. Chem. 362, 522–528 (1998)

[101] Van der Horst, C., Silwana, B., Iwuoha, E., Somerset, V.: Stripping voltammetric determination of palladium, platinum and rhodium in fresh water and sediment samples from South African water resources. J. Environ. Sci. Health 47, 2084–2093 (2012)

[102] Dalvi, A.A., Satpati, A.K., Palrecha, M.M.: Simultaneous determination of Pt and Rh by

catalytic adsorptive stripping voltammetry, using hexamethylene tetramine (HMTA) as complexing agent. Talanta 75, 1382–1387 (2008)

[103] Ulakhovich, N.A., Budnikov, G.K., Medyantseva, E.P.: Preconcentration in voltammetric analysis of platinum metal materials. Zh. Anal. Khim. 47, 1546–1566 (1992)

[104] Dybczyński, R.: Neutron activation analysis and its contribution to inorganic trace analysis. Chem. Anal. (Warsaw) 46(133–160) (2001)

[105] Alfassi, Z.B., Probst, T.U., Rietz, B.: Platinum determination by instrumental neutron activation analysis with special reference to the spectral interference of scandium-47 on the platinum indicator nuclide gold-199. Anal. Chim. Acta 360, 243–252 (1998)

[106] Garuti, G., Meloni, S., Oddone, M.: Neutron activation analysis of platinum group elements and gold in reference materials: a comparison of two methods. J. Radioanal. Nucl. Chem. 245, 17–23 (2000)

[107] Chajduk-Maleszewska, E., Dybczyn'ski, R.: Selective separation and preconcentration of trace amounts of Pd on Duolite ES 346 resin and its use for the determination of Pd by NAA. Chem. Anal. (Warsaw) 49(281–297) (2004)

[108] Chai, Z., Mao, X., Hu, Z., Zhang, Z., Chen, C., Feng, W., Hu, S., Ouyang, H.: Overview of the methodology of nuclear analytical techniques for speciation studies of trace elements in the biological and environmental science. Anal. Bioanal. Chem. 372, 407–411 (2002)

[109] Wong, E., Giandomenico Ch, M.: Current status of platinum-based antitumor drugs. Chem. Rev. 99, 2451–2466 (1999)

[110] Galanski, M., Jakupec, M.A., Keppler, B.K.: Update of the preclinical situation of anticancer platinum complexes: novel design strategies and innovative analytical approaches. Curr. Med. Chem. 12, 2075–2094 (2005)

[111] Küng, A., Pieper, T., Keppler, B.K.: Investigations into the interaction between tumorinhibiting Ru(Ⅲ) complexes and nucleotides by capillary electrophoresis. J. Chromatogr. B759, 81–89 (2001)

[112] Timerbaev, A.R., Hartinger, C.G., Aleksenko, S.S., Keppler, B.K.: Interactions of antitumor metallodrugs with serum proteins: advances in characterization using modern analytical methodology. Chem. Rev. 106, 2224–2248 (2006)

[113] Reisner, E., Arion, V.B., Keppler, B.K., Pombeiro, A.J.M.: Electron-transfer activated metal-based anticancer drugs. Inorg. Chim. Acta 361, 1569–1583 (2008)

[114] Goodisman, J., Hagrman, D., Tacka, K.A., Souid, A.K.: Analysis of cytotoxicities of platinum compounds. Cancer Chemother. Pharmacol. 57, 257–267 (2006)

[115] Todd, R.C., Lippard, S.J.: Inhibition of transcription by platinum antitumor compounds. Metallomics 1, 280–291 (2009)

[116] Esteban-Fernández, D., Moreno-Gordaliza, E., Cañas, B., Palacios, M.A., Gómez-Gómez, M.M.: Analytical methodologies for metallomics studies of antitumor Pt-containing drugs. Metallomics 2, 19–38 (2010)

[117] Crider, S.E., Holbrook, R.J., Franz, K.J.: Coordination of platinum therapeutic agents to met-rich motifs of human copper transport protein. Metallomics 2, 74–83 (2010)

[118] Groessl, M., Terenghi, M., Casini, A., Elviri, L., Łobiński, R., Dyson, P.J.: Reactivity of anticancer metallodrugs withserum proteins: new insights from size exclusion chromatography-ICP-MS and ESI-MS. J. Anal. At. Spectrom. 25, 305–313 (2010)

[119] Brouwers, E.E.M., Tibben, M.M., Rosing, H., Hillebrand, M.J.X., Joerger, M., Schellens, J.H.M., Beijnen, J.H.: Sensitive inductively coupled plasma mass spectrometry assay for the determination of platinum originating from cisplatin, carboplatin, and oxaliplatin in human plasma ultrafiltrate. J. Mass Spectrom. 41, 1186–1194 (2006)

[120] Morrison, J.G., White, P., McDougall, S., Firth, J.W., Woolfrey, S.G., Graham, M.A., Greenslade, D.: Validation of a highly sensitive ICP-MS method for the determination of platinum in biofluids: application to clinical pharmacokinetic studies with oxaliplatin. J. Pharm. Biomed. Anal. 24, 1–10 (2000)

[121] Bjorn, E., Nygren, Y., Nguyen, T.T.T.N., Ericson, C., Nojd, M., Naredi, P.: Determination of platinum in human subcellular microsamples by inductively coupled plasma mass spectrometry. Anal. Biochem. 363, 135–142 (2007)

[122] Brouwers, E.E.M., Tibben, M.M., Pluim, D., Rosing, H., Boot, H., Cats, A., Schellens, J.H.M., Beijnen, J.H.: Inductively coupled plasma mass spectrometric analysis of the total amount of platinum in DNA extracts from peripheral blood mononuclear cells and tissue from patients treated with cisplatin. Anal. Bioanal. Chem. 391, 577–585 (2008)

[123] Bettinelli, M.: ICP-MS determination of Pt in biological fluids of patients treated with antitumor agents: evaluation of analytical uncertainty. Microchem. J. 79, 357–365 (2005)

[124] Rudolph, E., Hann, S., Stingeder, G., Reiter, C.: Ultra-trace analysis of platinum in human tissue samples. Anal. Bioanal. Chem. 382, 1500–1506 (2005)

[125] Begerow, J., Turfeld, M., Dunemann, L.: Determination of physiological noble metals in human urine using liquid-liquid extraction and Zeeman electrothermal atomic absorption spectrometry. Anal. Chim. Acta 340, 277–283 (1997)

[126] Milačič, R., Čemažar, M., Serša, G.: Determination of platinum in tumor tissues after cisplatin therapy by electrothermal atomic absorption spectrometry. J. Pharm. Biomed. Anal. 16, 343–348 (1997)

[127] Ivanova, E., Adams, F.: Flow injection on-line sorption preconcentration of platinum in a knotted reactor coupled with electrothermal absorption spectrometry. Fresenius J. Anal. Chem. 361, 445–450 (1998)

[128] Kern, W., Braess, J., Bottger, B., Kaufmann, C.C., Hiddemann, W., Schleyer, E.: Oxaliplatin pharmacokinetics during a four-hour infusion. Clin. Cancer Res. 5, 761–765 (1999)

[129] Kloft, C., Appelius, H., Siegert, W., Schunack, W., Jaehde, U.: Determination of platinum complexes in clinical samples by a rapid flameless atomic absorption spectrometry assay. Therap. Drug Monit. 21, 631–637 (1999)

[130] Meerum, T.J.M., Tibben, M.M., Welbank, H., Schellens, J.H.M., Beijnen, J.H.: Validated method for the determination of platinum from a liposomal source (SPI-77) in human plasma using graphite furnace Zeeman atomic absorption spectrometry. Fresenius J. Anal. Chem. 366, 298–302 (2000)

[131] Tibben, M.M., Rademaker-Lakhai, J.M., Rice, J.R., Stewart, D.R., Schellens, J.H.M., Beijnen, J.H.: Determination of total platinum in plasma and plasma ultrafiltrate, from subjects dosed with the platinum-containing N-(2-hydroxypropyl)methacrylamide copolymer AP5280, by use of graphite-furnace Zeeman atomic-absorption spectrometry. Anal. Bioanal. Chem. 373, 233–236 (2002)

[132] Vouillamoz-Lorenz, S., Bauer, J., Lejeune, F., Decosterd, L.A.: Validation of an AAS method for the determination of platinum in biological fluids from patients receiving the oral platinum derivative JM216. J. Pharm. Biomed. Anal. 25, 465–475 (2001)

[133] Schnurr, B., Heinrich, H., Gust, R.: Investigations on the decompositions of carboplatin in infusion solutions II. Effect of 1,1-cyclobutanedicarboxylic acid admixture. Microchim. Acta 140, 141–148 (2002)

[134] Zimmermann, S., Messerschmidt, J., von Bohlen, A., Sures, B.: Determination of platinum, palladium and rhodium in biological samples by electrothermal atomic absorption spectrometry as compared with adsorptive cathodic stripping voltammetry and total-reflection X-ray fluorescence analysis. Anal. Chim. Acta 498, 93–104 (2003)

[135] Brouwers, E.E.M., Tibben, M.M., Joerger, M., van Tellingen, O., Rosing, H., Schellens, J.H.M., Beijnen, J.H.: Determination of oxaliplatin in human plasma and plasma ultrafiltrate by graphite-furnace atomic absorption spectrometry. Anal. Bioanal. Chem. 382, 1484–1490 (2005)

[136] Schierl, R., Rohrer, B., Hohnloser, H.: Long term platinum excretion in patients treated with cisplatin. Cancer Chemother. Pharmacol. 36, 75–78 (1995)

[137] Gelevert, T., Messerschmidt, J., Meinardi, M.T., Alt, F., Gietema, J.A., Franke, J.P., Sleijfer, D.T., Uges, D.R.A.: Adsorptive voltammetry to determine platinum levels in plasma from testicular cancer patients treated with cisplatin. Therap. Drug Monit. 23, 169–173 (2001)

[138] Zimmermann, S., Menzel, C.M., Berner, Z., Eckhardt, J.D., Stüben, D., Alt, F., Messerschmidt, J., Taraschewski, H., Sures, B.: Trace analysis of platinum in biological samples: a comparison between sector field ICP-MS and adsorptive cathodic stripping voltammetry following different digestion procedures. Anal. Chim. Acta 439, 203–209 (2001)

[139] Trebert, H.S., Lux, F., Karl, J., Spruss, T., Schönenberger, H.: Determination of platinum and biologically important traceelements in structure-activity relationship studies on platinum-containing anticancer drugs. Special procedures for removing phosphorus-32 as well as for the estimation of molybdenum-99 and gold-199. J. Radioanal. Nucl. Chem. 113, 461–467 (1987)

[140] Taskaev, E., Karaivanova, M., Grigorov, T.: Determination of platinum and gold in biological

materials by neutron activation analysis. J. Radioanal. Nucl. Chem. 120, 75-82 (1988)

[141] Rietz, B., Krarup-Hansen, A., Rørth, M.: Determination of platinum by radiochemical neutron activation analysis in neural tissues from rats, monkeys and patients treated with cisplatin. Anal. Chim. Acta 426, 119-126 (2001)

[142] Einhäuser, T.J., Galanski, M., Keppler, B.K.: Determination of platinum in protein-bound CDDP and DPD by inductively coupled plasma optical emission spectrometry and electrothermal atomic absorption spectrometry. J. Anal. At. Spectrom. 11, 747-750 (1996)

[143] Deforce, D.L.D., Kokotos, G., Esmans, E.L., De Leenheer, A.P., Van den Eeckhout, E. G.: Preparativecapillary zone electrophoresis in combination with off-line graphite furnace atomic absorption for the analysis of DNA complexes formed by a new aminocoumarine platinum (II) compound. Electrophoresis 19, 2454-2458 (1998)

[144] Sommer, L., Vlašánková, R.: Survey of the potential of high-performance liquid chromatography and capillary zone electrophoresis for the determination of platinum and platinumgroup metals. Chromatographia 52, 692-702 (2000)

[145] Warnke, U., Gysler, J., Hofte, B., Tjaden, U.R., van der Greef, J., Kloft, C., Schunack, W., Jaehde, U.: Separation and identification of platinum adducts with DNA nucleotides by capillary zone electrophoresis and capillary zone electrophoresis coupled to mass spectrometry. Electrophoresis 22, 97-103 (2001)

[146] Timerbaev, A.R., Küng, A., Keppler, B.K.: Capillary electrophoresis of platinum-group elements. Analytical, speciation and biochemical studies. J. Chromatogr. A 945, 25-44 (2002)

[147] Hann, S., Stefánka, Z., Lenz, K., Stingeder, G.: Novel separation methodfor highly sensitive speciation of cancerostatic platinum compounds by HPLC-ICP-MS. Anal. Bioanal. Chem. 381, 405-412 (2005)

[148] Stokvis, E., Rosing, H., Beijnen, J.H.: Liquid chromatography-mass spectrometry for the quantitative bioanalysis of anticancer drugs. Mass Spectrom. Rev. 24, 887-917 (2005)

[149] Bytzek, A.K., Reithofer, M.R., Galanski, M., Groessl, M., Keppler, B.K., Hartinger, C. G.: The first example of MEEKC-ICP-MS coupling and its application for the analysis of anticancer platinum complexes. Electrophoresis 31, 1144-1150 (2010)

[150] Ito, H., Yamaguchi, H., Fujikawa, A., Tanaka, N., Furugen, A., Miyamori, K., Takahashi, N., Ogura, J., Kobayashi, M., Yamada, T., Mano, N., Iseki, K.: A full validated hydrophilic interaction liquid chromatography-tandem mass spectrometric method for the quantification of oxaliplatin in human plasma ultrafiltrates. J. Pharm. Biomed. Anal. 71, 99-103 (2012)

[151] Groessl, M., Hartinger, C.G.: Anticancer metallodrug research analytically painting the "omics" picture-current developments and future trends. Anal. Bioanal. Chem. 405, 1791-1808 (2013)

[152] Yang, Z., Sweedler, J.V.: Application of capillary electrophoresis for the early diagnosis of cancer. Anal. Bioanal. Chem. 406, 4013-4031 (2014)

[153] Falta, T., Heffeter, P., Mohamed, A., Berger, W., Hann, S., Koellensperger, G.: Quantitative determination of intact free cisplatin in cell models by LC-ICP-MS. J. Anal. At. Spectrom. 26, 109–115 (2011)

[154] Bell, D.N., Liu, J.J., Tingle, M.D., McKeage, M.J.: Specific determination of intact cisplatin and monohydrated cisplatin in human plasma and culture medium ultrafiltrates using HPLC on-line with inductively coupled plasma mass spectrometry. J. Chromatogr. B 837, 29–34 (2006)

[155] Smith, C.J., Wilson, I.D., Abou-Shakra, F.R., Payne, R., Grisedale, H., Long, A., Roberts, D., Malone, M.: Analysis of a [^{14}C]-labelled platinum anticancer compound in dosing formulations and urine using a combination of HPLC-ICPMS and flow scintillation counting. Chromatographia 55, S151–S155 (2002)

[156] Smith, C.J., Wilson, I.D., Abou-Shakra, F., Payne, R., Parry, T.C., Sinclair, P., Roberts, D.W.: A comparison of the quantitative methods for the analysis of the platinum-containing anticancer drug {cis-amminedichloro(2-methylpyridine)]-platinum(II)} (ZD0473) by HPLC coupled to either a triple quadrupole mass spectrometer or an inductively coupled plasma mass spectrometer. Anal. Chem. 75, 1463–1469 (2003)

[157] Nygren, Y., Hemström, P., Åstot, C., Naredi, P., Björn, E.: Hydrophilic interaction liquid chromatography (HILIC) coupled to inductively coupled plasma mass spectrometry (ICPMS) utilizing a mobile phase with a low-volatile organic modifier for the determination of cisplatin, and its monohydrolyzed metabolite. J. Anal. At. Spectrom. 23, 948–954 (2008)

[158] Koellensperger, G., Stefanka, Z., Meelich, K., Galanski, M., Keppler, B.K., Stingeder, G., Hann, S.: Species specific IDMS for accurate quantification of carboplatin in urine by LC-ESI-TOFMS and LC-ICP-QMS. J. Anal. At. Spectrom. 23, 29–36 (2008)

[159] Esteban-Fernández, D., Montes-Bayón, M., Blanco, G.E., Gómez-Gómez, M.M., Palacios, M.A., Sanz-Medel, A.: Atomic (HPLC-ICP-MS) and molecular mass spectrometry (ESI-Q-TOF) to study cis-platin interactions with serum proteins. J. Anal. At. Spectrom. 23, 378–384 (2008)

[160] Mandal, R., Sawyer, M.B., Li, X.F.: Mass spectrometry study of hemoglobin-oxaliplatin complexes in colorectal cancer patients and potential association with chemotherapeutic responses. Rapid Commun. Mass Spectrom. 20, 2533–2538 (2006)

[161] Oe, T., Tian, Y., O'Dwyer, P.J., Roberts, D.W., Bailey, C.J., Blair, I.A.: Determination of the platinum drug cis-amminedichloro(2-methylpyridine)platinum(II) in human urine by liquid chromatography-tandem mass spectrometry. J. Chromatogr. B 792, 217–227 (2003)

[162] Hartinger, C.G., Ang, W.H., Casini, A., Messori, L., Keppler, B.K., Dyson, P.J.: Mass spectrometric analysis of ubiquitin-platinum interactions of leading anticancer drugs: MALDI versus ESI. J. Anal. At. Spectrom. 22, 960–967 (2007)

[163] Stefanka, Z., Hann, S., Koellensperger, G., Stingeder, G.: Investigation of the reaction of cisplatin with methionine in aqueous media using HPLC-ICP-DRCMS. J. Anal. At. Spectrom.

19, 894-898 (2004)

[164] Hann, S., Zenker, A., Galanski, M., Bereuter, T.L., Stingeder, G., Keppler, B.K.: HPIC-UV-ICP-SFMS study of the interaction of cisplatin with guanosine monophosphate. Fresenius J. Anal. Chem. 370, 581-586 (2001)

[165] Iijima, H., Patrzyc, H.B., Dawidzik, J.B., Budzinski, E.E., Cheng, H.C., Freund, H.G., Box, H.C.: Measurement of DNA adducts in cells exposed to cisplatin. Anal. Biochem. 333, 65-71 (2004)

[166] Garcia, S.D., Montes-Bayon, M., Blanco, G.E., Sanz-Medel, A.: Speciation studies of cisplatin adducts with DNA nucleotides via elemental specific detection (P and Pt) using liquid chromatography-inductively coupled plasma-mass spectrometry and structural characterization by electrospray mass spectrometry. J. Anal. At. Spectrom. 21, 861-868 (2006)

[167] Hartinger, C.G., Schluga, P., Galanski, M., Baumgartner, C., Timerbaev, A.R., Keppler, B.K.: Tumour-inhibiting platinum(II) complexes with aminoalcohol ligands: comparison of the mode of action by capillary electrophoresis and electrospray ionization-mass spectrometry. Electrophoresis 24, 2038-2044 (2003)

[168] Timerbaev, A.R., Alexenko, S.S., Połeć'-Pawlak, K., Ruzik, R., Semenova, O., Hartinger, C.G., Oszwałdowski, S., Galanski,M., Jarosz,M., Keppler, B.K.: Platinum metallodrug-protein binding studies by capillary electrophoresis-inductively coupled plasma-mass spectrometry: characterization of interactions between Pt(II) complexes and human serum albumin. Electrophoresis 25, 1988-1995 (2004)

[169] Rudnev, A.V., Aleksenko, S.S., Semenova, O., Hartinger, C.G., Timerbaev, A.R., Keppler, B.K.: Determination of binding constants and stoichiometrics for platinum anticancer drugs and serum transport proteins by capillary electrophoresis using the Hummel-Dreyer method. J. Sep. Sci. 28, 121-127 (2005)

[170] Schluga, P., Hartinger, C.G., Galanski, M., Meelich, K., Timerbaev, A.R., Keppler, B.K.: Tumour-inhibiting platinum(II) complexes with aminoalcohol ligands: biologically important transformations studied by micellar electrokinetic chromatography, nuclear magnetic resonance spectroscopy and mass spectrometry. Analyst 130, 1383-1389 (2005)

[171] Huang, Z., Timerbaev, A.R., Keppler, B.K., Hirokawa, T.: Determination of cisplatin and its hydrolytic metabolite in human serum by capillary electrophoresis techniques. J. Chromatogr. A 1106, 75-79 (2006)

[172] Hartinger, C.G., Keppler,B.K.: CE in anticancer metallodrug research-an update. Electrophoresis 28, 3436-3446 (2007)

[173] Brüchert, W., Krüger, R., Tholey, A., Montes-Bayon, M., Bettmer, J.: A novel approach for analysis of oligonucleotide-cisplatin interactions by continuous elution gel electrophoresis coupled to isotope dilution inductively coupled plasma mass spectrometry and matrixassisted laser desorption/ionization mass spectrometry. Electrophoresis 29, 1451-1459 (2008)

[174] Allardyce, C.S., Dyson, P.J.: Ruthenium in medicine: current clinical uses and future pros-

pects. Platin. Met. Rev. 45, 62-69 (2001)

[175] Egger, A.E., Hartinger Ch, G., Renfrew, A.K., Dyson, P.J.: Metabolization of [Ru(η^6-$C_6$$H_5CF_3$)(pta)$Cl_2$]: a cytotoxic RAPTA-type complex with a strongly electron withdrawing arene ligand. J. Biol. Inorg. Chem. 15, 919-927 (2010)

[176] Nazarov, A.A., Hartinger, C.G., Dyson, P.J.: Opening the lid on piano-stool complexes: an account of ruthenium(II)-arene complexes with medicinal applications. J. Organomet. Chem. 751, 251-260 (2014)

[177] Singh, A.K., Pandey, D.S., Xu, Q., Braunstein, P.: Recent advances in supramolecular and biological aspects of arene ruthenium(II) complexes. Coord. Chem. Rev. 270-271, 31-56 (2014)

[178] Kratz, F., Hartmann, M., Keppler, B.K., Messori, L.: The binding properties of two antitumor ruthenium(III) complexes to apotransferrin. J. Biol. Chem. 269, 2581-2588 (1994)

[179] Pongratz, M., Schluga, P., Jakupec, M.A., Arion, V.B., Hartinger, C.G., Allmaier, G., Keppler, B.K.: Transferrin binding and transferring-mediated cellular uptake of the ruthenium coordination compound KP1019, studied by means of AAS, ESI-MS and CD spectroscopy. J. Anal. At. Spectrom. 19, 46-51 (2004)

[180] Sulyok, M., Hann, S., Hartinger, C.G., Kepler, B.K., Stingeder, G., Koellensperger, G.: Two dimensional separation schemes for investigation of the interaction of an anticancer ruthenium(III) compound with plasma proteins. J. Anal. At. Spectrom. 20, 856-863 (2005)

[181] Ravera, M., Baracco, S., Cassino, C., Colangelo, D., Bagni, G., Sava, G., Osella, D.: Electrochemical measurements confirm the preferential bonding of the antimetastatic complex [ImH][$RuCl_4$(DMSO)(Im)] (NAMI-A) with proteins and the weak interaction with nucleobases. J. Inorg. Biochem. 98, 984-990 (2004)

[182] Dömötör, O., Hartinger, C.G., Bytzek, A.K., Kiss, T., Keppler, B.K., Enyedy, E.A.: Characterization of the binding sites of the anticancer ruthenium(III) complexes KP1019 and KP1339 on human serum albumin via competition studies. J. Biol. Inorg. Chem. 18, 9-17 (2013)

[183] Timerbaev, A.R., Rudnev, A.V., Semenova, O., Hartinger, C.G., Keppler, B.K.: Comparative binding of antitumor indazolium [*trans*-tetrachlorobis(1*H*-indazole)ruthenate(III)] to serum transport proteins assayed by capillary zone electrophoresis. Anal. Biochem. 341, 326-333 (2005)

[184] Połeć-Pawlak, K., Abramski, J.K., Semenova, O., Hartinger, C.G., Timerbaev, A.R., Kepler, B.K., Jarosz, M.: Platinum group metallodrug-protein binding studies by capillary electrophoresis-inductively couples plasma-mass spectrometry: a further insight into the reactivity of a novel antitumor ruthenium(III) complex toward human serum proteins. Electrophoresis 27, 1128-1135 (2006)

[185] Cetinbas, N., Webb, M.I., Dubland, J.A., Walsby, C.J.: Serum-protein interactions with anticancer Ru(III) complexes KP1019 and KP418 characterized by EPR. J. Biol. Inorg. Chem.

15, 131-145 (2010)

[186] Heffeter, P., Böck, K., Atil, B., Reza Hoda, M.A., Körner, W., Bartel, C., Jungwirth, U., Keppler, B.K., Micksche, M., Berger, W., Koellensperger, G.: Intracellular protein binding patterns of the anticancer ruthenium drugs KP1019 and KP1339. J. Biol. Inorg. Chem. 15, 737-748 (2010)

[187] Casini, A., Gabbiani, C., Michelucci, E., Pieraccini, G., Moneti, G., Dyson, P.J., Messori, L.: Exploring metallodrug-protein interactions by mass spectrometry: comparisons between platinum coordination complexes and an organometallic ruthenium compound. J. Biol. Inorg. Chem. 14, 761-770 (2009)

[188] Groessl, M., Tsybin, Y.O., Hartinger, C.G., Keppler, B.K., Dyson, P.J.: Ruthenium versus platinum: interactions of anticancer metallodrugs with duplex oligonucleotides characterized by electrospray ionization mass spectrometry. J. Biol. Inorg. Chem. 15, 677-688 (2010)

[189] Barefoot, R.R.: Distribution and speciation of platinum group elements in environmental matrices. Trends Anal. Chem. 18, 702-707 (1999)

[190] Schäfer, J., Hannker, D., Eckhardt, J.D., Stüben, D.: Uptake of traffic-related heavy metals and platinum group elements (PGE) by plants. Sci. Total Environ. 215, 59-67 (1998)

[191] Niemelä, M., Perämäki, P., Piispanen, J., Poikolainen, J.: Determination of platinum and rhodium in dust and plant samples using microwave-assisted sample digestion and ICP-MS. Anal. Chim. Acta 521, 137-142 (2004)

[192] Djingova, R., Heidenreich, H., Kovacheva, P., Markert, B.: On the determination of platinum group elements in environmental materials by inductively coupled plasma mass spectrometry and microwave digestion. Anal. Chim. Acta 489, 245-251 (2003)

[193] Beinrohr, E., Lee, M.L., Tschöpel, P., Tölg, G.: Determination of platinum in biotic and environmental samples by graphite furnace atomic absorption spectrometry after its electrodeposition into a graphite tube packed with reticulated vitreous carbon. Fresenius J. Anal. Chem. 346, 689-692 (1993)

[194] Akrivi, A.A., Tsogas, G.Z., Giokas, D.L., Vlessidis, A.G.: Analytical determination and bio-monitoring of platinum group elements in roadside grass using microwave assisted digestion and electrothermal atomic absorption spectrometry. Anal. Lett. 45, 526-538 (2012)

[195] Locatelli, C., Melucci, D., Torsi, G.: Determination of platinum-group metals and lead in vegetable environmental bio-monitors by voltammetric and spectroscopic techniques: critical comparison. Anal. Bioanal. Chem. 382, 1567-1573 (2005)

[196] Desimoni, E., Brunetti, B., Bacchella, R.: Cathodic stripping voltammetric determination of platinum in some foods and beverages at ng/g level under statistical control. Electroanalysis 14, 459-461 (2002)

[197] León, C., Emons, H., Ostapczuk, P., Hoppstock, K.: Simultaneous ultratrace determination of platinum and rhodium by cathodic stripping voltammetry. Anal. Chim. Acta 356, 99-104 (1997)

[198] Kołodziej, M., Baranowska, I., Matyja, A.: Determination of platinum in plant samples by voltammetric analysis. Electroanalysis 19, 1585–1589 (2007)

[199] Weber, G., Jakubowski, N., Stuewer, D.: Speciation of platinum in plant material. A combination of chromatography, elemental mass spectrometry and electrochemistry. In: Zereini, F., Alt, F. (eds.) Anthropogenic platinum-group element emissions. Their impact on men and environment, pp. 183–190. Springer, Berlin (2000)

[200] Jakubowski, N., Thomas, C., Klueppel, D., Stuewer, D.: Speciation of metals in biomolecules by use of inductively coupled plasma mass spectrometry with low and high mass resolution. Analysis 26, M37-M43 (1998)

[201] Alt, F., Weber, G., Messerschmidt, J., von Bohlen, A., Kastenholz, B., Guenther, K.: Bonding states of palladium in phytosystems: first results for endive. Anal. Lett. 35, 1349–1359 (2002)

[202] Messerschmidt, J., Alt, F., Tölg, G.: Platinum species analysis in plant material by gel permeation chromatography. Anal. Chim. Acta 291, 161–167 (1994)

[203] Ward, N.I., Dudding, L.M.: Platinum emissions and levels in motorway dust samples: influence of traffic characteristics. Sci. Total Environ. 334–335, 457–463 (2004)

[204] Gómez, B., Palacios, M.A., Gómez, M., Sanchez, J.L., Morrison, G., Rauch, S., McLeod, C., Ma, R., Caroli, S., Alimonti, A., Petrucci, F., Bocca, B., Schramel, P., Zischka, M., Pettersson, C., Wass, U.: Levels and risk assessment for humans and ecosystems of platinum-group elements in the airborne particles and road dust of some European cities. Sci. Total Environ. 299, 1–19 (2002)

[205] Rauch, S., Lu, M., Morrison, G.M.: Heterogeneity of platinum-group metals in airborne particles. Environ. Sci. Technol. 35, 595–599 (2001)

[206] Zaray, G., Ovari, M., Salma, I., Steffan, I., Zeiner, M., Caroli, S.: Determination of platinum in urine and airborne particulate matter from Budapest and Vienna. Microchem. J. 76, 31–34 (2004)

[207] Vanhaecke, F., Resano, M., Pruneda-Lopez, M., Moens, L.: Determination of platinum and rhodium in environmental matrices by solid sampling electrothermal vaporization-inductively coupled plasma mass spectrometry. Anal. Chem. 74, 6040–6048 (2002)

[208] Rauch, S., Morrison, G.M., Moldovan, M.: Scanning laser ablation-ICP-MS tracking of platinum group elements in urban particles. Sci. Total Environ. 286, 243–251 (2002)

[209] Alsenz, H., Zereini, F., Wiseman, C.L.S., Puttmann, W.: Analysis of palladium concentrations in airborne particulate matter with reductive co-precipitation, He collision gas, and ID-ICP-Q-MS. Anal. Bioanal. Chem. 395, 1919–1927 (2009)

[210] Meeravali, N.N., Kumar, S.J., Jiang, S.J.: An acid induced mixed-micelle mediated cloud point extraction for the separation and pre-concentration of platinum from road dust and determination by inductively coupled plasma mass spectrometry. Anal. Methods 2, 1101–1105 (2010)

[211] Limbeck, A., Rudolph, E., Hann, S., Koellensperger, G., Stingeder, G., Rendl, J.: Flow

injection on-line pre-concentration of platinum coupled with electrothermal atomic absorption spectrometry. J. Anal. At. Spectrom. 19, 1474-1478 (2004)

[212] Limbeck, A., Rendl, J., Puxbaum, H.: ET AAS determination of palladium in environmental samples with online preconcentration and matrix separation. J. Anal. At. Spectrom. 18, 161-165 (2003)

[213] Tokalioglu, S., Oymak, T., Kartal, S.: Determination of palladium in various samples by atomic absorption spectrometry after preconcentration with dimethylglyoxime on silica gel. Anal. Chim. Acta 511, 255-260 (2004)

[214] Tsogas, G.Z., Giokas, D.L., Vlessidis, A.G., Evmiridis, N.P.: On the re-assessment of the optimum conditions for the determination of platinum, palladium and rhodium in environmental samples by electrothermal atomic absorption spectrometry and microwave digestion. Talanta 76, 635-641 (2008)

[215] Probst, T.U., Rietz, B., Alfassi, Z.B.: Platinum concentrations in Danish air samples determined by instrumental neutron activation analysis. J. Environ. Monit. 3, 217-219 (2001)

[216] Giaveri, G., Rizzio, E., Gallorini, M.: Preconcentration and preseparation procedure for platinum determination at trace levels by neutron activation analysis. Anal. Chem. 73, 3488-3491 (2001)

[217] Fariseo, P., Speziali, M., Herborg, C., Orvini, E.: Platinum determination by NAA in two road dust matrices. Microchem. J. 79, 43-47 (2005)

[218] Schierl, R., Fruhman, G.: Airborne platinum concentrations in Munich city buses. Sci. Total Environ. 182, 21-23 (1996)

[219] Wei, C., Morrison, G.M.: Platinum in road dusts and urban river sediments. Sci. Total Environ. 146-147, 169-174 (1994)

[220] Alt, F., Bambauer, A., Hoppstock, K., Mergler, B., Tölg, G.: Platinum traces in airborne particulate matter. Determination of whole content, particle size distribution and soluble platinum. Fresenius J. Anal. Chem. 346, 693-696 (1993)

[221] Sako, A., Lopes, L., Roychoudhury, A.N.: Adsorption and surface complexation modeling of palladium, rhodium and platinum in surficial semi-arid soils and sediment. Appl. Geochem. 24, 86-95 (2009)

[222] Schäfer, J., Puchelt, H.: Platinum-group-metals (PGM) emitted from automobile catalytic converters and their distribution in roadside soils. J. Geochem. Explor. 64, 307-314 (1998)

[223] Beccaloni, E., Coccia, A.M., Musmeci, L., Stacul, E., Ziemacki, G.: Chemical and microbial characterization of indigenous topsoil and mosses in green urban areas of Rome. Microchem. J. 79, 271-289 (2005)

[224] Motelica-Heino, M., Rauch, S., Morrison, G.M., Donard, O.F.X.: Determination of palladium, platinum and rhodium concentrations in urban road sediments by laser ablation-ICP-MS. Anal. Chim. Acta 436, 233-244 (2001)

[225] Barefoot, R.R.: Determination of platinum at trace levels in environmental and biological mate-

rials. Environ. Sci. Technol. 31, 309–314 (1997)

[226] Barefoot, R.R., Van Loon, J.C.: Recent advances in the determination of the platinum group elements and gold. Talanta 49, 1–14 (1999)

[227] Meisel, T., Moser, J.: Reference materials for geochemical PGE analysis: new analytical data for Ru, Pd, Os, Ir, Pt and Re by isotope dilution ICP-MS in 11 geological reference materials. Chem. Geol. 208, 319–338 (2004)

[228] Sutherland, R.A.: Platinum-group element concentrations in BCR-723: a quantitative review of published analyses. Anal. Chim. Acta 582, 201–207 (2007)

[229] Bouma, M., Nuijen, B., Jansen, M.T., Sava, G., Picotti, F., Flaibani, A., Bult, A., Beijnen, J.H.: Development of a LC method for pharmaceutical quality control of the antimetastatic ruthenium complex NAMI-A. J. Pharm. Biomed. Anal. 31, 215–228 (2003)

第14章
挥发性有机化合物的浓缩和分析

14.1 概述

现代分析方法使人们能够全面、精确地分析各种基体中存在的所有化合物。为了自然生态系统的正常运行,对污染物的测定显得尤为重要,因为污染物会不可逆转地破坏生态系统。减少有害气体的排放,停止一些释放有害物质的生产是十分必要的。随着工业的不断扩张,不可避免地对生态系统造成一些直接或间接的危害。直接危害包括对水、土壤和大气等基体的污染,间接危害包括对食物、动植物和人类的影响。为了避免自然环境的恶化,需要对生态系统中发生的一些变化进行定期的监控和检测。

现代分析方法是基于物理化学现象和自然过程的变化,其不断发展有效地降低了化合物的检测限,提高了检测的准确度。特殊分析手段的可适用性不仅与被测物质的种类和性质有关,还与样品处理阶段的分离和浓缩方法的选择性和再现性有关。根据文献可知,样品的制备是定性和定量测定中最重要的步骤,完成样品的制备所需要的时间占到完成整个分析过程所需要的时间的60%~65%。样品制备方法具有多样性,基于表面现象的制备方法是日常痕量分析的主要方法,如吸附、解吸和萃取,其化学过程往往发生在气-液、气-固、液-液或液-固相边界上。这些技术可以直接与气相色谱(GC)或高效液相色谱(HPLC)等物理化学检测方法相结合,实现自动化过程,从而获得较好的数据再现性和较高的测量精度,还能获得令人满意的单次分析的成本效益。在各种常规分析中,仪器分析法具有操作过程简单、分析程序简单的显著优势。研发的新型分析方法和分析程序与现代高质量的设备联用,构成了现代分析的基础。

合适的色谱分离测量条件取决于待测物的物理化学性质(图14-1)。根据W. Giger定义,除了需要考虑分子量、化学性质和极性之外,还必须考虑分析物的可挥发性[1],这一特点与气相色谱法适用于高蒸汽压条件,并且测定的分子量达1000Da相吻合。

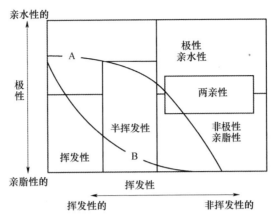

图 14-1 分析各种物理化学性质的分离方法范围
A—GC 限;B—HPLC 限;C—ZE 限(根据 W. Giger[1])。

从整个分析范围来看,挥发性有机化合物(VOC)这一类重要组分是可以鉴别的。根据美国国家环境保护局(EPA)的定义,蒸汽压高于 0.1013kPa 是将化合物纳入挥发性有机化合物的标准。在这一标准下,甲烷气体、其他含 4 个碳原子的碳氢化合物及其衍生物、芳香族分子和结构中含有 10 个以上碳原子的物质,都不属于 VOC 的范畴。

有些 VOC 是有毒的和致癌的。VOC 的主要来源分自然的和人为的两种。自然来源包括某些生物体的植被过程、同化过程、森林火灾、火山或间歇泉眼活动和每年 3000 万~6000 万吨的天然气释放。人为来源包括化石燃料开采和燃烧过程、原油加工、冶金、有机化学工业、溶剂生产和应用、食品加工工业、农业、固体废物处置、公路、空中和海洋运输[2]。

根据 Pauling 的研究,在 10000 种 VOC 中有 200 种是由活体器官释放出来的,这些物质可以对病人呼出的空气中的特殊气味物质进行鉴别。这类物质包括烷烃、甲基化烷烃或芳香类化合物及其功能化衍生物等[3]。目前,医学界对这一课题越来越感兴趣。

从这些简单的例子可以看出,对 VOC 这类物质化学监测的基本手段是采用现代分析方法对其痕量浓度进行测定。VOC 可以很容易地透过空气循环系统、植物和动物细胞的内外部活动等生物屏障进行共轭反应,主要有酶的氧化反应、自动氧化、起始、繁殖和终止等,从而产生新的代谢产物。这些代谢产物不仅能够确定环境中发生的某些现象与大气化学之间的关系,还对生物体的功能起重要作用如当自由基存在时发生的不可控的新反应(图 14-2)[4]。

目前,利用色谱法与光谱法相结合的技术,正确选择样品的前处理方法,是快速、全面、相对便宜以及准确测定各种环境和生物基体中 VOC 的唯一途径。

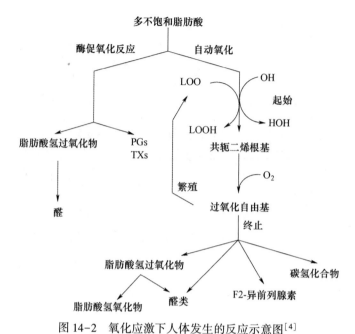

图 14-2 氧化应激下人体发生的反应示意图[4]

14.2 样品制备

1. 液-液微萃取

液-液微萃取(μ-LLE)的主要优势在于它是一种不需要特殊设备的简单操作程序,但它的一个根本问题是需要易燃或有毒的溶剂。现在许多与传统液-液微萃取有关的问题已被成功解决。

1978 年,Karlberg 和 Thelander[5]发现了流动注射萃取(FIE)技术。1979 年,Murray[6]改进了微萃取技术,使溶剂用量减少到 200μL,但是这些方法的主要问题是必须使用复杂的设备。Jeannot 和 Cantwel[7]以及 Hee 和 Lee[8]各自研究了一种较简单的微萃取方法——单滴微萃取法(SDME)。他们设计了一个微型反应器,在聚四氟乙烯管中加入 8μL 的正辛烷(图 14-3)[7]。研究者通过对扩散系数和系统动力学进行测量,提出了质量传递模型。

He 和 Lee[8]开发了一种微萃取方法,将有机液滴悬挂在气相色谱注射器的顶端。这个想法是在动态和静态两个修正实验中提出的。在动态实验中,使用传统的色谱注射器作为提取容器;在静态实验中,将 1μL 有机溶剂悬挂在针的顶端,采用配有电子捕获检测器(ECD)的气相色谱仪对提取液进行分析。Buszewski 等[9]

采用静态SDME法,滴加己烷和甲苯来提取氯化物。典型的尖角针气相色谱微量注射器已用于这方面的研究,其优势在于在试验过程中可使悬挂的液体保持稳定(图14-4)[10]。

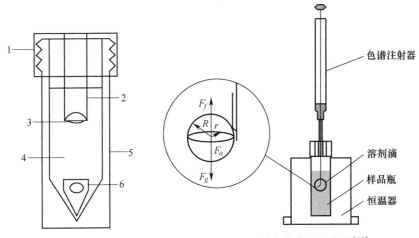

图14-3 微型反应器示意图
1—盖;2—聚四氟乙烯管;3—有机相;
4—水相;5—管道;6—搅拌棒[7]。

图14-4 单滴微萃取示意图[10]

使用标准曲线进行校准方法简便、重复性好、成本低,可用于自来水中三卤甲烷的常规测定。SDME法还与固相微萃取(SPME)、吹扫和捕集(P&T)和直接注射法(DAI)进行了比较[10],通过比较可知,SDME法的准确度可与P&T和DAI的准确度相媲美。相对于传统的LLE法,SDME法更加准确,另外,与DAI和P&T不同的是,SDME法不需要特殊设备。目前,SDME法已应用于提取氯酚[11]、杀虫剂[12-13]、战剂[14]、丁酮衍生物[15]以及食品副产物的控制[16]。由于SDME法不需要使用昂贵的仪器且具有成本低廉、操作简单、提取时间短等优势,特别适用于对水中有机污染物样品的初步分析。因为SDME法可作为SPME法的有效替代方案。

2. 分散液-液微萃取

上述SDME法虽然大大降低了溶剂消耗量,但是存在萃取效率低、达到平衡缓慢等缺点。在许多情况下,可以通过使用分散系统来提高萃取效率,例如,在水样中使用有机溶剂乳液作为分散剂。在分散液-液微萃取(DLLME)中,将萃取溶剂和分散剂的混合物注射到水样中。萃取溶剂是一种不溶于水的非极性液体,如甲苯、氯仿、二氯甲烷、四氯化碳和二硫化碳。分散剂是一种水溶性的极性溶剂,常见的分散剂有乙腈、丙酮、异丙醇和甲醇。两者的混合物中萃取剂的浓度通常在1%

~3%之间。由于有机相在水中的高分散性,混合液被快速地注入水样后,立即形成混浊的溶液。待测物从水相到有机相的扩散比在水中扩散更有效。乳化液通过离心来分离非混相溶剂,有机相沉积在锥形容器的底部,然后,将有机提取物注入色谱仪中。该方法快速高效、操作相对简单,具有较好的回收率和较高的富集因子。

分散液-液微萃取技术由 Rezaee 等发现,能够应用于多环芳烃、农药和烷基苯的萃取[17],紫外过滤[18]以及水样中氯酚的萃取[19]。分散液-液微萃取技术的应用可以参考相关的综述性文章[20-24]。其中值得一提的是,在这些应用中提到一个有趣的概念,即引入离子液体作为萃取剂[25-28]。

使用比水轻的溶剂是对分散液-液微萃取这项技术的一个改进。丙酮(分散剂)和己烷的混合溶液可用于 16 种多环芳烃的提取和测定。重要的一点是,在分离阶段不需要离心,只需再添加一部分丙酮。提取液测量重复性的相对标准偏差小于 11%[29]。采用 20nm 的纳米金颗粒作为提取剂,将纳米颗粒溶液和样品以 1400r/min 转速离心,然后在分离的固体中加入戊硫醇和辛烷混合物,再经过短时间的振动和离心(13400r/min),可以将多环芳烃提取出来[30],这是一种很有前景的多环芳烃提取方法。

3. 顶空萃取

顶空萃取法(HS)是一种常用的样品制备方法,在许多实验室中,特别是在工业生产中经常使用。顶空萃取技术涉及气相和液体或固体样品之间的平衡,在此技术中,将气相样品收集并引入气相色谱中。有两种分析类型:静态分析和动态分析。在静态分析中,当达到平衡时,气相被注入气相色谱中。在动态分析中,挥发性物质被气流带走。然而,由于基体效应的影响,导致某些物质的灵敏度降低,尤其是极性和亲水性样品。Kolb[31]出版了一本讲述顶空萃取技术的综合性书籍。

为了增加气相中分析物的浓度,本节进行了以下修正。

(1) 提高基质温度;

(2) 添加盐类化合物;

(3) 调节 pH。

静态顶空技术常用于测定食物[32]、尿液[34]、血液[35]和游泳池中的水[36]等复杂基体中的 VOC,这种方法也经常用于对制药工业残留溶剂的分析[37]。目前,已有许多类型的自动化静态顶空系统(图 14-5)[38]。

4. 吹扫捕集

动态顶空技术通常称为吹扫捕集法(P&T)。首先用惰性气体(一般是氦气或氮气)吹扫样品,VOC 会被不断地提取和转移到吸附剂中,然后加热捕集器并用载气冲洗,释放的挥发物就被转移到气相色谱中。此外,VOC 的解吸伴随着挥发物进入柱前低温聚焦同时发生[39]。流经样品的气流增加了系统中的传递速度。与

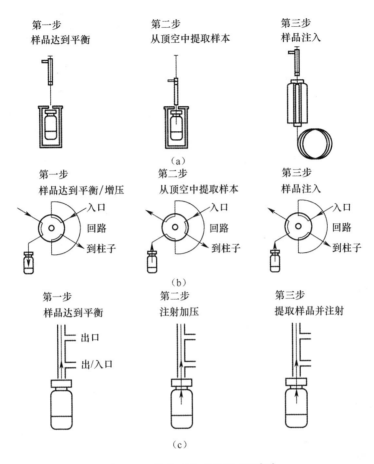

图 14-5 静态顶空系统示意图[38]

静态顶空技术相比,动态顶空技术具有更高的速度和提取效率。其提取效率与以下因素有关。

(1) 物质的极性及其在水中的溶解度;
(2) 水溶液的温度和离子强度;
(3) 用于萃取的气体体积;
(4) 吸附剂在捕集器中的穿透体积。

P&T 技术经常用于测定食品和饮料[40-41]、植物[42]、天然水和废水[43-44]、沉淀物[45]以及牛奶[46]中的挥发物,这项技术已被 EPA 广泛认可和推荐。

5. 闭环回路气提法

总体来说,闭环回路气提法(CLSA)和 P&T 法类似,不同点是在 CLSA 中的气体是通过提取器和含有吸附剂的捕集器在闭环回路中被抽走的。挥发性物质通常吸附在活性炭上,经过吹扫之后,用二硫化碳等有机溶剂从活性炭捕集器中人工洗脱[47]。闭环回路气提法的最大优势是具有超高灵敏度,目前这种方法已应用于地下水中挥发性有机物[48]以及烟草类[49]的提取和测定。

6. 热解吸

热解吸是另一种引入气相色谱的系统,专门用于对挥发性有机物的分析,但是,热解吸的目标分析物必须具有热稳定性,否则会发生分解。该方法主要用于测定空气中的挥发性组分,需要将样品收集到固体吸附剂上,然后对分析物进行解吸和气相色谱分析。通常情况下可用活性炭作吸附剂,然后用二硫化碳进行萃取。然而,溶剂解吸涉及 VOC 的再稀释,所以会在一定程度上对富集效应产生不利影响。因此,采样方法是利用含有某种吸附剂的吸附管将气体样品抽入管中,使 VOC 得到浓缩,然后将样品管置于热脱硫炉中,通过高温和载气的流动,将分析物从吸附剂中释放出来。与此同时,分析物被重新聚集在一个低温捕集器中,然后引入气相色谱柱中。这种两步式热解吸过程会在色谱柱的顶部形成一个窄的色谱带。

热解吸效率受温度、解吸时间和气流量的影响。另外,吸附剂的热稳定性限制了热解吸过程中的最高温度。通过在捕集器中加入少量的吸附剂,可以进一步提高冷捕集效率。

热解吸的一个重要问题是使用合适的吸附剂。吸附剂能捕集目标化合物,然后有效地释放它。选择合适的吸附剂取决于以下特性[50]。

(1) 热稳定性;
(2) 比表面积;
(3) 高穿透体积(BTV);
(4) 与水的亲和力低;
(5) 分析物的分子量和挥发性。

有许多吸附剂可以用于热解吸,表 14-1 列出了一些常用的吸附剂及其性能[51-52]。

许多多孔有机聚合物是从填充气相色谱柱的固定相中衍生来的。苯基对苯醚就是这样的例子,它是一种从二苯基对苯醚中获得的大孔聚合物。通常情况下,苯基对苯醚是疏水性的,不吸水。然而,它具备吸附极性化合物的性能。由于苯基对苯醚的比表面积较低($30m^2/g$),因此吸附能力有限,挥发性很强的化合物不能够被其吸附。因此,苯基对苯醚适用于吸附含有 4 个以上碳原子的较重的化合物。市场上有一种由石墨碳黑和苯基对苯醚(占 23%~77%)组成的苯基对苯醚吸附剂,这种吸附剂结合了两种材料的优点,其吸附效率大约是苯基对苯醚 TA 的 2 倍[50]。

表 14-1 常用的吸附剂及其性能

吸附剂	典型应用	温度上限/℃	比表面积/(m^2/g)	备注
Tenax TA™ Tenax GR	n-C_7 ~ n-C_{20}	350	15~25	吸附能力低,应用范围仅限于相对较高的浓度,常用于多层结构中
Chromosorb106	n-C_5 ~ n-C_{12}	250	750	稳定性不如 Tenax,不适合半挥发性化合物和多层结构
Carbotrap C™	n-C_8 ~ n-C_{30}	400	12	适用于多层结构,有相对疏水性,可用于半挥发性化合物和烷基苯(BTEX)
Carbopack B™	n-C_5 ~ n-C_{12}	400	100	有相对疏水性,可用于吸附酮和醇
Spherocarb	n-C_1 ~ n-C_5	400	1200	多层结构的末端,能有效跟踪,针对挥发性强的有机化合物(VVOCs),不适用于潮湿的样品
CarbosieveS Ⅲ Carboxenes	绝大部分气体	400	800	对大分子化合物有超强吸附或不可逆吸附

Tenax 也适用于热脱硫剂的低温捕集。但是它的缺点之一是在氧化条件下能够分解,例如,环境中存在臭氧和氮氧化物时,苯基对苯醚与臭氧发生反应生成苯乙酮和苯甲醛,与氮氧化物发生反应生成 2,6-二苯对苯醌。

填充了一种吸附剂的吸附管通常不适合同时分析具有不同挥发性和极性的化合物。解决这一问题的方法是采用含有不同层吸附剂(多层捕集器)的吸附管[53]。这种捕集器通常有两层,吸附管前端的第一层是吸附能力较弱的吸附剂,能吸附较重化合物;第二层含有吸附能力较强的吸附剂,可以吸附具有更强挥发性的化合物。在吸附过程中,分析物沿吸附强度增加的方向通过吸附管,最不易挥发的物质被吸附在位于管道入口的最弱的吸附剂上,较易挥发的物质被吸附在较强的吸附剂上,而最易挥发的物质被最强的吸附剂附着在吸附管的末端。解吸作用发生在与气体流动相反的方向,以免质量重的分析物滞留在最强的吸附剂上[54]。

热解吸可应用于空气样品中挥发分的测定和半挥组分的测定[55-58]。该方法已应用于昆虫信息素[59]和尿液中药物[60]的研究。相同的原理也可适用于固体材料或半固体材料,如土壤、沉淀物、制药原料、奶油、软膏、聚合物等。

位于吸附管中的样品在载气流中加热,释放出 VOC,同时挥发性组分集中在低温捕集器上,热解吸后引入气相色谱中。热解吸不仅适用于空气污染物的分析,也适用于环境颗粒物的分析[61],是一项重要的样品制备技术。

7. 针捕集装置

针捕集装置(NTD)的开发是将其应用在气体样品分析中所使用的微型吸附捕集器上,位于针内的吸附剂层作为捕集器。聚二甲基硅氧烷、羧酸、苯基对苯醚、二甲基苯等许多材料都被用作吸附剂。

在取样过程中,气体样品通过含有吸附剂层的针进行抽吸,一般是通过移动注射器的柱塞进行抽吸,随后,柱塞向下移动,通过针将样品拉出。经过几次循环后,挥发性化合物从高温气相注射器内的填充针中释放出来[62],注射器中的惰性气体增强了分析物在吸附剂中的解吸和转移。针捕集装置示意如图 14-6 所示,该系统可以进行手动或自动操作,由自动采样器完成。

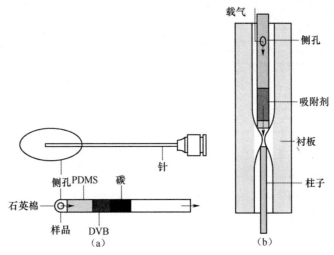

图 14-6 针捕集装置示意图
(a)针;(b)解吸内衬[63]。

NTD 的主要应用在气体样品分析和生物医学分析上[64-68]。其重要发展方向是针内衍生化的研究,以此来提高装置的灵敏度[69]。NTD 结合了固相微萃取技术的优点和传统吸附剂捕集器的灵敏度,其主要缺点是针垫的重复性较差。

8. 固相微萃取

在众多的样品制备方法中,固相微萃取(SPME)是最常用的方法之一。SPME 可用于测定气体、液体和固体样品中的 VOC。该方法的最大优点是能够在一个步骤内完成化合物的分离和富集,避免了有机溶剂的使用。

固相微萃取技术是一种可以替代其他提取方法的简单而廉价的方法。固相微萃取法是由 Arthur 和 Pawliszyn 在 20 世纪 80 年代末开发的[70]。

固相微萃取技术是在固定支架上使用涂上固定相薄层的熔融石英纤维。在提取过程中,纤维暴露在样品中,分析物被吸附在固定相上并被浓缩。经过一定时间的提取后,从支架中取出石英纤维,然后在气相色谱注射器中热脱除分析物。目前,市面上有多种类型的涂料,如聚二甲基硅氧烷(PDMS)、聚酰胺、聚乙二醇DVB、羧基PDMS和PDMS-DVB,也可以使用由3种材料制备的涂层(如DVB-羧酸-PDMS)[71]。纤维的选择主要基于"相似相溶"的原则,例如,PDMS吸附剂适用于提取碳氢化合物,吸附剂应是极性的以萃取醇类或酮类。涂层的厚度决定了纤维的吸附能力。调节液体试样的温度、pH或离子强度会影响分布常数,从而增加萃取效率。

由于某些分析物的物理化学性质及其很低的浓度水平,因此需要将分析物转化成极性较小、挥发性和热稳定性较高的衍生物,这一过程能够提高提取效率和检测精度。有几种衍生化方法可与固相微萃取技术相结合,其中纤维上的衍生化似乎是最适合的。在提取前,纤维中含有衍生剂,然后将纤维暴露在样品中,分析物在被提取的同时转化为衍生物。例如,甲醛的分析过程中会形成一种强亲和力的肟基[72]。用于将羰基衍生化的典型试剂有2,4-二硝基苯基肼、邻-(2,3,4,5,6-五氟苯基)-羟基苯胺和2,3,4,5,6-五氟唑苯胺。

为了提高提取的选择性和效率,研究者在痕量分析中研发了新的固定相,并对环糊精、石墨碳黑、聚吡咯和聚苯胺等导电聚合物进行了研究[73]。对溶胶-凝胶聚合物的研究很有意义,这些聚合物的一个重要特点是高热稳定性[74]。最近,生物样品分析的应用已经在文献[75-81]中提及。一些综述性文章也列出了近年来在方法学、固相萃取技术[82-84]和新型涂层方面的进展[85-86]。

9. 动态固相萃取

动态固相萃取技术(SPDE)类似于针捕集技术,即在针的内壁涂有聚合物。Lipinski引入了这一概念,并将一小段涂有典型聚二甲基硅氧烷聚合物的金属毛细管柱附着在气密注射器上[87]。动态固相萃取装置示意如图14-7所示[88]。

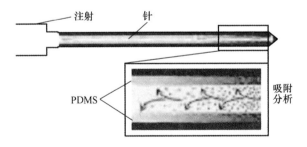

图14-7 动态固相萃取装置示意图

顶空样品被抽进注射器后,通过注射器柱塞向下移动,将样品吸入针头,重复操作几次后,通过针头吸取的分析物被PDMS层吸收。提取后,将注射器移至高温气相色谱进样器,并对分析物进行热解吸。

将针放入之后,在高温气相色谱中用氦气解吸,注射器中的气体会促进氦气的解吸,导致色谱带变窄。文献[88]介绍了一种利用该技术提取 VOC 的动力学模型,与固相微萃取技术相比,动态固相萃取技术的主要优点是具有更高的聚合物耐持久性和更好的灵敏度。最近这项技术已经成功地实现了自动化和商业化,一些研究型文章对这项技术进行了详细的描述[89-91]。

10. 搅拌棒吸附萃取

该技术的发展是为了解决在固相微萃取技术中纤维涂层材料使用量非常少的问题。搅拌棒吸附萃取(SBSE)是由 Baltussen 提出的[92],使用一种涂有聚合物吸附剂的磁棒。该磁棒的长度为 1cm 或 2cm,聚合物薄膜的厚度为 500μm 或 1000μm。搅拌棒提供了大量的聚合物涂层,其吸附容量远高于传统的固相微萃取方法。在这项技术中,吸附搅拌棒直接与液体样品接触。当吸附达到平衡时,将搅拌棒放置在热解吸装置中,分析物就会被释放到气相色谱中。

SBSE 技术已成功应用于环境样品[93-97]和食品的分析[98-99]。聚二甲基硅氧烷是最常用的聚合物,主要是因为它具有良好的热稳定性和持久性。通过对不同试剂的衍生化应用,对 SBSE 技术进行了改进[100-104],使该方法适用于需要衍生化的化合物的提取。采用多步衍生化方法提取多种萃取元素(每个反应在不同的搅拌棒上进行),可对酚类、类固醇、胺类、噻唑类、酮类等不同官能团的化合物进行高效的提取、解吸和色谱分析。乙酸酐、氯甲酸乙酯、四乙基硼烷和二甲基三氟丙酮酰胺钠均可以用这种技术萃取,并在萃取的同时进行衍生化[105]。

11. 膜技术

人们将膜技术用于样品制备的研究始于 20 世纪 80 年代。选择性萃取使膜技术成为 20 世纪 90 年代典型样品富集方法的替代方法。研究者们设计了不同的膜系统,并将其引入分析实践中,常见的膜系统有聚合物膜萃取(PME)、微孔膜液-液萃取(MMLLE)和支撑液膜萃取(SLME)[106-107]。膜辅助溶剂萃取(MASE)与 GC-MS 联用,可以分析环境样品中的有机污染物[108-111]。

吸附剂界面膜萃取(MESI)是萃取装置的一种,是与气相色谱连接的最有效的系统。在这种方法中,供体相是气体或液体样品,受体相是气体。挥发性物质被不断地吸附在吸附剂上,然后解吸到气相色谱中[112]。另一种方法是将离线的 GC-MESI 与一个低温捕集装置组合在一起,该捕集装置能够现场取样,并在送往实验室后进行气相色谱分析[113-114]。MESI 技术可以提取挥发性和相对非极性的化合物。

MESI 的另一种应用是将膜引入质谱系统(MIMS),这可以选择性地提取 VOC,并直接进行质谱分析,而无须色谱分离。文献[115-118]中描述了气态物质通过高疏水膜直接引入质谱仪的应用。

14.3 挥发性有机化合物测定方法

挥发性有机化合物的分离通常是由气相色谱进行的。当今的气相色谱毛细管柱具有良好的性能和较高的耐热性。VOC 的分离是通过含有有效交联聚合物的高热稳定性色谱柱进行的,这种聚合物与毛细管表面通过化学反应相结合。

14.3.1 气相色谱质谱法

在许多情况下,色谱峰的定性比对不足以对分析物进行正确的鉴别。气相色谱与质谱联用为待测物的鉴别分析提供了更多的信息。GC-MS 是一种众所周知的常用技术,将气相色谱具有的高效分离性能与质谱具备的高灵敏度和高选择性相结合。此外,仪器的质量分析器发展了不同的类型,如离子肼、四极杆和飞行时间质谱等,四极杆和飞行时间质谱大大提高了分析能力。不同种类的质谱仪见表 14-2[119]。

质谱检测器的另一个便利条件是商业化谱图数据库的广泛使用(如 NIST、Wiley),这对复杂混合物的定性分析很有帮助。人们普遍认为,GC-MS 是鉴别不同基体中挥发性有机化合物最合适的仪器。

表 14-2 普通质谱仪的选择参数

选择参数	四极杆(Q)	离子肼(IT)	飞行时间(TOF)
质量范围/Da	2~4000	10~2000	没有限制
分辨率	中	中	高
质量精度	1Da	1Da	<1ppm
检出限/g	$10^{-13} \sim 10^{-12}$	$10^{-13} \sim 10^{-12}$	—
线性范围	4~5	3~4	3~5
扫描速度/Hz	50	50	1000

实际应用中对样品制备过程简化的需求促使了新技术的发展,例如质子-转移-反应质谱(PTR-MS)以及选择离子流动管质谱-离子迁移谱联用(MS-SIFT IMS),这些技术可以在 1s 内实时测量,能够快速分析工艺过程中的气体成分,监测室内空气,以及检测活的有机体释放出的挥发性有机化合物。

14.3.2 质子-转移-反应质谱法

质子-转移-反应质谱法是一种新型的直接分析 VOC 的技术。最初在 1995 年由 Hansel 研究小组对混合气体进行直接分析时使用[120]。该方法的原理是将蒸汽引入电离区,形成反应物离子(H_3O^+)。这些离子是由空心阴极放电或 ^{241}Am 同位素 α 衰变产生的[121]。高纯度的反应物离子有助于提高该方法的灵敏度,并且无须在进入反应区之前使用四极杆来选择反应离子。此过程中可能出现共存的 H_3O^+、O_{2+} 离子,这是由 H_3O^+ 和 O_2 之间的电荷转移而形成的。

反应离子进入载气引入样品区域,在质子转移反应的基础上,H_3O^+ 离子与分析物分子 R 快速反应,这个过程会生成分析物阳离子(RH^+)。RH^+ 离子进入漂移区,在电场中根据其质荷比(m/z)进行分离:

$$H_3O^+ + R \rightarrow RH^+ + H_2O \tag{14-1}$$

也可以使用 NH_4^+ 离子作为反应离子。然而,大多数 VOC 都与 H_3O^+ 质子有超强的结合的能力。这也是有效碰撞较多,并能生成反应产物的阳离子的原因。低分子量的烷烃不具备这样的特点,因为其与质子的亲和力低。在这些条件下,发生质子转移分解,生成反应产物混合物[122]。例如,七氟醚($C_4H_3F_7O$)的电离生成多种产物,而不同产物的产率不同:

$$C_4F_7OH_2^+ + H_2O + H_2 (79\%) \tag{14-2}$$
$$H_3O^+ + C_4H_3F_7O \rightarrow C_4F_6H_2OH^+ + H_2O + HF (12\%) \tag{14-3}$$
$$CHFOH^+ + H_2O + C_3F_6H_2 (9\%) \tag{14-4}$$

离开漂移区,离子进入质量分析器,图 14-8 所示为质子-转移-反应质谱法的示意图[123]。该法在分析复杂混合物方面存在不足,因为同分异构体通过质子转移的准分子离子电离会得到相同的质荷比,这意味着质子-转移-反应质谱法在化学物质定性中的应用是有限的,不能用于鉴别未知物质。因此,在某物质被其他方法确认存在的情况下,质子-转移-反应质谱法是一种很好的用来直接监测该物质的方法[124]。

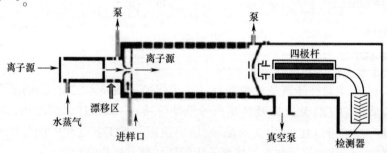

图 14-8 质子转移反应质谱法的示意图

将质子-转移-反应质谱法应用于气体混合物的分析,由于方法灵敏度高,避免了富集、预浓缩和预分离过程。质子-转移-反应质谱法可用于分析呼出气体中的挥发性有机化合物,虽然样品中含有高浓度的 CO_2、N_2、O_2、H_2O 等物质,但它们的存在并不干扰测量。质子-转移-反应质谱法的优点是分析时间短,通常只需几秒钟。在过去的 15 年里,已经有许多文献报道了使用质子-转移-反应质谱法对呼出空气进行直接分析。通过细菌培养之后可用该方法研究咖啡、肉类和水果中释放的挥发性有机物的扩散变化[125]。

14.3.3 选择离子流动管质谱法

选择离子流动管质谱法(SIFT-MS)是一种用于气体混合物中 VOC 的直接测定和定量测定分析技术。选择离子流动管质谱法是由 N. G. Adams 和 D. Smith 于 1976 年提出的,该方法可对一些物质进行实时监测[126]。选择离子流动管质谱装置如图 14-9 所示[127],其对气体混合物分析的原理是基于在特定时间内试剂离子与分析物分子的反应,采用化学电离法,在离子源中生成与合适的电离气体结合的试剂离子。在得到的所有离子中,只有具有理想质荷比 m/z 的阳离子被四极过滤器选择。通过四级杆的离子被注入惰性载气(通常是氦气)的反应流之后进入漂移管。漂移距离为 30~100cm,压力约为 100Pa,热力学温度为 300K。样品通过毛细管进入长管,样品导入速度需要适当的调整。为了防止毛细管中分析物的凝结,毛细管需要加热到 100℃。

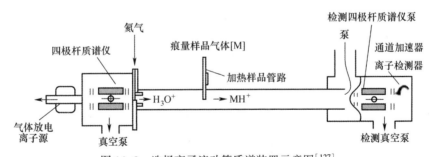

图 14-9 选择离子流动管质谱装置示意图[127]

前驱离子如 H_3O^+、NO^+ 或 O_2^+ 与分析物分子反应的速度非常快。通过这个过程,待测化合物生成特定的阳离子(MH^+):

$$H_3O^+ + M \rightarrow MH^+ + H_2O \tag{14-5}$$

随着分析物分子数量的增加,反应离子的数量减少。准分子离子的数量用动力学方程表示:

$$[MH^+]_t = [H_3O^+]k[M]t \tag{14-6}$$

式中：$[MH^+]_t$为分析物在t时间被质子化阳离子的浓度；t为前驱离子与分析物分子反应的时间；$[M]$为被分析物在气相中的浓度；k为前驱离子与分析物分子反应的速率常数。

试验测定了不同的分子与前驱离子H_3O^+、NO^+、$O_2^+ \cdot$反应的速率常数。为了计算分析物$[M]$的分压，必须知道每个反应的速率常数k和$[MH^+]$的浓度。该方法还可以同时测量几个不同分析物的浓度。研究表明，H_3O^+和NO^+离子与各种有机化合物的反应速度都比较快，但O_2^+只能与诸如NO、NO_2或NH_3等小分子迅速反应。对于含水量较大的样品（如呼吸样品或水溶液），除了H_3O^+离子外，还可能形成$H_3O^+(H_2O)_{1,2,3}$等簇离子。这些簇离子与分析物分子发生反应，提供阳离子，生成$MH^+(H_2O)_{1,2}$：

$$H_3O^+ + (H_2O)_{1,2,3} + M \rightarrow MH^+(H_2O)_{1,2} + H_2O \qquad (14-7)$$

反应过程中生成的各种分析物阳离子和反应离子撞击质谱分析器，并按质荷比（m/z）进行分离，通过测量，得到待测化合物的质谱。

使用选择离子流动管质谱可以在四极杆质谱的全扫描模式中记录质量谱线，也可以在离子以一定的质荷比（m/z）通过四极杆时在多离子监测模式中记录质量谱线。全扫描模式是在指定的时间周期内，在特定的质荷比（m/z）范围内分析物的全部质量谱线。多离子监测模式是对试剂离子和分析物在反应过程中形成的前驱离子和选定的离子进行扫描。

选择离子流动管质谱法用于分析空气样品中的挥发性有机化合物[127]，如检测ng/g级水平的氨、异戊二烯、丙酮、乙醇、乙腈等物质。此外，选择离子流动管质谱法也可用于分析血液[126]、尿液[128]、食物[129]、肺癌细胞顶空[130]等物质。其测定呼出空气中挥发性有机化合物的质谱图如图14-10所示。

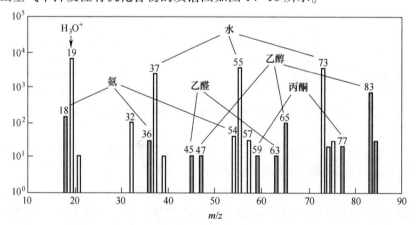

图14-10 选择离子流动管质谱法测定呼出空气中挥发性有机化合物的质谱图
（H_3O^+离子用作前驱离子[127]）

14.3.4 离子迁移谱法

离子迁移谱法(IMS)通常与气相色谱联用于 VOC 的分析。离子迁移谱法通过气相中的离子速度来鉴别化学物质,这些离子在外加电压(100~350V/cm)的影响下移动[131]。该方法具有单次分析成本低,获得测量结果所需的时间短(20~50ms)和对某些类型化合物的检出限低等优点,对醛、酮、胺和酯的检测在 ng/dm^3 级水平上。值得注意的是,使用离子迁移谱法很难检测到与质子亲和力低的挥发性有机化合物(如烷烃和芳烃)。离子迁移谱检测器操作简单、方便,但是对复杂基体中分析物的选择性差。因此,在实际应用中,多数情况下会与多分子柱层析法联用进行待测物的初步分离。筛选出的物质从毛细管柱进入离子源,在氮或合成空气等载气中经过 β 衰变发生电离(^{63}Ni、^3H);然后分析物离子通过载气离子与分析物分子发生碰撞。离子迁移谱检测器离子分离示意如图 14-11 所示[132]。

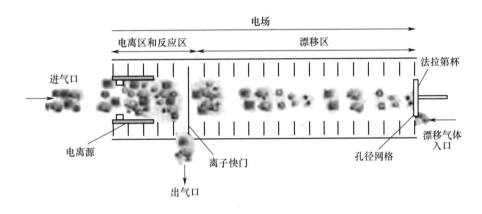

图 14-11 (见彩图)离子迁移谱检测器离子分离示意图[132]

另一种电离方法是利用激光或紫外光进行电晕放电和光电离。电离区与漂移区之间有一个隔断,这个隔断是周期性打开的。当隔板打开时,离子进入漂移区,并根据其质荷比进行分离。如果离子舱是封闭的,离子就会从系统中移除。最先到达探测器(如法拉第杯)的离子是电荷最大的、最轻的离子,其次是较重的离子。离子被收集在法拉第杯上,信号被放大并传输到记录仪上。氮气通入漂移区末端,将结合粒子转移到电离区。分析结果为漂移时间的谱线,该谱线是检测器检测到的电流与分离离子的漂移时间之间的关系。漂移时间是特定条件下离子的特性。如果测量系统由用于分析的气相色谱和作为检测器的离子迁移谱组成,则被测物的保留时间和漂移时间两个参数可以被表征。

在离子迁移谱检测器中会用到正极性或负极性。在第一步中,电离过程中的载气产生离子,然后与待测物分子发生碰撞发生电离。首先形成一个较弱的簇离子,在解离一个水分子后,会形成一个更稳定的离子,即反应产物。当被测样品中含有大量水蒸气时,会形成更多的团簇离子。

正极化作用:

$$\underset{\text{分析产物分子}}{M} + \underset{\text{试剂离子}}{H^+(H_2O)_n} \rightarrow \underset{\text{族离子}}{MH^+(H_2O)_{n-x}} + x\,H_2O \quad (14\text{-}8)$$

负极化作用:

$$\underset{\text{待测物分子}}{M} + \underset{\text{试剂离子}}{O^{2-}(H_2O)_n} \rightarrow \underset{\text{产物}}{MO^{2-}(H_2O)_{n-x}} + xH_2O \quad (14\text{-}9)$$

如果待测物浓度过高,形成的离子产物(单体)与非离子化的分析物分子发生碰撞的可能性就很高,从而导致二聚体的形成。当被测物质的浓度很高时,可能会有很少的反应离子,因此一些待测物分子不会被电离:

$$\underset{\text{质子化单体}}{MH^+(H_2O)_n} + M \rightarrow \underset{\text{二聚体}}{M_2H^+(H_2O)_{n-x}} + xH_2O \quad (14\text{-}10)$$

气相色谱-离子迁移谱联用技术有很多优点,可用于制药行业质量控制[131-132]、微量爆炸物检测、战剂和药物[133-135]检测,该系统也可用于监测周围空气[136]和直接分析呼出空气[137]。

14.4 小结

气相色谱法是挥发性有机物分析的首选技术。在化合物定性分析中,气相色谱与质谱的组合是不可缺少的。TOF-MS 和全二维气相色谱(GC×GC)的发展使高分辨质谱仪的准确性得到很大的提升。样品制备是分析之前的关键性阶段,PTR-MS、SIFT-MS 和 IMS 等实时测量方法的发展,对于仪器分析具有非常重要的价值。此外,样品浓缩技术特别是极性化合物的浓缩技术有待于进一步提升。另外,还应关注对于纳米颗粒、铁磁吸附剂、分子聚合物和铁磁离子液体的应用等。

参考文献

[1] Giger, W.: Hydrophilic and amphiphilic water pollutants; using advanced analytical methods for classic and emerging contaminants. Anal. Biochem. Chem. 393, 37-44 (2009)

[2] Baird, C., Cann, M.: Environmental chemistry, 3rd edn, p. 72. W.H. Freeman, New York (2005)

[3] Bajtarevic, A., Ager, C., Pienz, M., Kleiber, M., Schwarz, K., Ligor, M., Ligor, T., Filipi-

ak, W., Denz, H., Fiegl, M., Hilbe, W., Weiss, W., Lukas, P., Jamning, H., Hackl, M., Haidenberger, A., Buszewski, B., Miekisch, W., Schubert, J., Amann, A.: Noninvasive detection of lung cancer by analysis of exhaled breath. BMC Cancer 9, 348–364 (2009)

[4] Kinter, M.: Analytical technologies for lipid oxidation products analysis. J. Chromatogr. B 671, 223–236 (1995)

[5] Karlberg, B., Thelander, S.: Extraction based on the flow-injection principle: Part I. Description of the extraction system. Anal. Chim. Acta 98, 1–7 (1978)

[6] Murray, D.A.J.: Rapid micro extraction procedure for analyses of trace amounts of organic compounds in water by gas chromatography and comparisons with macro extraction method. J. Chromatogr. A 177, 135–140 (1979)

[7] Jeannot, M.A., Cantwell, F.F.: Solvent microextraction into a single drop. Anal. Chem. 68, 2236–2240 (1996)

[8] He, Y., Lee, H.K.: Liquid-phase microextraction in a single drop of organic solvent by using a conventional microsyringe. Anal. Chem. 69, 4634–4640 (1997)

[9] Ligor, T., Buszewski, B.: Extraction of trace organic pollutants from aqueous samples by a single drop method. Chromatographia 51, S279–S282 (2000)

[10] Buszewski, B., Ligor, T.: Single-drop extraction versus solid-phase microextraction for the analysis of VOCs in water. LC-GC Europe 2, 2–6 (2002)

[11] Ponnusamy, V.K., Ramkumar, A., Jen, J.F.: Microwave assisted headspace controlled-temperature single drop microextraction for liquid chromatographic determination of chlorophenols in aqueous samples. Microchim. Acta 179, 141–148 (2012)

[12] Tian, F., Liu, W., Fang, H., An, M., Duan, S.: Determination of six organophosphorus pesticides in water by single-drop microextraction coupled with GC-NPD. Chromatographia 77, 487–492 (2014)

[13] Fernandes, V.C., Subramanian, V., Mateus, N., Domingues, V.F., Delerue-Matos, C.: The development and optimization of a modified single-drop microextraction method for organochlorine pesticides determination by gas chromatography-tandem mass spectrometry. Microchim. Acta 178, 195–202 (2012)

[14] Park, Y.K., Chung, W.Y., Kim, B., Kye, Y., Shin, M., Kim, D.: Ion-pair single-drop micro-extraction determinations of degradation products of chemical warfare agents in water. Chromatographia, 76, 679–685 (2013)

[15] Hu, M., Chen, H., Jiang, Y., Zhu, H.: Headspace single-drop microextraction coupled with gas chromatography electron capture detection of butanone derivative for determination of iodine in milk powder and urine. Chem. Pap. 67(10), 1255–1261 (2013)

[16] Enteshari, M., Mohammadi, A., Nayebzadeh, K., Azadniya, E.: Optimization of headspace single-drop microextraction coupled with gas chromatography-mass spectrometry for determining volatile oxidation compounds in mayonnaise by response surface methodology. Food Anal. Methods, 7, 438–448 (2014)

[17] Rezaee, M., Assadi, Y., Milani Hosseini, M.R., Aghaee, E., Ahmadi, F., Berijani, S.: Determination of organic compounds in water using dispersive liquid-liquid microextraction. J. Chro-

matogr. A 1116, 1-9 (2006)

[18] Negreira, N., Rodríguez, I., Rubí, E., Cela, R.: Dispersive liquid-liquid microextraction followed by gas chromatography-mass spectrometry for the rapid and sensitive determination of UV filters in environmental water samples. Anal. Bioanal. Chem. 398, 995-1004 (2010)

[19] Wang, K.D., Chen, P.S., Huang, S.D.: Simultaneous derivatization and extraction of chlorophenols in water samples with up-and-down shaker-assisted dispersive liquid-liquid microextraction coupled with gas chromatography/mass spectrometric detection. Anal. Bioanal. Chem. 406, 2123-2131 (2014)

[20] Bosch Ojeda, C., Rojas, F.S.: Separation and preconcentration by dispersive liquid-liquid microextraction procedure. Chromatographia 69, 1149-1159 (2009)

[21] Liang, P., Xu, J., Li, Q.: Application of dispersive liquid-liquid microextraction and high-performance liquid chromatography for the determination of three phthalate esters in water samples. Anal. Chim. Acta 609, 53-58 (2008)

[22] Saraji, M., Boroujeni, M.K.: Recent developments in dispersive liquid-liquid micro-extraction. Anal. Bioanal. Chem. 406, 2027-2066 (2014)

[23] Vinas, P., Campillo, N., López-García, I., Hernández-Córdoba, M.: Dispersive liquid-liquid microextraction in food analysis. A critical review. Anal. Bioanal. Chem. 406, 2067-2099 (2014)

[24] Ojeda, C.B., Rojas, F.S.: Separation and preconcentration by dispersive liquid-liquid microextraction procedure: recent applications. Chromatographia 74, 651-679 (2011)

[25] Escudero, L.B., Grijalba, A.C., Martinis, E.M., Wuilloud, R.G.: Bioanalytical separation and preconcentration using ionic liquids. Anal Bioanal Chem 405, 7597-7613 (2013)

[26] Zhao, Q., Anderson, J.L.: Task-specific microextractions using ionic liquids. Anal. Bioanal. Chem. 400, 1613-1618 (2011)

[27] Zhou, Q., Ye, C.: Ionic liquid for improved single-drop microextraction of aromatic amines in water samples. Microchim. Acta 162, 153-159 (2008)

[28] He, L., Luo, X., Xie, H., Wang, C., Jiang, X., Lu, K.: Ionic liquid-based dispersive liquid-liquid microextraction followed high-performance liquid chromatography for the determination of organophosphorus pesticides in water sample. Anal. Chim. Acta 655, 52-59 (2009)

[29] Guo, L., Lee, H.K.: Low-density solvent-based solvent demulsification dispersive liquid-liquid microextraction for the fast determination of trace levels of sixteen priority polycyclic aromatic hydrocarbons in environmental water samples. J. Chromatogr. A 1218, 5040-5046 (2011)

[30] Wang, H., Yu, S., Campiglia, A.D.: Solid-phase nano-extraction and laser-excited time-resolved Shpolskii spectroscopy for the analysis of polycyclic aromatic hydrocarbons in drinking water samples. Anal. Biochem. 385, 249-256 (2009)

[31] Kolb, B., Ettre, L.S.: Static headspace-gas chromatography: theory and practice, 2nd edn. Wiley, Hoboken, NJ (2006). ISBN 978-0-471-74944-8

[32] Russo, M.V., Goretti, G., Liberti, A.: Direct headspace gas chromatographic determination of dichloromethane in decaffeinated green and roasted coffee. J. Chromatogr. 465, 429-433 (1989)

[33] Liu, M., Li, H., Zhan, H.: A novel method for the determination of the ethanol content in soy sauce by full evaporation headspace gas chromatography. Food Anal. Methods 7, 1043–1046 (2014)

[34] Ljungkvist, G., Larstad, M., Mathiasson, L.: Specific determination of benzene in urine using dynamic headspace and mass-selective detection. J. Chromatogr. B 721, 39–46 (1999)

[35] Schroers, H.J., Jermann, E.: Determination of physiological levels of volatile organic compounds in blood using static headspace capillary gas chromatography with serial triple detection. Analyst 123, 715–720 (1998)

[36] Montesinos, I., Gallego, M.: Headspace gas chromatography-mass spectrometry for rapid determination of halonitromethanes in tap and swimming pool water. Anal. Bioanal. Chem. 402, 2315–2323 (2012)

[37] Penton, Z.: Determination of residual solvent in pharmaceutical preparations by static headspace GC. J. High Resol. Chromatogr. 15, 329–331 (1992)

[38] Technical guide. A technical guide for static headspace analysis using GC, Restek, Bellefonte, PA, USA, Application note (2000)

[39] Grob, R.L., Barry, E.F. (eds.): Modern practice of gas chromatography, 4th edn, pp. 790–794. Wiley, Hoboken, NJ (2004)

[40] Jiemin, L., Ning, L., Meijuan, W., Guibin, J.: Determination of volatile sulfur compounds in beverage and coffee samples by purge-and-trap on-line coupling with a gas chromatography-fiame photometric detector. Microchim. Acta 148, 43–47 (2004)

[41] Beltran, J., Serrano, E., López, F.J., Peruga, A., Valcarcel, M., Rosello, S.: Comparison of two quantitative GC-MS methods for analysis of tomato aroma based on purge-and-trap and on solid-phase microextraction. Anal. Bioanal. Chem. 385, 1255–1264 (2006)

[42] Zang, L.H., Liu, Y.L., Liu, J.Q., Tian, Q., Xiang, F.N.: An improved method for testing weak VOCs in dry plants with a purge and trap concentrator coupled to GC-MS. Chromatographia 68, 351–356 (2008)

[43] Zhao, R.S., Cheng, C.G., Yuan, J.P., Jiang, T., Wang, X., Lin, J.M.: Sensitive measurement of ultratrace phenols in natural water by purge-and-trap with in situ acetylation coupled with gas chromatography-mass spectrometry. Anal. Bioanal. Chem. 387, 687–694 (2007)

[44] Barco-Bonilla, N., Plaza-Bolanos, P., Fernández-Moreno, J.L., Romero-González, R., Frenich, A.G., Vidal, J.L.M.: Determination of 19 volatile organic compounds in wastewater effluents from different treatmentsby purge and trap followed by gas-chromatography coupled to mass spectrometry. Anal. Bioanal. Chem. 400, 3537–3546 (2011)

[45] Han, D., Ma, W., Chen, D.: Determination of biodegradation process of benzene, toluene, ethylbenzene and xylenes in seabed sediment by purge and trap gas chromatography. Chromatographia 66, 899–904 (2007)

[46] Larreta, J., Bilbao, U., Vallejo, A., Usobiaga, A., Arana, G., Zuloaga, O.: Multisimplex optimization of the purge-and-trap preconcentration of volatile fatty acids, phenols and indoles in cow slurries. Chromatographia 67, 93–99 (2008)

[47] Barcelo, D.: Environmental analysis. Elsevier, Amsterdam (1993)

[48] Buszka, P.M., Zaugg, S.D., Werner, M.G.: Determination of trace concentrations of volatile organic compounds in ground waterusing closed-loop stripping. Bull. Environ. Contamin. Toxicol. 45, 507–515 (1990)

[49] Meruva, N.K., Penn, J.M., Farthing, D.E.: Rapid identification of microbial VOCs from tobacco molds using closed-loop stripping and gas chromatography/time-of-flight mass spectrometry. J. Ind. Microbiol. Biotechnol. 31, 482–488 (2004)

[50] Harper, M.: Sorbent trapping of volatile organic compounds from air. J. Chromatogr. A 885, 129–151 (2000)

[51] Dettmer, K., Knobloch, T., Engewald, W.: Stability of reactive low boiling hydrocarbons on carbon based adsorbents typically used for adsorptive enrichment and thermal desorption. Fresenius J. Anal. Chem. 366, 70–78 (2000)

[52] Palluau, F., Mirabel, P., Millet, M.: Influence of relative humidity and ozone on the sampling of volatile organic compounds on carbotrap/carbosieve adsorbents. Environ. Monit. Assess. 127, 177–187 (2007)

[53] Dettmer, K., Engewald, W.: Adsorbent materials commonly used in air analysis for adsorptive enrichment and thermal desorption of volatile organic compounds. Anal. Bioanal. Chem. 373, 490–500 (2002)

[54] Dettmer, K., Bittner, T., Engewald, W.: Adsorptive enrichment and thermal desorption of low-boiling oxygenated compounds possibilities and limitations. Chromatographia Suppl. 53, S322-S326 (2001)

[55] Massold, E., Bahr, C., Salthammer, T., Brown, S.K.: Determination of VOC and TVOC in air using thermal desorption GC-MS-practical implications for test chamber experiments. Chromatographia 62, 75–85 (2005)

[56] Ras, M.R., Marcé, R.M., Borrull, F.: Volatile organiccompounds in air at urban and industrial areas in the Tarragona region by thermal desorption and gas chromatography-mass spectrometry. Environ. Monit. Assess. 161, 389–402 (2010)

[57] Bahrami, A.R., Fam, I.M., Donaldson, J.: Development of a thermal desorption method for the analysis of particle associated polycyclic aromatic hydrocarbons in ambient air. Int. J. Environ. Sci. Technol. 1, 165–169 (2004)

[58] Juillet, Y., Dubois, C., Bintein, F., Dissard, J., Bossée, A.: Development and validation of a sensitive thermal desorption-gas chromatography-mass spectrometry (TD-GC-MS) method for the determination of phosgene in air samples. Anal. Bioanal. Chem. 406, 5137–5145 (2014)

[59] Drijfhout, F.P., Beek, T.A., Visser, J.H., Groot, A.: On-line thermal desorption-gas chromatography of intact insects for pheromone analysis. J. Chem. Ecol. 26, 1383–1392 (2000)

[60] van Hout, M.W.J., de Zeeuw, R.A., Franke, J.R., de Jong, G.J.: Solid-phase extraction-thermal desorption-gas chromatography with mass selective detection for the determination of drugs in urine. Chromatographia 57, 221–225 (2003)

[61] Schnelle-Kreis, J., Orasche, J., Abbaszade, G., Schäfer, K., Harlos, D.P., Hansen, A.D.A., Zimmermann, R.: Application of direct thermal desorption gas chromatography time-of-flight mass spectrometry for determination of non-polar organics in low-volume samples from ambient

particulate matter and personal samplers. Anal. Bioanal. Chem. 401, 3083–3094 (2011)

[62] Lord, H.L., Zhan, W., Pawliszyn, J.: Fundamentals and applications of needle trap devices. A critical review. Anal. Chim. Acta 677, 3–18 (2010)

[63] Wang, A., Fang, F., Pawliszyn, J.: Sampling and determination of volatile organic compounds with needle trap devices. J. Chromatogr. A 1072, 127–135 (2005)

[64] Ueta, I., Mizuguchia, A., Fujimura, K., Kawakubo, S., Saito, Y.: Novel sample preparation technique with needle-type micro-extraction device for volatile organic compounds in indoor air samples. Anal. Chim. Acta 746, 77–83 (2012)

[65] Ueta, I., Samsudin, E.L., Mizuguchi, A., Takeuchi, H., Shink, T., Kawakubo, S., Saito, Y.: Double-bed-type extraction needle packed with activated-carbon-based sorbents for very volatile organic compound. J. Pharm. Biomed. Anal. 88, 423–428 (2014)

[66] Trefz, P., Kischkel, S., Hein, D., James, E.S., Schubert, J.K., Miekisch, W.: Needle trap micro-extraction for VOC analysis: effects of packing materials and desorption parameters. J. Chromatogr. A 1219, 29–38 (2012)

[67] Ueta, I., Mizuguchi, A., Okamoto, M., Sakamaki, H., Hosoe, M., Ishigurod, M., Saito, Y.: Determination of breath isoprene and acetone concentration with a needle-type extraction device in gas chromatography-mass spectrometry. Clin. Chim. Acta 430, 156–159 (2014)

[68] Ueta, I., Saito, Y., Hosoe, M., Okamoto, M., Ohkita, H., Shirai, S., Tamura, H., Jinno, K.: Breath acetone analysis with miniaturized sample preparation device: in-needle preconcentration and subsequent determination by gas chromatography-mass Spectroscopy. J. Chromatogr. B 877, 2551–2556 (2009)

[69] Saito, Y., Ueta, I., Ogawa, M., Jinno, K.: Simultaneous derivatization/preconcentration of volatile aldehydes with a miniaturized fiber-packed sample preparation device designed for gas chromatographic analysis. Anal. Bioanal. Chem. 386, 725–732 (2006)

[70] Arthur, C.L., Pawliszyn, J.: Solid phase microextraction with thermal desorption using fused silica optical fibers. Anal. Chem. 62, 2145–2148 (1990)

[71] Pawliszyn, J.: Solid phase microextraction: theory and practice. Wiley, New York (1997)

[72] Martos, P.A., Pawliszyn, J.: Sampling and determination of formaldehyde using solid phase microextraction with on-fiber derivatization. Anal. Chem. 70, 2311–2320 (1998)

[73] Djozan, D.J., Bahar, S.: Solid-phase microextraction of aliphatic alcohols based on polyaniline coated fibers. Chromatographia 59, 95–99 (2004)

[74] Chong, S.L., Wang, D., Hayes, J.D., Wilhite, B.W., Malik, A.: Sol-gel coating technology for the preparation of solid-phase microextraction fibers of enhanced thermal stability. Anal. Chem. 69, 3889–3898 (1997)

[75] Mahdi Moein, M., Said, R., Bassyouni, F., Abdel-Rehim, M.: Solid phase microextraction and related techniques for drugs in biological samples. J. Anal. Methods Chem. 1(1–24) (2014)

[76] Ulrich, S.: Solid-phase microextraction in biomedical analysis. J. Chromatogr. A 902, 167–194 (2000)

[77] Mills, G.A., Walker, V.: Headspace solid-phase microextraction procedures for gas chromato-

graphic analysis of biological fluids and materials. J. Chromatogr. A 902, 267–287 (2000)

[78] Theodoridis, G., Koster, E.H.M., de Jong, G.J.: Solid-phase microextraction for the analysis of biological samples. J. Chromatogr. B 745, 49–82 (2000)

[79] Augusto, F., Valentey, A.L.P.: Applications of solid-phase microextraction to chemical analysis of live biological samples. Trends. Anal. Chem. 21(428–438) (2002)

[80] Ligor, T., Ligor, M., Amann, A., Bachler, M., Ager, C., Bachler, M., Dzien, A., Buszewski, B.: The analysis of healthy volunteers' exhaled breath by use of solid phase microextraction and GC-MS. J. Breath Res. 2, 046006 (2008)

[81] Buszewski, B., Ulanowska, A., Ligor, T., Denderz, N., Amann, A.: Analysis of exhaled breath from smokers, passive smokers and non-smokers by solid-phase microextraction gas chromatography/mass spectrometry. Biomed. Chromatogr. 23, 551–556 (2008)

[82] Ouyang, G., Pawliszyn, J.: A critical review in calibration methods for solid-phase microextraction. Anal. Chim. Acta 627(184–197) (2008)

[83] Lord, H., Pawliszyn, J.: Evolution of solid-phase microextraction technology. J. Chromatogr. A 885, 153–193 (2000)

[84] O'Reilly, J., Wang, Q., Setkova, L., Hutchinson, J.P., Chen, Y., Lord, H.L., Linton, C.M., Pawliszyn, J.: Automation of solid-phase microextraction. J. Sep. Sci. 28, 2010–2022 (2005)

[85] Dietz, C., Sanz, J., Camara, C.: Recent developments in solid-phase microextraction coatings and related techniques. J. Chromatogr. A 1103(183–192) (2006)

[86] Kumar, A., Ashok, G., Malik, K., Kumar Tewary, D., Singh, B.: A review on development of solid phase microextraction fibers by sol-gel methods and their applications. Anal. Chim. Acta 610, 1–14 (2008)

[87] Lipinski, J.: Automated solid phase dynamic extraction-Extraction of organics using a wall coated syringe needle. Fresenius J. Anal. Chem. 369, 57–62 (2001)

[88] Van Durme, J., Demeestere, K., Dewulf, J., Ronsse, F., Braeckman, L., Pieters, J., Van Langenhove, H.: Accelerated solid-phase dynamic extraction of toluene from air. J. Chromatogr. A 1175, 145–153 (2007)

[89] Laaks, J., Jochmann, M.A., Schmidt, T.C.: Solvent-free microextraction techniques in gas chromatography. Anal. Bioanal. Chem. 402, 565–571 (2012)

[90] Laaks, J., Letzel, T., Schmidt, T.C., Jochmann, M.A.: Fingerprinting of red wine by headspace solid-phase dynamic extraction of volatile constituents. Anal. Bioanal. Chem. 403, 2429–2436 (2012)

[91] Nerín, C., Salafranca, J., Aznar, M., Batlle, R.: Critical review on recent developments in solventless techniques for extraction of analytes. Anal. Bioanal. Chem. 393, 809–833 (2009)

[92] Baltussen, E., Sandra, P., David, F., Cramers, C.: Stir bar sorptive extraction (SBSE), a novel extraction technique for aqueous samples: theory and principles. J. Microcol. Sep. 11, 737–747 (1999)

[93] David, F., Tienpont, B., Sandra, P.: Stir-bar sorptive extraction of trace organic compounds

from aqueous matrices. LC GC North America 21, 108-118 (2003)

[94] Baltussen, E., Cramers, C.A., Sandra, P.: Sorptive sample preparation - a review. Anal. Bioanal. Chem. 373, 3-22 (2002)

[95] Tienpont, B., David, F., Bicchi, C., Sandra, P.: High capacity headspace sorptive extraction. J. Microcol. Sep. 12, 577-584 (2000)

[96] Popp, P., Bauer, C., Weinrich, L.: Application of stir bar sorptive extraction in combination with column liquid chromatography for the determination of polycyclic aromatic hydrocarbons in water samples. Anal. Chim. Acta 436, 1-9 (2001)

[97] Rita, A., Silva, M., Nogueira, J.M.F.: Stir-bar-sorptive extraction and liquid desorption combined with large-volume injection gas chromatography-mass spectrometry for ultratrace analysis of musk compounds in environmental water matrices. Anal. Bioanal. Chem. 396, 1853 - 1862 (2010)

[98] Ruan, E.D., Aalhus, J.L., Jua'rez, M., Sabik, H.: Analysis of volatile and flavor compounds in grilled lean beef by stir bar sorptive extraction and thermal desorption gas chromatography mass spectrometry. Food Anal. Methods 8, 363-370 (2015)

[99] Niu, Y., Yu, D., Xiao, Z., Zhu, J., Song, S., Zhu, G.: Use of stir bar sorptive extraction and thermal desorption for gas chromatography-mass spectrometry characterization of selected volatile compounds in Chinese liquors. Food Anal. Methods (2015).doi: 10.1007/s 12161-014-0060

[100] Bonet-Domingo, E., Grau-González, S., Martín-Biosca, Y., Medina-Hernández, M.J., Sagrado, S.: Harmonized internal quality aspects of a multi-residue method for determination of forty-six semivolatile compounds in water by stir-bar-sorptive extraction-thermal desorption gas chromatography-mass spectrometry. Anal. Bioanal. Chem. 387, 2537-2545 (2007)

[101] Iparraguirre, A., Prieto, A., Navarro, P., Olivares, M., Fernández, L.Á., Zuloaga, O.: Optimisation of stir bar sorptive extraction and in-tube derivatisation-thermal desorption-gas chromatography-mass spectrometry for the determination of several endocrine disruptor compounds in environmental water samples. Anal. Bioanal. Chem. 401, 339-352 (2011)

[102] Kawaguchi, M., Ito, R., Sakui, N., Okanouchi, N., Saito, K., Seto, Y., Nakazawa, H.: Stir-bar-sorptive extraction, with in-situ deconjugation, and thermal desorption with in-tube silylation, followed by gas chromatography-mass spectrometry for measurement of urinary 4-nonylphenol and 4-tert-octylphenol glucuronides. Anal. Bioanal. Chem. 388, 391-398 (2007)

[103] Ferreira, A.M.C., Möder, M., Laespada, M.E.F.: GC-MS determination of parabens, triclosan and methyl triclosan in water by in situ derivatisation and stir-bar sorptive extraction. Anal. Bioanal. Chem. 399, 945-953 (2011)

[104] Badoil, L., Benanou, D.: Characterization of volatile and semivolatile compounds in waste landfill leachates using stir bar sorptive extraction-GC/MS. Anal. Bioanal. Chem. 393, 1043-1054 (2009)

[105] Van Hoeck, E., Canale, F., Cordero, C., Compernolle, S., Bicchi, C., Sandra, P.: Multi-

residue screening of endocrine-disrupting chemicals and pharmaceuticals in aqueous samples by multi-stir bar sorptive extraction-single desorption-capillary gas chromatography/mass spectrometry. Anal. Bioanal. Chem. 393, 907-919 (2009)

[106] Hyötyläinen, T.: Critical evaluation of sample pretreatment techniques. Anal. Bioanal. Chem. 394, 743-758 (2009)

[107] Farajzadeh, M.A., Sorouraddin, S.M., Mogaddam, M.R.A.: Liquid phase microextraction of pesticides: a review on current methods. Microchim. Acta 181, 829-851 (2014)

[108] Vincelet, C., Rousse, J.M., Benanou, D.: Experimental designs dedicated to the evaluation of a membrane extraction method: membrane-assisted solvent extraction for compounds having different polarities by means of gas chromatography-mass detection. Anal. Bioanal. Chem. 396, 2285-2292 (2010)

[109] Kuosmanen, K., Hyötyläinen, T., Hartonen, K., Jönsson, J.A., Riekkola, M.L.: Analysis of PAH compounds in soil with on-line coupled pressurised hot water extraction-microporous membrane liquid-liquid extraction-gas chromatography. Anal. Bioanal. Chem. 375, 389 - 399 (2003)

[110] Lüthje, K., Hyötyläinen, T., Riekkola, M.L.: On-line coupling of microporous membrane liquid-liquid extraction and gas chromatography in the analysis of organic pollutants in water. Anal. Bioanal. Chem. 378, 1991-1998 (2004)

[111] Iparraguirre, A., Navarro, P., Prieto, A., Rodil, R., Olivares, M., Ferna'ndez, L.A., Zuloaga, O.: Membrane-assisted solvent extraction coupled to large volume injection-gas chromato-graphy-mass spectrometry for the determination of a variety of endocrine disrupting compounds in environmental water samples. Anal. Bioanal. Chem. 402, 2897-2907 (2012)

[112] Pratt, K.F., Pawliszyn, J.: Gas extraction kinetics of volatile organic with a hollow fiber membrane. Anal. Chem. 64, 2101-2106 (1992)

[113] Matz, G., Kibelka, G., Dahl, J., Lenneman, F.: Experimental study on solvent-less sample preparation methods - membrane extraction with a sorbent interface, thermal membrane desorption application and purge-and-trap. J. Chromatogr. 830, 365-376 (1999)

[114] Juan, S., Guo, X., Mitra, S.: On-site and on-line analysis of chlorinated solvents in ground water using pulse introduction membrane extraction gas chromatography (PIME-GC). J. Sep. Sci. 24, 599-605 (2001)

[115] Kotiaho, T.: On-site environmental and in situ process analysis by mass spectrometry. J.Mass Spectrom. 31, 1-15 (1996)

[116] Viktorova, O.S., Kogan, V.T., Manninen, S.A., Kotiaho, T., Ketola, R.A.,Dubenskii, B.M., Parinov, S.P., Smirnov, O.V.: Utilization of a multimembrane inlet and a cyclic sudden sampling introduction mode in membrane inlet mass spectrometry. J. Am. Soc. Mass Spectrom. 15, 823-831 (2004)

[117] Li, X., Xia, L., Yan, X.: Application of membrane inlet mass spectrometry to directly quantify denitrification in flooded rice paddy soil. Biol. Fertil. Soils 50, 891-900 (2014)

[118] Beckmann, K., Messinger, J., Badger, M.R., Wydrzynski, T., Hillier, W.: On-line mass spectrometry: membrane inlet sampling. Photosynth. Res. 102, 511–522 (2009)

[119] Masucci, J.A., Caldwell, G.W.: Techniques for gas chromatography/mass spectrometry. In: Grob, R.L., Barry, E.F. (eds.) Modern practice of gas chromatography, pp. 339–401. Wiley, Hoboken, NJ (2004)

[120] Hansel, A., Jordan, A., Holzinger, R., Prazeller, P., Vogel, W., Lindinger, W.: Proton transfer reaction mass spectrometry - online trace gas analysis at the ppb level. Int. J. Mass Spectrom. Ion Proc. 150, 609–619 (1995)

[121] Cao, W., Duan, Y.: Current status of methods and techniques for breath analysis. Crit. Rev. Anal. Chem. 37, 3–13 (2007)

[122] Lindinger, W., Hansel, A., Jordan, A.: On-line monitoring of volatile organic compounds at pptv levels by means of proton-transfer-reaction mass spectrometry (PTR-MS), medical applications, food control and environmental research. Int. J. Mass Spectrom. Ion. Proc. 173, 191–241 (1998)

[123] Zhan, X., Duan, J., Duan, Y.: Recent developments of proton-transfer reaction mass spectrometry (PTR-MS) and its applications in medical research. Mass Spectrom. Rev. 32, 143–165 (2013)

[124] Blake, R.S., Whyte, C., Monks, P.S., Ellis, A.M.: Proton transfer reaction time-of-flight mass spectrometry: a good prospect for diagnostic breath analysis. In: Amann, A., Smith, D. (eds.) Breath analysis for clinical diagnosis and therapeutic monitoring, p. 45. World Scientific, Toh Tuck Link, Singapore (2005)

[125] Moser, B., Bodrogi, F., Eibl, G., Lechner, M., Rieder, J., Lirk, P.: Mass spectrometric profile of exhaled breath—field study by PTR-MS. Resp. Physiol. Neurobiol. 145, 295–300 (2005)

[126] Španěl, P., Diskin, A.M., Abbott, S.M., Wang, T., Smith, D.: Quantification of volatile compounds in the headspace of aqueous liquids using selected ion flow tube mass spectrometry. Rapid Commun. Mass Spectrom. 16, 2148–2153 (2002)

[127] Smith, D., Španěl, P.: Selected ion flow tube mass spectrometry, SIFT-MS, for on-line trace gas analysis. In: Amann, A., Smith, D. (eds.) Breath Analysis for Clinical Diagnosis and Therapeutic Monitoring, pp. 3–34. World Scientific Publishing, Toh Tuck Link, Singapore (2005)

[128] Španěl, P., Smith, D.: On-line measurements of the absolute humidity of air, breath and liquid headspace samples by selected ion flow tube mass spectrometry. Rapid Commun. Mass Spectrom. 15, 563–596 (2001)

[129] Španěl, P., Smith, D.: Selected ion flow tube mass spectrometry: detection ad real-time monitoring of flavours released by food products. Rapid Commun. Mass Spectrom. 13, 585–597 (1999)

[130] Smith, D., Wang, T., Sule-Suso, J., Španěl, P., Haj, A.E.: Quantification of acetaldehyde

released by lung cancer cells in vitro using selected ion flow tube mass spectrometry. Rapid Commun. Mass Spectrom. 17, 845-850 (2003)

[131] Baumbach, J.I.: Process analysis using ion mobility spectrometry. Anal. Bioanal. Chem. 384, 1059-1070 (2006)

[132] Li, F., Xie, Z., Schmidt, H., Sielemann, S., Baumbach, J.I.: Ion mobility spectrometer for online monitoring of trace compounds. Spectrochim. Acta Part B 57, 1563-1574 (2002)

[133] Ewing, R.G., Atkinson, D.A., Eiceman, G.A., Ewing, G.J.: A critical review of ion mobility spectrometry for the detection of explosives and explosive related compounds. Talanta 54, 515-529 (2001)

[134] Rearden, P., Harrington, P.B.: Rapid screening of precursor and degradation products of chemical warfareagents in soil by solid-phase microextraction ion mobility spectrometry (SPME-IMS). Anal. Chim. Acta 545, 13-20 (2005)

[135] Wang, Y., Nacson, S., Pawliszyn, J.: The coupling of solid-phase microextraction/surface enhanced laser desorption/ionization to ion mobility spectrometry for drug analysis. Anal. Chim. Acta 582, 50-54 (2007)

[136] O'Donnell, R.M., Sun, X., Harrington, P.B.: Pharmaceutical applications of ion mobility spectrometry. Trends Anal. Chem. 27, 44-53 (2008)

[137] Miekisch, W., Schubert, J.K.: From highly sophisticated analytical techniques to life-saving diagnostics: technical developments in breath analysis. Trends Anal. Chem. 25, 665-673 (2006)

第15章
放射性核素分析

15.1 概述

15.1.1 放射性及其伴随现象

放射性是由物质的原子核衰变而产生的,释放出 α 射线、β 射线和 γ 射线。放射性同位素的原子核是不稳定的,可能发生以下转变。

(1) α 衰变(α):由原子核放射出 α 粒子,α 粒子是由两个质子和两个中子组成的一个氦核。

(2) 负 β 衰变($β^-$):在转变过程中释放电子,核内由于中子过剩,其中的一个中子转变为质子。

(3) 正 β 衰变($β^+$):在转变过程中释放正电子,核内由于质子过剩,其中的一个质子转变为中子。

(4) 电子俘获(EC):原子核从 K 层或 L 层轨道俘获电子;被俘获的电子与质子结合产生中子、中微子、X 射线、γ 射线、俄歇电子和内部"韧致辐射"。

除了上述转变之外,在原子核中还可能发生以下伴随现象。

(1) 发射 γ 射线(γ);
(2) 内部转换;
(3) 发射 X 射线[1]。

15.1.2 天然和人工放射性

有三位科学家与放射性发现有关:亨利·贝克勒尔,其在 1896 年发现了这一现象[2];玛丽亚·居里将这个过程命名为"放射性";以及她的丈夫皮埃尔·居里[3]。他们证实铀会发射电离射线,除此之外,玛丽亚还发现钍会发射同样的射线。她发现辐射不是分子相互作用的结果,而是来自原子,这个发现绝对是革命性的。玛丽亚和皮埃尔发现了前两种放射性元素:钋和镭。在自然环境中大约有20种放射性元素和50种放射性核素。

第一个人造放射性核素^{30}P是在1934年由费雷德克里和伊伦·约里奥·居里(居里夫人的女儿和女婿)通过加速器中的质子轰击铝发现的[4]。目前,特别是在发现并应用^{235}U和^{239}Pu核裂变反应之后,已经有2000多种人工放射性核素被制造并得到确认。

随时间变化的放射性核子的数量是较少的。半衰期是每种放射性核素的特征参数,定义为放射性核数目衰变到一半所需要的时间。每个放射性核素进行衰变,表达如下:

$$A_0 = A_t \cdot \exp\left[\frac{t \cdot \ln 2}{T}\right] \quad (15-1)$$

式中:A_0为$t=0$时刻的放射性核素的活度;A_t为t时刻后的放射性核素的活度;T为放射性核素半衰期;t为$0\sim t$的时间。

放射性核素的活度是指单位时间内衰变的量。如今,放射性活度的单位为Bq(1Bq=1次衰变/s),但在过去是用居里来表示的($1\text{ Ci} = 3.7\times10^{10}\text{ Bq}$)。

环境中存在的放射性核素分为天然来源或人工来源[1]:

1) 天然放射性核素

(1) 陆生放射性核素,如^{40}K和^{87}Rb;

(2) 宇生放射性核素,如^{3}H、^{14}C、^{32}Si和^{36}Cl;

(3) 原生放射性核素:这些长期存在的放射性核素自形成以来就普遍存在于地球上,放射性核素^{238}U、^{232}Th和^{235}U分别是铀、钍和锕放射性衰变链的母体元素。

2) 人工放射性核素

(1) 中子活化产物,如^{22}Na、^{54}Mn、^{55}Fe、^{60}Co、^{63}Ni、^{64}Cu、^{65}Zn、^{110m}Ag、^{124}Sb、^{125}Sb;

(2) ^{235}U和^{239}Pu裂变放射性核素,如^{90}Sr、^{95}Zr、^{131}I、^{132}I、^{132}Te、^{137}Cs、^{144}Ce;

(3) 超铀元素,如^{238}Pu、^{239}Pu、^{240}Pu、^{241}Pu、^{241}Am、^{243}Am。

考虑到它们的半衰期、衰变类型和放射性毒性,自然环境中最重要的放射性核素是^{210}Po、^{210}Pb、^{222}Rn、^{226}Ra、^{234}U、^{235}U和^{238}U;最重要的人工放射性核素有^{55}Fe、^{60}Co、^{63}Ni、^{90}Sr、^{137}Cs、^{238}Pu、^{239}Pu、^{240}Pu、^{241}Pu。在自然环境中,这些放射性核素以微量的形式出现[1]。

15.1.3 辐射测量方法

探测器是辐射测量仪的一个非常重要的组成部分,用于测量气体、液体和固体中的电离辐射,以下列举了几种用于测量电离辐射的探测器[5]。

(1) 气体探测器:电离室、正比计数器(测量 α 粒子和 β 粒子),以及盖格计数器(专门针对 β 粒子)。

(2) 闪烁探测器:晶体、塑料和液体闪烁体。在闪烁体中,电离辐射与物质作用产生光量子。

(3) 半导体探测器:不仅可以对电离粒子进行计数,还可以作为谱测量仪器来测量能谱,由单晶硅和锗制备的探测器用来测量 α 粒子和 γ 射线。

低水平 α 能谱法、β 能谱法和 γ 能谱法用于测定环境中低活度水平的放射性物质。考虑到 α 能谱和 γ 能谱的非线性特征,放射性核素识别是基于测量放射性同位素发出的 γ 射线或 α 粒子的能量,故在 γ 能谱和 α 能谱中可以同时检测和鉴定一些放射性核素。在这些方法中,可以通过符合和反符合测量系统降低地球化学本底。与 α 能谱和 γ 能谱相比,β 电子发射的能谱是线性的,以平均能量 E_{av} 和最大能量 E_{max} 来表征。因此,用 β 电子发射的能谱无法同时检测和鉴定某些放射性核素,尤其是同一元素的放射性同位素[1]。

15.1.4　γ 能谱法

γ 辐射是一种光子流,其能量介于几万电子伏特和千万亿电子伏特之间,环境中典型 γ 辐射的范围为 40keV~3MeV[5]。

对 γ 能谱的精确研究需要保证测量区域免受污染,特别是要使用新鲜空气,因为氡及其衰变产物,以及含有高密度元素或高序数金属,如铅、钨、铜、铁的材料中会含有少量的放射性物质,特别是 U、Ra、Th、K 及其衰变产物。通过增大光栅宽度来减小谱仪的本底值,并不能保证其检出限的降低,还需要将样品中的辐射电压降至最低,并将 X 射线荧光的感应区转移到谱仪未注册的能量区[5]。

由于自然地球化学背景的影响很大,γ 能谱法定量测定自然环境中低活度放射性核素的应用有限,因此该技术常用于气载放射性活度的研究[6-7]。初级 γ 能谱法主要用于测定如 ^{40}K、^{60}Co、^{137}Cs、^{210}Pb 等核素的活性(图 15-1)。

15.1.5　β 能谱法

测量 β 辐射谱最精确的方法是电磁法,利用交叉磁场和电场以及机械光栅来选择电子或正电子的能量。电磁谱仪对环境辐射的分析效果不佳,在这种情况下,如气体计数器和液体闪烁计数器等非能谱测量设备,常用来测量低能 β 放射性核素。将含有放射性核素的试剂与有机闪烁液混合,在闪烁计数器中进行分析。目前,该方法是测定环境样品中放射性核素的最著名、最常见、最通用和最完善的方

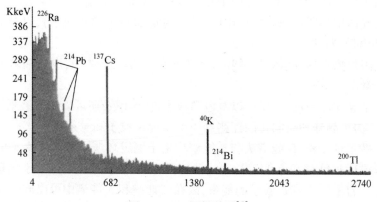

图 15-1 γ 辐射谱图[5]

法。过去,闪烁试剂是含有甲苯或苯的 2,5-二苯基噁唑(PPO)和一种能改变波谱发射光波长的物质 1-4-双-2-(5-苯基噁唑基)苯(POPOP),如今,使用的是高化学稳定性的非燃烧闪烁剂(如超金堪培拉或希法菲 3 华拉克)[5]。采用 5~10cm 厚的铅盖和一种使用塑料(如 Wallac Guardian)、液体(如 Wallac Quantulus)或铋-锗酸盐晶体闪烁体(如 Packard-Canberra 模式 2770)的附加系统,可以减少测量过程中的地球化学本底值。通过分析闪烁过程的时间信息可分辨 α 辐射和 β 辐射信号[8]。最近,有研究者尝试利用半导体探测器来测量 β 辐射[9]。使用液体闪烁光谱法(Wallac 1414-003 Guardian)测量获得的 β 辐射能谱图如图 15-2 所示。

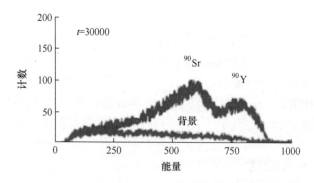

图 15-2 放射性同位素 ^{90}Sr-^{90}Y 的 β 辐射谱图[5]

某些放射性同位素的活度,特别是 ^{90}Sr-^{90}Y,也可以通过切伦科夫现象用液晶光谱法检测[10-11]。切伦科夫效应用于测定发射 β 射线的 E_{max} 在 500keV 以上的 β 粒子[12]。利用切伦科夫效应测定 β 放射性活度的主要优点是,使用分析制剂进行另一种化学分析(如计算回收率)。此外,附加低能 β 辐射或 α 辐射不会干扰测量,从而降低了分析成本。但是,这种方法的缺点是降低了可回收性和减少了有关β能谱的真实信息[13]。环境样品中β同位素的测定非常困难,需要进行化学分离。样品类型和化学分析的时间决定了分析方法的选择。与此同时,样品被污染和采集之间的时间也很重要,近期被污染的样品所用的方法与已经发生了短时间的放射性核素衰变的旧样品所用的方法是不同的[1,5]。

15.1.6 α能谱法

α能谱法是测量天然和人工放射性核素发射的α粒子的最重要的放射测量技术之一。α能谱仪包含活性表面为 $100\sim600mm^2$ 的低能级半导体探测器,该探测器被放置在与1024多道分析仪相连的真空室中。所用的探测器是带有薄金层的硅二极管。这些探测器的效率在25%~40%,分辨率在20~35keV。α能谱仪能够测量能量范围为 $3\sim8MeV$ 的α粒子,并能测定自然环境中大多数α放射性核素活度[1]。考虑到α粒子具有低穿透性,在对待分析的放射性核素进行化学分离之前,应先采用α能谱法测量。分析环境样品通常使用热的浓酸(HF、HNO_3、HCl、$HClO_4$),采用干法或湿法进行消解[1,14]。

15.1.7 中子活化分析法

中子活化分析法(NAA)是一种灵敏的多元素分析技术,可用于主要元素、次要元素、微量元素和稀有元素的定性和定量分析。中子活化分析法是通过中子轰击样品,将稳定核素转化为放射性核素,然后测量其辐射,特别是γ辐射。目前,样品中的人工放射性核素可以用两种方法来测量[15-16]。

(1)破坏性方法(放射化学分离中子活化分析,RNAA):在化学分离的基础上,将放射性元素分解成不同的组分,每一组分都含有一些放射性核素[17-20]。

(2)非破坏性方法(仪器中子活化分析,INAA):采用高分辨率γ能谱法对样品进行放射性活度测量,利用放射性核素衰变速率的微分进行分析[15]。

中子活化分析法有许多优点[15]。

①对许多元素具有低检出限;

②无空白样品或最小化空白样品;

③多元素分析;

④可对干燥、固体或粉状样品进行无损分析;

⑤元素分析的高准确性。

中子活化分析法注重分析一些特殊的样品,例如陨石、标准物质定值、分析化学质量保证等[21]。"确定的方法"基于样品的活度,用柱(离子交换)色谱选择分离待分析的放射性核素,并用γ能谱法进行测量[15-16,19]。在"确定的方法"中,放射性化学分离、共沉淀法、萃取法和离子交换法保证了待分析的放射性核素的高纯度和高回收率。放射性核素的分离通常是通过添加非活性载体来进行的,回收率是通过重量法、滴定法或比色法来确定的。在确定的方法中,分离方法在数量和质量上必须具有高度的选择性,因为这些方法仅用于测定一种元素[16]。

中子活化分析法在痕量分析中的应用涉及核反应堆,核反应堆是中子流的来源。中子活化分析法最重要的优点是非破坏性,这使得其对于分析土壤、沉积物、岩石、生物样品和食物非常有利。中子活化分析法也用于犯罪学和司法毒理学,研究存在于杀虫剂和除草剂及海洋生物中的无机化合物(As、Hg)或有机化合物,例如卤代(或氯代、溴代、碘代)有机化合物的毒性[22-23]。

15.2 放射分析方法在环境中的应用:分析方面

15.2.1 ^{40}K 活度测定

同位素^{40}K 是一种天然放射性核素,具有较长的半衰期($T=1.26\times10^9$年),在天然钾中含量为 0.0118%。^{40}K 经 β 辐射转变成为^{40}Ca(89.05%),通过电子俘获转化成^{40}Ar(10.95%)[24]。利用 1461keV 能谱,可以在干燥均匀的环境样品中进行^{40}K 活度的测定。在天然水样中,可以采用切伦科夫计数法测定^{40}K[25-26],也可以采用热电离质谱法(TIMS)测定环境样品中的^{40}K[27]。

15.2.2 活化产物^{55}Fe、^{60}Co、^{63}Ni 的活度测定

在核武器试验、核电厂的后处理厂和反应堆运行以及核科学研究中,由于活化而产生了一些人工放射性核素。现代放射分析技术使环境中的^{22}Na、^{51}Cr、^{54}Mn、^{55}Fe、^{60}Co、^{63}Ni、^{65}Zn、^{110}Ag、^{124}Sb 等活化产物得以检测[28-29]。含铁、镍、钴的不锈钢是核反应堆的重要材料,用于建造核试验装置或其支撑结构[30-31]。在中子活化钴稳定同位素期间,产生放射性同位素^{60}Co($T=5.27$ 年),然后发生 β 衰变,释放出两个 γ 量子(1.173MeV 和 1.333MeV)之后,衰变成 E_{max} 为 0.314MeV 的^{60}Ni[24],用 γ 能谱法测定^{60}Co 在环境样品中的活度[1]。中子活化稳定同位素铁产生放射性核素^{55}Fe($T=2.685$a),通过电子俘获衰变,它是一个具有 6.9 keV 能量的 β 发射体[24]。放射性镍的同位素(^{59}Ni、^{63}Ni 和^{65}Ni)通过中子活化产生,^{63}Ni 同位素是三者中最重要的核素,因为^{59}Ni/^{63}Ni 的活度比仅为 0.01。^{63}Ni 同位素也

是一种β粒子发射体,半衰期为100.1年[32]。环境样品中^{55}Fe和^{63}Ni的分析方法包括以下步骤。

（1）^{55}Fe、^{63}Ni在天然水中与氢氧化铁共沉淀；

（2）矿物（沉积物和土壤）和生物样品的灰化；

（3）铁(^{55}Fe)与铁络合物分离,镍(^{63}Ni)与二甲基乙二肟络合物$(DMG)_2$Ni分离；

（4）用阴离子交换树脂（Dowex）纯化铁、镍馏分；

（5）在抛光铜盘上电解^{55}Fe和^{63}Ni。

用原子吸收光谱法测量电沉积前后稳定的铁和镍的放射化学产额。使用位于盖革-弥勒区的反符合屏蔽无窗低水平β粒子流气式计数器测量^{55}Fe和^{63}Ni的活度。工作气体由氩气（99%）和异丁烯（1%）组成[28,33]。被^{55}Fe和^{63}Ni污染的盐类,可以采用闪烁计数器与高分辨离子色谱联用技术[34-35],或^{63}Ni在Chelex 100树脂上吸附并与$(DMD)_2$Ni络合纯化等辐射测量方法[36]。除此之外,还可以利用加速质谱法（AMS）测定环境样品中的^{63}Ni。

15.2.3 ^{137}Cs活度测定

^{137}Cs是人工放射性核素中最重要的长寿命（T=30.17年）裂变产物和常见污染物之一。它释放出1176 keV（6%）和514 keV（94%）两种能量的β辐射,激发2.55 min ^{137}Ba*的平衡体。这一平衡体通过发射661.66 keV的单能γ射线来激发自身的能量[24]。在平衡状态下,^{137}Cs和^{137}Ba*的活度是相同的,^{137}Cs的活度可以直接用β能谱法测定,也可以用γ能谱法间接测定^{137}Ba*的活度。β能谱法直接测定^{137}Cs需要将环境样品消解,或用磷钼酸铵（AMP）吸附铯将其分离,随后通过阳离子交换树脂（Dowex 1）对^{137}Cs进行提纯,并用六氯铂酸铯(Cs_2PtCl_6)进行共沉淀,再用低水平流气式β计数器测量^{137}Cs的活度[38]。间接法包括对分析样品进行干燥和均匀化,并用γ能谱法测量^{137}Cs(^{137}Ba*)的活度。直接法是最精确的,但也是最费时的,需要校准^{137}Cs的活度和Cs_2PtCl_6沉淀物的质量以及作为化学示踪剂的CsCl的质量[1]。校准通常使用活度为1kBq的^{137}Cs标准溶液,0.6mg~30mg CsCl载体,50mg~200mg六氯铂酸(H_2PtCl_6)。Cs_2PtCl_6沉积物质量与^{137}Cs活度之间的校正系数按下式计算：

$$\eta = \frac{A}{CPM_{100\%}} \quad (15-2)$$

式中：η为校准系数Bq·计数$^{-1}$·min^{-1}（一般范围为0.03~0.06）；A为^{137}Cs标准活度；$CPM_{100\%}$为根据铯回收率100%计算出来的值。

计算样品中^{137}Cs的活度：

$$C = \frac{\text{CPM}_\eta \cdot \eta \cdot 100\%}{m \cdot Y} \quad (15-3)$$

式中：C为^{137}Cs的比活度(Bq/g)；m为样品质量(g)，η为校准系数；CPM_η为在β计数器(无背景)中测量的^{137}Cs的计数率；Y为回收率(%)。

^{137}Cs活度的标准偏差(SD)由下式计算：

$$\delta = \sqrt{\frac{\text{CPM}_b}{t_P} + \frac{\text{CPM}_t}{t_t}} \quad (15-4)$$

式中：CPM_b为β计数器测样时的计数率(含本底)；CPM_t为β计数器测样的本底计数率；t_p为样品计数时间；t_t为本底计数时间。

15.2.4 ^{90}Sr活度测定

^{90}Sr是毒性最强和最危险的放射性同位素之一，也是一个纯β发射体(E_{max} = 546keV)，能够衰变成另一个纯β发射体^{90}Y(E_{max} = 2283.9keV)[24]。

环境样品(水、土壤、沉积物和生物群)中^{90}Sr的放射化学测定方法是基于测定平衡状态或分离后仅有^{90}Sr的放射性活度[5,39]。分析方法的选择取决于样品的种类和化学成分、数据采集和分析的时间、辐射测量技术的类型(闪烁计数器、切伦科夫计数器或反符合计数器)以及分离方法(锶或钇分离)[5]。放射性化学污染与样品采集之间的时间对于^{90}Sr的分析非常重要，根据Fischer[40]的研究，在新采集的样品(除^{90}Sr之外)中存在以下放射性核素：^{89}Sr、^{91}Sr、^{92}Sr、^{91}Y、^{92}Y和^{93}Y。在两三年后，它们的活度变得很小，在此之后，短寿命放射性核素与^{90}Sr的活度比小于20。对于新采集样品(最多14天)，应采用快速测定^{90}Sr的方法，因为分析样品中还存在^{89}Sr、^{90}Sr和^{90}Y同位素。对于老样品的分析，可以采用去除化学污染物和放射性化学污染物的准确方法[5]。

测定环境样品中^{90}Sr的分析方法分为快速法和缓慢法两种。快速法测定^{90}Sr，是在污染后的几天收集新鲜土壤和沉淀物样品，基于不同填充剂(Sr-Spec@，Amberlit树脂)在色谱柱上的吸附和随后锶的提取，在液体闪烁计数器中测定^{90}Sr的活度[41-42]。Mateos等[43]提出了一种顺序注入分析法，以确定在分离两种放射性核素30min后，矿泉水样品中^{90}Sr/^{90}Y的活度比。此外，Friberg[40,44]还提出了用浓HNO$_3$从固体样品中提取放射性核素，然后用HDEHP[乙二酸(2-乙基硅氧烷)]分离^{90}Sr和^{90}Y，并在切伦科夫计数器中测定^{90}Sr的活度。^{90}Sr与^{90}Y分离24h后，^{90}Sr在有机相中的活度占放射性锶初始活度的22%。对于新采集的样品，^{90}Sr分析的总时间为50天[40]。

缓慢法测定环境样品中^{90}Sr活度的误差最小,因此样品的采集、制备和消解时间是非常重要的[39,45]。对于土壤、沉积物、骨头、肉类和植物样品,建议在450~600℃的温度下焚烧4~24h。在^{90}Sr分离过程中,常用HF、H_2SO_4、HNO_3混合酸进行消解[44,46],然后用HNO_3或HCl浸取,也常使用NaOH、HCl[39,47]、65% HNO_3[44,48-50]或发烟HNO_3[39,51-52]浸取。在这些方法中,用γ射线能谱法测定样品中作为活性示踪剂的^{85}Sr的活度,用于计算分析产额。此外,用能量色散X射线荧光光谱仪(ED-XRF)或原子吸收光谱法(AAS)测定样品消解前后的稳定锶,用于计算回收率[53-54]。土壤样品的分析产额为2.7%~58%,取决于样品中锶的含量[53-54]。

对于^{90}Sr的痕量分析(如天然水样)应使用载体,最常见的载体是离子类,如Sr^{+2}、Ba^{+2}、Ca^{+2}、Fe^{+3}[39]。钙离子与锶离子非常相似,应根据钙与硝酸锶在浓HNO_3或发烟HNO_3中溶解度的不同从样品中去除。Chen等[55]提出,当样品中含有50g以上Ca,且Ca/Sr比为250时,应以$Ca(OH)_2$的形式从样品中分离钙离子。此外,利用冠醚提取^{90}Sr[50,52,56]或用离子交换树脂分离也是十分重要的[47,50]。许多放射性核素,特别是Th、U、Ra、Ac、^{210}Po、^{212}Pb、^{214}Pb和污染物样品,应该从分析物中去除。对于镭,使用$Ba(Ra)Cl_2$或$Ba(Ra)CrO_4$共沉淀可使近100%的锶从镭中分离出来[55]。另外,可通过离子交换树脂Dowex 1x8或Dowex 50Wx8从^{90}Sr与^{90}Y分离后的溶液中去除天然矿石中的Th、U、Pa、Ra、Ac[57]。

放射化学分析过程中最重要的阶段是将^{90}Sr与其衰变产物^{90}Y分离。所采用的方法是与$SrSO_4$、SrC_2O_4或$SrCO_3$共沉淀[39,50-52,55],液-液萃取(如使用HDEHP)[44,48,53]以及固相萃取[41-43,58-62]。

对于^{90}Sr的分析方法,^{90}Sr与其衰变产物^{90}Y之间的放射性平衡状态非常重要,这种状态是在放射性锶分离12天后达到平衡[62]。^{90}Sr测定结果的可靠性取决于最小可检测活度(MDA)[5],每个分析样本均应计算MDA。通常使用发烟HNO_3分离^{90}Sr,然后与镭、铅、钡通过共沉淀法形成铬酸盐,可用于植物、土壤、灰尘过滤器和水样的分析。熔融产物(如^{137}Cs)由氢氧化物共沉淀去除,之后转化为草酸钇,在低浓度比例计数器中测量^{90}Y的活度。分析之前,通过测量每个样品中添加的^{85}Sr(γ发射体)作为活度测量的内部示踪剂来控制聚变产物的产额[1,46]。使用标准物质(CRM)验证分析结果的准确性。

液体闪烁计数器测量后,根据下式计算^{90}Sr的活度[1]:

$$A = \frac{29.55 \cdot N \cdot 100\%}{2 \cdot t \cdot \text{eff} \cdot Y \cdot m} \quad (15-5)$$

式中:A为^{90}Sr的活度(Bq/g干重);N为计数率;t为计数时间(s);Y为γ测量的^{85}Sr示踪剂的回收率(%);m为样品的质量(g);29.55为^{90}Y和^{90}Sr-^{90}Y能谱之间

的计数比例系数;2 为^{90}Sr 或^{90}Y 的活度值(二者处于平衡状态);eff 为 β 辐射的有效系数(通常为 0.90~1.00)。

Eff 计算公式为

$$\text{eff} = \frac{A_2}{A} \qquad (15-6)$$

式中:A_2 为样品中^{90}Y 的活度;A 为实际的^{90}Y 的活度。

^{90}Sr 的检出限 L_d 的计算公式为

$$L_d = 2.86 + 4.78\sqrt{(B + 1.36)} \qquad (15-7)$$

式中:B 为本底值。

15.2.5 天然和人工放射性核素活度的测定

1. 镭(^{226}Ra)

^{226}Ra(半衰期为 1602 年)是^{238}U 自然衰变产生的放射性同位素。基于公共卫生方面的考虑,地球学家、海洋学家和环境学家经常分析天然样品中的^{226}Ra,特别是在水生环境中,^{226}Ra 是地球化学过程的有效示踪剂。在自然、公共和饮用水给水中镭的检测已备受关注,因为镭是内照射中最危险的元素之一[63-65]。在环境样品中,天然水体是迄今为止通过液体闪烁法[64-66]、切伦科夫计数法[67]、α 能谱法和 γ 能谱法以及电感耦合等离子体质谱(ICP-MS)法[68]检测镭的最常见的样品。能谱法是灵敏度最好的方法,因为其可以直接测量样品中^{226}Ra 的活度。其他方法,特别是液体闪烁法、切伦科夫计数法和 γ 能谱法,都是测量^{226}Ra 的衰变产物^{222}Rn 的测量方法[69]。在天然水中,镭与二氧化锰(MnO_2)共沉淀,也会与矿物、沉积物和土壤样品(消解后)中的微晶 Ba(Ra)SO_4共沉淀。利用 α 能谱法测定^{226}Ra 的活度[70-71],用^{133}Ba(γ 发射体)或^{224}Ra(α 发射体)作为内部示踪剂计算回收率。

2. 氡(^{222}Rn)

氡是惰性气体,在自然环境中最重要的同位素是^{222}Ra 和^{220}Rn。^{222}Rn(半衰期为 3.8 天)是^{226}Ra 的直系子核,而^{220}Rn(半衰期为 55s)是^{224}Ra 的衰变产物[24]。从放射学角度来看,^{222}Rn 是最重要的,测量其在空气中的活度非常有意义[5]。空气样品中氡的浓度可以通过电离室或闪烁室实时测量,也可以通过活性炭对^{222}Rn 的吸附来测量[5]。^{222}Rn 的衰变产物(^{214}Bi 和^{214}Pb 处于放射性平衡状态)的活度可以用 γ 能谱法或闪烁计数法来测量(以及^{222}Rn 的活性),其他衰变产物,如^{218}Po 和^{214}Po 被吸附在玻璃过滤器上,并通过被动式植入平面硅(PIPS)探测器进行测量。最近,已经研发了基于活性炭吸附^{222}Rn 并测量其 α、β 和 γ 衰变产物的方法[72-73]。3h 后,用 γ 能谱法测定 ^{214}Bi 和 ^{214}Pb 的活度,其活度与 ^{222}Rn 在活性炭和空气中的活度成正比。该方法需要对镭腔内氡浓度的探测器进行校准[73]。另一种方法是液体闪烁测量法,探测器是一个 30mL 的圆柱形塑料容器(Pico-Rad)[5],在容器内部有用于吸附氡的活性炭和用于吸湿的硅胶,在空气中暴露 48h(吸附 95%氡)后,加入闪烁液(二甲苯或甲苯),8h 后,氡转移到闪烁液中,在液体闪烁计数器中测量其活度。这种方法(Pico-Rad 检测器)检测 ^{222}Rn 的限值在 10 Bq/m^3 以上,误差为±10%[74]。

Zikovsky 和 Roireau 已经研发出用正比计数器测量水中氡的简单方法[75]。该方法是用氩气从水中吹出氡,通过水样吹出,然后对准计数管,其检出限为 0.02Bq/L,优于其他测量水中氡活度的方法,如液体闪烁法、切伦科夫计数法和发光分析法[75-78]。

3. 钋(^{210}Po)、铅(^{210}Pb)、铀(^{234}U、^{235}U、^{238}U)和钚(^{238}Pu、$^{239+240}$Pu、^{241}Pu)

许多天然和人工放射性核素是或者可能是研究自然环境中地球化学和生物过程的指标,但它们的浓度都非常低。

自然环境中发现的钋,存在于铀和钍矿石中。在钋的 7 种天然放射性核素中,^{210}Po 是最重要的天然放射性核素。它是一个 α 发射体,能量为 5.305MeV,半衰期为 138.376 天[24]。钋是一种具有放射性的有毒元素,在陆地和水生生物体内具有很强的生物积累[1]。

天然铀含有 3 种放射性核素,分别为 ^{238}U(99.2745%)、^{235}U(0.7200%)和 ^{234}U(0.0055%),它们都是寿命较长的放射性同位素,^{234}U 的半衰期为 2.455×10^5 年,^{235}U 的半衰期为 7.037×10^8 年,^{238}U 的半衰期为 4.468×10^9 年。铀同位素的 α 辐射能量介于 4.040~4.776MeV[24]。1Bq ^{238}U 的活度相当于环境样品中 81.6μg 的总铀。此外,同位素 ^{235}U 可裂变,可以用作核能燃料[1]。

钚属于人类活动产生的一组人工元素,包含 3 个发射 α 射线的放射性核素 ^{238}Pu、^{239}Pu、^{240}Pu 和发射 β 射线的 ^{241}Pu 同位素。这些放射性核素由于具有高放射毒性、长半衰期、高化学反应性和在环境系统中的长期存在,因此在放射学上是很重要的。^{238}Pu 的半衰期为 87.7 年,^{239}Pu 为 2.411×10^9 年,^{240}Pu 为 6583 年,^{241}Pu 为 13.2 年[24]。钚同位素的 α 辐射能量介于 4.755MeV~5.499MeV[24]。^{239}Pu 同位素是最重要的,因为其可以发生裂变,可用来制造核弹。环境中钚的主要来源(约 5t)是核武器试验产生的大气沉降物[1],可利用能谱法对环境样品中钚同位素进行放射化学测定[79-81]。

环境样品中天然(^{210}Po、^{210}Pb、^{234}U、^{235}U 和 ^{238}U)和人工(^{238}Pu、$^{239+240}$Pu 和 ^{241}Pu)放射性同位素测定其活度的化学方法基于以下步骤[1,14]。

(1) 天然水样中放射性核素与二氧化锰的共沉淀;
(2) 沉积物、土壤和生物样品的消解;

(3) 离子交换树脂上放射性核素的连续分离纯化；

(4) 在银盘上电沉积钋，在钢盘上电解铀和钚；

(5) α 能谱法测定钋(^{210}Po)、铀(^{234}U 和 ^{238}U)和钚(^{238}Pu、$^{239+240}$Pu 和 ^{241}Pu)的活度。

在放射性化学分析之前,在每个样品中加入可回收示踪剂(5mBq~50mBq 的 ^{209}Po、^{232}U 和 ^{242}Pu)[14]。样品中的钋共沉淀和消解后,在0.5mol/L HCl 溶液中的在银盘上自发电沉积 4h[82],在钋沉积后,该溶液可用于测定放射性铅(^{210}Pb)、铀和钚。考虑到发射的 β 粒子能量较低,在天然样品中直接测量^{210}Pb 的活度比较困难,所以,^{210}Pb 的活度是通过测量由^{210}Pb 衰变产生的^{210}Po 的活度间接计算出来的[14]。Pu(Ⅳ)在浓酸溶液中(8mol/L HNO$_3$ 和 10mol/L HCl)包含阴离子络合物[Pu(NO$_3$)$_6$]$^{2-}$和[PuCl$_6$]$^{2-}$,它们被吸附在阴离子交换树脂上(如 Dowex),而 Pu(Ⅲ)以 Pu^{3+}阳离子的形式存在[83]。用碘化铵将吸附的 Pu(Ⅳ)阴离子络合物还原,并使其分解为 Pu(Ⅲ)。除了钋、镎、钍之外,8mol/L 硝酸溶液中还含有阴离子络合物[Th(NO$_3$)$_6$]$^{2-}$和[Np(NO$_3$)$_6$]$^{2-}$,这些络合物被吸附在阴离子交换树脂上,但铀(UO$_2^{2+}$)、钋(Po^{4+})和铁(Fe^{3+})通过了吸附柱。另外,在 10mol/L HCl 溶液中,铀和铁以阳离子 UO$_2$Cl$_{42-}$ 和 FeCl^{4-} 的形式存在,但钍、镅、锔难以形成阴离子络合物,因此不会被吸附在阴离子交换树脂上。正是由于以上这些特性,才使得钚的提纯能够消除其他放射性元素的干扰[1,14]。

U(Ⅳ)(包括 Fe、Co、Cu、Zn 和 Cd)在 10mol/L 的 HCL 溶液中以铀酰阴离子 UO$_2$Cl$_4^{2-}$ 的形式存在,其可以吸附在阴离子交换树脂上[84-85]。在硫酸溶液中能够从其他元素中分离和提纯铀,当 H$_2$SO$_4$(aq)的浓度大于 0.01mol/L 时,铀以阴离子络合物 UO$_2$(SO$_4$)$_2^{2-}$ 和 UO$_2$(SO$_4$)$_3^{4-}$ 的形式存在。与铀不同的是,其他元素(Fe、Co、Cu、Zn)在硫酸溶液中不形成阴离子络合物[1,14]。

分离纯化后,将铀和钚的纯馏分电镀在抛光的不锈钢盘上,用能谱法测定其放射性核素的活度。α 探测器的分辨率为 17~20keV,这点非常重要。两个放射性核素(^{239}Pu 和 ^{240}Pu)可以同时测量,因为它们的粒子能量差小于 15keV[1,14]。图 15-3~图 15-5 所示分别为用于测定钋、铀和钚的 α 能谱图[1]。

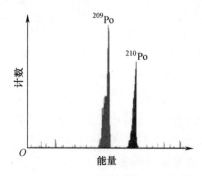

图 15-3　放射性钋(^{209}Po,^{210}Po)的 α 能谱图[9]

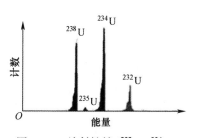

图 15-4 放射性铀(^{232}U, ^{234}U, ^{235}U, ^{238}U)的 α 能谱图[9]

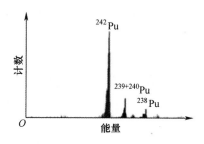

图 15-5 放射性钚(^{238}Pu, $^{239+240}$Pu, ^{242}Pu)的 α 能谱图[9]

4. 钋、铅、铀、钚放射性活度的计算

^{210}Po、^{234}U、^{235}U、^{238}U、^{238}Pu、$^{239+240}$Pu 活度的计算式如下[1,14]：

$$A_i = \frac{I_i}{e_i \cdot Y_i} \pm SD_i \quad (15-8)$$

式中：i 为 ^{210}Po、^{234}U、^{235}U、^{238}U、^{238}Pu 或 $^{239+240}$Pu；A_i 为放射性核素的活度(Bq)；I_i 为样本(无本底)计数率，定义为 N_i/t_p 的比值，其中 N_i 是 ^{210}Po、^{234}U、^{235}U、^{238}U、^{238}Pu 或 $^{239+240}$Pu 的计数率(无本底)，t_p 是 ^{210}Po、^{234}U、^{235}U、^{238}U、^{238}Pu 或 $^{239+240}$Pu 的计数时间(s)；e_i 为探测器效率；Y_i 为 ^{210}Po、^{234}U、^{235}U、^{238}U、^{238}Pu 或 $^{239+240}$Pu 的回收率；SD_i 为被测样品活度的标准偏差(不确定度)。

^{210}Po、^{234}U、^{235}U、^{238}U、^{238}Pu、$^{239+240}$Pu 的回收率[1,14]为

$$Y_i = \frac{I_i}{e_i \cdot s} \quad (15-9)$$

式中：Y_i 为回收率；I_i 为回收示踪剂(^{209}Po、^{232}U、^{242}Pu)的计数率；s 为放射性化学分析前添加的 ^{209}Po、^{232}U、^{242}Pu 的活度(Bq)；e_i 为探测器效率。

钋、铀、钚的活度的 SD 计算公式为[1,14]

$$SD = \frac{A_0}{A_t} \cdot \frac{\sqrt{\dfrac{I_p}{t_P} + \dfrac{I_t}{t_t}}}{Y \cdot e} \quad (15-10)$$

式中：A_0 为电沉积时 ^{210}Po 的活度(Bq)；A_t 为计数时 ^{210}Po 的活度(Bq)；I_p 为样本计数率(含本底)；I_t 为本底计数率；t_p 为样本计数时间(s)；t_t 为本底计数时间(s)；Y 为回收率；e 为检测器效率。

如果本底计数率非常低，当分析时间 $A_0 = A_t$ 时(铀和钚)，根据下式计算 SD[1,14]：

$$SD = \frac{\sqrt{N_P}}{Y \cdot e \cdot t_P} \tag{15-11}$$

式中：N_P 为铀的(^{234}U、^{235}U、^{238}U)计数率。

根据银盘上钋的电沉积时间，环境样品中 ^{210}Po 的活度公式为[1,14]：

$$A_0 = A_t \cdot \exp\left[\frac{t \cdot \ln 2}{T}\right] = A_t \cdot \exp(0.00502 \cdot t) \tag{15-12}$$

式中：A_0 为电沉积银盘时 ^{210}Po 的活度(Bq)；A_t 为计数时 ^{210}Po 的活度(Bq)；t 为电沉积与 ^{210}Po 计数(天数)之间的时间间隔；T 为 ^{210}Po 的半衰期(138.4 天)。

用间接法测量放射性铅的活度，在提纯并储存几个月(最多 2 年)后，根据 ^{210}Po 的增长估算出 ^{210}Pb 的活度，根据式(15.13)计算采样时的 ^{210}Pb 的活度[1,14]：

$$A_0(^{210}\text{Pb}) = \frac{A_2(^{210}\text{Po})}{1 - \exp[-k(t_2 - t_1)]} \tag{15-13}$$

式中：$A_0(^{210}\text{Pb})$ 为样品采集时 ^{210}Pb 的活度；$A_2(^{210}\text{Po})$ 为第二次电沉积后 ^{210}Pb 衰变产生的 ^{210}Po 的活度；t_1 为收集到第一次 ^{210}Po 计数的时间；t_2 为收集到第二次 ^{210}Po 计数的时间；k 为 ^{210}Po 的衰变常数。

切尔诺贝利事故中钚在环境样品中的影响[1,14]可以表达为

$$F_{\text{ch}} = \frac{R_{\text{obs}} - R_n}{R_{\text{ch}} - R_n} = \frac{R_{\text{ch}} - 0.04}{0.56} \tag{15-14}$$

式中：R_{obs} 为样品中 ^{238}Pu/$^{239+240}$Pu 的活度比；R_n 为全球大气沉降中的 ^{238}Pu/$^{239+240}$Pu 的活度比(0.04)，R_{ch} 为切尔诺贝利事故中的 ^{238}Pu/$^{239+240}$Pu 的活度比(0.60)。

15.2.6 ^{241}Pu 活度测定

^{241}Pu 是低能发射体，$E_{\max}$ 为 21 keV，半衰期为 13.2 年。^{241}Pu 可以直接通过 β 正比计数器和液体闪烁计数器测量得到(^{241}Pu 含量相对较高的样品)[86-87]，也可间接通过其子体放射性核素 ^{241}Am 的 α 能谱测量得到[88-89]。

这种测量是基于 ^{241}Am 的增长，只有在长时间的增长之后，也就是 4~20 年才能进行测量。4 年后，^{241}Am/^{241}Pu 活度比仅为 1:166，因此，使用正比计数器直接测量 ^{241}Pu 的检出限约为 10mBq，而 ^{241}Am 的检测需要累积到 200mBq[90]。所得到的钚的 α 能谱必须与 4~20 年前获得的相同能谱进行比较，这样才可以根据 ^{241}Am 的 5.49MeV 峰值的增加来估算 ^{241}Pu 的含量，其中已经考虑了切尔诺贝利事故的环境样品中存在的 ^{238}Pu。^{241}Pu 的活度的计算公式为[88]

$$A_{\text{Pu}_0} = 31.11 \cdot \frac{A_{^{241}\text{Am}} \cdot e^{\lambda_{\text{Am}} \times t}}{(1 - e^{-\lambda_{\text{Pu}} \cdot t})} \tag{15-15}$$

式中：A_{Pu_0} 为 ^{241}Pu 在取样时的活度；$A_{^{241}Am}$ 为 ^{241}Am 在测量 4 年后的活度；λ_{Pu} 为 ^{241}Pu 的衰变量（0.050217/年）；λ_{Am} 为 ^{241}Am 的衰变量（0.001604/年），31.11 是 $\lambda_{Pu}/\lambda_{Am}$ 的比值，t 是从抽样到 ^{241}Am 的测量时间（4~20 年）。

15.2.6.1 放射性分析方法的误差估计

在测量钋、铀和钚的活度时，固有误差可以通过国际实验室试验中测定 CRM 中的放射性核素和使用国际原子能机构（IAEA）生产的 CRM 来评估。^{210}Po 的准确度估计小于 9%；^{234}U、^{235}U、^{238}U 小于 7.5%；$^{239+240}$Pu 小于 6%，但精密度估计在 10% 以下。化学回收率：^{210}Po 为 95%~99%，^{238}U 为 85%~95%，$^{239+240}$Pu 为 50%~85%[1,14]。α 能谱测量的 MDA 值计算公式为

$$MDA = 2.71 + 4.65\sqrt{B/t}\eta Y \tag{15-16}$$

式中：B 为本底计数；t 为计数时间；η 为计数效率；Y 为 α 能谱测量法的回收率[5]。

在 3~10 天的测量中，^{210}Po、^{238}U 和 $^{239+240}$Pu 放射性核素的 MDA 为 0.08~0.15 mBq[1,14]。

参考文献

[1] Skwarzec, B.: Determination of radionuclides in aquatic environment. In: Namieśnik, J.I., Szefer, P. (eds.) Analytical Measurement in Aquatic Environments, pp. 241-258. Taylor & Francis PE, London (2009)

[2] Becquerel, H.: Sur les radiations invisibles emises par les sels d'uranium. C. R. Acad. Sci. Paris 122, 689-694 (1896)

[3] Skłodowska-Curie, M.: Rayons emis par les composes de l'uranium et du thorium. C. R. Acad. Sci. Paris 126, 1101-1103 (1898)

[4] Curie, I., Joliot, M.F.: Un nouveau type de radioactivite. C. R. Acad. Sci. Paris 198, 254-256 (1934)

[5] L'Annunziata, M.: Handbook of Radioactivity Analysis, 2nd edn. Academic Press, Amsterdam (2003)

[6] Bem, H., Bem, E.M., Krzemin'ska, M., Ostrowska, M.: Determination of radioactivity in air filters in alpha and gamma spectrometry. Nukleonika 47(2), 87-91 (2002)

[7] Kozak, K., Mietelski, J.W., Jasińska, M., Gaca, P.: Decreasing of the natural background counting-passive and active method. Nukleonika 46(4), 165-169 (2001)

[8] Oikari, T., Kajola, H., Nurmi, I., Kaihola, L.: Simultaneous counting of low alpha-and beta-particle activities with liquid scintillation spectrometry and pulse-shape analysis. Appl. Radiat. Isot. 38A, 875-878 (1987)

[9] Courti, A., Goutelard, F., Burger, P., Blotin, E.: Development of beta-spectrometer using PIPS technology. Appl. Radiat. Isot. 53, 101–108 (2000)

[10] Cherenkov, P.A.: Visible grow of pure liquids under the influence of γ-rays. C. R. Acad. Sci. URSS (Dokłady Akademii NaukSSSR) 2, 451–454 (1934)

[11] Elrick, R.H., Parker, R.P.: The use of Cerenkov radiation in the measurement of beta-emitting radionuclides. Appl. Radiat. Isot. 19, 263–271 (1968)

[12] Aguilar-Benitez, M., Barnett, R. M., Crawford, R. L., Eichler, R. A., et al.: Particle Properties Data Booklet (from Review of particle properties, Phys. Rev., 170B). North Holland, Amsterdam (1986)

[13] Mietelski, J.W., LaRosa, J.J., Ghods, A.: Results of ^{90}Sr, and ^{238}Pu, $^{239+240}$Pu, ^{241}Am, measurements in some samples of mushrooms and forest soil from Poland. J. Radioanal. Nucl. Chem. 170(1), 243–258 (1993)

[14] Skwarzec, B.: Radiochemical methods for the determination of polonium, radiolead, uranium and plutonium in environmental samples. Chem. Anal. 42, 107–115 (1997)

[15] Dybczyński, R.: The position of NAA among the methods of inorganic trace analysis in the past and now. In: Proceedings of the Conference NEMEA-2, Bucharest, 20–23 October 2004, EUR Report (2005)

[16] Dybczyński, R.: Two philosophies in designing radiochemical separation intended for neutron activation analysis. Chem. Anal. (Warsaw) 54, 123–134 (2009)

[17] Dybczyński, R., Samczyński, Z.: The use of amphoteric ion exchange resin for the selective separation of cadmium from other radioelements present in neutron irradiated biological materials. J. Radioanal. Nucl. Chem. 150, 143–153 (1991)

[18] Dybczyński, R.: Neutron activation analysis and its contribution to inorganic trace analysis. Chem. Anal. (Warsaw) 46, 133–160 (2001)

[19] Samczyn'ski, Z., Danko, B., Dybczyński, R.: Application of Chelex 100 ion exchange resin for separation and determination of palladium, platinum and gold in geological and industrial materials by neutron activation analysis. Chem. Anal. (Warsaw) 45, 843–857 (2000)

[20] Chajduk-Maleszewska, E., Dybczyński, R.: Ion exchange preconcentration and selective separation of trace amounts of Pd on chelating resin Duolite ES 346 and its determination by NAA. Chem. Anal. (Warsaw) 49, 281–297 (2004)

[21] Dybczyński, R., Danko, B., Polkowska-Matrenko, H.: NAA study on homogeneity of reference materials and their suitability for microanalytical techniques. J. Radioanal. Nucl. Chem. 245, 97–104 (2000)

[22] Kawano, M., Falandysz, J., Wakimoto, T.: Instrumental neutron activation analysis of extractable organohalogens (EOX) in Arctic marine organisms. J. Radioanal. Nucl. Chem. 255(5), 235–237 (2003)

[23] Kawano, M., Falandysz, J., Morita, M.: Instrumental neutron activation analysis of extractable organohalogens in marine mammal, harbour porpoise (*Phocoena phocoena*) and itsfeed, Atlantic

herring (*Clupea harengus*), from the Baltic Sea. J. Radioanal. Nucl. Chem. 278(2), 263–266 (2008)

[24] Browne, E., Firestone, F.B.: Table of radioactive isotopes. In: Shirley, V.S. (eds.). Wiley, New York (1986)

[25] Pullen, B.P.: Cerenkov counting of ^{40}K in KCl using a liquid scintillation spectrometer. J. Chem. Educ. 63, 971 (1986)

[26] Rao, D.D., Sudheendren, V., Baburajan, A., Chandramouli, S., Hegde, A.G., Mishra, U.C.: Measurement of high energy gross beta and ^{40}K by Cerenkov counting in liquid scintillation analyzer. J. Radioanal. Nucl. Chem 243(3), 651–655 (2000)

[27] Fletcher, I.R., Maggi, A.L., Rosman, K.J., McNaughton, N.J.: Isotopic abundances of K and Ca using a wide-dispersion multi-collector mass spectrometer and low fractionation ionization techniques. Int. J. Mass Spectrom. 163, 1–17 (1997)

[28] Holm, E., Oregoni, B., Vas, D., Pettersson, H., Rioseco, J., Nilsson, U.: Nickel-63: radiochemical separation and measurement with an ion implanted silicon detector. J. Radioanal. Nucl. Chem. 138, 111–116 (1990)

[29] Holm, E.: Source and distribution of anthropogenic radionuclides in marine environment. In: Holm, E. (ed.) Radioecology, pp. 63–83. World Scientific, Singapore (1994)

[30] Holm, E., Roos, P., Skwarzec, B.: Radioanalytical studies offallout ^{63}Ni. Appl. Radiat. Isot. 43, 371 (1992)

[31] Skwarzec, B., Holm, E., Roos, P., Pempkowiak, J.: Nickel-63 in Baltic fish and sediment. Appl. Radiat. Isot. 54, 609–611 (1994)

[32] Martin, J.E.: Physics for Radiation Protection. Wiley-Interscience, Hoboken, NJ (2000)

[33] Skwarzec, B., Holm, E., Strumin'ska, D.I.: Radioanalytical determination of ^{55}Fe and ^{63}Ni in the environmental samples. Chem. Anal. (Warsaw) 46, 23–30 (2001)

[34] Geckeis, H., Hentschel, D., Jansen, D., Görzen, Kerner, N.: Determinationof Fe-55 and Ni-63 using semi-preparative ion chromatography-a feasibility study. Fresenius J. Anal. Chem. 357, 864–869 (1997)

[35] Nouri, J., Alloway, B.J., Peterson, P.J.: The chemical extractability and mobility of ^{63}Ni in soil and sludge. J. Biol. Sci. 1(10), 976–985 (2001)

[36] Scheuerer, C., Schupfner, R., Schuttelkopf, H.: A very sensitive LSC procedure to determine Ni-63 in environmental samples, steel and concrete. J. Radioanal. Nucl. Chem. 193(1), 127–131 (1995)

[37] McAninch, J.E., Hainsworth, L.J., Marchetti, A.A., Leivers, M.R., Jones, P.P., Dunlop, A.E., Mauthe, R., Vogel, J.S., Proctor, I.D., Straume, T.: Measurement of ^{63}Ni and ^{59}Ni by accelerator mass spectrometry using characteristic projectile X-rays. In: 7th International Accelerator Mass Spectrometry Conference, May 20–24, Tuscon, AZ, USA (1996)

[38] Folsom T.R., Sreekumaran C.: Some reference methods for determining radioactive and natural ceasium for marine studies. Technical reports series No.118. Reference methods for marine radio-

activity studies. IAEA, Vienna (1970)

[39] HASL-300: The procedures manual of the environment measurement laboratory. In: Ericson, M. D., Chiecko, N. A. (eds.), Edition EML, 28 edn. U. S. Department of Energy, Sr-03-RC (1997)

[40] Friberg, I.: Determination of ^{90}Sr and the transuranium elements in even of nuclear accident. Doctoral thesis, Goteborg (1998)

[41] Fliss, M., Botsch, W., Handl, J., Michel, R.: A fast method for the determination of strontium-89 and strontium-90 in environmental samples and its application to the analysis of strontium-90 in Ukrainian soils. Radiochim. Acta 83, 81–92 (1998)

[42] Grahek, Z., Eskinja, I., Kosutic, K., Cerjan-Stefanowic, S.: Isolation of yttrium and strontium from soil samples and rapid determination of ^{90}Sr". Croatica Chem. Acta 73(3), 795–807 (2000)

[43] Mateos, J.J., Gomez, E., Garcias, F., Casas, M., Cerda, V.: Rapid ^{90}Sr–^{90}Y determination in water samples using a sequential injection method. Appl. Radiat. Isot. 53, 139–144 (2000)

[44] Friberg, I.: Development and application of a method for the determination of ^{90}Sr in environmental samples. J. Radioanal. Nucl. Chem. 226(12), 55–60 (1997)

[45] IAEA: Measurement of radionuclides in food and the environment. A guidebook. IAEA, Vienna (1989)

[46] Gaca, P., Skwarzezc, B., Mietelski, J.: Geographical distribution of ^{90}Sr contamination in Poland. Radiochim. Acta 94, 174–179 (2006)

[47] Stella, R., Ganzerli, M.T., Valentini, Maggi, L.: A novel approach to Ca-Sr separation in the determination of ^{90}Sr using inorganic exchangers: an application to environmental samples. J. Radioanal. Nucl. Chem. 161(2), 413–419 (1992)

[48] Zhu, S., Ghods, A., Veselsky, J.C., Mirna, A., Schelenz, R.: Interference of yttrium-91 with the rapid determination of strontium-90 originating from the Chernobyl fallout debris. Radiochim Acta 51, 195–198 (1990)

[49] Bojanowski, R., Knapińska-Skiba, D.: Determination of low-level ^{90}Sr in environmental materials: a novel approach to the classical method. J. Radioanal. Nucl. Chem. 138, 207–218 (1991)

[50] Gómez, E., Garcías, F., Casas, M., Cerdá, V.: Determination of 137Cs and 90Sr in calcareous soils: geographical distribution on the island of Majorca. Appl. Radiat. Isot. 48(5), 699–704 (1997)

[51] Bunzl, K., Kracke, W., Meyer, H.: A simple radiochemical determination of ^{90}Sr in brines. J. Radioanal. Nucl. Chem. 212(2), 143–149 (1996)

[52] Pimpl, M.: ^{89}Sr/^{90}Sr-Determination in soils and sediments using crown ethers for Ca/Sr-separation. J. Radioanal. Nucl. Chem. 194(2), 311–318 (1995)

[53] Solecki, J., Chibowski, S.: Studies of soilsample mineralization conditions preceding the determination of ^{90}Sr. J. Radioanal. Nucl. Chem. 247, 165–169 (2001)

[54] Melquiades, F.L., Appoloni, C.R.: Application of XRF and field portable XRF for environmen-

tal analysis. J. Radioanal. Nucl. Chem. 262(2), 533–541 (2004)

[55] Chen, Q.J., Hou, X.L., Yu, Y.X., Dahlgaard, H., Nielsen, S.P.: Separation of Sr from Ca, Ba and Ra by means of Ca(OH)2 and Ba(Ra)Cl2 or Ba(Ra)SO4 for the determination of radiostrontium. Anal. Chim. Acta 466, 109–116 (2002)

[56] Mikulaj, V., Hlatky, J., Vašekova, L.: An emulsion membrane extraction of strontium and its separation from calcium utilizing crowns and picric acid. J. Radioanal. Nucl. Chem. 101(1), 51–57 (1986)

[57] Alhassanieh, O., Abdul-Hadi, A., Ghafar, M., Aba, A.: Separation of Th, U, Pa, Ra and Ac from natural uranium and thorium series. Appl. Radiat. Isot. 51(5), 493–498 (1999)

[58] Blasius, E., Klein, W., Schön, U.: Separation of strontium from nuclear waste solutions by solvent extraction with crown ethers. J. Radioanal.Nucl. Chem. 89(2), 389–398 (1985)

[59] Gjeci, E.: Analysis of ^{90}Sr in environmental and biological samples by extraction chromatography using a crown ether. J. Radioanal. Nucl. Chem. 213(3), 165–174 (1996)

[60] Vajda, N., Ghods-Esphahan, A., Cooper, E., Danesi, P.R.: Determination of radiostrontium in soil samples using a crown ether. J. Radioanal. Nucl. Chem. 162(2), 307–323 (1992)

[61] Goutelard, F., Nazard, R., Bocquet, C., Coquenlorge, N., Letessier, P., Calmet, D.: Improvement in ^{90}Sr measurements at very low levels in environmental samples. Appl. Radiat. Isot. 53 (1–2), 145–151 (2000)

[62] Lee, M.H., Chung, K.H., Choi, G.K., Lee, C.W.: Measurement of ^{90}Sr in aqueous samples using liquid scintillation counting with full spectrum DPM method. Appl. Radiat. Isot. 57(2), 257–263 (2002)

[63] Aupiais, J., Fayolle, C., Gilbert, P., Dacheux, N.: Determination of Ra-226 in mineral drinking water by alpha liquid scintillation with rejection of beta emitters. Anal. Chem. 70, 2353–2359 (1998)

[64] Burnett, W.C., Tai, W.-C.: Determination of radium in natural waters by alpha liquid scintillation. Anal. Chem. 64, 1691–1697 (1992)

[65] Higuchi, H., Uesugi, M., Satoh, K., Ohashi, N.: Determination of radium in water by liquid scintillation counting after preconcentration with ion exchange resin. Chem. Anal. 56, 761–763 (1984)

[66] Bem, E.M., Bem, H., Majchrzak, I.M.: Comparison of two methods for ^{226}Ra determination in mineral mater. Nukleonika 43(4), 59–68 (1998)

[67] Blackburn, R., Al-Masri, M.S.: Determination of radium-226 in aqueous samples using liquid scintillation counting. Analyst 117, 1949–1951 (1992)

[68] Becker, J.K., Dietze, H.-J.: Determination of long-lived radionuclides by double focusing sector field ICP mass spectrometry. Adv. Mass Spectrom. 14, 681–689 (1998)

[69] IAEA: The environmental behaviour of radium, Technical report 310, Vol. 1, Part 3, pp. 145–319, Vienna (1991)

[70] Sill, C.W.: Determination of radium-226 in ores, nuclear waste and environmental samples by

high-resolution alpha spectrometry. Nucl. Chem. Waste Manag. 7, 239–256 (1987.)
[71] Kowalewska, G.: ^{226}Ra in water and sediment of the southern Baltic Sea. Oceanologia 23, 65–76 (1989)
[72] Cohen, B.L., Cohen, E.S.: Theory and practice of radon monitoring with charcoal adsorption. Health Phys. 45, 501–508 (1983)
[73] Cohen, B.L.: Liquid scintillation versus gamma ray counting in radon measurement with charcoal. Am. J. Publ. Health, 81(12) (1991)
[74] Fleischer, R.L.: Radon in the environment-opportunities and hazards. Nucl. Trans. Radiat. Meas. 14(4), 421–435 (1988)
[75] Zikovsky, L., Roireau, N.: Determination of radon in water by argon purging and alpha counting with a proportional counter. Appl. Radiat. Isot. 41, 679–681 (1990)
[76] Salonen, L.: Measurement of low levels of ^{222}Rn in water with different commercial liquid scintillation counters and pulse shape analysis. In: Noakes, J.E., Schonhofer, F., Polach, H.A. (eds.) Advances in Liquid Scintillation Spectrometry, pp. 361–371. Radiocarbon Publishers, University of Arizona, Tucson, AZ (1992)
[77] Blackburn, R., Al-Masri, M.S.: Determination of radon-222 and radium-226 in water samples by Cerenkov counting. Analyst 118, 873–876 (1993)
[78] Homma, Y., Murase, Y., Takiue, M.: Determination of ^{222}Rn by air luminescence method. J. Radioanal. Nucl. Chem. Lett. 119, 457–465 (1987)
[79] Pietrzak-Flis, Z., Chrzanowski, E., Dembin'ska, S.: Intake of ^{226}Ra, ^{210}Pb, ^{210}Po with food in Poland. Sci. Total Environ. 203, 157–165 (1997)
[80] Pietrzak-Flis, Z., Kamińska, I., Chrzanowska, E.: Uranium isotopes in waters and bottom sediments of rivers and lake in Poland. Nukleonika 49(2), 69–76 (2004)
[81] Suplińska, M.: Vertical distribution of ^{137}Cs, ^{210}Pb, ^{226}Ra and $^{239+240}$Pu in bottom sediments from the Southern Baltic Sea in the years 1998–2000. Nukleonika 47(2), 45–52 (2002)
[82] Bagnall, K.W.: The Chemistry of Selenium, Tellurium and Polonium. Elsevier, Amsterdam (1966)
[83] Niesmiejanow, A.N.: Radiochemia, pp. 488–505. Wydawnictwo PWN, Warszawa (1975)
[84] Strelow, F.W., Bohme, E.: Anion exchange and selectivity scale for elements in surface acid media with a strongly basic resin. Anal. Chem. 39, 595–599 (1967)
[85] Sing, N.P., Wrenn, W.: Determination of alpha-emitting uranium isotopes in soft tissue by solvent extraction and alpha-spectrometry. Talanta 30, 271–274 (1973)
[86] Moreno, J., LaRosa, J.J., Deseni, P.R., Burns, K., DeRegge, P., Vajda, N., Sinojmeri, M.: Determination of ^{241}Pu by liquid-scintillation counting in the combined procedure for Pu radionuclides, ^{241}Am and ^{90}Sr analysis in environmental samples. Radioact. Radiochem. 9, 35–44 (1998)
[87] Mietelski, J.W., Dorda, J., Was, B.: Pu-241 in samples of forest soil from Poland. Appl. Radiat. Isot. 51, 435–447 (1999)

[88] Strumin'ska, D. I., Skwarzec, B., Mazurek-Pawlukowska, M.: Plutonium isotopes ^{238}Pu, $^{239+240}$Pu and ^{241}Pu in Baltic Sea ecosystem. Nukleonika 50, S45–S48 (2005)

[89] Strumińska, D.I., Skwarzec, B.: ^{241}Pu concentration in southern Baltic Sea ecosystem. J. Radioanal. Nucl. Chem. 268(1), 59–63 (2006)

[90] Rosner, G., Hötzl, H., Winkler, R.: Determination of 241Pu by low level beta proportional counting, application to Chernobyl fallout samples and comparison with the ^{241}Am build-up method. J. Radioanal. Nucl. Chem. 163, 225–233 (1992)

著者列表

Maria Balcerzak 华沙理工大学，化学学院分析化学系，波兰 华沙
Irena Baranowska 西里西亚理工大学 无机分析化学和电化学系，波兰 格利维策
Jolanta Borkowska-Burnecka 波兰佛罗茨瓦夫理工大学，分析化学部化学系，波兰 佛罗茨瓦夫
Bogusław Buszewski 哥白尼大学 化学学院 环境化学与生物分析系主任，波兰 托伦
Rajmund S. Dybczyński 核化学与技术研究所 核分析技术实验室，波兰 华沙
Małgorzata Grembecka 格但斯克医科大学 药学院食品科学系，波兰 格但斯克
Adam Hulanicki 华沙大学化学系，波兰 华沙
Maciej Jarosz 波兰华沙科技大学，化学院分析化学系主任，波兰 华沙
Maria Kała 波兰克拉科夫法医学研究所
Roman Kaliszan 格但斯克医科大学 药学院 药剂学和药效学系，波兰 格但斯克
Piotr Konieczka 格但斯克理工大学 化学学院 分析化学系，波兰 格但斯克
Marta Koper 西里西亚理工大学 分析化学系，波兰 格利维策
Paweł Kościelniak 雅吉洛尼亚大学 化学学院 分析化学系，波兰 克拉科夫
Bartosz Kowalski 西里西亚理工大学 化学学院 无机分析和电化学系，波兰 格利维策
Joanna Kozak 雅吉洛尼亚大学 化学学院 分析化学系，波兰 克拉科夫
Anna Leśniewicz 弗罗茨瓦夫理工大学 化学学院 分析化学专业，波兰 佛罗茨瓦夫
Tomasz Ligor 哥白尼大学 化学学院 环境化学与生物分析系主任，波兰 托伦
Sylwia Magiera 西里西亚理工大学 化学学院 无机分析和电化学系，波兰 格利维策

Barbara Marciniec 卡罗尔·马辛考夫斯基医科大学 药物化学系,波兰 波兹南

Michał J. Markuszewski 格但斯克医科大学 生物制药和药物动力学系,波兰 格但斯克

Henryk Matusiewicz 波兹南理工大学 分析化学系,波兰 波兹南

Katarzyna Pawlak 华沙理工大学,化学学院 分析化学系主任,波兰 华沙

Mikołaj Piekarski
卡罗尔·马辛考夫斯基医科大学 药物化学系,波兰 波兹南
加拿大 BC 癌症管理局北方药学研究中心,加拿大 海利纳 乔治王子城

Polkowska-Motrenko 核化学与技术研究所 核分析技术实验室,波兰 华沙

Lena Ruzik 华沙理工大学 化学学院分析化学系主席,波兰 华沙

Iwona Rykowska 亚当米基维奇大学 化学学院,波兰 波兹南

Bogdan Skwarzec 格但斯克大学 化学学院 化学与环境放射化学系主席,波兰 格但斯克

Maciej Stawny 卡罗尔·马辛考夫斯基医科大学 药物化学系,波兰 波兹南

Wiktoria Struck-Lewicka 格但斯克医科大学 生物制药和药效学系,波兰 格但斯克

Piotr Szefer 格但斯克医科大学 药学院 食品科学系,波兰 格但斯克

Agnieszka Ulanowska 哥白尼大学 化学学院 环境化学与生物分析系主任,波兰 托伦

Wiesław Wasiak 亚当米基维奇大学 化学学院,波兰 波兹南

Marcin Wieczorek 雅吉洛尼亚大学 化学学院分析化学系,波兰 克拉科夫

Janina Zięba-Palus Prof. Dr.Jan Sehn 法医学研究所,波兰 克拉科夫

Wiesław Żyrnicki 弗罗茨瓦夫理工大学 化学学院 分析化学专业,波兰 弗罗茨瓦夫